ULTRA WIDEBAND – CURRENT STATUS AND FUTURE TRENDS

Edited by **Mohammad Abdul Matin**

Ultra Wideband – Current Status and Future Trends

http://dx.doi.org/10.5772/2588
Edited by Mohammad Abdul Matin

Contributors

M. A. Matin, Rshdee Alhakim, Kosai Raoof, Emmanuel Simeu, Xubo Wang, Anh Dinh, Daniel Teng, Huanan Yu, Shuxu Guo, Carlos Pomalaza-Ráez, Attaphongse Taparugssanagorn, H. Lu, T. Xu, H. Nikookar, Xian Ling Liang, D. Tran, N. Haider, S. E. Valavan, I. E. Lager, A. Szilagyi, O. Yarovyi, L. P. Ligthart, Danial Nezhad Malayeri, Ali Imran Najam, Yvan Duroc, Smail Tedjini, Y.F. Weng, S.W. Cheung, T.I. Yuk, L. Liu, L. Liu, S.W. Cheung, Y.F. Weng, T.I. Yuk, Mongkol Meeloon, Sarawuth Chaimool, Prayoot Akkaraekthalin, Ahmed Bazil Taha, Miguel Calvo Ramon, João M.B. Oliveira, Luís M. Pessoa, Diogo Coelho, Henrique M. Salgado, Jorge C.S. Castro

Published by InTech

Janeza Trdine 9, 51000 Rijeka, Croatia

Publishing Process Manager Mirna Cvijic

Typesetting InTech Prepress, Novi Sad

Cover InTech Design Team

Ultra Wideband – Current Status and Future Trends, Edited by Mohammad Abdul Matin
p. cm.
ISBN 978-953-51-0781-1

For EU product safety concerns:
IN TECH d.o.o., Prolaz Marije Krucifikse Kozulić 3, 51000 Rijeka, Croatia,
info@intechopen.com or visit our website at intechopen.com.

Meet the editor

Dr. Mohammad A Matin is currently working at EEE department, Institut Teknologi Brunei (ITB) as an Associate Professor. He obtained his MSc degree in digital communication from Loughborough University, UK and a PhD degree in wireless communication from Newcastle University, UK. He has taught several courses in communications, electronics and signal processing at North South University, KUET, Khulna University, and BRAC University during his career. Dr. Matin was the visiting academic staff at the National University of Malaysia (UKM) and university of Malaya (UM). He has published over 50 refereed journals and conference papers. He is the author of three academic books and few book chapters. He is currently serving as a member of the editorial board of several international journals such as IET Wireless Sensor Systems (IET-WSS), IJCTE, JECE, etc. His current research interests include UWB communication, wireless sensor networks, cognitive radio, EM modeling, and antenna engineering.

Contents

Preface

Ultra wideband (UWB) has advanced and merged as a technology, and many more people are aware of the potential for this exciting technology. The current UWB field is changing rapidly with new techniques and ideas where several issues are involved in developing the systems. Among UWB system design, the UWB RF transceiver and UWB antenna are the key components. Recently, a considerable amount of researches has been devoted to the development of the UWB RF transceiver and antenna for its enabling high data transmission rates and low power consumption. Our book attempts to present current and emerging trends in research and development of UWB systems as well as future expectations.

Chapter 1 presents RF transceiver architecture for ultra-wideband (UWB) data communications which can fabricate in a small chip area. Actually, the transceiver design offers a huge challenge of this UWB technology as it requires low cost, low power consumption and single chip transceiver architecture. This chapter highlights the current design issues of RF transceiver which will empower readers to make the further developments in UWB transceiver.

Chapter 2 presents an IR-UWB timing synchronization algorithm for achieving rapid, accurate and low-complexity synchronization. The suggested approach improves the estimation performance of the original dirty template algorithms.

Chapter 3 talks about the implementation of the low-power low-cost CMOS biomedical radar using UWB pulse for heart rate monitoring with improved accuracy and reliable performance compared with conventional heart rate measurement devices built using discrete components. This work will contribute in the development of next generation of non-invasive real-time medical monitoring methods.

Chapter 4 approaches a pre-filtering method for UWB channel estimation based on the theory of compressed sensing (CS), whose measurement matrix is just a Toeplitz matrix. The presented approach improves the channel estimation accuracy.

Chapter 5 describes the features of the UWB channel that should be taken into account when it is being considered for medical applications, in particular in hospital scenarios. These scenarios include cases where the human body is in motion and the mathematical models of the channel have been developed based on actual measurements.

Chapter 6 presents a design of a cooperative wavelet scheme for the multi-relay, multi-scale and multi-lag diversity, which can be widely applied in the ultra-wideband communications and underwater acoustic communications.

Chapter 7 introduces the-state-of-art of the conventional and modern UWB antennas. First, it talks about some of the conventional UWB antennas and then discusses and summarizes some novel types of omni-directional UWB antennas and wideband techniques, mainly concentrate on two types: the planar monopole UWB antenna and the printed monopole UWB antenna. Several novel types of directional UWB antenna and wideband techniques are also summarized including the UWB printed wide-slot antenna and the UWB dielectric resonator antennas (DRA). Finally, to avoid the interference between the UWB system and other exist communication systems, various notch function techniques for UWB antenna design are introduced.

Chapter 8 presents the concept and the design methodology of a generic SWB antenna architecture (SWBA). The antenna architecture provides powerful isolated-parameters to control the antenna characteristics, such as resonance-shifting, resonance matching, bandwidth broadening, diffraction reduction, and SWB pattern maintaining.

Chapter 9 describes the design of active integrated antenna for UWB (ultra-wideband) application.

Chapter 10 presents several UWB-MIMO antennas and a critical description of relevant characteristics need to take into account for the design and a review of presented designs in the literature classified according to the techniques applied to improve the mutual coupling and enhance isolation. Some new designs of UWB-MIMO antennas, based on simple radiating elements, are also introduced, analyzed and compared in this chapter.

Chapter 11 presents several UWB antennas with single, dual, triple and quadruple band-notched characteristics using different resonator structures including the CPW resonators,$\lambda/4$-resonators and the MLs. The bandwidths and center frequencies of all these notches can be controlled independently by adjusting the dimensions of the resonators which are also discussed in this chapter.

Chapter 12 discusses the cable effects on measuring compact planner monopole antennas for UWB applications. The small ground plane of the antennas causes the currents to flow back on the surface of the feeding cable, results in secondary radiation. This leads to discrepancies between the simulated and measured performances and uncertainties in the design of the antenna.

Chapter 13 presents advanced UWB-bandpass filters based on slotted linear tapered-line resonator (SLTR) and slotted step-impedance resonator (SSIR) structures for size reduction and improved stopband performances. A comprehensive treatment of slotted resonators and both ends of the resonator with interdigital coupled lines is described. The design concept is demonstrated using two filter examples including one with an SLTR and another one with an SSIR. These filters are compact in size and provide a wider upper stopband resulting from resonator bandstop characteristics.

Chapter 14 presents a study on UWB coexistence with the 3G and 4G Cellular Systems. The methodology used to account for the impact of UWB interference on the coverage range and capacity of the interfered systems has been explained in detail. Finally, it has been applied in a set of study cases in scenarios involving the 3G and 4G selected systems.

Chapter 15 discusses the performance of the transmission of UWB signals over two distinctive optical based networks. A low-cost RoF solution based on direct modulation of a VCSEL and two polymer optical fibers are presented and the UWB signal transmission performance is assessed. It is considered two high performance PF-GI POFs with diameters of 62.5 mm and 120 mm and with bandwidth distance products higher than 1 GHz_Km. Experimental measurements of packet error rate (PER) and minimum transmitted powers to achieve the maximum allowed PER showed that it is possible to have a viable transmission at data rates of 480 Mbps, 200 Mbps and 53.3 Mbps over 100, 150 and 200 meters of PF-GI-POF, respectively, preceded by a 1 meter wireless link.

It is believed that this book serves as a comprehensive reference for graduate students who are wishing to greatly enhance their knowledge of all aspects of UWB systems.

Mohammad Abdul Matin
Institut Teknologi Brunei
Brunei Darussalam

Chapter 1

Ultra-Wideband RF Transceiver

M. A. Matin

Additional information is available at the end of the chapter

http://dx.doi.org/10.5772/49095

1. Introduction

Ultra-wideband (UWB) technology has developed rapidly over the past several years due to its high date rate with small current consumption in short range communication. According to Shannon-Hartley theorem, the maximum rate of clean (or arbitrarily low bit error rate) date through an AWGN (Additive white Gaussian Noise) channel, is limited to

$$C = BW \cdot \log_2(1 + SNR) \tag{1}$$

Where, C is the channel capacity, BW is the bandwidth, SNR is the ratio of average received signal power to the noise spectral density. It can be seen from (1), channel capacity increases linearly with bandwidth but only logarithmically with SNR which means capacity increases as a function of BW faster than as a function of SNR and with a wide bandwidth, high data rate can be achieved with a low transmitted power. Its main applications include imaging systems, vehicular radar systems and communications and measurement systems. Ever since, the FCC released unlicensed spectrum of 3.1-10.6 GHz for UWB application in 2002, UWB has received significant interest from both industry and academia. Mutli-Band OFDM (MB-OFDM) and Direct-Sequence UWB (DS-UWB) are two existing competing proposals for UWB; each gained multiple supports from industry. The MB-OFMD divides the 3 ~ 10 GHz UWB spectrum into fourteen sub-bands which has a 528 MHz bandwidth. Due to incompatible of these two proposals, it experiences huge difficulties in commercialization of UWB technology. On the other hand, Impulse Radio UWB (IR-UWB) has become a hot research area in academia due to its low complexity and low power.

2. UWB modulation

As UWB pulse itself does not contain information, we must add digital information to the analog pulse through modulation. The MB-OFDM systems are dealing with continuous ultra-wideband modulated signals while DS-UWB systems are transmitting discrete short pulses which cover ultra-wide bandwidth. On the other hand, IR-UWB is a carrier-less pulse-based system which means IR-UWB and DS-UWB are the two different categorizes of pulse based UWB. Pulse modulation scheme includes OOK (On Off Keying), BPSK (Binary Phase Shift Keying) and PPM (Pulse Position Modulation). OOK modulation is performed by generating transmitted pulses only while transmitting '1' symbols. BPSK modulation generates 180° phase-shifted pulses while transmitting baseband symbols '1' and '0'. PPM modulation is performed by generating pulses where each pulse is delayed or sent in advance of a regular time scale. Thus a binary communication system can be established with a forward and backward shift in time. By specifying specific time delays for each pulse, an M-ary system can be created. BPSK has an advantage over other modulation types due to an inherent 3 dB increase in separation between constellation points (Wentzloff & Chandrakasan, 2006); however, BPSK modulation is not suitable for some receiver architectures, e.g., noncoherent receivers.

3. UWB transceivers

Both MB-OFDM (Ranjan & Larson, 2006; Zheng H et al., 2007; Bergervoet et al., 2007; Beek et al., 2008) and DS-UWB (Zheng Y. et al., 2007, 2008) are carrier-modulated systems, where a mixer is used to up/down convert the radio frequency (RF) signal, therefore it requires local oscillator (LO) synthesis. On the other hand, IR-UWB (Yang, C. et al., 2005 ; Xia L. et al., 2011) is a carrier-less pulse-based system, therefore, we can eliminate the fast hopping LO synthesis, thus reducing the complexity and power consumption of the entire radio. Furthermore, since the signal of a pulse-based UWB system is duty-cycled, the circuits can be shut down between pulses intervals which would lead to an even lower power design.

There are a number of different fabrication options for UWB transceivers; CMOS is mainly compelling due to its low cost, low power consumption and single chip transceiver architecture with few external components. Poor passive components and lower operating voltages associated with process scaling pose significant problems for the radio architect and designer. Moreover, the design of UWB transceivers faces the following issues such as - 1) broadband circuits and matching; 2) the low-noise amplifier (LNA) with reasonable noise figure (NF) and impedance matching 3) broadband transmit/receive switch. Narrowband interference imposes some extra issues- the linearity and dynamic range. Even though some important issues that impact the receiver design are given above, there are many other factors that affect the receiver design and choice. For example, the modulation that is used at the transmitter impacts the receiver design. If the transceiver complexity and cost are the primary concerns, a scheme that enables noncoherent demodulation (OOK, positive PAM, PPM, and M-ary PPM) can be considered. On the other hand, some other modulations like BPSK, M-ary PAM, and QAM have the potential to provide better performance and require coherent demodulation since the information is embedded in the polarities of the pulses.

3.1. UWB transmitter/Pulse generator

In principle, all the pulses with the spectra (≥ 500 MHz) falling into the UWB band can be used as signals. However, for practical purposes, the pulses which are simple to generate, controlled, and have low power-consumption (no direct component), are selected to generate UWB signals. The proper selection of the source pulse can maximize the radiated power within the UWB band and meet the required emission limits without filters before the transmitting antennas while minimizing anticipated inter-symbol (and in the case of DS-UWB, inter-chip) interference and providing spectral flexibility as a method to coexist with other radio systems.

In the transmitter, the binary information stream from devices such as PC, PDA or DVD player is passed to the front end of the transmitter and mapped from bits to symbols if higher order modulation schemes are to be used. Each symbol representing multiple bits is then mapped to analog pulse shape which is generated by pulse generator. The mapping of information into waveforms is referred to data mapping or modulation. The generated pulse then can be optionally amplified before being passed to transmitting antenna. Typical IR-UWB use transition generators with edge rates designed to occupy 3 GHz of bandwidth or more while other systems use various forms of gated frequency generators, where the edge rates are selected to spread the energy around the fundamental frequency of the generator.

3.2. UWB receiver

It is necessary to have an optimal receiving system same as generating signal with the desired spectral characteristics. The optimal receiving technique often used in UWB is a correlation receiver. The correlator in the receiver multiplies the received signal with the template waveform. It is critical to note that the mean value of the correlator is zero. Thus, for in-band noise signals received by a UWB radio, the correlator's output has an average value of zero. Moreover, the standard deviation or rms of the correlator output is related to the power of those in-band noise signals. The level of hardware implementation and computational complexity plays an important role in determining which modulation to be used in what application.

The receiver sensitivity is generally defined by the signal level required to gain the given signal-to-noise (S/N) ratio. This means sensitivity is increased when there is less noise. The following formula shows the factors used to define receiver sensitivity.

$$S\,(\text{dBm}) = -174 + NF + 10\log B + 10\log(S/N) \qquad (2)$$

Where, S (dBm) is the receiver sensitivity, NF is the noise figure, B is the bandwidth and S/N is the signal to noise ratio.

If communication is established by QPSK with 8 dB of S/N ratio and 6 dB of total circuit NF, receiver sensitivity with MBOFDM receiver will become –73dBm, when the bandwidth, B= 528 MHz for the data rate 480 Mb/s, To raise the data rate from 54 to 480 Mb/s, the channel bandwidth B need to increase from 20 to 178 MHz. MB-OFDM derives the receiver sensitivity requirements ranging from -80.8 dBm (for 54 Mb/s) to -73 dBm (for 480 Mb/s) at different data rates. If the required SNR is 2.4 dB, the receiver noise figure is 11.7 dB and the channel bandwidth is 1.32 GHz, the receiver sensitivity with a DS-UWB receiver will become –76.5 dBm at 220 Mb/s.

4. RF transceiver for IR-UWB

The transmitter for IR-UWB integrates amplitude and spectrum tunability, thereby providing adaptable spectral characteristics for different data rate transmission. The receiver employs noncoherent architecture because of its low complexity and low power. A 3-5 GHz fully integrated IR-UWB transceiver is presented as shown in Fig. 1 (Xia et al., 2011). IR-UWB transceiver is implemented in a 0.13 μm 1P8M CMOS technology. The transceiver die microphotograph is shown in Fig. 2. The die area is 2 mm×2 mm. The chip is bonded to the 4-layer FR-4 PCB with chip-on-board (COB) assembly. With a supply voltage of 1.2 V, the power consumption of the transmitter is only 1.2 mW and 2.2 mW when transmitting 50 Mb/s and 100 Mb/s baseband signals, respectively; the power consumption of the receiver is 13.2 mW.

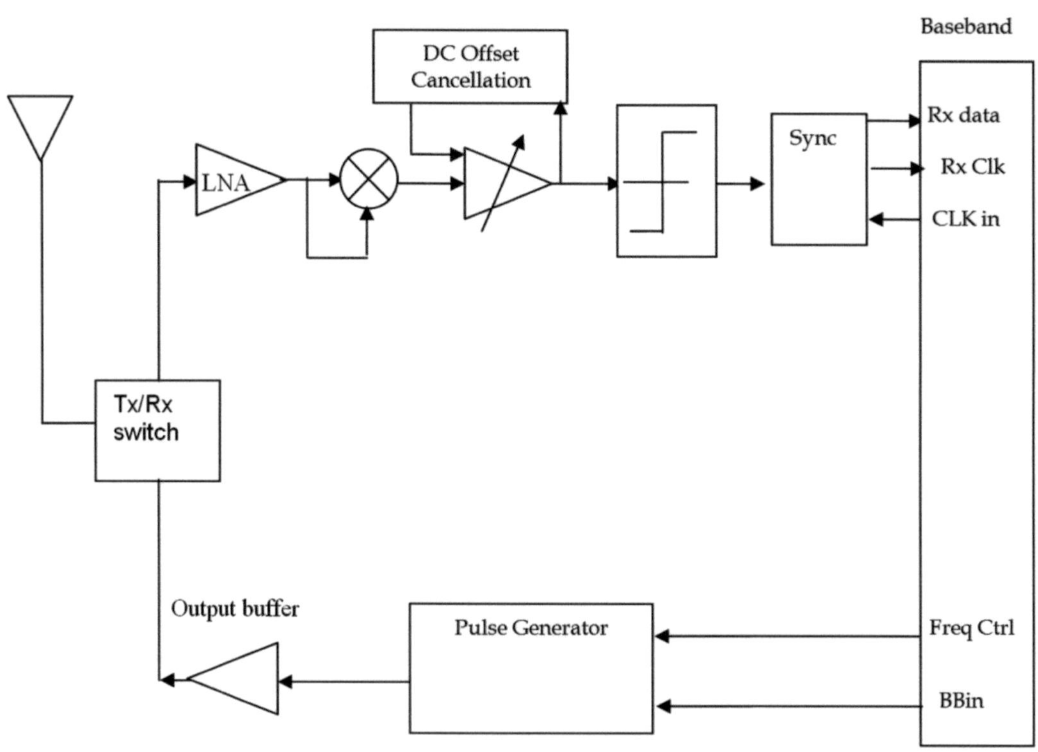

Figure 1. The proposed IR-UWB transceiver system architecture with OOK modulation

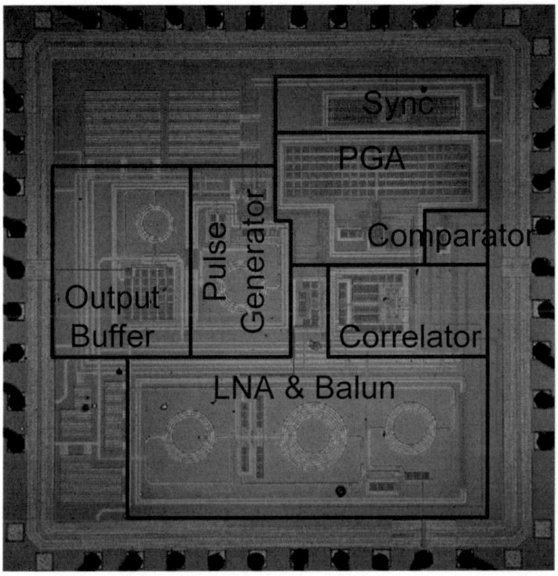

Figure 2. Microphotograph of IR-UWB transceiver

In fact, most companies are diving head-on into DS-CDMA and MB-OFDM to form the foundation for most of the coming UWB products though the impulse approach is the hot research area in academia.

5. DS-UWB scheme and RF transceiver

Direct-sequence spread-spectrum (DSSS) technique is a powerful multiple access (MA) technique that could be combined with UWB modulation to provide robustness against interference. In DS-UWB, the data to be transmitted is modulated using bipolar modulation, based upon a certain spreading code. Modulation is either phase-shift keying (PSK) or PPM. DS-UWB transmitters are super simple and use very low power, but the receiver and its complex correlation recovery circuits are somewhat more of a challenge. DS-UWB has many attractive properties, including low peak-to-average power ratio and robustness to multiple access interference (MAI) [Win et al., 1997].

The basic transmitted CDMA waveform of user k is given by

$$x_k(t) = \sum_{j=0}^{N-1} C_j^k w\left(t - jT_c\right) \tag{3}$$

Where, $w\ (t)$ represents the transmitted monocycle and C_j^k denotes jth spreading chip of the pseudo-random noise (PN) Sequence. N is the number of pulses of the PN sequences to be used for each user.

The transmission signal format is shown in Fig. 3. The encoded data of each user are considered as a data symbol, which is multiplied by the transmitted CDMA code.

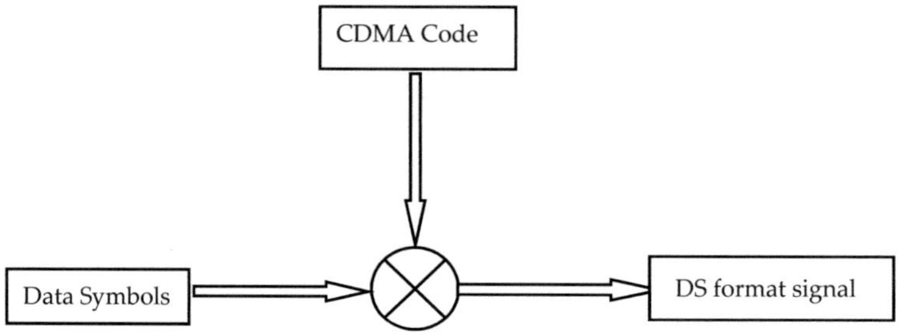

Figure 3. Transmission signal format

Let, T_f be the symbol period and T_c be the chip period such that $T_f = NT_c$. Hence, a typical DS format of the *k*th impulse radio transmitter output signal is given by

$$S_k(t) = \sqrt{p_k} \sum_m d_m^k \, x_k(t - mT_f) \tag{4}$$

Where d_m^k represents the data symbols and p_k is the transmitted power corresponding to the *k*th user. It is important to note that even an ideal channel and antenna system modify the shape of the transmitted monocycle $w(t)$ to $w_{rec}(t)$ at the output of the receiving antenna, where $w_{rec}(t)$ is the derivatives of a Gaussian function.

As indicated in [Ge et al., 2002, Wu et al., 2002, Wang et al., 2007], the DS-UWB system performance is severely downgraded by inter-symbol and multiple access interferences. Hence, researches on reducing effects of inter-symbol interference (ISI) and MAI is of great importance in designing of the transceiver for DS-UWB [Nassar et al. 2003]. The transceiver architecture of DS-UWB is shown in Fig.2 and the building blocks have bee presented in the following subsections.

5.1. Low noise amplifier

The primary factors in choosing a low noise amplifier (LNA) scheme are noise figure, dynamic range, linearity and power consumption. LNA is not the first block of a receiver circuit. It is followed by a band pass filter, a switch, or a duplexer that has to be implemented as a first block of the receiver chain in front of Low noise amplifier (LNA). As the band-pass filter is constructed with LC-tank in the first stage mixer to perform filtering of out-of-band interference and this block has signal loss characteristics instead of signal amplification, LNA requires providing a reasonable noise figure (NF) and impedance matching. By using a darlington topology a high gain can be achieved over the entire operating band. The design of a UWB LNA is more challenging than a traditional narrow-band LNA. Detailed design and consideration of the LNA can be found in [Hu et al., 2010; Lee H-J, 2006].

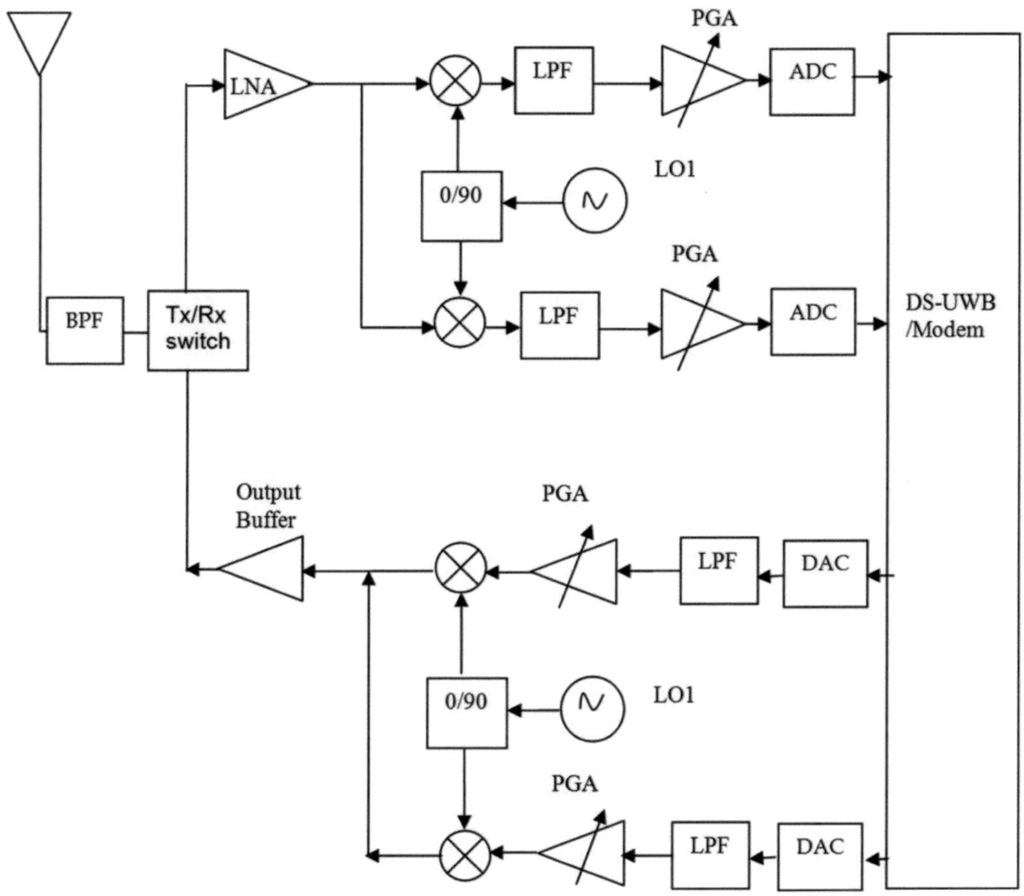

Figure 4. DS-UWB transceiver

5.2. Mixer

A combined mixer is proposed for both RF down-conversion in RX and for the RF up-conversion in the TX. In the receiver, it needs to synchronize the received pulse with local controlling signals to down-converted first. For superheterodyne transceivers, it is further down converted to baseband signal by a quadrature mixer. Because of the two stage frequency translation, local oscillator leakage does not have a significant impact on the receiver. In case of direct conversion transceiver, the RF signal is directly down-converted to baseband signal without any intermediate frequency. Therefore, the cost and size of the overall transceiver are reduced. The double-balanced Gilbert-type mixer topology has been widely used due to its low oscillator leakage and low even-order distortion products at the output.

5.3. Bandpass filter

The UWB filters are required to have a specified a small bandpass filter (BPF) with a notched band in the UWB passband (for DS- UWB) in order to avoid being interfered by the

5–6 GHz for IEEE 802.11a wireless local area networks (WLANs). To avoid the frequency use of WLAN radio signals, the direct sequence ultra-wideband (DS–UWB) specifications for wireless personal area networks (WPANs) need further to divide into a low band of 3.1–4.9 GHz and a high band of 6.2–9.7 GHz [IEEE.15 Working Group].

5.4. Variable Gain Amplifier

Variable gain amplifier (VGA) is an essential block at the front end of ultra-wideband transceiver to maximize the dynamic range of the receivers. VGAs are also used in the transmitter part of ultra-wideband transceivers to control the transmission signal power. The VGA is typically implanted in an automatic gain control amplifier (AGC) loops to provide constant output signal regardless the variations in the input signal. The variable gain amplifier suppresses even harmonics, rejects common-mode noises and provides good linearity and wideband performance regardless of the control voltage.

6. Multiband OFDM (MB-OFDM) scheme and transceiver architecture

According to (Batra et al., 2003, 2004a, 2004b), Multiband OFDM (MB-OFDM) scheme divides the available band into 14 sub-bands of 528 MHz each, as illustrated in Fig. 5. Each subband contains 128 subcarriers of which 10 are used for guard tones and can be used for various purposes, 12 are dedicated to the pilot signals and 100 are for information. It can be seen from the figure, each band group being made from three consecutive sub-bands, except for the fifth one which encompasses only the last two sub-bands. A WiMedia compatible device uses only one out of these six defined channels. Initially, most of the studies done in the literature have been performed on the first band group from 3.1 to 4.8 GHz.

The MB-OFDM system can transmit information at different data rates varying from 53.3 to 480 Mbps, listed in Table 1. These data rates are obtained through the use of different convolutional coding rates, frequency-domain spreading (FDS) and time-domain spreading (TDS) techniques. FDS consists in transmitting each complex symbol and its conjugate symmetric within the same OFDM symbol. It is used for the modes with data rates of 53.3 and 80 Mbps. With the TDS, the same information is transmitted during two consecutive OFDM symbols using a time-spreading factor of 2. It is applied to the modes with data rates between 53.3 and 200 Mbps.

In MB-OFDM, quadrature phase-shift keying (QPSK) and dual carrier modulation (DCM) are used for data modulation. For data rates lower than 320 Mbps, the constellation applied to the different subcarriers using quadrature phase-shift keying (QPSK) and for data rates of 320 Mbps and higher, the binary data is mapped onto two different 16-point constellations using a dual-carrier modulation (DCM) technique.

As illustrated the MB-OFDM transmitter in Fig. 6, the first LO1 signal down-converts RF signals to a fixed IF which is further down converted to another IF by LO2. As a unique feature of MB-OFDM transceiver, a single RF mixer is proposed for both RF to IF down conversion in the receiver and IF to RF up conversion in the transmitter.

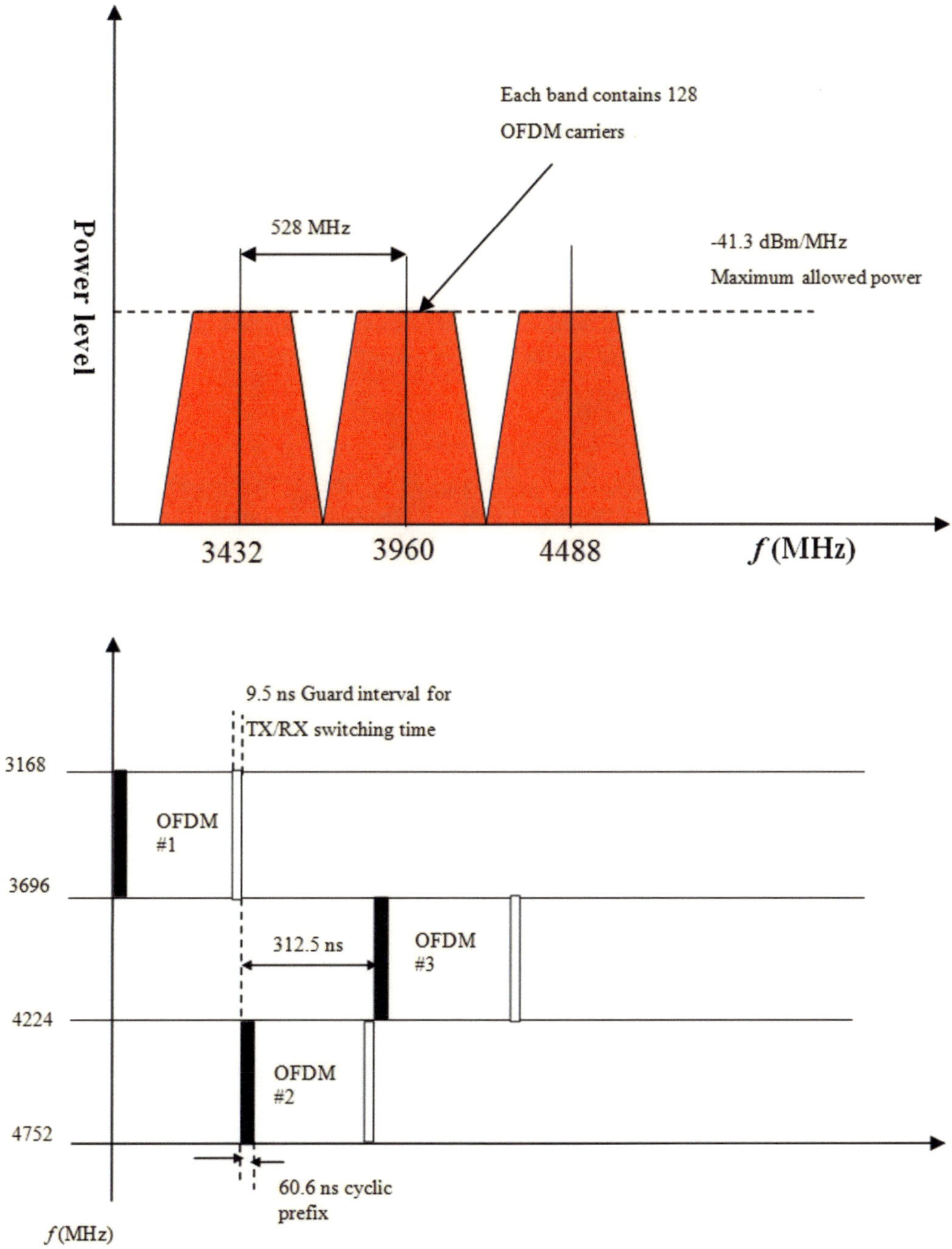

Figure 5. MB-OFDM system

Info. Data Rate (Mbps)	Modulation/ Constellation	FFT Size	Coding Rate (K=7)	Spreading rate
53.3	OFDM/QPSK	128	1/3	4
55	OFDM/QPSK	128	11/32	4
80	OFDM/QPSK	128	1/2	4
106.7	OFDM/QPSK	128	1/3	2
110	OFDM/QPSK	128	11/32	2
160	OFDM/QPSK	128	1/2	2
200	OFDM/QPSK	128	5/8	2
320	OFDM/QPSK	128	1/2	1
400	OFDM/QPSK	128	5/8	1
480	OFDM/QPSK	128	3/4	1

Data Rate =640Mbps*Coding Rate/Spreading

Table 1. Data rate dependent parameters [26]

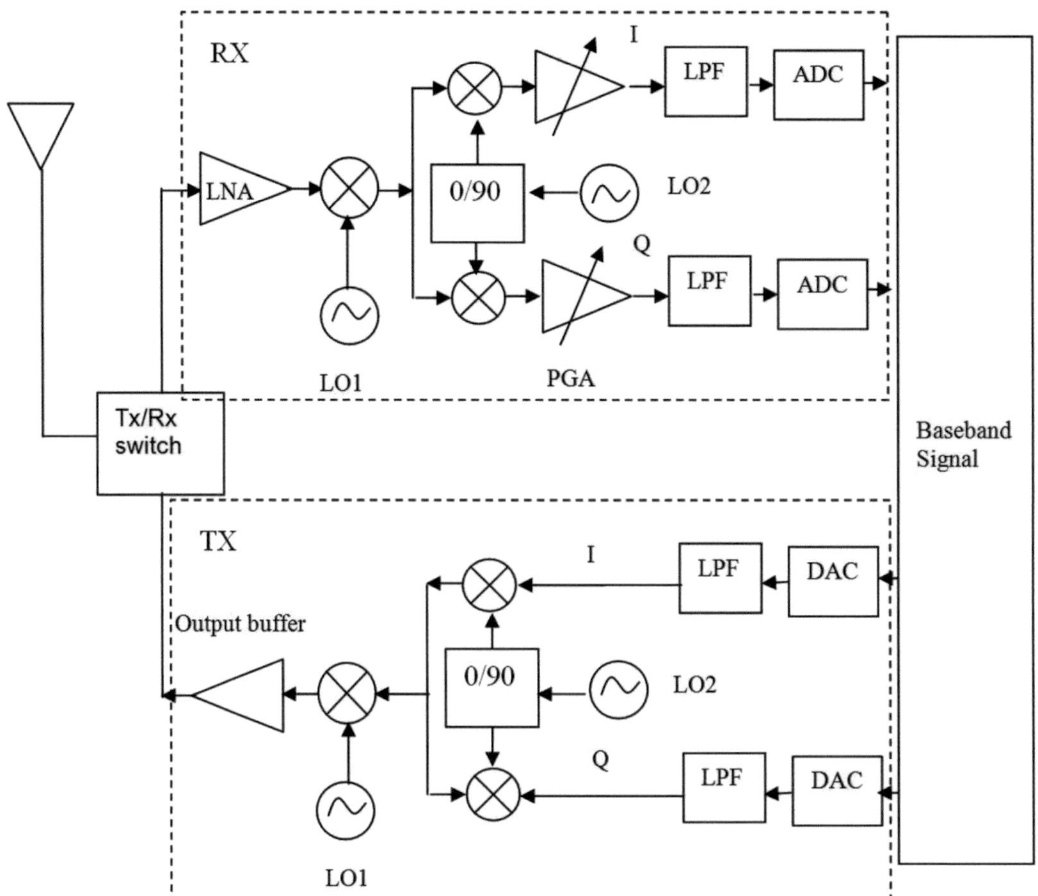

Figure 6. MB-OFDM Transceiver

7. UWB antennas

An UWB communication system requires transmitter and receiver with a wideband antenna. Antennas are the fundamental component of a communication system, both at the receiver and at the transmitter, subject to performance requirements while at the same time supporting demand constraints to incorporate it in terminals or network access points. The contradiction between requirements and constraints make the selection or the design of an antenna something difficult in the ultra-wideband (UWB) case as the large bandwidth places additional needs in comparison to narrowband radio. The second problem for the antenna designer is the lack of tools to evaluate the performance of an antenna embedded in a radio system, apart from tools intended to determine the antenna input impedance, gain, efficiency, and its radiation patterns. These tools are obviously quite important in order to describe where the direction of the radiation would go or would be received, and what signal power can be lost due to antenna losses, but nothing about the "matching" between the antenna and the channel. It is well known that matched filtering is a necessary requirement for optimal signal reception, therefore since both antennas and channels are filters that are involved in the transfer function between the signal to be transmitted and the signal at the receiver output. This means, we should analyze antenna performance and channel properties in a correlated manner, if optimization of the radio link performance is a goal to reach.

As Antennas are considered to be the largest components of integrated wireless systems; antenna miniaturization is necessary to achieve an optimal design. The printed antennas present good solution because of providing several advantages compared to the conventional microwave antennas. The main advantages are: lightweight, small volume, low-profile, planar configuration, compact, can be made conformal to the host surface, easy integrated with printed-circuit technology and with other MICs on the same substrate, low cost, allow both linear polarization and circular polarisation. Monopole disc antennas, with circular, elliptical and trapezoidal shapes, have simpler two-dimensional geometries and are easier to fabricate compared to the traditional UWB monopole antennas with three-dimensional geometries such as spheroidal, conical and teardrop antennas. These disc monopole antennas can be designed to cover existing and upcoming UWB communication applications, (Honda et al., 1992) & (Hammoud & Colomel, 1993).

In the last few years, circular monopole antennas have been studied extensively for UWB communications systems because of some appealing features (easy fabrication, feedgap optimization alone gives wide impedance matching and omnidirectional radiation patterns). One of the strongest competitors in terms of good impedance bandwidth, radiation efficiencies, and omnidirectional radiation patterns are the circular disc monopole (CDM) and elliptical antennas (Abbosh & Bialkowski, 2008; Allen et al., 2007; Antonino et al., 2003, Liang et al., 2004; Powell, 2004; schartz, 2005; Srifi et al., 2009). There is great demand for UWB antennas that offer miniaturized planar structure, so the vertical disc monopole is still not suitable for integration with a PCB. This drawback limits its practical application. For this reason, a printed structure of the UWB disc monopole is well desired, which consist on printed radiator disc on substrate. Printed CDM antennas can be fed simple microstrip line, coplanar waveguide (CPW), or slotted structures.

Figure 7. The prototype of simple fed CDM antenna

8. Conclusions

The objective of this chapter is to provide the fundamentals of UWB transceiver systems so that the general readers can be able to easily grasp some of the ideas in transceiver design for ultra-wideband communications. The chapter briefly describes signaling and modulation techniques, UWB transceiver system architecture, UWB antennas. Devices used for this exciting technology have become small, low power and low cost which in turn will accelerate their widespread use in indoor communications.

Author details

M. A. Matin

Institut Teknologi Brunei, Brunei Darussalam

9. References

Wentzloff, D.D. & Chandrakasan, A.P. (2006). Gaussian pulse generators for subbanded ultra-wideband transmitters, *IEEE Transactions on Microwave Theory and Techniques*, Vol. 54, No. 4, April 2006, pp. 1647-1655

Ranjan, M. & Larson, L. (2006). A sub-1mm2 dynamically tuned CMOS MB-OFDM 3-to-8GHz UWB receiver front-end, *IEEE International Solid-State Circuits Conference*, 2006, pp. 128-129

Zheng, H.; Lou, S.; Lu, D. et al. (2007). A 3.1-8.0GHz MB-OFDM UWB transceiver in 0.18μm CMOS, *IEEE Custom Integrated Circuits Conference*, 2007, pp. 651-654

Bergervoet, J.R.; Harish, K.S.; Lee, S. et al. (2007). A WiMedia-compliant UWB transceiver in 65nm CMOS, *IEEE International Solid-State Circuits Conference*, 2007, pp. 112-113.

Beek, R.; Bergervoet J.; Kundur, H. et al. (2008). A 0.6-to-10GHz receiver front-end in 45nm CMOS, *IEEE International Solid-State Circuits Conference,* 2008, pp. 128-129

Xia, L.; Shao, K.; Chen, H. et al. (2011). 0.15-nJ/b 3-5-GHz IR-UWB system with spectrum tunable transmitter and merged-correlator noncoherent receiver, *IEEE Transactions on Microwave Theory and Techniques,* Vol. 59, No. 4, April 2006, pp. 1147-1156

Xia, L.; Changhui Hu and Patrick Chiang (2011). Ultra Wideband RF Transceiver Design in CMOS Technology, Ultra Wideband Communications: Novel Trends - System, Architecture and Implementation, Mohammad Matin (Ed.), ISBN: 978-953-307-461-0, InTech.

Zheng, Y.; Wong, K.W.; Asaru, M.A. et al. (2007). A 0.18μm CMOS dual-band UWB transceiver, *IEEE International Solid-State Circuits Conference,* 2007, pp. 114-115

Zheng, Y.; Arasu, M.A; Wong, K.W. et al. (2008). A 0.18μm CMOS 802.15.4a UWB transceiver for communication and localization, *IEEE International Solid-State Circuits Conference,* 2008, pp. 118-119.

Yang, C.; Chen, K. & Chiueh, T. (2005). A 1.2V 6.7mW impulse-radio UWB baseband transceiver, *International Solid-State Circuits Conference,* 2005, pp. 442-443

Win, M. Z.; Scholtz, R. A.; and Barnes, M.A. (1997) "Ultra-Wide Bandwidth Signal Propagation for Indoor Wireless Communications," *Proc. IEEE ICC '97,* Vol. 1, pp. 56–60, June 1997.

Ge, L.; Yue, G.; and Affes, S. (2002) "On the BER performance of the Pulse Position Modulation UWB Radio in Multipath Channels, "*in Proc. of IEEE conference on Ultra Wideband Systems and Technologies, UWBST 2002,* Maryland, USA, pp. 231-234, May 2002.

Wu, Z.; Zhu, F.; Nassar, C.R. (2002) "High Performance, High Throughput UWB via Novel Pulse Shape and Frequency Domain Processing," *IEEE Ultra Wideband Systems and Technologies Conference,* Baltimore, MD, pp. 53-58, May 2002.

Wang, H; Ji, F; and Jiang, S. (2007) "Stability Analysis Of Timing Acquisition Methods With Multi-hypothesis in Indoor UWB Multipath And MAI Channel," *Eighth ACIS International Conference on Software Engineering, Artificial Intelligence, Networking, and Parallel/Distributed Computing*. IEEE, pp.478-483, 2007.

Nassar, C. R; Zhu, F. and Wu, Z. (2003) "Direct Sequence Spreading UWB Systems: Frequency Domain Processing for Enhanced Performance and Throughput," *in Proceedings of IEEE International Conference on Communications 2003, ICC 2003,* vol. 3, pp. 2180-2184, May 2003.

Hu, R.; Jou, C.F. (2010) "Complementary UWB LNA Design Using Asymmetrical Inductive Source Degeneration" *IEEE Microwave and wireless Components letters,* vol. 20, no.7, pp. 402-404.

Lee H-J; Ha, D.S.; Choi, S.S. (2006) "A 3 to 5GHz CMOS UWB LNA with Input Matching using Miller Effect" *2006 IEEE International Solid-State Circuits Conference,* pp. 731 – 740.

IEEE.15 Working Group for Wireless Personal Area Networks, "Detailed DS-UWB simulation results," Tech. Rep., IEEE 802.15, 2004, p. 802.

A. Batra et al. *"Multi-band OFDM Physical Layer Proposal for IEEE 802.15 Task Group 3a,"* IEEE P802.15 Working Group for Wireless Personal Area Networks (WPANs), doc:IEEE P802.15-03/268r4, 2004, 78 p.

A. Batra, J. Balakrishnan, G. R. Aiello, J. R. Foerster, and A. Dabak, "Design of a Multiband OFDM system for Realistic UWB Channel Environments," *IEEE Trans. on Microwave theory and Tech.*, vol. 52 No. 9, pp. 2123-2138, Sept. 2004.

Abbosh, A.M.; and Bialkowski, M.E. "Design of Ultrawideband Planar Monopole Antennas of Circular and Elliptical Shape," *IEEE Trans. On Ant. and Prop.*, vol. 56, no. 1, pp. 17–23, Jan. 2008.

Liang, J.; Chiau, C. C.; X. Chen, and C. G. Parini, Printed circular disc monopole antenna for ultra-wideband applications, *Electronics Letters*, vol. 40, no. 20, pp. 1246–1248, Sep. 2004.

Powell, J. Antenna design for ultra-wideband radio, *Massachusetts institute of technology*, May 2004

Schartz, H.; The Art and Science of Ultrawideband Antennas. *Artech House*, Inc., 2005.

Srifi, M. N.; Podilchak, S.K.; Essaaidi, M. and Y.M.M. Antar, Planar Circular Disc Monopole Antennas Using Compact Impedance Matching Networks for Ultra-Wideband (UWB) Applications, *Proc. of the IEEE Asia Pacific Microwave Conference*, pp. 782–785, Dec. 2009.

Hammoud, P. P. and Colomel, F., "Matching the input impedance of a broadband disc monopole," Electronics Letters, vol. 29, pp. 406-407, Feb. 1993.

Honda, S.; Ito, M.; Seki, H. and Jinbo, Y., "A disc monopole antenna with 1:8 impedance bandwidth and omni-directional radiation pattern," Proc. ISAP, pp. 1145-1148, Sep. 1992.

Chapter 2

Timing Synchronisation for IR-UWB Communication Systems

Rshdee Alhakim, Kosai Raoof and Emmanuel Simeu

Additional information is available at the end of the chapter

http://dx.doi.org/10.5772/48663

1. Introduction

The interest for Ultra Wide Band (UWB) technology is growing fast especially in the short-range indoor wireless communication, for example, in wireless personal area networks (WPAN). The basic concept is to transmit and receive baseband impulse waveform streams of very low power density and ultra-short duration pulses (typically at nanosecond scale). These properties of UWB give rise to fine time resolution, rich multipath diversity, low probability of detection, enhanced penetration capability, high user-capacity, and potential spectrum compatibility with existing narrowband systems [1]. However, one of the most critical challenges in enabling the unique benefits of UWB transmissions is timing synchronization, because the transmitted pulses are narrow and have low power density under the noise floor [2].

Timing synchronization in wireless communication systems typically depends on the sliding correlator between the received signal and a transmit-waveform template (Clean Template). In Impulse-Radio Ultra-Wideband (IR-UWB) devices however, this approach is not only sub-optimum in the presence of rich resolvable multipath channel, but also incurs high computational complexity and long synchronization time [2, 3]. Some research for improving the synchronization performance for IR-UWB systems has been reported in [4-9]. Each of these approaches requires one or more of the following assumptions: 1) the absence of multipath; 2) the absence of time-hopping (TH) codes; 3) the multipath channel is known; 4) high computational complexity and long synchronization time; and 5) degradation of bandwidth and power efficiency. Timing with Dirty Templates (TDT) is an efficient synchronization approach proposed for IR-UWB, introduced in [10-13]. This technique is based on correlating the received signal with "dirty template" extracted from the received waveforms. This template is called dirty; because it is distorted by the unknown channel and by the ambient noise. TDT allows the receiver to enhance energy capture even when the multipath channel and the Time-Hopping (TH) spreading codes are both unknown. Consequently, TDT approach contributes to enhance synchronization performance for IR-UWB and to reduce receiver structure complexity [14-15].

In general, the synchronization system consists of three units: signal detection, timing acquisition and tracking. Figure 1 depicts metaphorically the synchronization system block. Signal detection is the first unit, for deciding if the signal received is desired UWB signal or noise only [16]. Timing Acquisition unit is, a coarse synchronization, employed to find approximately a starting point of each received symbol and to reduce the timing error to within a fraction of UWB pulse duration [10-12]. The third step is a tracking to maintain and lock the satisfactory synchronization in the presence of timing offset variations in the received waveform, as a result of oscillator drifts or transmitter-receiver motion (Doppler effects) [17-18].

This chapter is organized as follows. Section 2 introduces the UWB signal model and presents the dirty template (DT) technique. Section 3 describes the DT detection of UWB signals in the presence of ambient noise and dense multipath channel. The Neyman-Pearson (NP) theorem is applied to set an optimal threshold and decide if the UWB signal is present or not. Then, the performance of the dirty template detector is evaluated in terms of the detection property and the false alarm property versus the threshold for different values of signal-to-noise ratio (SNR) and data-aided bit number. Section 4 extracts a timing acquisition estimator based on the dirty template algorithms. The performance of this estimator will be improved by modifying the structure of the cross-correlation operation. The focus of Section 5 is on the third unit of the synchronization system: Tracking. The delay locked loop DLL is a fundamental tracking technique used for maintaining the satisfactory synchronization and reducing timing error in UWB systems. DLL structure is derived and its loop parameters are selected to optimize noise and transient performance while taking Doppler effects into account. Section 6 ends the chapter with some conclusions.

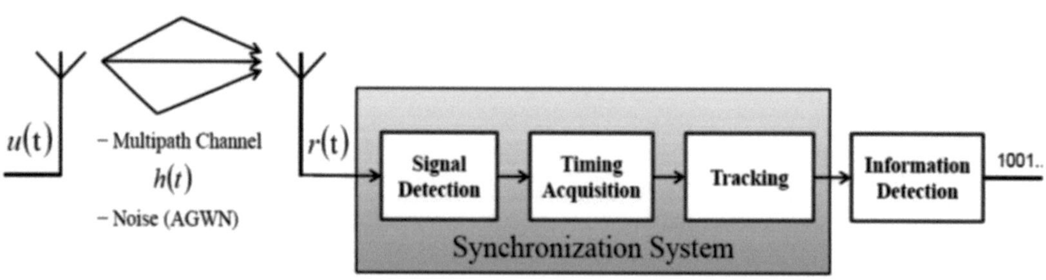

Figure 1. Synchronization System Block

2. UWB signal modelling

Consider an impulse radio UWB-IR system, where every symbol is transmitted over T_s period that consists of N_f pulses over N_f frames (one pulse per frame). Every frame of duration T_f contains N_c chips. The symbol waveform of duration $T_s := T_f N_f$ is $p_T(t) =$

$\sum_{j=0}^{N_f-1} p(t - jT_f - c_jT_c)$, where $p(t)$ is un ultra-short pulse, that has duration $T_p(\ll T_f)$, and $T_c := T_f/N_c$ is the chip duration with pseudo-random time-hopping (TH) codes $\{c_j\} \in [0, N_c - 1]$, $j \in [0, N_f - 1]$ (see Figure 2).

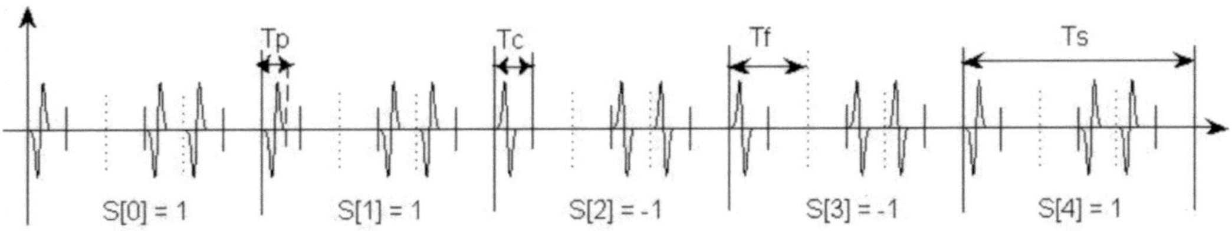

Figure 2. TH-UWB signal with PAM modulation ($N_f = 3, N_c = 2$, TH codes=[0, 1,0])

We consider that the symbol waveform has unit energy ($\int p_T{}^2(t)dt = 1$). By focusing on pulse amplitude modulation (PAM), where the information-bearing symbols $s[n] \in \{\pm 1\}$ are modelled as binary independent and identically distributed (i.i.d.) with energy $\mathcal{E}_s$ spread over N_f frames. The transmitted UWB waveform is then given by [15]:

$$u(t) = \sqrt{\mathcal{E}_s} \sum_{n=0}^{\infty} s[n] p_T(t - nT_s) \tag{1}$$

The signal $u(t)$ propagates through a multipath channel, whose impulse response $h(t) := \sum_{l=0}^{L-1} \alpha_l \delta(t - \tau_l)$ has coefficients α_l and delays τ_l, obeying $\tau_l < \tau_{l+1}$.The timing offset τ_0 refers to the first arrival time. To isolate τ_0, we define $\tau_{l,0} := \tau_l - \tau_0$ as the relative time delay of each channel tap, where $\tau_{L-1,0}$ is channel delay spread. To avoid inter-symbol interference (ISI), T_f is selected to satisfy the following condition: $T_f \geq \left(c_{N_f-1} - c_0\right) T_c + \tau_{L-1,0} + T_p$ [2]. The received pulse within each frame is $p_r(t) := \sum_{l=0}^{L-1} \alpha_l p(t - \tau_{l,0})$; The waveform in the output of the receiver antenna is:

$$r(t) = \sqrt{\mathcal{E}_s} \sum_{n=0}^{\infty} s[n] p_R(t - nT_s - \tau_0) + w(t) \tag{2}$$

where $p_R(t)$ is the received waveform of each symbol:

$$p_R(t) := \sum_{j=0}^{N_f-1} p_r\left(t - jT_f - c_jT_c\right) = \sum_{l=0}^{L-1} \alpha_l p_T\left(t - \tau_{l,0}\right) \tag{3}$$

and $w(t)$ represents the bandpass filtered zero-mean additive white Gaussian noise (AGWN) with power spectral density (PSD) $N_0/2$ and with double sided bandwidth B $(\gg 1/T_s)$ [11]. The timing offset τ_0 could be represented by: $\tau_0 = n_sT_s + \tau$, where $n_s = \lfloor \tau_0/T_s \rfloor \geq 0$ denotes the symbol-level timing offset, $\lfloor \cdot \rfloor$ represents the floor operation, and $\tau \in [0, T_s)$ the fine-timing offset [2]. By substituting τ_0 in (2), the received signal can be expressed by:

$$r(t) = \sqrt{\mathcal{E}_s} \sum_{n=0}^{\infty} s[n] p_R(t - nT_s - n_sT_s - \tau) + w(t) \tag{4}$$

Under mistiming ($\tau_0 \neq 0$), any T_s-long received segment of $r(t)$ can be represented by parts of two consecutive symbols (see Figure 3, the received UWB pulses are represented by triangles to illustrate the rich multipath effects of the channel), as bellow:

$$x(t+nT_s) = w(t+nT_s) + \begin{cases}\sqrt{\mathcal{E}_s}s[n-n_s-1]p_R(t+T_s-\tau): t\in[0,\tau)\\ \sqrt{\mathcal{E}_s}s[n-n_s]p_R(t-\tau) \qquad\qquad : t\in[\tau,T_s)\end{cases} \tag{5}$$

where $x(t)$ is the received segment of duration T_s.

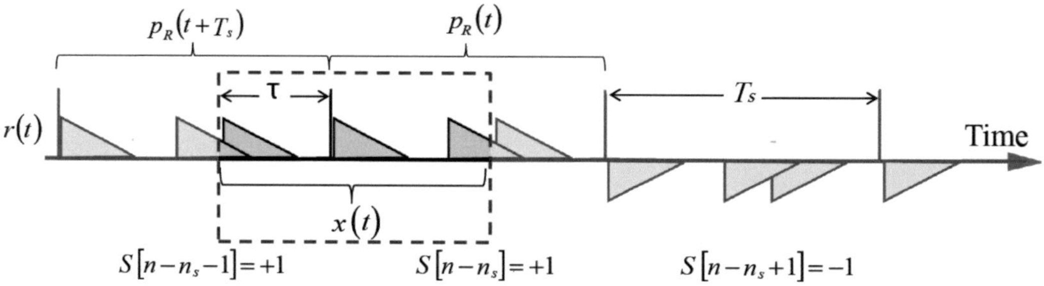

Figure 3. T_s-long observed received Symbol $x(t)$

The original TDT method proposed in [2], is defined as the cross-correlation between the pairs of successive received segments of duration T_s. That means: one segment in each pair of successive received symbols serves as the reference template signal for the other segment, as shown below:

$$R_{x,x}[n] = \int_0^{T_s} x(t+nT_s)x(t+(n+1)T_s)\,dt \tag{6}$$

$$= \int_0^{T_s}\{w(t+nT_s) + \sqrt{\mathcal{E}_s}\textstyle\sum_{m=0}^{1} s[n-n_s-m]p_R(t+mT_s-\tau)\} \times \{w(t+(n+1)T_s) + \sqrt{\mathcal{E}_s}\textstyle\sum_{m=0}^{1} s[n-n_s-m+1]p_R(t+mT_s-\tau)\}dt$$

$$= \breve{\omega}[n] + A\int_0^{T_s} p_R^2(t+T_s-\tau)\,dt + B\int_0^{T_s} p_R^2(t-\tau)\,dt$$

where the sampled noise $\breve{\omega}[n]$ explained in [11], is composed of three terms, two of them are the result of correlation between the symbol and the noise, and the third term is between shifted noises (noise-cross-noise). These terms could be treated approximately as mutually uncorrelated zero mean Gaussian noises. So $\breve{\omega}[n]$ could represent as zero-mean Gaussian noise, and its statistic properties are given as [11]:

$$\breve{\omega}[n] \sim \mathcal{N}\left(0, \sigma_w{}^2 = \frac{N_0{}^2}{2}BT_s + N_0\underbrace{\mathcal{E}_s\mathcal{E}_{max}}_{\mathcal{E}_r}\right) \tag{7}$$

Where: $\mathcal{E}_{max} = \int_0^{T_s} p_R(t)^2\,dt$ represents T_s-long received segment energy. The dirty template sample $R_{x,x}[n]$ is thus given as:

$$R_{x,x}[n] = \breve{\omega}[n] + A\int_{T_s-\tau}^{T_s} p_R^2(t)\,dt + B\int_0^{T_s-\tau} p_R^2(t)\,dt \tag{8}$$

$$where: \begin{cases}A = \mathcal{E}_s s[n-n_s-1].s[n-n_s]\\ B = \mathcal{E}_s s[n-n_s].s[n-n_s+1]\end{cases}$$

The possible values of A and B are exhibited in Table 1.

$s[n-1]$	+1	+1	-1	-1	+1	-1	+1	-1
$s[n]$	+1	-1	-1	+1	+1	-1	-1	+1
$s[n+1]$	-1	-1	+1	+1	+1	-1	+1	-1
A	+1	-1	+1	-1	+1	+1	-1	-1
B	-1	+1	-1	+1	+1	+1	-1	-1

Table 1. Possible values of A and B in (8)

The TDT technique can be applied to UWB receiver even in the presence of time hopping TH codes or Inter-Frame Interference (IFI), because the received segment and its dirty template contain the same TH codes and IFI properties regardless of the unknown channel characteristics (e.g. unknown time offset), but with the condition of the absence of ISI. Moreover, this method exploits the rich multipath diversity provided by UWB channels, and does not need to neither estimate the propagation channel nor generate clean correlation template at the receiver. Consequently, it reduces receiver complexity with high received energy capture.

3. Signal detection

At the receiver, detecting the received signal $r(t)$ and identifying the symbol-level offset n_s are achieved by using the dirty template data-aided (DA) algorithms. Supposing that we send M_1 training symbols, have the same value $\{s[n]=1\}_{n=0}^{M_1-1}$ (or all training bits are equal -1), they are thus received during $[\tau_0, \tau_0 + M_1 T_s]$. With the presence of M_1 training symbols, the successive received symbols have the same values. In this case, by looking on the Table 1 and taking the corresponding values of A and B, then substituting them in (8), $R_{x,x}[n]$ becomes:

$$R_{x,x}[n] = \mathcal{E}_s \int_0^{T_s} P_R^2(t)\,dt + \widetilde{\omega_{d1}}[n] = \underbrace{\mathcal{E}_s \mathcal{E}_{max}}_{\mathcal{E}_r} + \widetilde{\omega_{d1}}[n] = \mathcal{E}_r + \widetilde{\omega_{d1}}[n] \tag{9}$$

where $\widetilde{\omega_{d1}}[n]$ is the dirty template noise in the case of signal detection. Assume that under the noise-only hypothesis $\mathcal{H}_0$, we observe M_1 (i.i.d.) samples of $R_{x,x}[n] = \widetilde{\omega_{d0}}[n]$ for $n = 0,1,\dots,M_1-1$, while under signal-present hypothesis $\mathcal{H}_1$, we observe $R_{x,x}[n] = \mathcal{E}_r + \widetilde{\omega_{d1}}[n]$ for $n = n_s, n_s+1, \dots, n_s+M_1-1$. The detection problem is to distinguish between these two hypotheses:

$$\begin{aligned} &\mathcal{H}_0: R_{x,x}[n] = \widetilde{\omega_{d0}}[n] && n = 0,1,\dots,M_1-1 \\ &\mathcal{H}_1: R_{x,x}[n] = \mathcal{E}_r + \widetilde{\omega_{d1}}[n] && n = n_s, n_s+1, \dots, n_s+M_1-1 \end{aligned} \tag{10}$$

From (7), we could find easily that: In the case of the hypothesis $\mathcal{H}_0$, the UWB signal is absent and the signal-noise terms become zeros. Thus:

$$\widetilde{\omega_{d0}}[n] \sim \mathcal{N}\left(0, \sigma_0{}^2 = \frac{N_0{}^2}{2} B T_s\right)$$

$$\widetilde{\omega_{d1}}[n] \sim \mathcal{N}\left(0, \sigma_1^2 = N_0 \mathcal{E}_r + \frac{N_0^2}{2} BT_s\right) \quad (11)$$

The Neyman-Pearson (NP) detector, mentioned in [16], decides $\mathcal{H}_1$ and declares the presence of $r(t)$ if the likelihood ratio exceeds a threshold : $(J(n) > \eta)$, where η is a threshold set by the desired probability of false alarm (FA).

$$J(n) = \frac{p(R_{x,x}[n]; \mathcal{H}_1)}{p(R_{x,x}[n]; \mathcal{H}_0)} > \eta \quad (12)$$

$$J(n) = \frac{\frac{1}{(2\pi\sigma_1^2)^{M_1/2}} exp\left[-\frac{1}{2\sigma_1^2} \sum_{m=n}^{n+M_1-1} (R_{x,x}[m] - \mathcal{E}_r)^2\right]}{\frac{1}{(2\pi\sigma_0^2)^{M_1/2}} exp\left[-\frac{1}{2\sigma_0^2} \sum_{m=n}^{n+M_1-1} (R_{x,x}[m])^2\right]} > \eta$$

$$J(n) = \left(\frac{\sigma_0}{\sigma_1}\right)^{M_1} exp\left[-\frac{1}{2\sigma_1^2} \sum_{m=n}^{n+M_1-1} (R_{x,x}[m] - \mathcal{E}_r)^2 + \frac{1}{2\sigma_0^2} \sum_{m=n}^{n+M_1-1} (R_{x,x}[m])^2\right] > \eta$$

$$J(n) = \left(\frac{\sigma_0}{\sigma_1}\right)^{M_1} exp\left[\sum_{m=n}^{n+M_1-1} \left\{\left(-\frac{1}{2\sigma_1^2} + \frac{1}{2\sigma_0^2}\right) R_{x,x}^2[m] + \frac{\mathcal{E}_r}{\sigma_1^2} R_{x,x}[m] - \frac{\mathcal{E}_r^2}{2\sigma_1^2}\right\}\right] > \eta$$

Taking the logarithm of both sides does not change the inequality, so that:

$$ln(J(n)) = M_1 ln\left(\frac{\sigma_0}{\sigma_1}\right) + \sum_{m=n}^{n+M_1-1} \left\{\left(-\frac{1}{2\sigma_1^2} + \frac{1}{2\sigma_0^2}\right) R_{x,x}^2[m] + \frac{\mathcal{E}_r}{\sigma_1^2} R_{x,x}[m] - \frac{\mathcal{E}_r^2}{2\sigma_1^2}\right\} > ln(\eta)$$

$$\sum_{m=n}^{n+M_1-1} \left\{\left(-\frac{1}{2\sigma_1^2} + \frac{1}{2\sigma_0^2}\right) R_{x,x}^2[m] + \frac{\mathcal{E}_r}{\sigma_1^2} R_{x,x}[m]\right\} > ln(\eta) - M_1 ln\left(\frac{\sigma_0}{\sigma_1}\right) + \frac{M_1 \mathcal{E}_r^2}{2\sigma_1^2}$$

$$\sum_{m=n}^{n+M_1-1} \left\{\frac{\sigma_1^2 - \sigma_0^2}{2\mathcal{E}_r \sigma_0^2} R_{x,x}^2[m] + R_{x,x}[m]\right\} > \left(ln(\eta) - M_1 ln\left(\frac{\sigma_0}{\sigma_1}\right) + \frac{M_1 \mathcal{E}_r^2}{2\sigma_1^2}\right) \frac{\sigma_1^2}{\mathcal{E}_r}$$

From (11) we have $(\sigma_1^2 - \sigma_0^2 = N_0 \mathcal{E}_r)$, then

$$\sum_{m=n}^{n+M_1-1} \left\{\frac{N_0}{2\sigma_0^2} R_{x,x}^2[m] + R_{x,x}[m]\right\} > \left(ln(\eta) - M_1 ln\left(\frac{\sigma_0}{\sigma_1}\right) + \frac{M_1 \mathcal{E}_r^2}{2\sigma_1^2}\right) \frac{\sigma_1^2}{\mathcal{E}_r}$$

By replacing the value of σ_0^2 from (11) into the last equation, the test statistic T will be:

$$T = \sum_{m=n}^{n+M_1-1} \{R_{x,x}^2[m] + N_0 BT_s R_{x,x}[m]\} > \underbrace{\left(ln(\eta) - M_1 ln\left(\frac{\sigma_0}{\sigma_1}\right) + \frac{M_1 \mathcal{E}_r^2}{2\sigma_1^2}\right) \frac{\sigma_1^2 N_0 BT_s}{\mathcal{E}_r}}_{\xi} \quad (13)$$

The Neyman-Pearson (NP) detector in (13) decides $\mathcal{H}_1$ and declares the presence of signal if the test statistic T exceeds a threshold $T > \xi$ .The architecture of the dirty template detector (13) is shown in Figure 4. To verify the performance of the detection, first the false alarm property P_{FA} and the detection property P_D are identified as following:

$$P_{FA} = Pr\{T > \xi; \mathcal{H}_0\} \quad (14)$$

$$P_D = Pr\{T > \xi; \mathcal{H}_1\} \quad (15)$$

But it is not easy to find the probability that the test statistic value T proposed in (13), exceeds a threshold ξ ($Pr\{T > \xi\}$) by analytical means, so we must exploit a Monte Carlo simulation. We plot the results of the Monte Carlo evaluation of $Pr\{T > \xi\}$ versus ξ in Figures 5-7 for different values of SNR and data-aided number. The solid curve represents false alarm property $Pr\{T > \xi; \mathcal{H}_0\}$ and the dashed curve represents detection property $Pr\{T > \xi; \mathcal{H}_1\}$. From these figures we could determine the optimal value of the threshold ξ where the difference between the detection property and the false property is maximized $(max(P_D - P_{FA})|_\xi)$. Furthermore, Figure 5 shows the interference area between the two curves, this area provides an indication on possible errors. When we reduce the false property by increasing ξ, the detection property will be simultaneously reduced. Figure 6-7 show that by increasing the SNR or data aided number, the two curves are clearly separated, which decreases the interference area. That is why the detection error property is reduced and the detection performance is enhanced, but these improvements are in the price of power efficiency or detector design simplification.

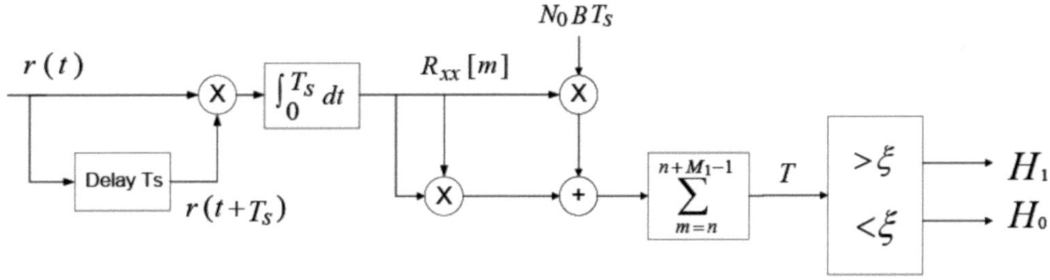

Figure 4. Block diagram of detection model for Dirty Template System

Another way of representing the detection performance of a Neyman-Pearson detector is to plot the detection property P_D versus the false alarm property P_{FA}. Each point on the curve corresponds to a value of (P_{FA}, P_D) for a given threshold ξ. As we have already found when ξ increases, P_{FA} and P_D decrease. This type of plot is called the receiver operating characteristic (ROC), shown in Figures 8-9. The ideal detector is when $P_D = 1$ for any value of P_{FA}. So it is clear that when the SNR or the data aided number increases the ROC curve approaches to the ideal case and the detection performance is improved.

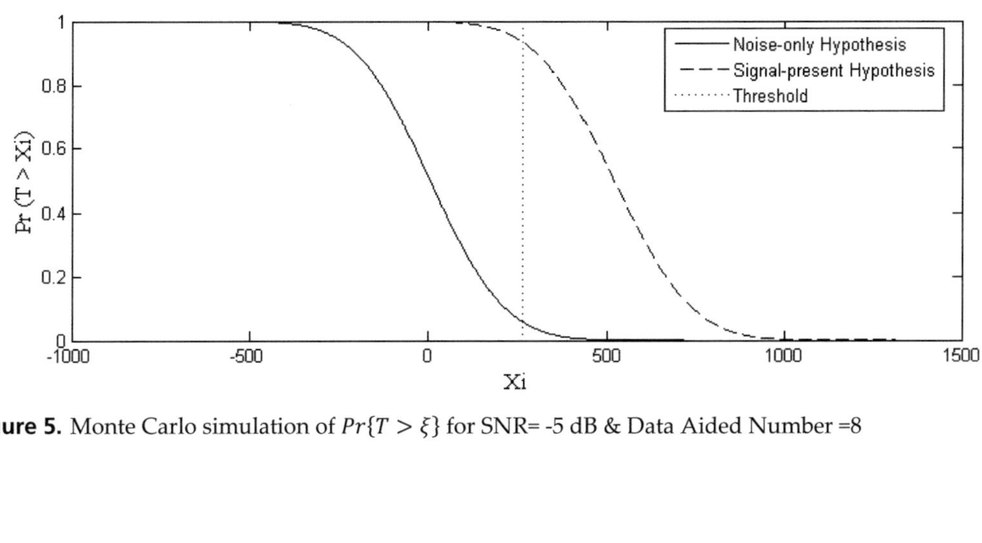

Figure 5. Monte Carlo simulation of $Pr\{T > \xi\}$ for SNR= -5 dB & Data Aided Number =8

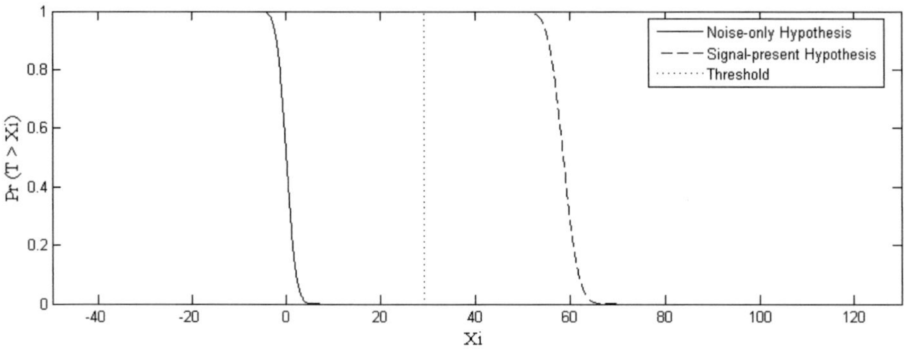

Figure 6. Monte Carlo simulation of $Pr\{T > \xi\}$ for SNR= -5 dB & Data Aided Number =32

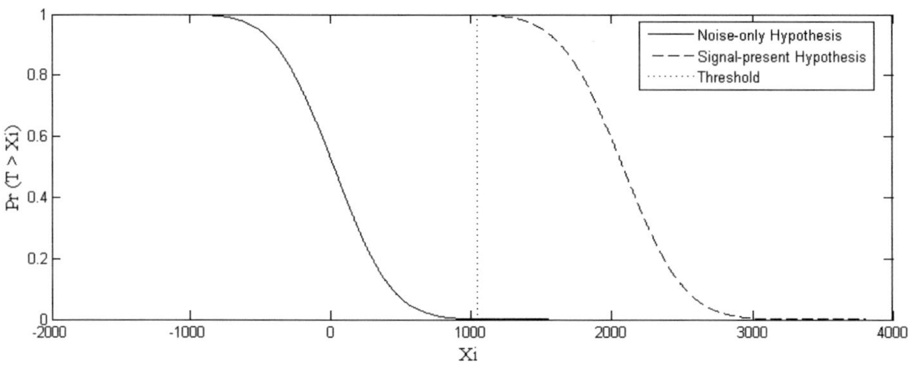

Figure 7. Monte Carlo simulation of $Pr\{T > \xi\}$ for SNR= 5 dB & Data Aided Number =8

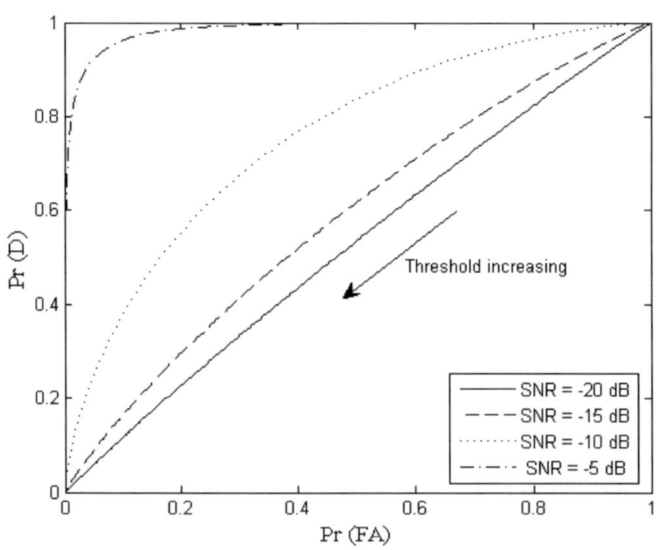

Figure 8. ROC for Data Aided Number =8

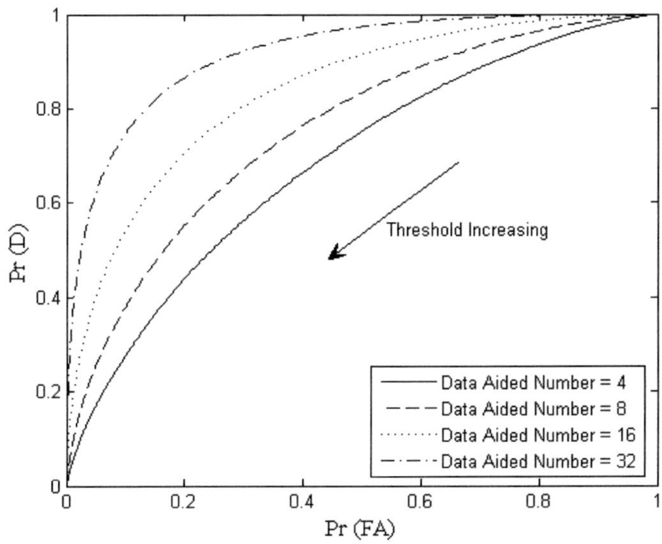

Figure 9. ROC for SNR= -10 dB

In this simulation, we select the pulse $p(t)$ as the third derivative of the Gaussian function with unit energy and duration $T_p \approx 1$ ns. Each symbol contains $N_f = 24$ frames with duration $T_f = 100$ ns. After the signal detection is done, the symbol-level timing offset $\hat{n}_s$ can be estimated via a line search to maximize the test statistic $T(n_s)$,as show below:

$$\hat{n}_s = arg\ max\ T(n_s); \qquad T(n_s) = \sum_{m=n_s}^{n_s+M_1-1}\left(R_{x,x}^2[m] + N_0 B T_s R_{x,x}[m]\right) \tag{16}$$

To avoid the influence of neighboring information-bearing symbols on $\hat{n}_s$ estimating, we separate between the data-aided symbols and information-bearing symbols by four-symbol pattern $(-1, -1,\ +1, +1)$. This pattern permit to exhibit a unique maximum peak at $n = \hat{n}_s$.

In this section, UWB signal detection has been carried out based on the Neyman-Pearson theory in (13) (Figure 4). Then, detection performance of the proposed method has been evaluated by Matlab simulation results (Figures 5-9). The analysis shows that the detection performance of the dirty template approach is improved by increasing the signal-to-noise ratio (SNR) or data-aided (DA) number. But these improvements reduce the power efficiency and complicate the receiver design. After detecting the UWB signal, the symbol-level timing offset estimation relies on searching the best statistics T in (16).

In practice, before the signal detection system runs for the first time, the value of ξ in (13) should be adjusted by sending train of data-aided symbols and by utilizing temporarily feedback loop, the optimal value of the threshold ξ is determined where the difference between the detection property and the false property is maximized $(max(P_D - P_{FA})|_{\xi})$. For the bandwidth B in (13), it is specified by The UWB standards approved by the FCC (e.g. $B \approx 5 - 7$ GHz for indoor applications) [19]. And for the background noise energy N_0 in (13), it could be estimated by listening to the environment during noise-only period.

4. Timing acquizition

Supposing that signal detection and symbol-level offset estimation are correctly achieved, the new time offset is confined within one symbol duration: $\tau_0 = \tau \in [0, T_s)$. Thus the objective of this section is to estimate the timing offset τ_0 based on dirty template algorithms in both data aided (DA) and non-data aided (NDA) modes. The original TDT acquisition method, proposed in [2], is based on searching a peak in the output of the correlation between the received signal and a dirty template, as shown below:

$$\hat{\tau}_0 = \underset{m}{arg\,max}\ \frac{1}{M_2}\sum_{n=0}^{M_2-1}\left(\int_0^{T_s} x(t+nT_s+\tau_m).x(t+(n+1)T_s+\tau_m)\,dt\right)^2 \qquad (17)$$

where M_2 is the size of the successive received segments, used for accomplishing the TDT acquisition operation, $\tau_m = m\Delta_t \in [0, T_s)$; Δ_t represents the size of the increment, and m denotes the number of increments. Since the synchronization accuracy can be improved by reducing Δ_t, but in the price of fast estimation timing. Practically, in DA mode, the form of training symbols used for achieving fast acquisition is as follows [2]:

$$s[n] = \begin{cases} -1, & if(n \bmod 4) = 0, or, 1 \\ +1, & if(n \bmod 4) = 2, or, 3 \end{cases} : n \in [0, M_2 - 1] \qquad (18)$$

where (X mod Y) denotes the modulo operation, where X and Y are both real. We can explain TDT in (17) by another way: when we do the cross-correlation between signals with its sliding replica, we obtain the unique maximum point at $\tau_0 = 0$. The TDT use the same principle to achieve the synchronization.

However, we can notice from Figure 10, in the UWB received signal, the presence of gaps between the received symbol and its dirty template. These gaps may cause multiple maxima points around the peak (optimal point) at the correlator output in (17), so the estimation error of the timing synchronization may increase. To avoid this problem, we modify the structure of the cross-correlation operation by adding a suitable window filter. In this case, the timing estimation in (17) can be achieved, as follows:

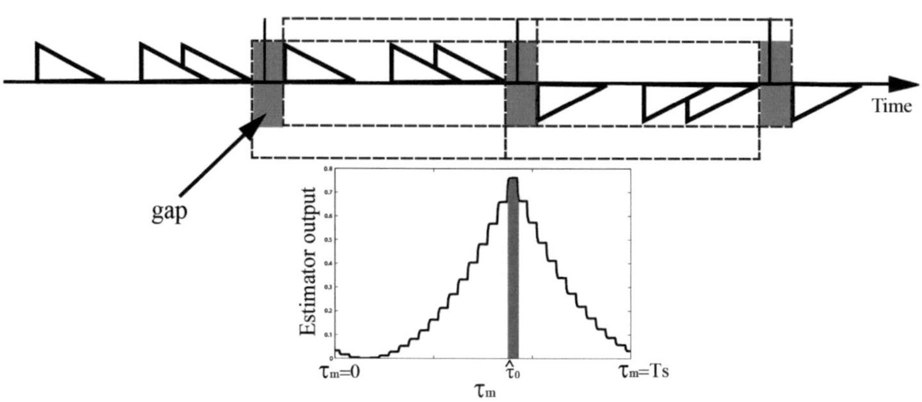

Figure 10. Timing reference illustration

$$\hat{\tau}_0 = \underset{m}{arg\,max}\ \frac{1}{M_2}\sum_{n=0}^{M_2-1}\left(\int_0^{T_s} x(t+nT_s+\tau_m).x(t+(n+1)T_s+\tau_m).W(t)dt\right)^2 \quad (19)$$

where the window $W(t)$ contains the information of spreading TH codes, as shown below:

$$W(t) = \sum_{j=0}^{N_f-1} c_j p\left(t - jT_f\right);\ p(t) = \begin{cases} 1 & si\ 0 \le t \le T_p + \tau_{L-1,0} \\ 0 & otherwise. \end{cases} \quad (20)$$

As the time of channel delay spread $\tau_{L-1,0}$ is unknown, we assume that the window width could be practically inside the range $\{T_c, 2T_c\}$. Figure 11 illustrates the structure of this window filter. We can notice that this window filter reduces the effect of these gaps at the UWB receiver. However, the window may slightly reduce the energy capture at the output of cross-correlation operation, but fortunately that doesn't impact on the estimation accuracy.

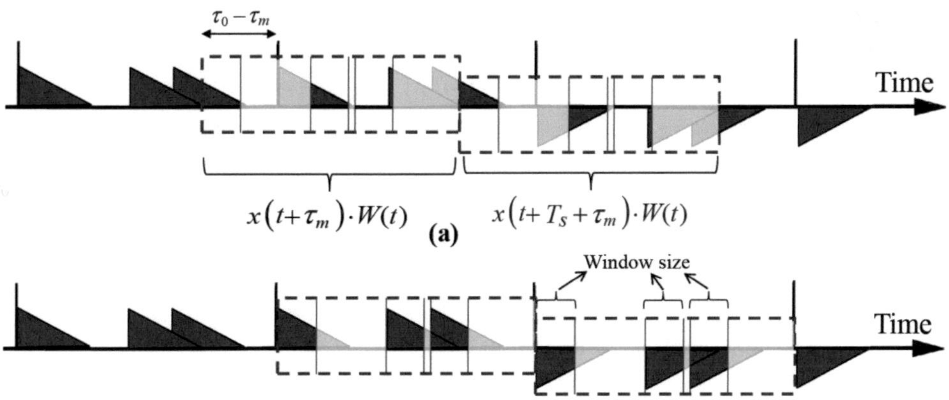

Figure 11. Window filter illustration, (a): $\tau_m \neq \tau_0$, (b): $\tau_m = \tau_0$

Figures 12-14 show the captured energy at the correlator output for various values of $\tau_m = m\Delta_t \in [0, T_s)$. Each of these figures has three graphs, one (*dashed line*) represents the DT estimator without any window, and the others (*dotted & solid lines*) represent the DT estimator with window size equal to $6T_c$ and $2T_c$ respectively. We have studied the UWB system without and with TH codes and with SNR= (20 or 3) dB. These figures show that the correlator output without window in (17) has multiple maxima points around the peak. Besides, they exhibit that decreasing the window size in (20) leads to decrease the maxima point number, until we reach a single maxima point (at $\tau_m \approx \tau_0$) in (19). Moreover, the window filter also contributes to reduce the unwanted noise effect on the timing estimation. For example, when window size equals $2T_c$, the received noise energy will be equal to (N_0) multiplied by the normalized window size $(2T_c/T_f)$. In practice, this window $W(t)$ may simplify the implementation of timing operation, because we don't need to take all sampled points of the symbol to calculate the cross-correlation, the sample points inside the windows are sufficient.

To evaluate the performance of the proposed DT synchronizers for DA & NDA modes, we run a simulation in Matlab. In this simulation, we select the pulse $p(t)$ as the third derivative of the Gaussian function with unit energy and duration $T_p \approx 1$ns. Each symbol contains $N_f = 24$ frames with duration $T_f = 100$ns. The simulations are performed in a Saleh-Valenzuela channel [20]. The parameters of this channel are chosen with (1/Λ, 1/λ, Γ, γ) = (2, 0.5, 30, 5) ns. The maximum channel delay spread of the channel is about 99 ns, and the inter-symbol interference (ISI) is negligible. We generate τ_0 randomly from a uniform distribution over $[0, T_s)$. We employ TH spreading codes of period N_f, which is generated from a uniform distribution over $[0, N_c - 1]$, with $N_c = 6$, and $T_c = 10$ns. We supposed the size of the increment Δ_t equal to $T_c/2$.

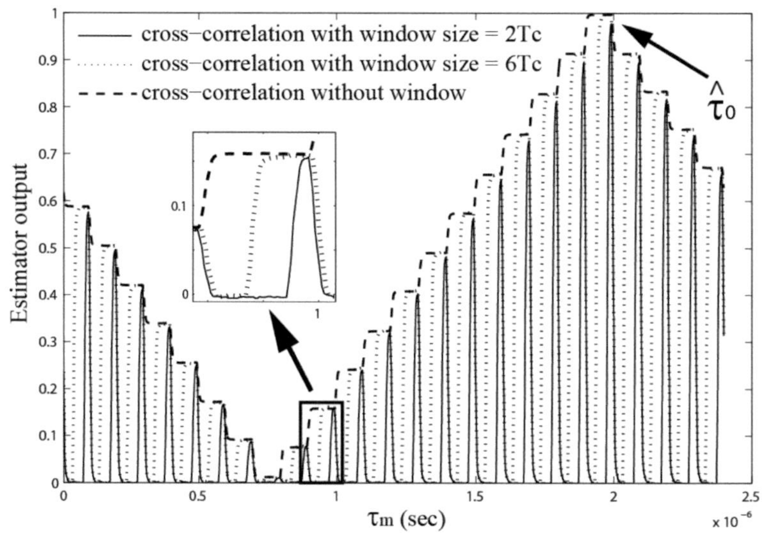

Figure 12. Correlator output energy (without TH codes, SNR=20 & M_2=16)

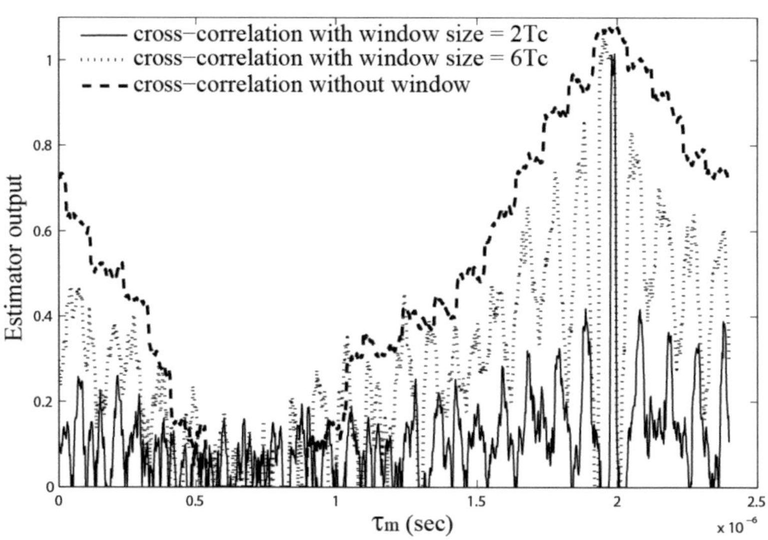

Figure 13. Correlator output energy (TH codes, SNR=3 & M_2=16)

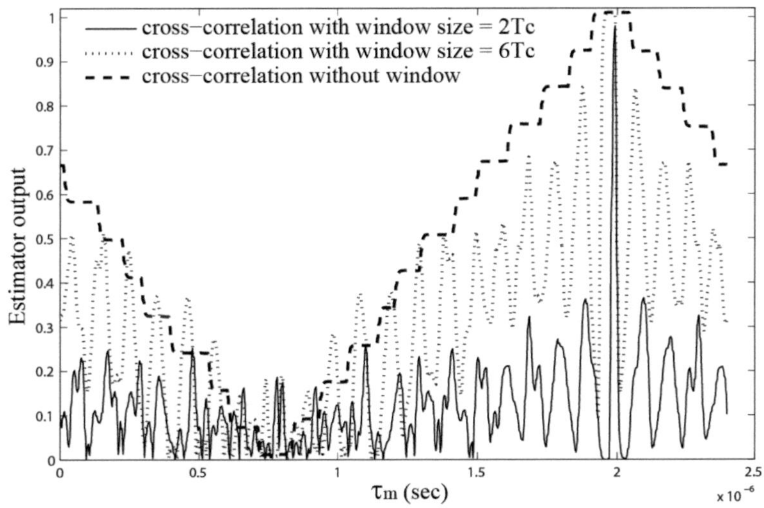

Figure 14. Correlator output energy (TH codes, SNR=20 & M_2=16)

We compare the accurate acquisition between the DT estimator without any window mentioned in (17) and the modified estimator with window size equal to $2T_c$ as mentioned in (19). Figures 15-16 show the comparison of mean square error (MSE) performance in DA & NDA modes with dirty templates for various values of M_2. In these figures, the MSE results are normalized by T_s^2, and plotted versus the signal-to-noise ratio (SNR) per pulse. The simulations confirm that, the DT estimator with window (*solid lines*) has higher timing estimation performance than the DT estimator without window (*dashed lines*). In general, as M_2 increases the normalized MSE decreases. Increasing SNR also helps to reduce the MSE. That means, when the observation number M_2 or the signal-to-noise ratio SNR is increased, the mean square error MSE is reduced and the estimation performance is improved. In addition, we can clearly notice that all curves with high SNR reach an error floor, which depends on the synchronization accuracy. This error floor can be reduced by decreasing the size of (Δ_t), but with increasing simulation time and computational complexity.

The results also show that the timing estimator for DA mode in Figure 15 has smaller MSE values and more accurate timing simulation than NDA mode in Figure 16 for the same SNR & M_2, but with less bandwidth efficiency. Moreover, Figure 15 shows that the DA estimator could be used with small training pattern size such as $M_2 = 4$, that helps to reduce the number of operations performed at the receiver as well as the synchronization time.

In this section, we present the original TDT algorithm used for achieving rapid, accurate and low-complexity timing acquisition. It relies on searching a peak in the output of the correlation between the received signal and a dirty template. But we have found the presence of multiple maxima points around the peak at the output of the correlator, and that may increase the complication to estimate the timing offset error (TOE). To avoid this problem, we modify the structure of the cross-correlation, by adding the suitable window filter. This window contains information of the TH codes. This modified approach guarantees that we obtain a single maximal peak at the outside of the estimator and that improve the estimation error performance. The simulation results show the estimation performances of DA and NDA modes of the proposed DT method, and confirm that for the same size of correlation sequence pattern, the DA mode has high performance and fast timing, compared to NDA mode, but that is in the price of the bandwidth efficiency.

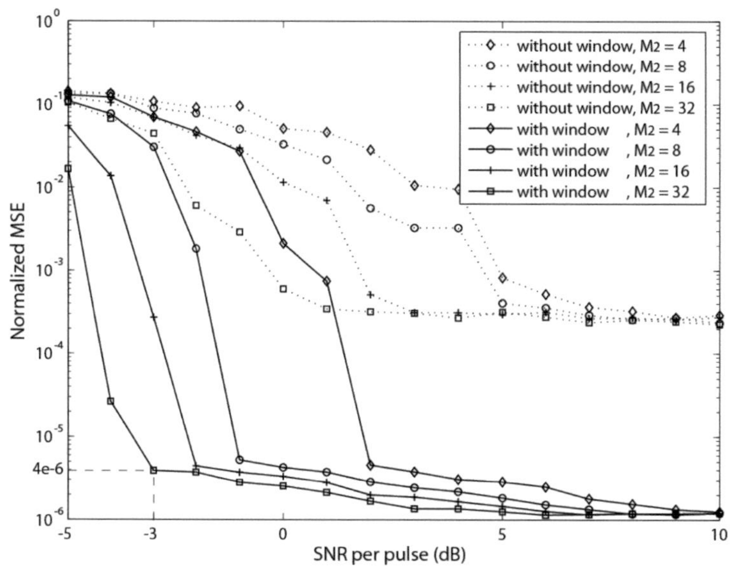

Figure 15. Normalized MSE vs. SNR per pulse for DA mode (window size = 2Tc)

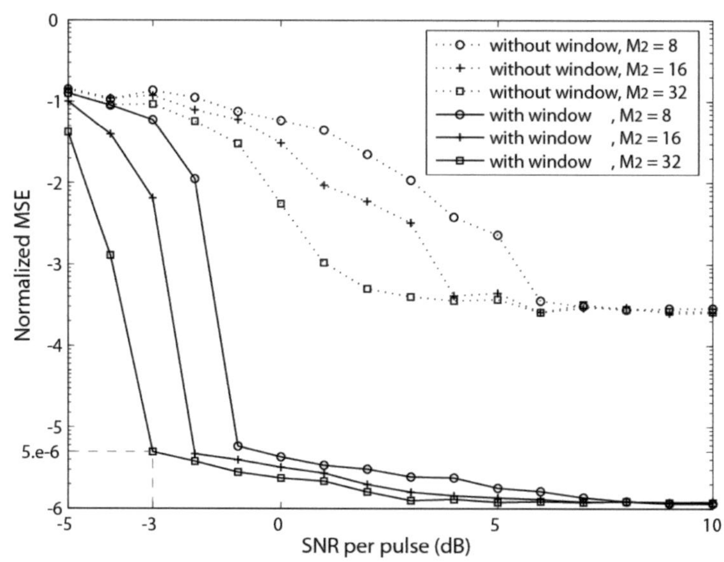

Figure 16. Normalized MSE vs. SNR per pulse for NDA mode (window size = 2Tc)

5. Tracking

The next step, after detecting UWB signals and estimating the timing offset, is to maintain and lock the satisfactory synchronization between the receiver and the received data signals. Tracking unit is used to alleviate the effects of timing offset variations due to transmitter-receiver motion (Doppler effects) and to maintain the transmission quality. For tracking purpose, we will use Delay-Locked Loop (DLL) approach which is considered as a fundamental tracking technique for UWB and spread-spectrum devices [21]. Several DLL schemes have been proposed to improve the tracking performance in UWB systems [22-24]. This section shows how to combine DLL with TDT, which enables adaptively following timing offset variations in the received signals and to enhance the BER performance.

Assuming that signal detecting and timing acquisition have been correctly achieved, so the timing error is confined within a fraction of UWB pulse duration and the tracking loop is *"in lock"* at the beginning of its operation. Figure 17 shows the proposed DLL method, known also as Early-Late Gate (ELG), consists of two correlator branches. The first branch is called *early correlator* because the received signal is cross-correlated with the advanced version of the reference signal $y(t+\Delta)$. And the second is called *late correlator*, where the received signal is correlated with the retarded version of the reference signal $y(t-\Delta)$. In other word, $y(t)$ is advanced and delayed in time by Δ. The correlation outputs are given by:

$$\begin{cases} R_e = \int_0^{T_s} r(t) y(t-\hat{\tau}+\Delta)\, dt \\ R_l = \int_0^{T_s} r(t) y(t-\hat{\tau}-\Delta)\, dt \end{cases} \tag{21}$$

where the reference signal $y(t)$ is extracted from the received waveform based on training (data-aided) sequence. We can employ the same training bits which have already been used to detect UWB signal (or even which have been used to acquire the timing). Assuming that M transmitted training bits equal one $\{s[n]=1\}_{n=1}^{M}$, in this case, the noise-free part of $r(t)$ in (2) consists of $p_R(t)$ replicas with spacing T_s. Hence, $y(t)$ can be approximately yield from $r(t)$ by averaging operation:

$$y(t) = \frac{1}{M}\sum_{n=1}^{M}\big(p_R(t-nT_s)+w_n(t)\big);\quad t \in [0,T_s) \tag{22}$$

Then:

$$y(t) \cong p_R(t) + \bar{w}(t):\quad \bar{w}(t) = \frac{1}{M}\sum_{n=1}^{M} w_n(t) \sim \mathcal{N}\left(0, \bar{\sigma}^2 = \frac{N_0}{2M}\right) \tag{23}$$

We can notice that an increase in the number of data-aided bits M lead to reduce the noise variance $\bar{\sigma}^2$ and improve the tracking performance. Another way to decrease the amount of noise term $\bar{w}(t)$ inside the reference template $y(t)$ in (23) is to pass $y(t)$ through the window filter $W(t)$, mentioned in (20). Assume that the window size equals $1.5T_c$, the noise energy $\bar{w}(t)$ will be divided by $(1.5\,T_c/T_f)$.

In the presence of oscillator drifts or transmitter-receiver motion, this will produce the mistiming $(\tau \neq 0)$ between the reference and received signals. Figure 17 shows that DLL tracking is performed by estimating the time offset τ between $r(t)$ and $y(t)$. Then, it compensates this offset by shifting the reference signal position to $y(t-\hat{\tau})$, where $\hat{\tau}$ denotes the DLL estimate of τ.

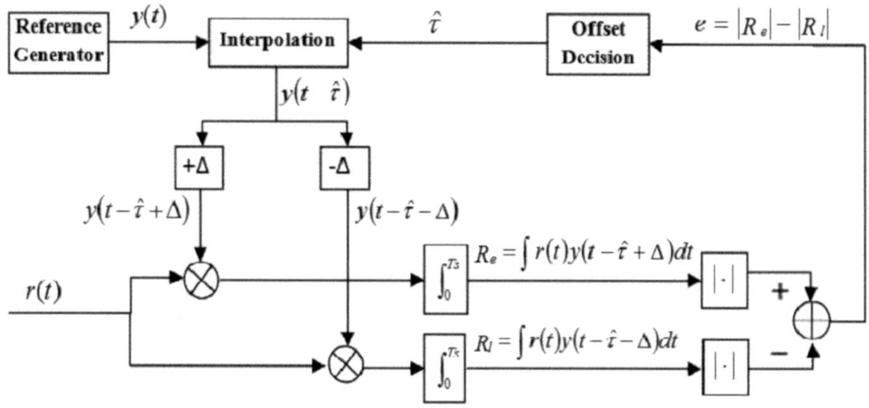

Figure 17. Delay-Locked Loop diagram

Supposing the time offset $\epsilon = \hat{\tau} - \tau$ is within UWB pulse duration (the tracking loop is *"in lock"*). Substituting (2) and (23) in (21), the cross-correlation outputs $R_{e,l}[\epsilon, n]$ become:

$$R_{e,l}[\epsilon,n] = \int_0^{T_s}\{s[n]p_R(t) + w_n(t)\} \times \{p_R(t-\epsilon \pm \Delta)\overline{w}(t)\}dt\,; \quad t \in [0,T_s) \tag{24}$$

Thus:

$$R_{e,l}[\epsilon,n] = s[n]\Gamma_{e,l}[\epsilon] + n_{e,l}[\epsilon,n] \tag{25}$$

where:

$$\Gamma_{e,l}[\epsilon] = \int_0^{T_s} p_R(t)p_R(t-\epsilon \pm \Delta)\,dt$$

Like as we have seen before in (6), the sample noise factors $n_{e,l}[\epsilon,n]$ is composed of three terms, two of them are the result of correlation between the symbol and the noise, and the third term is as noise-cross-noise [11]. $n_{e,l}[\epsilon,n]$ could represent as zero-mean Gaussian noises $n_{e,l} \sim \mathcal{N}(0, \sigma_{e,l}^2)$:

$$n_{e,l}[\epsilon,n] = \xi_1 + \xi_2 + \xi_3 \tag{26}$$

where:

$$n_{e,l}[\epsilon,n]: \begin{cases} \xi_1 = \int_0^{T_s} p_R(t-\epsilon \pm \Delta)w_n(t)\,dt \\ \xi_2 = \int_0^{T_s} s[n]p_R(t)\overline{w}(t)\,dt \\ \xi_3 = \int_0^{T_s} w_n(t)\overline{w}(t)\,dt \end{cases}$$

From Figure 17, the difference value (discriminator) between the correlator outputs in the ELG tracking serves as a timing error indicator. For example, if the discriminator (e) is close to zero, then $y(t)$ and $r(t)$ are synchronized perfectly ($\hat{\tau} \approx \tau$). Otherwise, $y(t)$ is too late or too early. The discriminator error e at the output of DLL tracking is expressed by:

$$e[\epsilon, n] = R_e[\epsilon, n] - R_l[\epsilon, n]. \tag{27}$$

Substituting (25) in (27), we get:

$$e[\epsilon, n] = s[n] \underbrace{(\Gamma_e[\epsilon] - \Gamma_l[\epsilon])}_{\mathbb{S}[\epsilon]} + \underbrace{n_e[\epsilon, n] - n_l[\epsilon, n]}_{\eta[n]} = s[n] \cdot \mathbb{S}[\epsilon] + \eta[n] \tag{28}$$

where $\eta[n] = n_e[\epsilon, n] - n_l[\epsilon, n]$ is the equivalent additive noise for DLL tracking system, and $\mathbb{S}[\cdot]$ is the loop discriminator characteristic (S-curve), which is the useful term for tracking τ. From (28), we notice that the noise-free part of discriminator error (e) depends on the time offset ϵ and also on the data sign $s[n]$. To avoid the influence of the received date $s[n] \in \{\pm 1\}$ on the tracking performance, we add an absolute value block inside the DLL diagram. The absolute value method is chosen to make the tracking error indicator independent of the unknown data sign, is as follows:

$$e[\epsilon, n] = (R_e[\epsilon, n] - R_l[\epsilon, n]) \times sign(R_e[\epsilon, n] + R_l[\epsilon, n]) \tag{29}$$

Hence, the discriminator error e in (28) becomes as follows:

$$e[\epsilon, n] = \mathbb{S}[\epsilon] + \eta[n] \tag{30}$$

After series of mathematical manipulation, we found that the noise $\eta[n]$ could be treated as a zero mean Gaussian noise:

$$\eta[n] \sim \mathcal{N}(0, \sigma_\eta^2) \quad : \sigma_\eta^2 = \frac{N_0(M+1)}{M} (\mathcal{E}_{max} - \Gamma_{rr}(2\Delta)) \tag{31}$$

where $\mathcal{E}_{max} = \int_0^{T_s} p_R(t)^2 \, dt$ represents T_s-long received segment energy, and $\Gamma_{rr}(2\Delta) = \int_0^{T_s} p_R(t) \, p_R(t - 2\Delta) dt$.

Assuming that the time offset ϵ is small to be inside the linear range of the S-curve $\mathbb{S}[\epsilon]$. In this case, the DLL behaves approximately like a linear filter ($\mathbb{S}[\epsilon] = A\epsilon$, where $A = \dot{\mathbb{S}}[0]$ is the gradient of the discriminator characteristic at $\epsilon = 0$). The DLL tracking system could be represented by the equivalent model shown in Figure 18. We focus now on designing a fit offset decision block, which is used to update the current estimate of time offset $\hat{\tau}$, to reduce then the offset error ϵ and to maintain timing synchronization. Let us consider the case where the dynamic input $\tau(t)$ has a ramp form:

$$\tau(t) = \tau_0 + mt \tag{32}$$

where τ_0 represents a constant time delay and m is a Doppler shift which is produced by a constant velocity relative to transmitter and receiver. For tracking efficiently a ramp input, the second-order DLL is selected [17]. Figure 18 depicts the suggested DLL structure with filter $F(s) = (1 + k_2 s)/(k_1 s)$, where the constants $\{k_1, k_2\}$ are loop parameters which will be found to optimize the tracking performance. Letting the closed loop function behave like lowpass filter with noise equivalent bandwidth $B_L \ll 1/T_s$, which can be found as (see [17, Table 3.3.1]):

$$B_L = \frac{\xi\omega_n}{2}\left(1+\frac{1}{4\xi^2}\right) \tag{33}$$

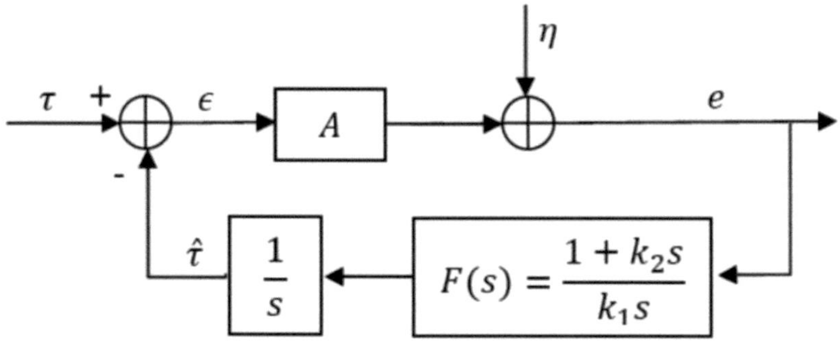

Figure 18. Linear Model of the second-order DLL

where the natural frequency $\omega_n = (A/k_1)^{1/2}$ and the loop damping ratio $\xi = (k_2\omega_n)/2$. With ramp input signal, DLL tracking system is subject to two error sources: additive noise error and transient error. So we are seeking to determine suitable values of the loop parameters $\{\omega_n, \xi\}$ for optimizing the performance criterion: minimizing noise energy effect as well as maintaining specified transient error. For selecting $\{\omega_n, \xi\}$, we apply Wiener-filter theory [25]. This optimization method allows us to minimize the following design criterion: $\sigma_\epsilon^2 + \lambda^2\epsilon_T^2$, where σ_ϵ^2 is the error energy (variance) due to noise, $\epsilon_T^2 = \int_0^\infty [\tau_0(t) - \tau(t)]^2 dt$ is the error energy due to transients, and λ is the Lagrange multiplier (considered as relative weight between noise and transient error energies). Based on Wiener-filter theory, the optimal second-order loop parameters are given as follows (see [17, Table 3.7.1]):

$$\omega_n^2 = m\,\lambda\,\sigma_\eta^{-1}, \quad \xi = \sqrt{2}/2 \tag{34}$$

Substituting (34) in (33), we get:

$$B_L = 0.53\omega_n \ll 1/T_s\ , \ k_2 = \sqrt{2}/\omega_n\ , \ k_1 = A/\omega_n^2 \tag{35}$$

This section has shown how to design a suitable UWB tracking system and to select its parameter values for maintaining the synchronization in the presence of ambient noise and Doppler effects. Figure 19 summarizes the synchronization system based on dirty template technique, which consists of three main blocks: signal detector, timing acquisition, and tracking. Relying on this synchronization system, the received symbols $\hat{s}[n]$ are demodulated, using:

$$\hat{s}[n] = \text{sign}\left[\int_0^{T_s} r(t+nT_s)y(t-\hat{\tau})dt\right] \tag{36}$$

In the presence of the noise and with the channel variations, the DLL might be derived out of lock. We have mentioned in (2) that the received UWB signal $r(t)$ consists of a large

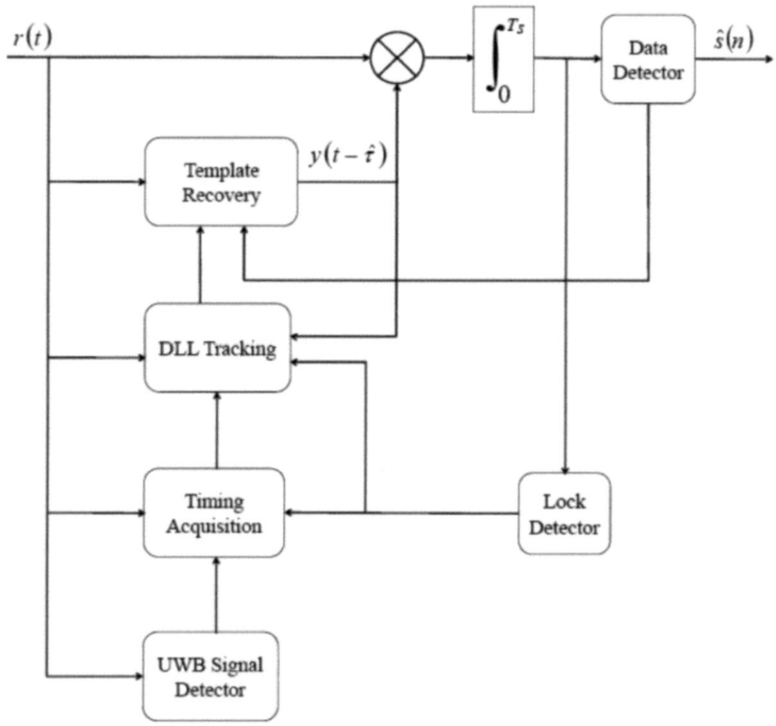

Figure 19. Block diagram of the synchronization system based on Dirty Template technique

number of channel paths. So when the DLL falls out of lock, it begins to track with a wrong path instead of the first one. It has been found in [24] that dirty template demodulator is able to perform efficiently symbol detecting, even if DLL is tracking another path close to the first one. In Figure 19, a lock detector is employed to measure capture energy at the correlator output and to compare it versus a threshold, for deciding if the DLL remains in lock or it is out of lock. Once the lose-lock signal is generated, the receiver turns back to the acquisition mode. After executing successfully the acquisition phase, the receiver enters again the tracking phase and DLL is brought back in lock. In UWB channel, due to the relative movement between the transmitter and the receiver (or object movement inside the signal propagation field), the multipath channel parameters vary slowly with time. Consequently, the reference signal $y(t)$, extracted in (22), becomes outdated and is not efficiently similar to T_s-long received segments. This leads to degrade gradually synchronization performance and increase BER. To avoid this problem, The dirty template $y(t)$, which is used for tracking the received signal and for estimating $\hat{s}[n]$, should be updated, as follows:

$$y(t) = \frac{1}{K}\sum_{n=1}^{K} \hat{s}[n]r(t + nT_s); \quad t \in [0,T_s) \tag{37}$$

where K denotes the averaging size. Update of the template $y(t)$, does not need to be continuous, only periodic with a frequency $(1/T_y)$, where update period T_y must be less than the channel coherence time $(T_y \ll T_{coh})$. On the other hand, by considering that UWB pulses are ideal candidates for indoor communication applications, where the expected speed of objects or users is typically in the order of $v = 1\, m/s$. With maximum carrier frequency for IR-UWB $f_c = 6$ GHz, the maximum Doppler spread is given as: $f_D = v/\lambda = (v/c) \times f_c = 20$ Hz, where c is the speed of light, and λ is the wavelength. As discussed in [26], the channel coherence time is about $T_{coh} = 0.423/f_c \approx 21$ msec. Thus, updating the DT reference $y(t)$ must be done periodically with T_y less than 21 msec.

For Matlab simulation, we select the pulse $p(t)$ as the second derivative of the Gaussian function with unit energy and duration $T_p \approx 1$ns. Each symbol contains $N_f = 10$ frames with duration $T_f = 100$ns, the symbol duration is $T_s = N_f T_f = 1$µs. We employ TH spreading codes of period N_f, which is generated from a uniform distribution over $[0, N_c - 1]$, with $N_c = 9$, and $T_c = 10$ ns. The DT reference signal $y(t)$ is enhanced by passing it through the filter $W(t)$ with window filter size $= 1.5T_c$. The simulations are performed in a Saleh-Valenzuela channel [20]. Figures 20-21 show the mean square error (MSE) for various values of natural frequency ω_n and data-aided bit number M. In these figures, the MSE results are normalized by T_p^2, and plotted versus the signal-to-noise ratio (SNR) per pulse. Doppler effects are neglected $m = 0$. The simulations confirm that an increase in M or a decrease in ω_n lead to reduce noise energy effect and improve tracking performance. We can notice from Figure 20 that for SNR≥ 5 dB the MSE lines at various M are matched. In order to evaluate the transient performance, the input signal time varies as follows $\tau(t) = mt$ with $m = T_p/100T_s$. We select $\lambda = 10^7$, and $\sigma_\eta^2 = 10^{-16}$; thus, the optimum natural frequency in (34) is $\omega_n = 0.1$ GHz. Since DLL operates digitally with sampling rate T_s, we have $\tau(n) = 0.01nT_p$. Figure 22 presents DLL behavior in the presence of input time variations $\tau(n)$ for different ω_n. It is seen that DLL follows the input signal $\tau(n)$ rapidly when the natural frequency increases. Another way of representing DLL transient performance is exhibited in

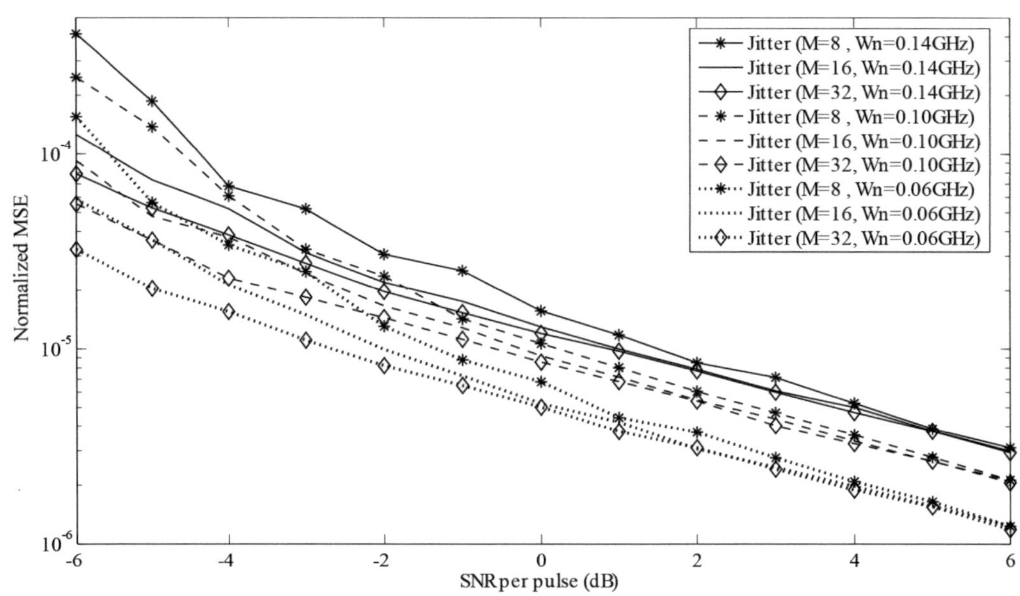

Figure 20. Normalized MSE vs. SNR per pulse for second-order DLL

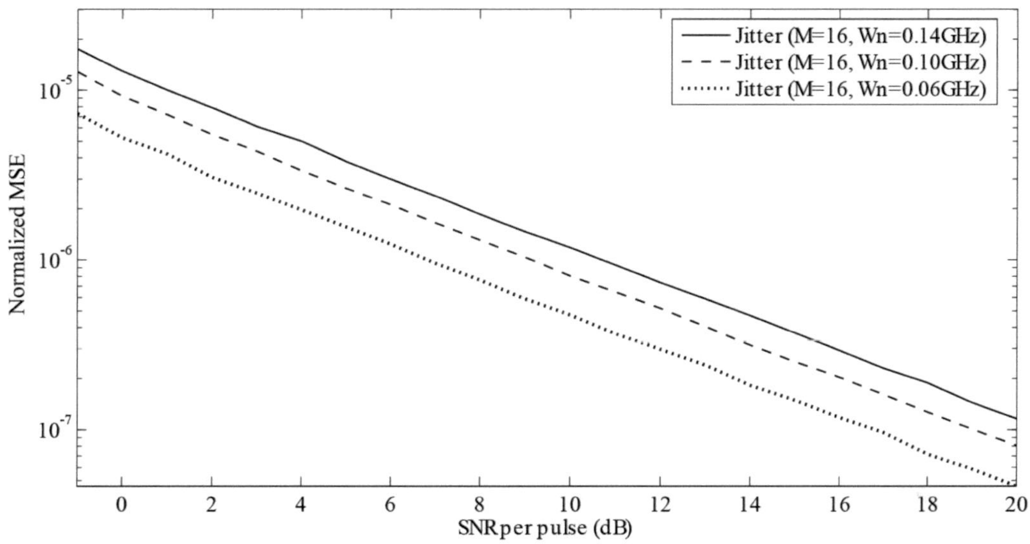

Figure 21. Normalized MSE vs. SNR per pulse for second-order DLL with data-aided bit number $M = 16$

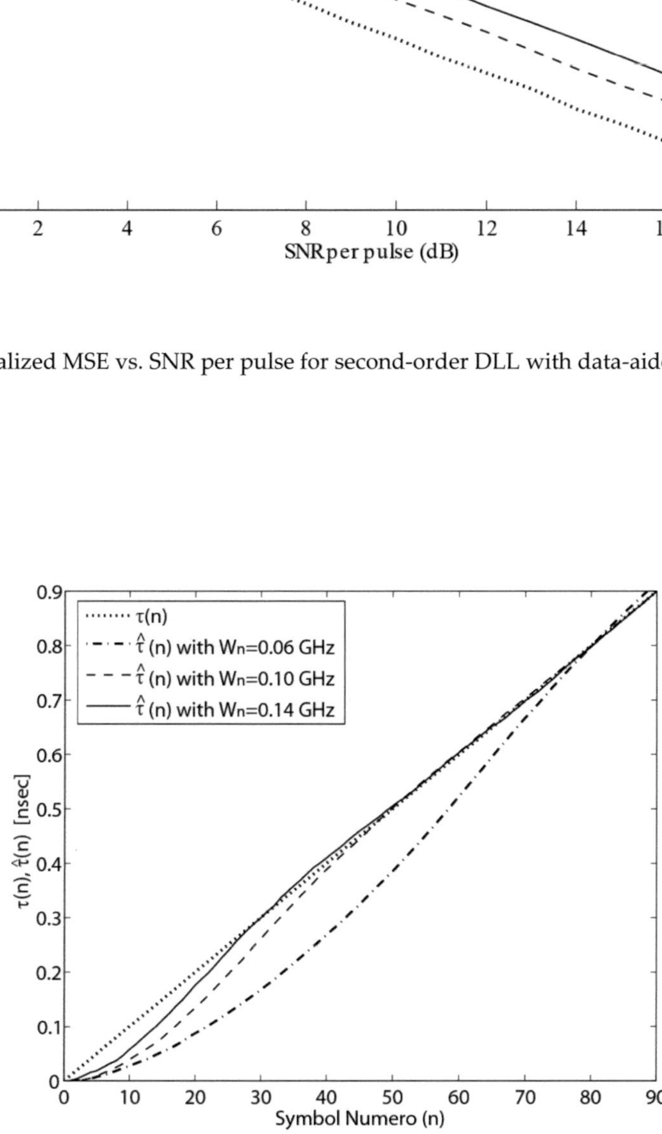

Figure 22. DLL transient reponse for ramp input (SNR=3 dB)

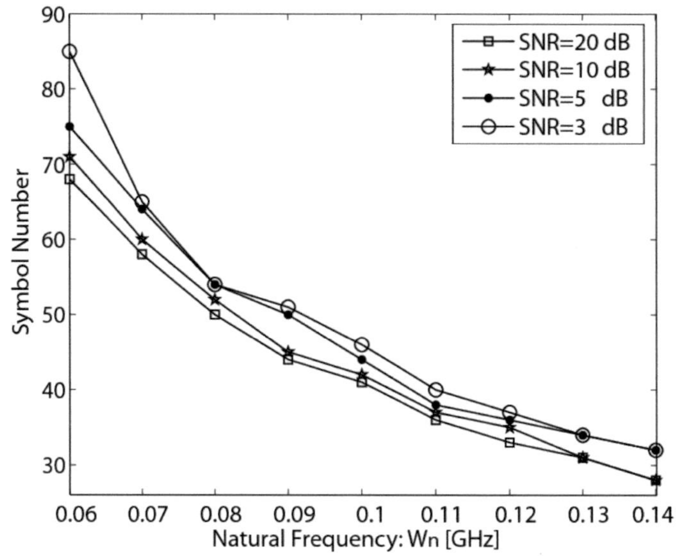

Figure 23. DLL transient performance (symbol number vs. natural frequency)

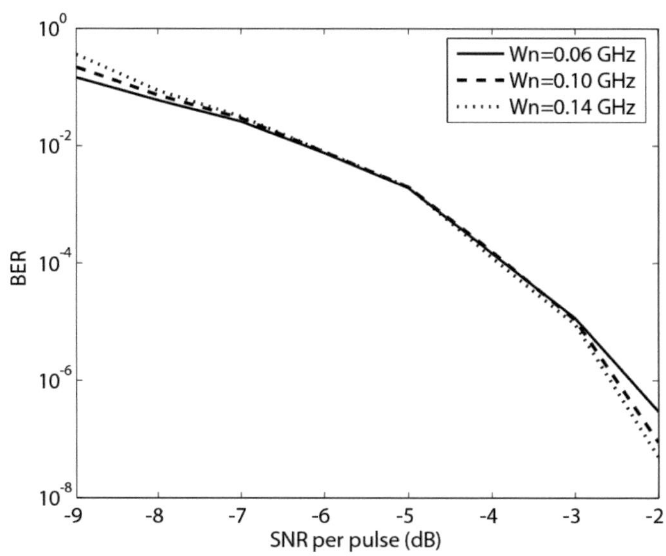

Figure 24. BER for the proposed DLL ($M = 16,\ \tau(n) = 0.01nT_p$)

Figure 23, which shows how many symbols (iteration steps) DLL requires to match efficiently $\tau(n)$. It is clear that when ω_n increases, the required number of symbols would decrease and transient response would be better. The BER comparisons for various ω_n are depicted in Figure 24. We observe that for high SNR values, an increase in ω_n (Simultaneously, B_L augments) leads to speed up the tracking operation and to improve BER performance. On the other hand, for small SNR values, the bandwidth B_L should be decreased for alleviating the noise effects, but that degrades the transient performance. Consequently, increasing ω_n helps to reduce the transient error effect, but in the price of noise handling ability.

In this section, we design a second-order DLL used for tracking timing offset variations in the received waveform, taking Doppler effects into account. We combine DLL with TDT, which enables enhancing the received energy capture even when the multipath channel and the TH codes are both unknown. Consequently, the proposed approach contributes to improve tracking performance for UWB systems and to reduce receiver structure complexity. For selecting the optimum parameter values for the proposed DLL, we apply Wiener-filter theory. Simulation results show the performance of the proposed DLL and conform that increasing ω_n helps to reduce the transient error effect, but in the price of noise handling ability.

6. Conclusion

In this chapter, we use TDT algorithms for carrying out low-complexity high-performance timing synchronization, which constitutes a major challenge in realizing the UWB communications. TDT technique is based on correlating the received signal with "dirty template" extracted from the received waveforms. This template is called dirty; because it is distorted by the unknown channel and by the ambient noise. TDT allows enhancing received energy capture and reducing receiver structure complexity. The described technique can be applied to UWB systems even in the presence of time hopping TH codes, Inter-Frame Interference (IFI) and rich multipath environment, where Inter-Symbol Interference (ISI) is absent.

TDT synchronization system consists of three main blocks: signal detection, timing acquisition and tracking. Each block of them is explained in a separated section of this chapter. In signal detection section: the dirty template detector is derived by applying the Neyman-Pearson theory. Then, Monte Carlo simulations are performed to find the probabilities of false alarm (P_{FA}) and detection (P_D). The results of simulation analysis show that the detection performance of the dirty template approach is improved by increasing the signal-to-noise ratio (SNR) or data-aided (DA) number. After detecting the UWB signal, the symbol-level timing offset estimation relies on searching the best statistics T.

Next section presents the timing acquisition in both DA and NDA modes. Then, we improve the timing estimator by adding a suitable window filter to the structure of the cross-correlation. Both the theoretical analysis and Matlab simulation results show the performance of the proposed TDT estimator, and confirm that DA mode has the high performance and fast timing, compared to NDA mode, but that is at the price of bandwidth efficiency.

In tracking section, we demonstrate tracking system design taking into consideration the relative motion between Transmitter and receiver. We combine DLL with TDT, which enables to enhance the received energy capture and to improve tracking performance. Simulation results illustrated that increasing natural frequency parameter ω_n helps to reduce transient error effect, but in the price of noise handling ability.

Author details

Rshdee Alhakim and Emmanuel Simeu
TIMA Laboratory, Grenoble University, Grenoble, France

Kosai Raoof
LAUM Laboratory, Maine University, Le Mans, France

7. References

[1] Pagani P, Talom F, Pajusco P, Uguen B (2007) Communications Ultra Large Bande: le canal de propagation radioélectrique. Hermes. 246 p.

[2] Tian Z, Giannakis G.B (2006) Timing Synchronization for UWB Impulse Radios. In: Shen X, Guizani M, Qiu R.C, Ngoc T.L, editors. UWB Wireless Commun. and Networks. John Wiley & Sons. pp. 53-81.

[3] Win M.Z, Scholtz R.A (1998) On the energy capture of ultrawide bandwidth signals in dense multipath environments. IEEE Commun. Letters. pp. 245-247.

[4] Homier E.A, Scholtz R.A (2002) Rapid acquisition of Ultra-Wideband signals in the dense multipath channel. Proc. IEEE Conf. Ultra Wideband Systems and Techno. pp. 105-110.

[5] Fleming R, Kushner C, Roberts G, Nandiwada U (2002) Rapid acquisition for Ultra-Wideband localizers. Proc. IEEE Conf. Ultra Wideband Systems and Techno. pp. 245-250.

[6] Yang L, Tian Z, Giannakis G.B (2003) Non-data aided timing acquisition of Ultra-Wideband transmissions using cyclostationarity. Proc. IEEE Conf. ASSP. pp. 121-124.

[7] Strohmer T, Emami M, Hansen J, Papanicolaou G, Paulraj A.J (2004) Application of time-reversal with MMSE equalizer to UWB communications. IEEE Global Telecommun. Conf. pp. 3123-3127.

[8] Zhang H, Goeckel D.L (2003) Generalized Transmitted-Reference UWB Systems. Proc. IEEE Conf. Ultra Wideband Systems and Techno. pp. 147-151.

[9] Pardinas-Mir J.A, Muller M, Lamberti R, Gimenes C (2011) TR-UWB detection and synchronization - Using the Time Delayed Sampling and Correlation detection method. Proc. 8th EuRAD Conf. pp.202-205.

[10] Alhakim R, Raoof K, Simeu E (2010) A Novel Fine Synchronization Method for Dirty Template UWB Timing Acquisition. Proc. 6th IEEE Conf. Wireless Commun. Networking and Mobile Computing. pp. 1-4.

[11] Alhakim R, Raoof K, Simeu E, Serrestou Y (2011) Cramer-Rao Lower Bounds and Maximum Likelihood Timing Synchronization for Dirty Template UWB Communications. Signal Image and Video Processing J. pp. 1-17.

[12] Alhakim R, Raoof K, Simeu E, Serrestou Y (2011) Data-aided timing estimation in UWB communication systems using dirty templates. IEEE Int. Conf. on Ultra Wideband. pp. 435-439.

[13] Alhakim R, Simeu E, Raoof K (2011) Internal model control for a self-tuning Delay-Locked Loop in UWB communication systems, Proc. 17th IEEE Conf. pp. 121-126.

[14] Yang L, Giannakis G.B (2005) Timing Ultra-Wideband Signals with Dirty Templates. IEEE Trans. On Commun. pp. 1952–1963.

[15] Yang L, Giannakis G.B (2003) Low-complexity training for rapid timing acquisition in ultra-wideband communications. Proc. Global Telecommun. Conf. pp. 769–773.

[16] Kay S.M (1998) Fundamentals of Statistical Signal Processing: Detection Theory. Prentice Hall. 672 p.

[17] Meyr H, Ascheid G (1990) Synchronization in Digital Communications: Phase-, Frequency-Locked Loops, and Amplitude Control. John Wiley & Sons. pp. 21-162.

[18] Meyr H, Moeneclaey M, Fechtel S.A (1997) Digital Communication Receivers: Synchronization, Channel Estimation, and Signal Processing. John Wiley & Sons. pp 61.206

[19] , FCC First Report and Order: In the matter of Revision of Part 15 of the Commission's Rules Regarding Ultra-Wideband Transmission Systems. FCC 02-48, Apr. 2002.

[20] Saleh A.A.M, Valenzuela R.A (1987) A statistical model for indoor multipath propagation. Selected Areas in Commun. j. pp. 128-137.

[21] Lee J.S, Miller L.E (1998) CDMA Systems Engineering Handbook. Artech House Publishers. pp. 677-834.

[22] Salem F, Pyndiah R, Bouallegue A (2005) Synchronization Using an Adaptive Early-Late Algorithm for IR-TH-UWB Transmission in Multipath Scenarios. 2nd Int. Symp. on Wireless Comm. Sys. pp. 268-271.

[23] Zhang W, Shen H, Kwak K.S (2007) Improved Delay-Locked Loop in a UWB Impulse Radio Time-Hopping Spread-Spectrum System. ETRI j. Available: http://dx.doi.org/10.4218/etrij.07.0506.0032.

[24] Farahmand S, Xiliang L, Giannakis G.B (2005) Demodulation and Tracking with Dirty Templates for UWB Impulse Radio: Algorithms and Performance. IEEE Trans. On Vehicular Technology. pp. 1595 – 1608.

[25] Jaffe R, Rechtin E (1955) Design and performance of phase-lock circuits capable of near-optimum performance over a wide range of input signal and noise levels. IRE Trans. Inf. Theory. pp. 66–76.

[26] Rappaport T.S (2001) Wireless Communications: Principles and Practice 2nd ed. Prentice Hall. pp. 139-196.

Chapter 3

Radar Sensing Using Ultra Wideband – Design and Implementation

Xubo Wang, Anh Dinh and Daniel Teng

Additional information is available at the end of the chapter

http://dx.doi.org/10.5772/48587

1. Introduction

Ultra-wideband (UWB) has received significant attention for applications in target positioning and wireless communications recently. The extremely short pulses in turn generate a very wide bandwidth and offer several advantages, such as large throughput, covertness, robustness to jamming, lower power, and coexistence with current radio services. UWB not only can transmit a huge amount of data over a short distance at very low power, but also has the capability to pass through physical objects that tend to reflect signals with narrow bandwidth.

The extremely narrow pulse (usually in order of few nanoseconds to few hundred picoseconds) makes it possible to build radar with much better spatial resolution (usually 0.1 to 1 ft) and very short-range capability compared to other conventional radars. Also, the large bandwidth allows the UWB radar to get more information about the possible surrounding targets and detect, identify, and locate only the most desired target among others. The fine resolution makes the UWB radar beneficial for medical applications. The properties of short pulse indicate that the UWB signal can penetrate a great variety of biological materials such as organic tissues, fat, blood, and bone. Experiment results show that the signals with low center frequencies achieve better material penetration. Compared to a radar system with a pulse-length of one microsecond, a short Gaussian or Gaussian monopole pulse of 200ps in width has a wavelength in free space of only 60 mm, compared to 300m. Since the pulse length in conventional radar is significantly longer than the size of the target of interest, the majority of the duration of the returned signal is an exact replica of the radiated signal. Thus, the returned signal provides little information about the nature of the target. However, since the UWB pulse length is in the same order of magnitude with the potential targets, UWB radar reflected pulses are changed by the target structure and electrical characteristics. Those changes in pulse waveform provide valuable information such as shape and material properties about the targets. Discrimination of target using higher order signal processing of impulse signals can distinguish between materials that would not be otherwise distinguishable by the narrowband signals.

To work as UWB radar, the UWB transmitter sends a narrow pulse toward a target and an UWB receiver detects the reflected signal. This is a very simple algorithm of radar sensing which has been widely used. For biomedical radar, the target is, for example, a human heart. When the UWB pulse in propagation encounters an boundary of two types of medium with different dielectric properties, a portion of the incident electromagnetic energy is reflected back to the original medium with a reflection angle θ_r (zero reflection angle if the incident wave path is parallel to the normal line), while the other portion continues propagating through the next medium. The analogy of the transmission of UWB pulse is shown in Figure 1.

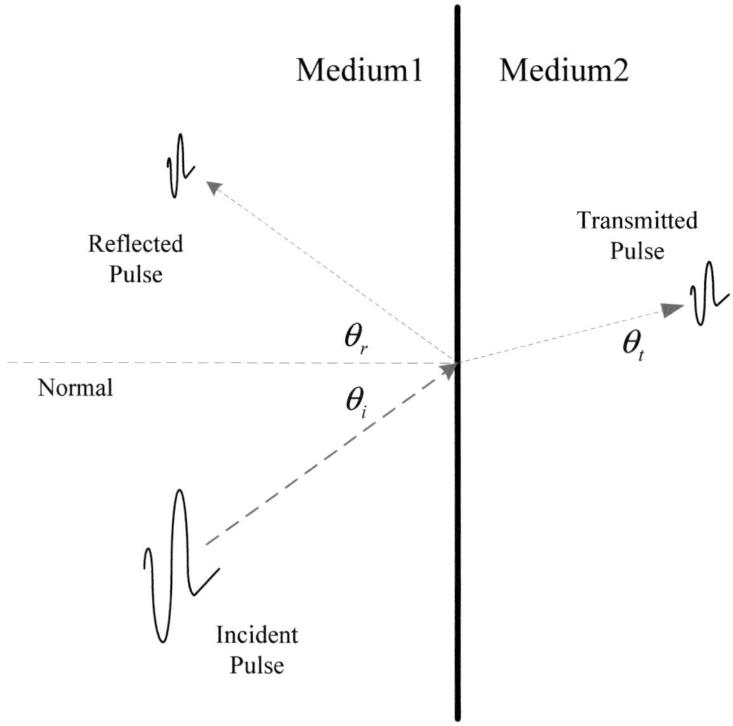

Figure 1. Pulse reflection and transmission diagram

Unlike ultrasound device, which is being widely used at the present time that requires direct skin contact, the UWB makes imaging internal organ movements without invasive surgical or direct skin contact possible. Another advantage in using UWB technology is that the UWB transceiver is simple and occupies a very small chip area as it does not require complicated frequency recovery system as in the narrow bandwidth transceiver. In addition, power consumption of the impulse based UWB systems is extremely low because the power is consumed only during pulse transmitting period.

2. UWB radar architecture

When designing a UWB radar transceiver system, two design aspects need to be considered: architecture and implementation. Different architecture set the fundamental performance capabilities of the design, and good implementation choices improves radar performances. Impulse radar detection range depends on radiated energy, transmitter and receiver design, target size, and signal processing. Among various UWB transceiver architectures, the impulse-based energy detection UWB transceiver architecture is discussed here.

An example of the impulse-based energy detection transceiver architecture is shown in Figure 2. In this architecture, the transmitter sends a pulse train toward the target. The interface between two medias produces a partial reflection. Then the receiver detects and samples this particular type of reflected pulse train, and the decision circuit makes the final decision. Pulses are diffracted and scattered by different tissue layers and organs in human body. Channel distortion and power loss easily destructs the reflected pulses and make them undistinguishable. The rang-gate is designed to look for the destined reflected pulse rather than wait and receive every reflected pulse from every location and try to identify the expected return pulse, which in many cases are very week and tangled with other return pulses. The receiver samples only the pulses arriving at the receiver during a very narrow time window after pulse transmission, as shown in Figure.3. By estimating the distance of the expected target, a delay time is chosen.

This proposed transceiver architecture enormously reduces the circuitry complexity and power consumption. The transmitter consists of a modulator, a pulse generator, and a variable gain amplifier (VGA) driver. An on-off keying (OOK) modulation scheme is used to modulate the pulse. The VGA and driver are used to amplify output and match output impedance. The receiver consists of a low noise amplifier (LNA), a correlator, an integrator, a clocked voltage comparator, and a delay controller. The input clock train and control signal are modulated to a sequence of clock pulse, which then enters the pulse generator to produce a pulse train. This pulse train is passed onto a driver amplifier and then to an UWB antenna. The reflected pulse is caught by the antenna in the non-coherent receiver and amplified by a LNA. The signal then is squared by a multiplier at the asynchronous receiver. The squared output is then fed into an integrator and clocked comparator to boost up the voltage and reconstruct the signals. The range controller uses logic gates to switch on/off the LNA and disable the sampling operation of the comparator for range finding.

3. UWB radar transmitter

Two classes of UWB signals are utilized to transmit symbols in UWB system: carrier-free impulse signal, and carrier-based short sinusoidal signal. The impulse UWB signal is often represented using Gaussian (different orders of Gaussian derivatives), Rayleigh, or Hermitian pulse. The advantages of impulse signal are that the impulse-based transceiver architecture often very simple and consume the least amount of power due to its low pulse repetition rate and low duty cycle. However, the drawback for impulse-based signal is the

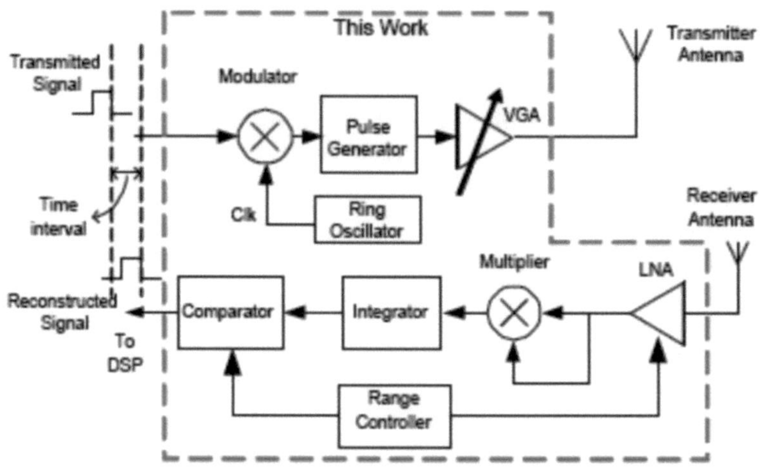

Figure 2. An example of the architecture of the UWB transceiver

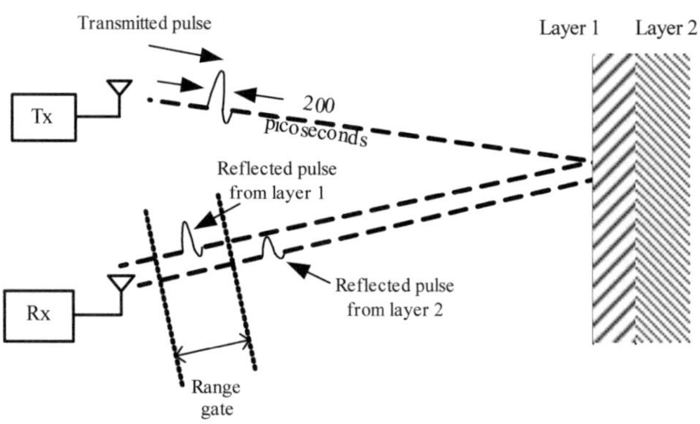

Figure 3. A range gate at receiver opens briefly after a fixed delay time to sample the reflected pulse.

frequency characteristic is largely determined by the pulse shape. Compared to the impulse-based signal, the carrier-based UWB uses sinusoidal wave instead of short pulse to represent signals, and these signals are easier to manage within the FCC spectrum and produce less distortion. Both two classes of UWB transmitter are discussed in following sections.

3.1. CMOS high-order pulse digital transmitter

In this section, a pulsed-UWB BPSK transmitter using higher order Gaussian pulse with digital circuit implementation is discussed. This transmitter outputs tenth-order Gaussian derivative BPSK-modulated pulses. By implementing the transmitter in digital circuit, the pulses can be switched on/off between very short pulse intervals. One way to reduce transmitter power is to avoid the use of circuit blocks that operate constantly or have long setting time and start-up times such as voltage controlled oscillator (VCO).

The proposed UWB high-order Gaussian pulse transmitter with pulse tuning capability is shown in Figure 4. The proposed transmitter consists of two Gaussian derivative pulse generators, a variable gain driver amplifier, and peripheral circuit. The input of the transmitter is a baseband signal. A D-type flip-flop as a frequency divider is used to expand the control signal for pulse generators. The baseband signal is also used to triggered the generation of the pulse generator. The baseband signals are BPSK modulated such that one pulse generator outputs positive Gaussian derivative pulses when the baseband signal is logic high, and the other pulse generator outputs negative Gaussian derivative pulses when the baseband signal is logic low.

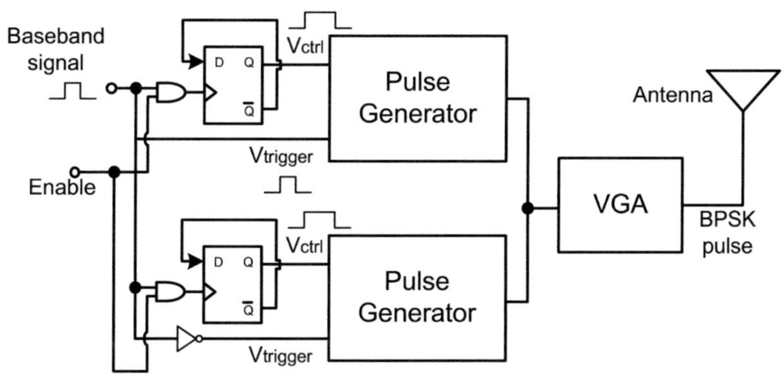

Figure 4. Proposed transmitter architecture.

As seen from the transmitter block diagram, the proposed transmitter is not using any clock signals. This greatly reduces the power consumption of the circuit at the tradeoff of lower pulse repetition frequency (PRF), which is not the ultimate concern for the bio-radar sensing applications. The average transmitter output power emission in the UWB band is limited to -41.3dBm/MHz. But gated UWB systems can also transmit at higher power levels and then sit quietly only if the average emission power density during the time period still satisfies the power limitation. The enable signal in the transmitter controls the gating timing. When the transmitter is configured to operate with gating control at 25 percent duty cycle during transmission of baseband signals from a transmitter to a target, the impulse UWB transmitter can achieve up to four times better performance. This gated signal can then achieve the same average transmit power as a continuous signal while occupying only a fraction of the channel time available for transmissions in the UWB system.

The core component of the pulse generator is the edge generator, as shown in Figure 5. The edge generator generates the basic element which is the Gaussian pulse. Each rising edge of the Vtrigger triggers positively-peaked pulses through a NOR gate and negatively-peaked pulse through a NAND gate. The pulse width and pulse amplitude are controlled through the delay elements. The pulse generation logic is that when one XOR input is slower than the other XOR input, the time difference between these input produces a positive normally distributed voltage pulse (usually represented by Gaussian distribution), and when the inverted NAND input is slower than the other NAND input, the time difference of these two input produces a negative normally distributed voltage pulse. The polarity of the edge generator output is control by Vctrl. The Vctrl select either the positive pulse or the negative pulse to pass to the output through the AND and OR logic gates.

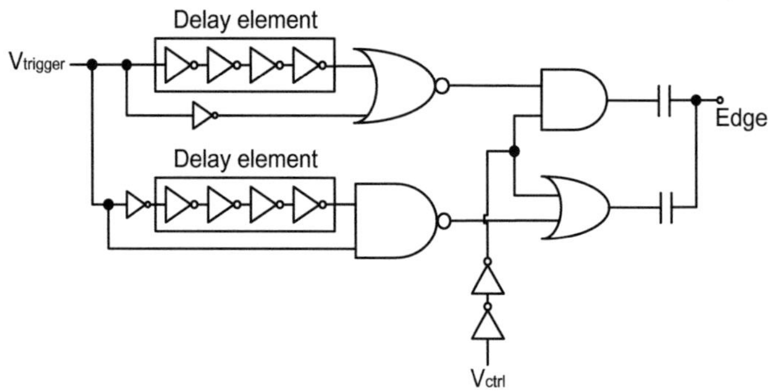

Figure 5. Edge generator circuit with edge tuning capability.

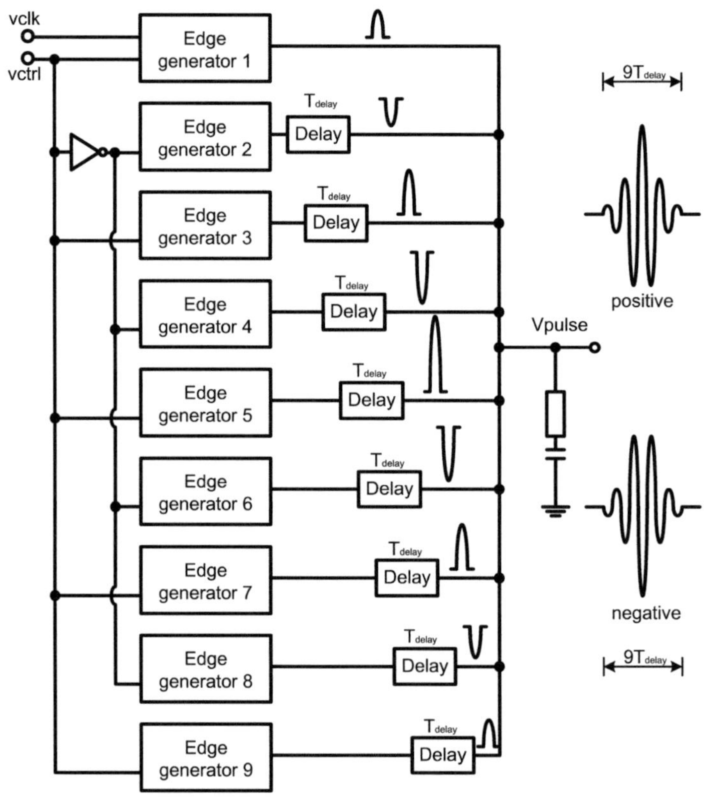

Figure 6. Block diagram of the pulse generator.

The Gaussian derivative pulse generator, as shown in Fig. 6, consists of an array of edge generators connected in parallel. Each edge generator produces a single pulse with different amplitude and the same pulse width T_{delay}. The delay in each edge generator is adjusted based on the shape factor of the tenth derivative of the Gaussian pulse. The control signal is inverted at the 2nd, 4th, 6th, and 8th edge generators to produce pulses with negative polarity at these locations. The trigger signal Vtrigger is delayed one pulse width Tdelay at the input of the 2nd edge generator, delayed two pulse width 2*Tdelay at the 3rd edge generator, 3*Tdelay at the 4th edge generator, 4Tdelay at the 5th edge generator, 5*Tdelay at the 6th edge generator, 6*Tdelay at the 7th edge generator, 7Tdelay at the 8th edge generator, and 8*Tdelay at the last stage. The final output of the tenth derivative Gaussian pulse is constructed based on each single edge generator output. The total pulse width is 9*Tdelay.

The design is a transmitter prototype. It not only can generate tenth-order Gaussian derivative, but any pulse combinations, either Gaussian derivatives or rectangular or Gaussian modulated sine pulses by controlling the delay elements in each edge generator and the number of edge generators used based on different orders of derivative of Gaussian pulse, the bandwidth, and the center frequency.

The transmitter prototype mentioned in last section was designed and simulated in TSMC 90nm CMOS with 1.0V voltage supply. The base-band signal was generated at 100Mbps to leave enough headroom for pulse generation. The positive and negative tenth derivative Gaussian pulses were generated, as shown in Figure 7. The pulse width was adjusted to 0.5ns through the delay components. The transmitter output peak-to-peak amplitude is 130mV, and the amplitude ranges from 35mV to 500mV.

A Monte Carlo simulation was performed to validate the robustness of the transmitter circuit against process voltage and temperature variations and mismatches. The Monte Carlos simulation results of the tenth-order Gaussian derivative pulses are shown in Figure 8. The simulated power spectral density plot of the transmitter output with simulated white noise added is shown in Figure 9.

The layout of the proposed transmitter is shown in Figure 10. The transmitter core occupies a chip area of 200um by 140um. Simulating on a 1V voltage supply, the transmitter draws average 9.037mW when generating the tenth-order Gaussian derivative pulses at a pulse repetition frequency of 50MHz. The pulse width is 0.6ns. The transmitter has output pulse energy of 3.1pJ/pulse.

3.2. CMOS Gaussian pulse transmitter

The proposed impulse-based UWB technique enormously reduces the circuitry complexity and power consumption. Figure 11 shows the block diagram of the transmitter which consists of a modulator, a pulse generator, and a variable gain amplifier (VGA) driver, and a ring oscillator. An on-off keying (OOK) modulation scheme is used to modulate the pulse. The VGA and driver are used to amplify and adjust the output pulse, as well as for impedance matching. The incoming data V_{data} is modulated with clock train generated by the ring oscillator, yielding a sequence of clock pulse, which then enters the pulse generator to produce a pulse train. This pulse train is passed onto a driver amplifier and then to an

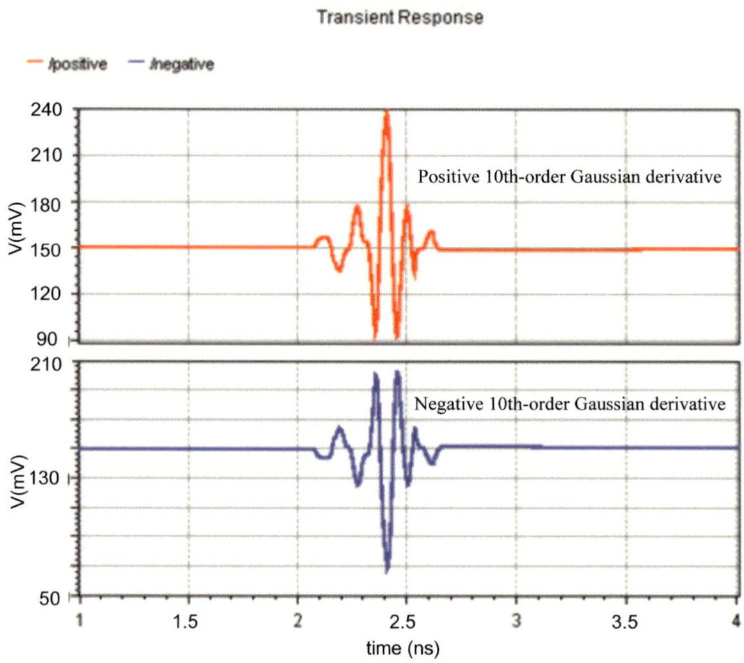

Figure 7. Simulated tenth order Gaussian derivative pulses.

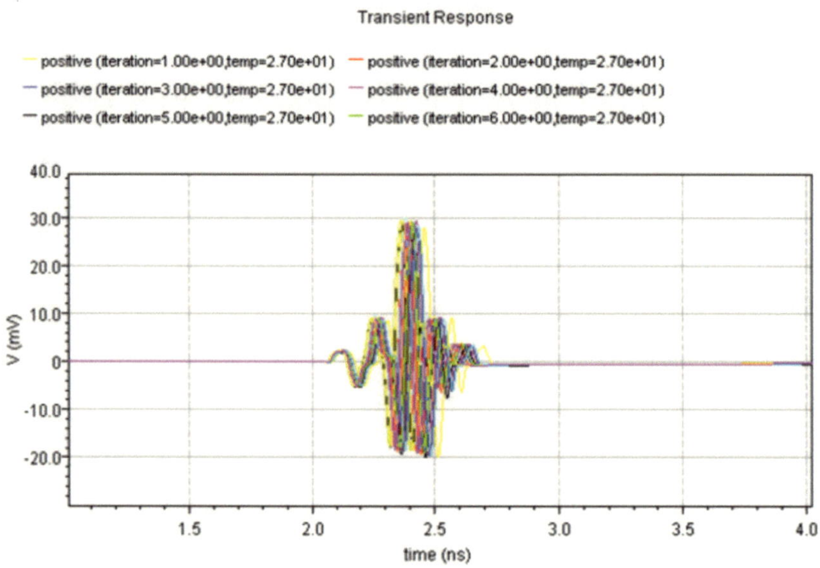

Figure 8. Monte Carlo simulation of the transmitter.

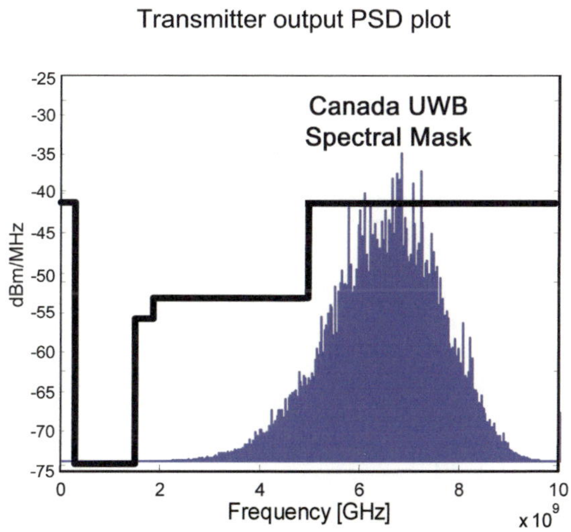

Figure 9. Simulated power spectral density of the pulse.

Figure 10. Layout of the proposed digital transmitter.

UWB antenna. The output pulse amplitude is adjustable through the variable gain amplifier in the driver. The range controller uses logic gates to switch on/off the VGA. The transmitter is fabricated using CMOS 90nm process, and the whole design consumes less than 0.5mA of static current.

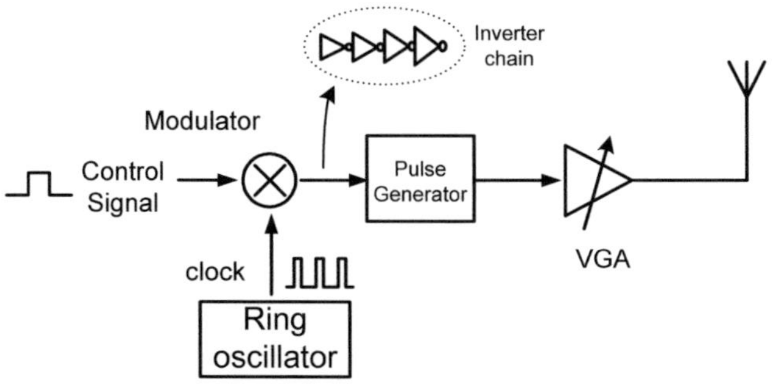

Figure 11. Proposed IR-UWB transmitter

Fig. 12(a) shows two critical components of the transmitter: the OOK modulator and the pulse generator. The clock signal Clk, generated by a ring oscillator, has a period of 1ns and a pulse width of 500ps. The clock train is modulated by the input data with OOK modulation by an AND gate. The number of pulses in one bit of data is determined by both the bit length and the clock frequency. For radar sensing purpose, it is the best interest of the energy-collecting receiver to include more pulses in a single pulse train sent to the target for a good SNR and easy detection. In communication cases, the number of pulses representing one bit is set less for higher data transmission rate. The clock rate is higher than the data rate to ensure reliable modulation and demodulation. Figure 12(b) illustrates the pulse generator and modulator in the transistor level.

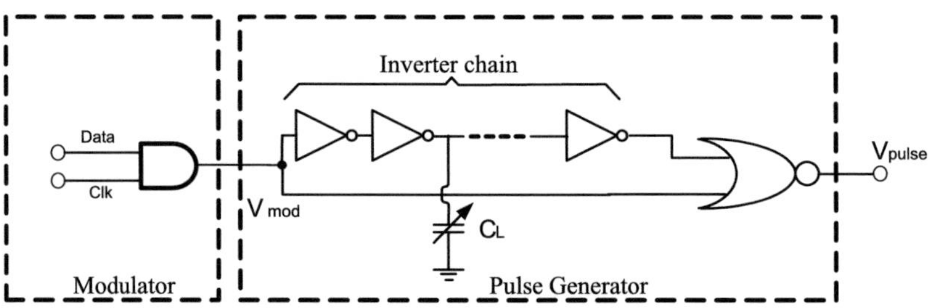

(a) Pulse generator and modulator

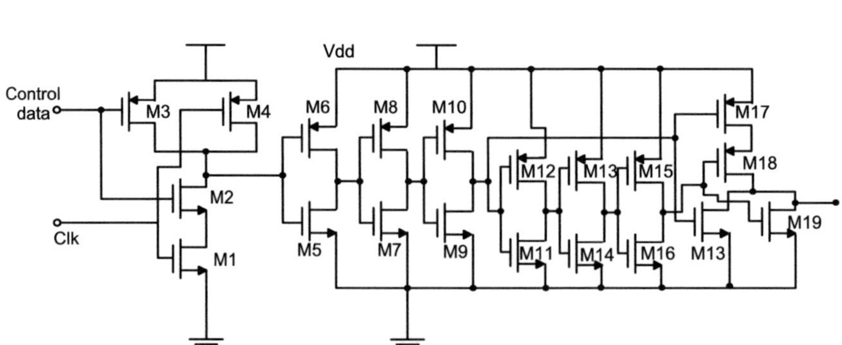

(b) Pulse generator circuit

Figure 12. Pulse generator and modulator circuit in transistor level

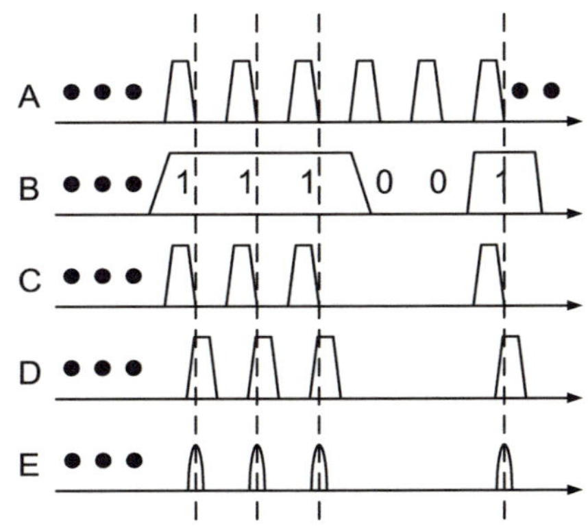

Figure 13. Timing waveforms of the transmitter

The signal flow at each block of the proposed transmitter is illustrated in Figure 13. The transmitter clock signal is represented by A. B is the input digital sequence. The modulated clock train C passes through an inverter chain to sharpen the rising and falling edge of the each clock signal. The modulated clock train is then split into two signal paths and fed into a NOR gate. Signals in one path is delayed and inverted, as shown by D. The NOR gate only outputs high when both inputs are low, and the time for both inputs be low is the delay time set by the inverter chain in the delay path. The signal E represents the output pulses. The output pulse width is determined by this delay time. The output pulse width can be adjusted by changing the delay of the inverters, which in turn, by varying the load capacitance of the inverters. Figure 14 shows the simulated pulses with various pulse widths when changing the load capacitance of the inverters C_L. This load capacitance C_L is implemented using a CMOS varactor.

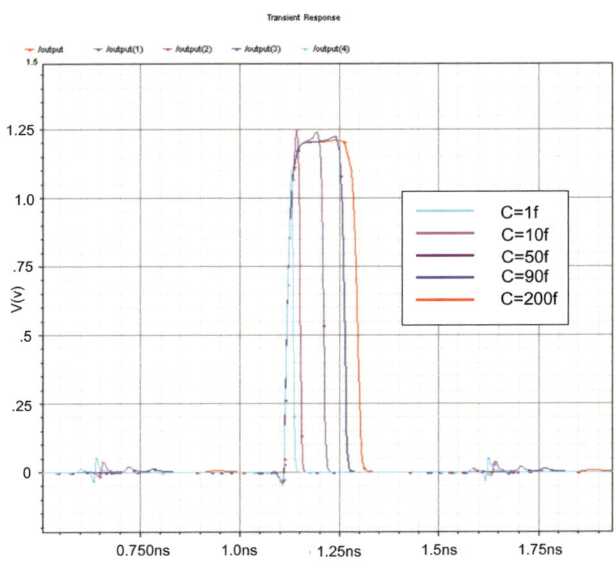

Figure 14. Simulation of different pulse widths

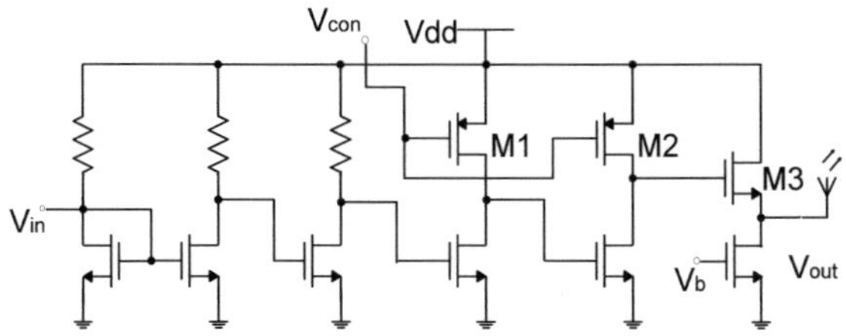

Figure 15. Driver amplifier and VGA

The driver amplifier is used to amplify and shape the spectrum of the out-going short pulses, to adjust the transmitting pulse amplitude, and to match output impedance. As shown in Figure 15, the variable gain driver amplifier consists of three stages. The first stage employs three cascade common-source amplifiers with load resistors. Each amplifier is designed with relatively low gain to achieve a high bandwidth. The second stage consists of two cascade common-source amplifiers with PMOS transistors load. By analyzing the high frequency operation of a cascaded common-source amplifier, the voltage gain can be shown as

$$A_V = \frac{-g_m Z_L}{1 + g_m Z_S} \tag{1}$$

For a MOS transistor biased in the linear region, a linear resistance is

$$r_{DS} \equiv \frac{v_{DS}}{i_D} = \frac{1}{\mu_n C_{OX} \frac{W}{L}(v_{GS} - V_t)} \tag{2}$$

The value of the resistance r_{DS} is controlled by varying the value of v_{GS}. The overall gain of the driver amplifier is therefore

$$gain = \frac{-g_m Z_L - g_m \left(\mu_n C_{OX} \frac{W}{L}(v_{GS} - V_t) \right)^{-1}}{1 + g_m Z_S} \tag{3}$$

The two PMOS transistors are tuned in the triode region and the equivalent resistances are controlled by the control voltage Vcon. Simulation shows that the change of Vcon from 0V to 0.15V corresponds to output pulse level from 30mV to 560mV, as shown in Figure 16. The last stage is an output buffer for 50ohms impedance matching purpose.

The impulse UWB transmitter was designed and implemented using standard ST 90nm CMOS technology with 1.2V power supply to verify the proposed idea. A chip photograph of the fabricated test circuit is shown in Figure 17. The active transmitter area measures $50\text{um} \times 100\text{um}$. The transmitter die is housed using CQFP44 package and the chip is mounted on a PCB for measurement. The fabricated impulse transmitter module is tested under normal operating environment. Figure 18 illustrates the measurement setup. For digital data input, a pattern generator is used to provide a 250Mb/s bit sequence data. The

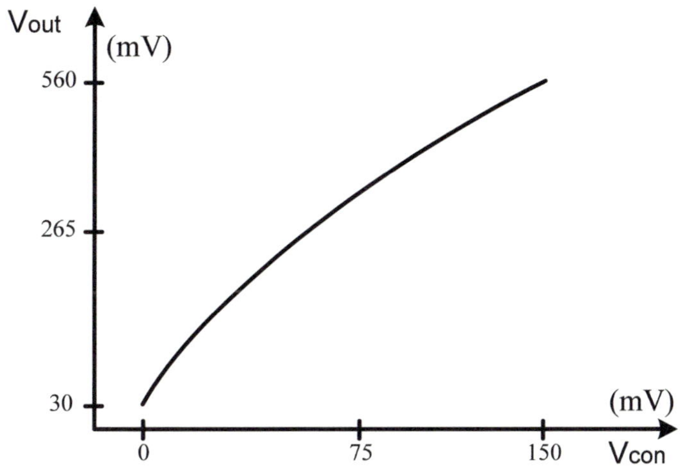

Figure 16. Output pulse amplitude vs. different control voltage

internal clock generator of the transmitter provides a 500MHz square wave clock signal. The output of the transmitter is connected to a high frequency oscilloscope through a SMA cable. The power supply of the chip package is 1.2V.

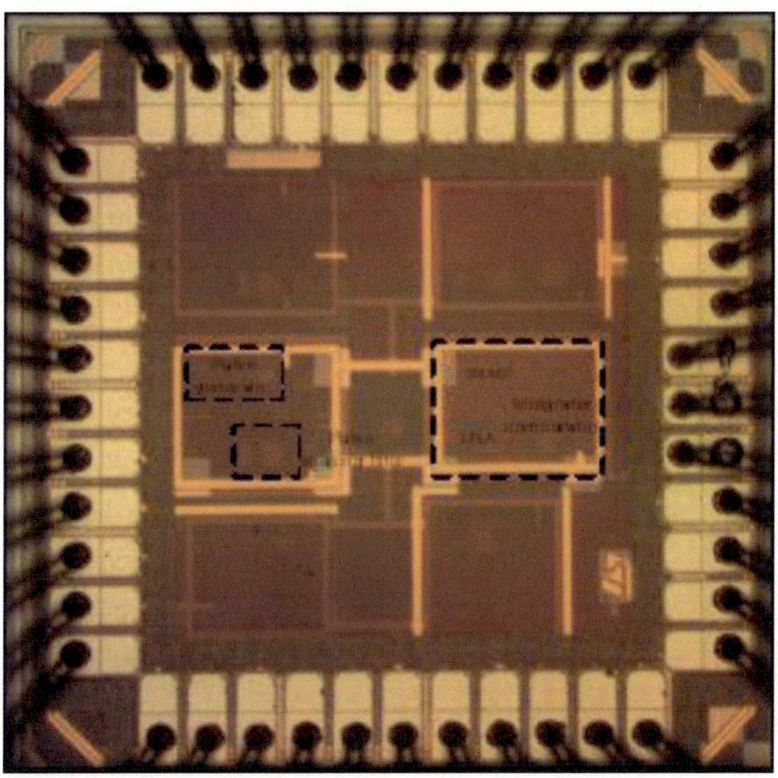

Figure 17. Photograph of the transceiver die

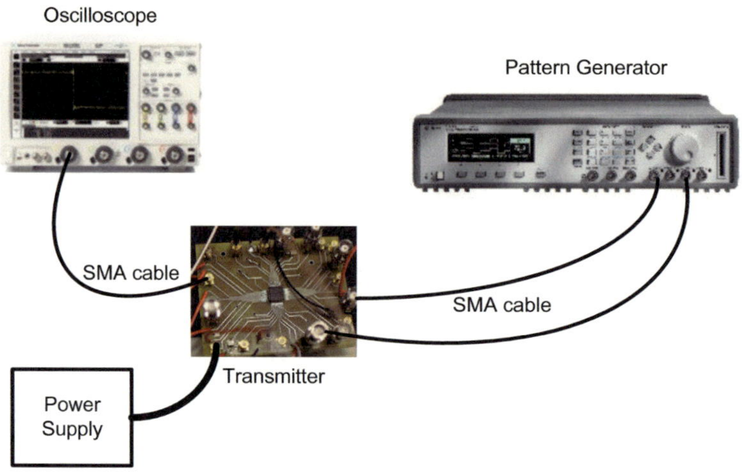

Figure 18. Measurement setup for the performance test of the proposed transmitter

In one of the tests, the pattern generator inputs a digital sequence with a pattern of 111001 at 250Mb/s, as shown in Figure 19, to the transmitter module. The corresponding transmitter output pulses are shown in Figure 20 with the scale doubled. The first derivative of the Gaussian pulse at the receiver side is shown in Figure 21. Each pulse-width is 1ns and has an amplitude of 177mV.

Figure 22 shows the measured output pulse waveforms of the implemented transmitter in another test where a single bit data is applied as an input. The pulse has a close-to-maximum amplitude of 521mV at a 50 Ω load and a pulse width of 1ns when the VGA control voltage was set at 0.14V. The output pulse has a minimum amplitude of 30mV.

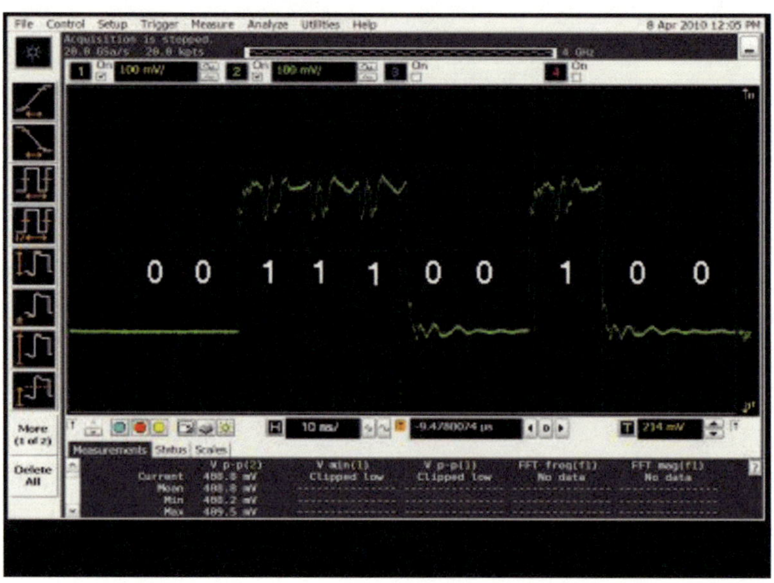

Figure 19. Waveform of input

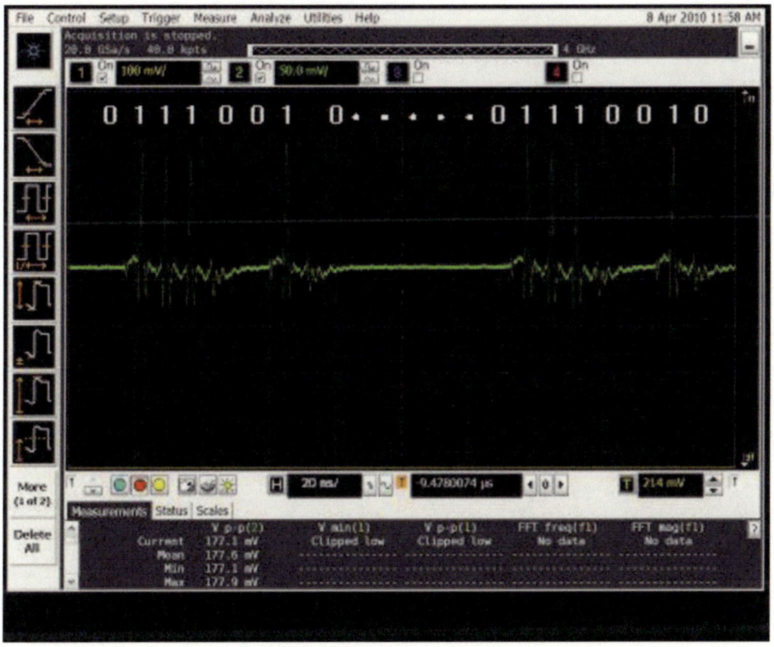

Figure 20. Measured transmitter output pulses

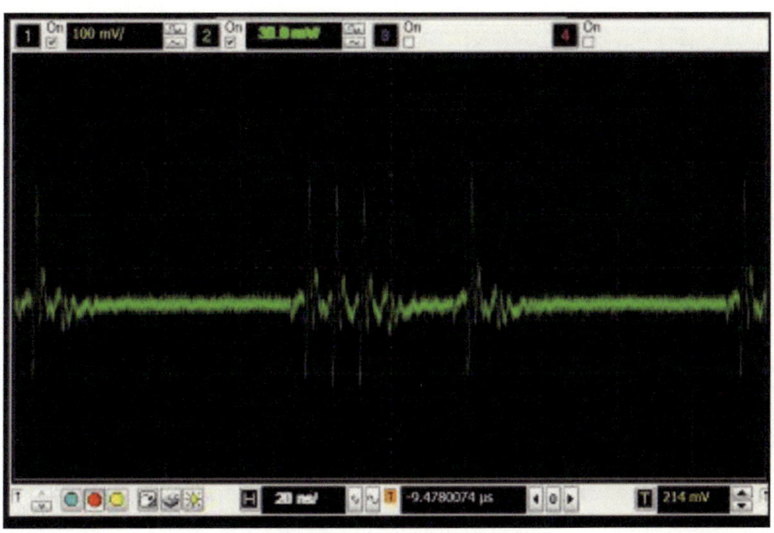

Figure 21. Measured first derivative of the pulses at receiver side

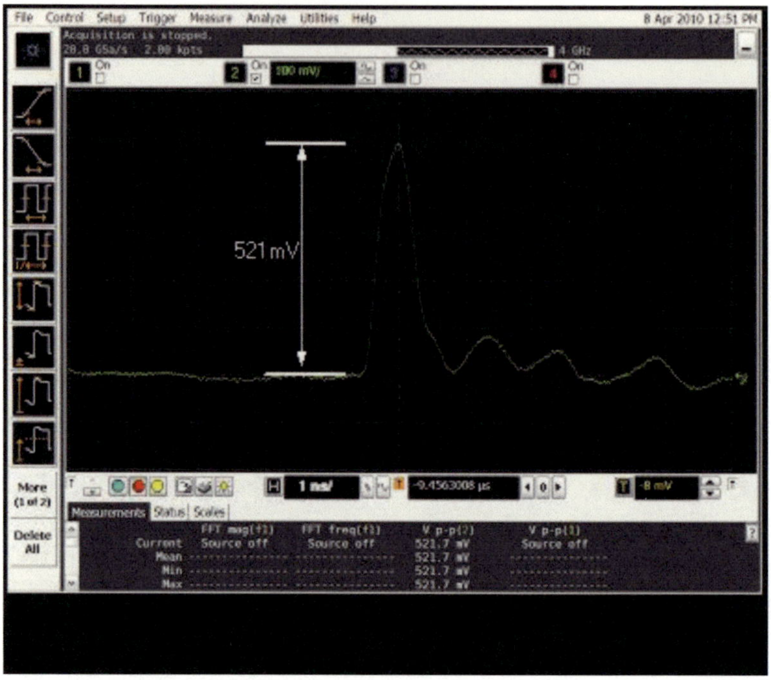

Figure 22. Single measured Gaussian pulse

The power spectral density of the transmitter modulated output pulse train is calculated using discrete Fourier transform on the measured UWB signal. Figure 23 shows the calculated power spectral density in dBm/MHz unit with the FCC spectral mask.

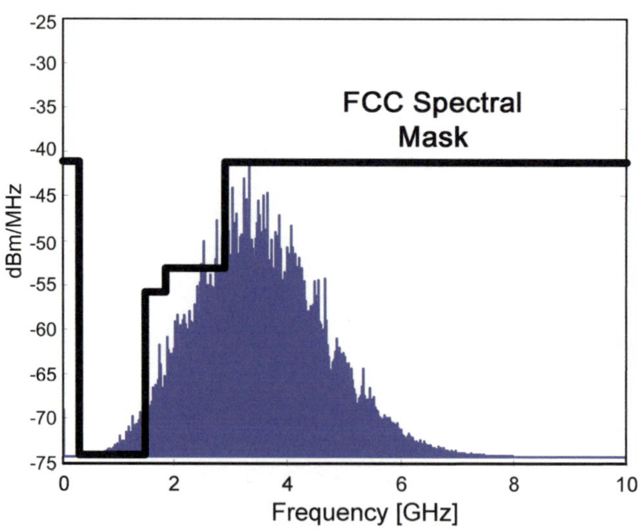

Figure 23. Power spectrum density of the output pulse

3.3. Other types of transmitter

In this section, the proposed system, which is shown in Figure 24, uses carrier-based continuous wave to represent information. The proposed non-coherent UWB transceiver transmits signals at different carrier frequency within the full 3-10 GHz band. This allows the radar to detect the targets with more details since more waves at different frequencies are reflected from dielectric boundaries favoring different frequencies. The signal at the transmitter is generated by a voltage controlled oscillator. The digital signal with specific envelop shape is multiplied with the output wave of the VCO to generate a modulated waveform consisting of multiple cycles of sinusoidal waves with small bandwidth. The carrier frequency can be adjusted over the total UWB frequency band by varying the capacitance value of the capacitor bank in the circuit. For radar sensing, a low repetition rate is desired. The transmitter employs a switching mechanism to reduce the power consumption and avoid the oscillation start-up transient delay and oscillation leakage. The proposed receiver contains a LNA, a down-conversion mixer, a filter and ADC. By using this approach, there is more flexibility on power spectrum control and output frequency selectivity, at the cost of higher power and more complexity compared to the first approach. Figure 25 shows the transmitter architecture in detail. It consists of a tunable LC oscillator, a voltage multiplier, a variable gain amplifier (VGA) driver, and a pulse generator. In this proposed design, the incoming data *Vdata* is split into two paths: one enters the LC oscillator and controls the on/off switch of the oscillation, the other path is fed into the pulse generator to trigger a square voltage pulse with a narrower pulse width. After the square pulse *Vdata* switches both the LC oscillator and the multiplier on and the oscillation reaches a steady state, the pulse generator output *Vpulse* with narrower time window is multiplied with the sinusoidal carrier to produce the transmitter output signal.

The voltage oscillator is realized based on the LC oscillation circuit using a cross-coupled transistor pair as shown in Figure 26. The oscillation frequency is determined by the resonance frequency of the inductor in each arm and the capacitor bank capacitance. The cross-coupled transistors create a negative impedance of value -1/gm at the drain of M1 and M2. The oscillator resonant frequency is set at 1/sqrt(LC), where L and C is the inductance and capacitance of L1 and capacitor bank, respectively. The signal Vdata turns on/off the oscillator by switching the PMOS M3 on/off. This signal also controls on/off of the multiplier, which is implemented using a Gibert Cell.

The signals at each block of the proposed transmitter are illustrated in Figure 27. The transmitter input signal Vdata controls signal C and B, and signal B generates signal D. The oscillator generates the carrier waveform with different center frequency within the 3-10GHz UWB band. The tuning is achieved through selecting different capacitors and adjusting the control voltage of the varactor in the capacitor bank. The six capacitors divide the 3-5GHz and 6-10GHz band into six sub-bands. By tuning voltage of the varactor in the capacitor bank, each single sub-band has a frequency tuning range of 500MHz. The overall tuning range of the capacitor bank covers the whole 7GHz UWB band. The simulated power spectral density of the transmitter output at different center frequency is shown in Figure 28.

Six transmitter outputs at frequency 3.1GHz, 4GHz, 4.8GHz, 6.3GHz, 8GHz, and 9GHz, tuned through the capacitor bank, are shown. The transmitter core area is 0.051 mm². The overall average power consumption of the transmitter at frequency of 9GHz is 1.85mW.

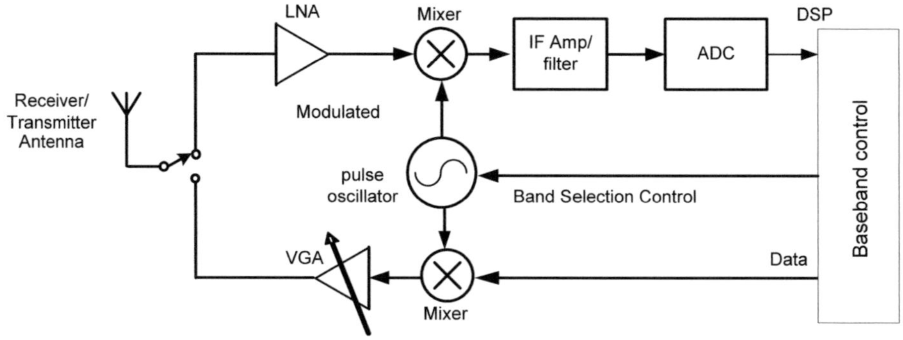

Figure 24. Proposed carrier-based UWB transceiver architecture

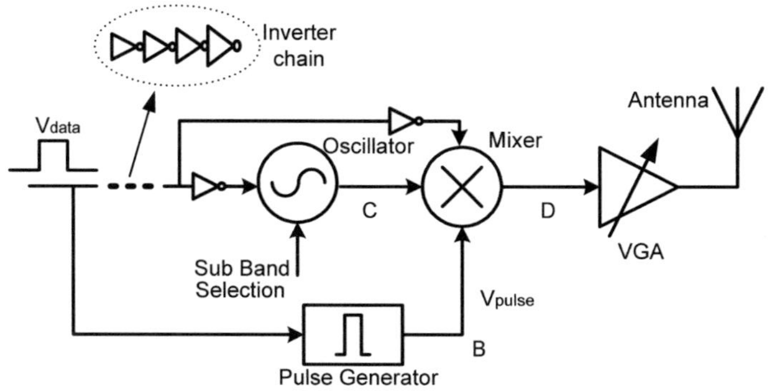

Figure 25. Proposed carrier-based transmitter

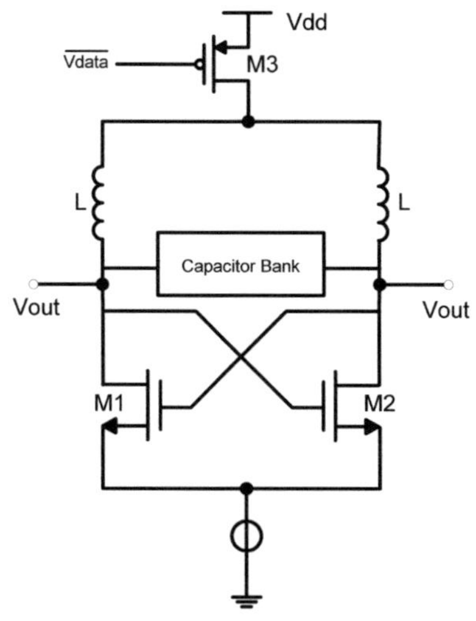

Figure 26. LC oscillator circuit

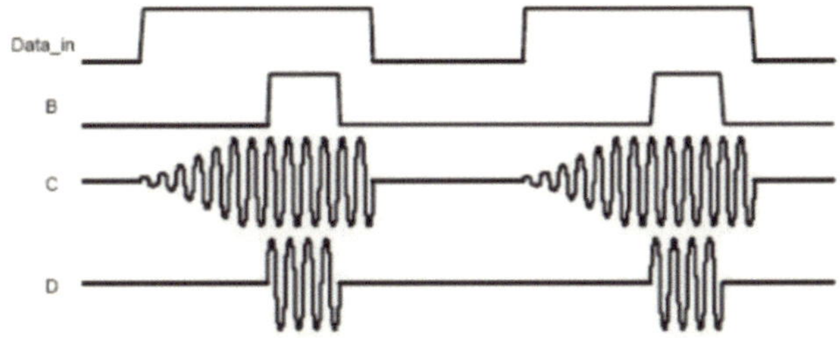

Figure 27. Waveforms of the carrier-based transmitter

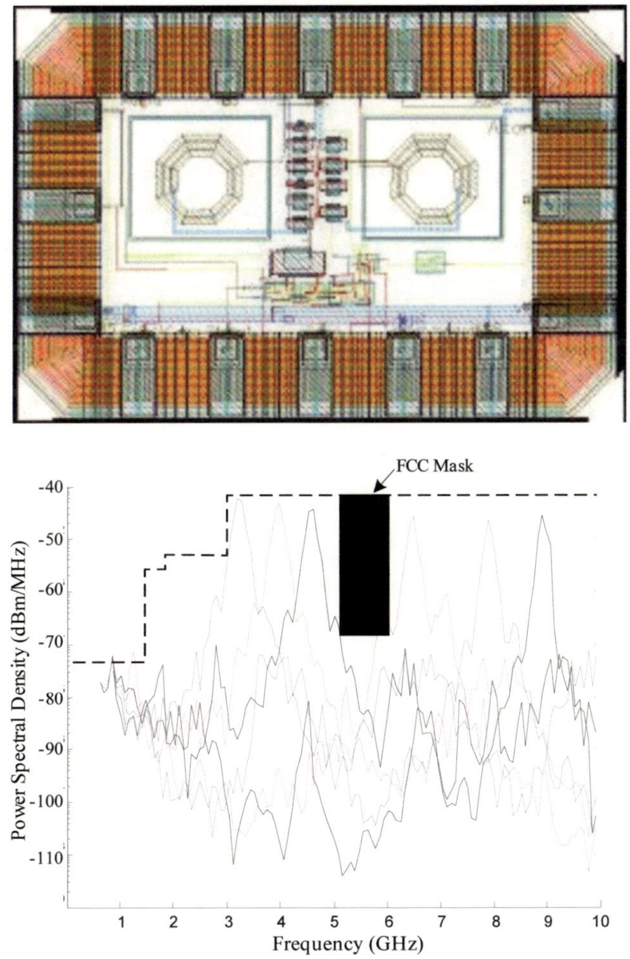

Figure 28. Transmitter layout and simulated PSD of the designed pulse generator output.

4. UWB radar energy detection receiver

Receiver design is another major challenge in UWB system. To detect the UWB signals, the goal of the UWB radar is to develop the method to maximize the signal-to-noise ratio (SNR).

Figure 29 shows an impulse-based UWB radar receiver. It consists of a LNA, a voltage multiplier, an integrator, and a clocked voltage comparator.

The LNA, shown in Figure 30, consists of two common-source, common-gate cascade amplifiers. The cascade configuration with inductive peaking (L2, L4) improves the reverse isolation and frequency response. The input matching network includes L1, C1, and L3. The inductors can be replaced by bonding wire to save chip area.

A Gilbert cell, as shown in Figure 31, is used to implement the voltage multiplication. For non-coherent receiver, the correlator uses the received signal as a signal template. The gates of M1, M3, and M6 are the correlator inputs. An analog integrator is used to collect the charges in a pulse or a pulse train. By integrating the pulse voltage, the voltage level of the received signal is much more distinguishable than noise. The integration time is adjusted through a voltage controlled capacitor C and R. A sense amplifier, as shown in Figure 32, is employed as a clocked comparator because of its sensitivity and low circuit complexity. Transistor M3, M4, M8, and M9 form a flow-through latch to sense the voltage difference between the input voltage and the reference voltage. Cascode input configuration is employed to minimize feedthrough of the clock signal. The range gating signal is combined with the comparator clock to control the sampling time of the comparator. The comparator is functioning only at the window when the range gating signal is on.

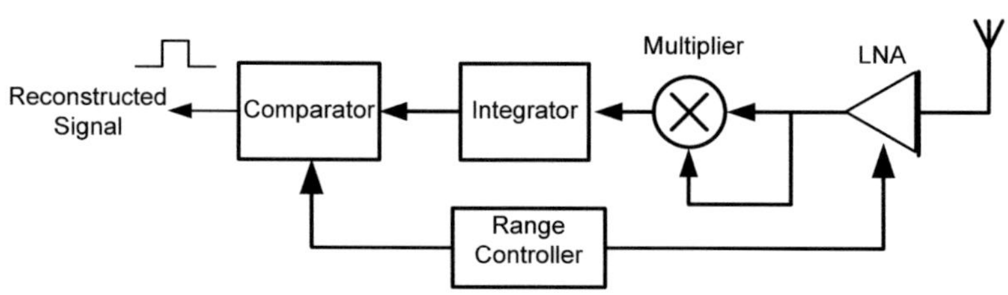

Figure 29. Impulse-based UWB receiver

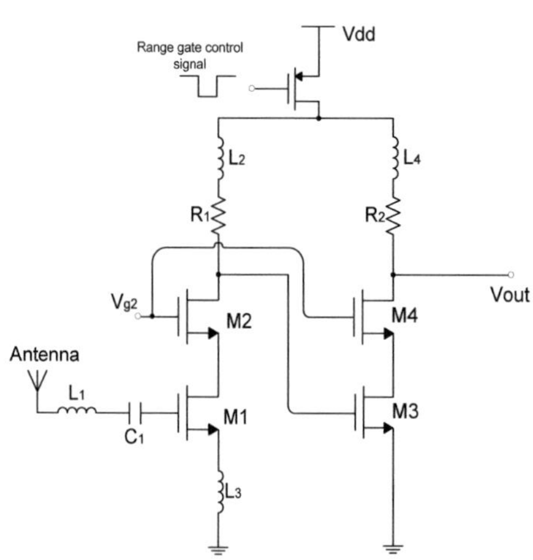

Figure 30. LNA circuit

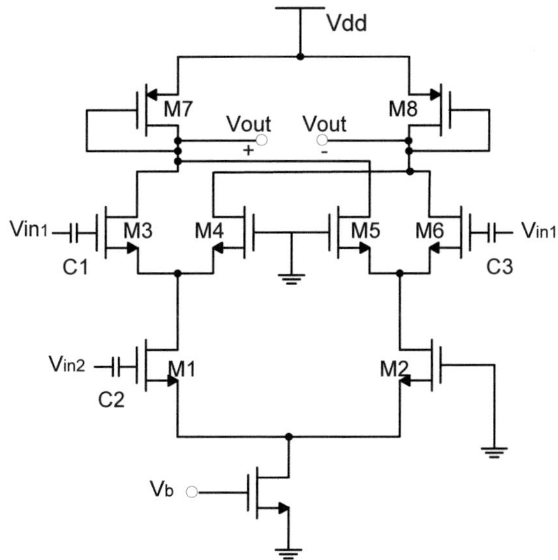

Figure 31. Multiplier circuit

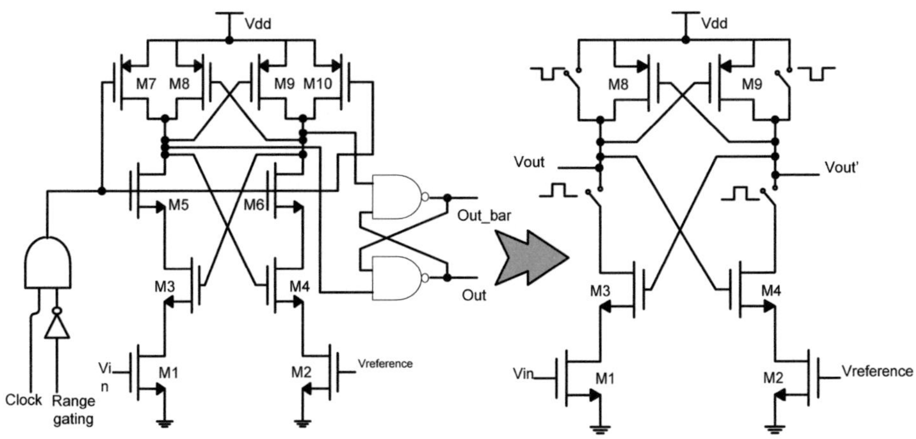

Figure 32. Comparator

Figure 33 shows the receiver for carrier-based UWB receiver. It consists of a wideband low noise amplifier, bandpass filters, square circuit, energy integrator, and comparator. The principle operation of the receiver involves energy correlation and detection at each different frequency sub-band. This energy detection topology is chosen because it reduces the sensitivity of the receiver required to immune noise and multi-path. The signal detected by antenna is amplified by the LNA and filtered by a bank of bankpass filters into 500 MHz sub-bands in UWB band. The filtered signal is squared and integrated over a fixed period of time t, and then quantized by a decision-making circuit. For non-coherent receiver, there is no need for synchronization mechanism, and thus reduce the complexity of the receiver circuit. For the case of OOK modulation scheme, there are two possible situations: signal energy collected or only noise energy collected. The greatly relaxes the required SNR for ADC operation.

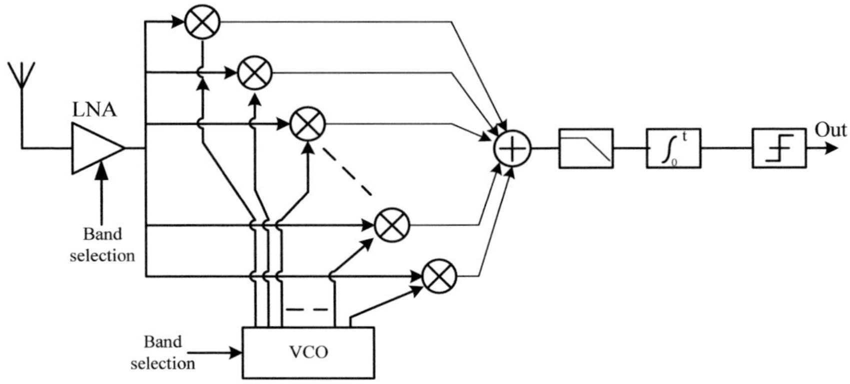

Figure 33. Carrier-based UWB receiver

Another advantage of this receiver topology is it can receive and process different frequency content at the same time, ie. The parallel receiver structure. This will increase the data rate by a factor of N, where N equals to the number of the filters in the filter bank.

5. Conclusion

Many published works have discussed the CMOS transceiver design for communications, but few demonstrated the CMOS transceiver design for medical radar sensing. This chapter demonstrates the design of biomedical radar sensing on the single CMOS integrated UWB transceiver. The advantage of using integrated CMOS UWB technology in biomedical sensing is that this technology provides ultra-low power, ultra-low cost, and ultra-low area solutions with much accurate and reliable performance. This chapter proposes an integrated radar system architecture which can achieve the radar sensing for heart rate monitoring, and explores and implements the integrated single chip radar transceiver circuit in CMOS IC. This chapter shows the implementation of the low-power low cost CMOS biomedical radar using UWB pulse for bio-monitoring.

This chapter can be expended further to apply in biomedical imaging using impulse radio radar. By characterizing the reflection properties of different tissues inside human body, an image of fluoroscopy of the human body can be generated under UWB radar scanning. The UWB radar will lead a technology breakthrough in the medical imaging area.

Author details

Xubo Wang, Anh Dinh and Daniel Teng
Electrical and Computer Engineering Department, University of Saskatchewan, Saskatoon, Canada

6. References

[1] Federal Communications Commission, *"FCC notice of proposed rule making, revision of part 15 of the commission's rules regarding ultra-wideband transmission systems,"* FCC, Washington DC, ET-docket 98-153.

[2] C. N. Paulson, J. T. Chang, C. E. Romero, J. Watson, F. J. Pearce, N. Levin, "Ultra-wideband Radar Methods and Techniques of Medical Sensing and Imaging," *SPIE International Symposium on Optics East*, Boston, MA, US, 2005.

[3] I. Oppermann, M. Hmlinen and J. Iinatti, *UWB : Theory and Applications*. John Wiley & Sons, September 2004.

[4] Faranak Nekoogar, *Ultra-Wideband Communications: Fundamentals and applications.* Prentice Hall Pearson Education, Inc., 2005.

[5] E. M. Staderini, "UWB radar in medicine," *Aerospace and Electronic System Magazine, IEEE*, Vol. 17, Issue 1, pp. 13-18, January 2002.

[6] A. Reisenzahn, T. Buchegger, D. Scherrer, S. Matzinger, S. Hantscher, C. Diskus, "A low-cost UWB radar system for sensing applications," *Ultra-Wideband, The 2006 IEEE International Conference*, pp. 299-302, September 2006.

[7] J. F. Holzrichter, G. C. Burnett, L. C. Ng, "Speech Articulator Measurements using Low Power EM-wave Sensor," The *Journal of the Acoustical Society of America*, Vol. 103, pp. 622-625, 1998.

[8] T. E. McEwan, "Body Monitoring and Imaging Apparatus and Method," *United States Patent* 5,766,208, June 16, 1998.

[9] T. McEwan, S. Azevedo, "Micropower Impulse Radar," *Science & Technology Review*, January 1996.

[10] Wireless2000, "Patient Assessment Monitor (PAM 3000)", *Wireless2000 RF & UWB Tech.*, Burnaby, BC, Canada, 2007, http://www.wireless2000.com/hrrm.htm.

[11] C. G. Bilich, "Bio-medical Sensing using Ultra Wideband Communications and Radar Technology: a feasibility study," *1st International Conference on Pervasive Computing Technologies for Healthcare*, 2006.

[12] I. Y. Immoreev, S. Samkov, and T. H. Tao, "Short Distance Ultra Wideband Radars," *Aerospace and Electronic Systems Magazine, IEEE*, Vol. 20, No. 6, pp. 9-14, June 2005.

[13] C. Y. Lee and C. Toumazou, "Ultra-low Power UWB for Real Time Biomedical Wireless Sensing," *IEEE International Symposium on Circuits and Systems*, pp. 57-60, May 2005.

[14] Immoreev I. Y. "Practical Application of Ultra-Wideband Radars," *IEEE Ultrawideband and Ultrashort Impulse Signals*, pp. 44-49, Septemeber 2006.

[15] X. Shen, M. Guizani, R. C. Qiu, T. Le-Ngoc, *Ultra-Wideband: Wireless Communicatioins and Networks*, John Wiley & Sons, 2006.

[16] T. Phan, V. Krizhanovskii, S. Lee, "Low-power CMOS energy detection transceiver for UWB Impulse Radio System," *IEEE Custom Intergrated Circuits Conference*, pp. 675-678, September 2007.

[17] Hilton MF, Bates RA, Godfrey KR. "Evaluation of Frequency and Time-Frequency Spectral Analysis of Heart Rate Variability as a Diagnostic Marker of the Sleep Apnea Syndrome," *Med. Biol. Eng. Comput.*, 37, (6), pp. 760-769, Novemeber 1999.

[18] John Proakis, *Digital Communications 2nd Edition*, McGraw-Hill, New York, 1989.

[19] James D. Taylor *Introduction to Ultra-Wideband Radar Systems*, CRC Press, 1994.

[20] Ehud Heyman, Benjamin Mandelbaum, Joseph Shiloh, *Ultra-Wideband Short-Pulse Electromagnetics*, Springer, 1999.

[21] Chang Y. J. "The NPAC Visible Human Viewer," *Syracuse University, NY.* www.npac.syr.edu/projects/vishuman/VisibleHuman.html.

[22] Gabriel C, "Compilation of the Dielectric Properties of Body Tissues at RF and Microwave Frequencies," *Physics Department, King's College London, Armstrong*

Laboratory, Occupational and Environmental Health Directorate, Radiofrequency Radiation Division, Brooks Air Force Base, TX, 78235-5102. Report: AL/OE-TR-1996-0037.

[23] Guillermo Gonzalez, *Microwave Transistor Amplifiers: Analysis and Design, 2nd* Edition, Rentice Hall, New Jersey, 1997.

[24] Sangyoub Lee, "Design and Analysis of Ultra-Wide Bandwidth Impulse Radio Receiver," *Doctor of Philosophy Dissertation, University of Southern California*, 2002.

[25] B. Razavi, *RF microelectronics*. Prentice Hall PTR, pp.147-175, 1998.

[26] T. H. Lee, *The Design of CMOS Radio-Frequency Integrated Circuits*, 1st ed., New York: Cambridge Univ. Press, pp. 272-377, 1998.

[27] T. Terada, S.Yoshizumi, Y. Sanada, and T. Kuroda, "Transceiver Circuits for Pulse-based Ultra-wideband," *IEEE ISCAS*, Vol. 4, pp. 349-352, May 2004.

[28] Y. Zheng, Y. Tong, J. Yan, and Y. Xu, W. Yeoh, F. Lin, "A low power noncoherent CMOS UWB transceiver ICs," *IEEE Radio Frequency Integrated Circuits (RFIC) Symposium*, pp. 347-350, June 2005.

[29] A. Bevilacqua, A. Niknejad, "An Ultrawideband CMOS Low Noise Amplifier for 3.1-10.6 GHz Wireless Receivers," *IEEE Journal of Solid-State Circuits*, Vol. 39, No. 12, pp. 2259-2268, December 2004.

[30] Bo-Yang Chang, Christina F. Jou, "Design of a 3.1-10.6GHz Low-Voltage, Low-Power CMOS Low-Noise Amplifier for Ultra-wideband Receivers," *IEEE APMC2005 Proceedings*, Vol. 2, 4 pages, December 2005.

[31] Bi Ngoc Pham, "A Dual Mode Q-enhanced RF Front-end Filter for 5GHz WLAN and UWB with NB Interference Rejection," *Master of Science Thesis, University of Saskatchewan*, 2007.

[32] P. K. Saha, N. Sasaki and T. Kikkawa, "Impulse-based UWB transmitter in 0.18um CMOS for Wireless interconnect in future ULSI," *Ext. Abst. of the Forth Hiroshima International Workshop on Nanoelectronics for Tera-bit Information Processing, Hiroshima University*, Japan, pp. 76-77, September 2005.

[33] B. Razavi, "Design Considerations for Direct-conversion Receivers," *IEEE Transactions on Circuits and Systems II:Analog and Digital Signal Processing*, Vol. 44. 6, pp. 428-435, June 1997.

Chapter 4

Compressed Sensing: Ultra-Wideband Channel Estimation Based on FIR Filtering Matrix

Huanan Yu and Shuxu Guo

Additional information is available at the end of the chapter

http://dx.doi.org/10.5772/48714

1. Introduction

Ultra-wideband (UWB) communication (Win & Scholtz, 1998; Yang & Giannakis, 2004a) is a fast emerging technology since the Federal Communication Commission released a spectral mask in the spring of 2002. The major reason for UWB technology to receive much attention is its promising ability to provide low-power consumption, high bit rate and multipath resolution, and coexist with the narrow-band system by trading bandwidth for a reduced transmits power. In the impulse radio UWB (IR-UWB) systems, the duration of pulse is ultra-short, typically on the order of nanoseconds. On one hand, the ultra-short impulses make it possible to resolve and combine signal echoes with path length differential down to 1 ft exploiting the diversity inherent in the multipath channel and improving the position accuracy. On the other hand, the new technical (Witrisal et al., 2009) challenges are posed: (1) analog-to-digital converters (ADCs) working at the Nyquist rate are in general very expansive and power demanding; (2) the synchronization which is accomplished at the scale of sub nanosecond duration is extremely complex; (3) capture a sufficient amount of the rich multipath diversity need accuracy channel estimation. Compare to the transmitter easily implement, the IR-UWB receiver are too complex.

The emerging theory of compressed sensing (CS) (Candès, et al., 2006; Donoho, 2006) provides new approaches for practical UWB receiver design. When the short duration pulses in the UWB system propagate through the multipath channels, the received signals remain sparse in time domain. The sampling rate can be reduced to sub-Nyqusit rate and the receiver can reconstruct the initial signal with high probability. Accordingly, there has been a growing interest in applying the CS theory to sparse channel estimation (Bajwa et al., 2010; Berger et al., 2010). The recent literature on sparse channel estimation can be found in (Bajwa et al., 2010; Berger et al., 2010) and in their references. It is proved that conventional channel estimation methods provide higher errors because they ignore the prior knowledge of the sparseness (Wan et al., 2010). The sparse channel estimation problem is faced in (Paredes et al., 2007) under a time domain sparse model point of view. In (Paredes et al., 2007) a suitable dictionary formed by delayed versions of the UWB transmitted pulse is defined in order to better match the UWB signal. However, the spike basis achieves maximal incoherence with the Fourier basis (Candès & Wakin, 2008) and is for that reason that seems more convenient to work with frequency domain. To ensure that every measurement counts, they propose to pre-modulate the input signal with a spread spectrum sequence before the Fourier transformation. As the IR-UWB signals have resolvable multipath with a sparse structure at the receiver, the application of CS theory to UWB channel estimation has also found wide interest in the UWB community. For the CS based UWB channel estimation, the main goal has been to estimate the sparse channel with reduced number of observations (Paredes et al., 2007; Liu & Lu, 2009; Naini et al., 2009).That is equivalent to reducing the sampling rate at the receiver. In (Paredes et al., 2007), a channel detection method based on the Matching Pursuit algorithm is proposed, where the path delays and gains are calculated iteratively. In (Liu & Lu, 2009), the authors combine the maximum likelihood (ML) approach with the CS theory. In (Naini et al., 2009), a spread spectrum modulation structure is placed before the measurement matrix to enhance the estimation performance. The common assumption of the studies in (Paredes et al., 2007; Liu & Lu, 2009; Naini et al., 2009) is that the UWB channels are sparse. However, depending on the environment (e.g., an industrial environment may have dense multipath), the sparsity assumption of the channels may not hold. And the receiver may be a little complex for the compressed sensing framework.

In this context a Finite Impulse Response (FIR) filtering matrix estimator for UWB channel based on the theory of CS is advanced. An FIR filter is introduced at the transmitter to get a quasi-Toeplitz measurement matrix. So the reconstruction accuracy using the CS framework is improved. Also, the receiver is simplified since a filter at the transmitter has been adopted in place of the measurement matrix at the receiver. The key point is to avoid the magnification of noise by the measurement matrix. Unlike the Generalized Likelihood Ratio Test (GLRT) detector design, the correlation detector for UWB signals employing the channel parameters estimated in this chapter needs no prior knowledge about the channel noise. In addition, the desired receiver performance calls for fewer measurements. Then both the Orthogonal Matching Pursuit (OMP) and the Basis Pursuit De-noising (BPDN) are compared to the Dantzig Selector (DS) for different signal noise ratio (SNR) to give the opinions for choosing suitable reconstruction algorithms. Realistic channel estimation is considered. Simulations discussed later indicate the efficiency of the proposed method.

This chapter is organized as follows. In section 1, the motivation and research status are introduced. In section 2, a brief description of compressed sensing and its application for UWB channels is introduced. In section 3, the FIR filtering matrix method for UWB channel estimation based on the CS theory is proposed. In section 4, the estimation results are used in the UWB signals detection. In section 5, the simulation results together with the analysis are given. In section 6, we offer the conclusions and discussions. The references are given in section 7.

2. Compressed sensing for UWB channel estimation

In essence, CS theory has shown that a sparse signal can be recovered with high probability from a set of random linear projections using nonlinear reconstruction algorithms. The sparsity of the signal can be in any domain (time domain, frequency domain, wavelet domain, etc.) and the number of random measurements, in general, is much smaller than the number of samples in the original signal, which leads to a reduced sampling rate and, hence, reduced use of ADCs resources.

In UWB impulse radio communications, an ultra-short duration pulse, typically on the order of nanoseconds, is used as the elementary pulse-shaping to carry information (Reed, 2005). Transmitting ultra-short pulses leads to several desirable characteristics: (1) simplicity is attained in the transmitter since a carry-less baseband signal is used for conveying information (Lottici et al., 2002); (2) the transmitted signal power is spread broadly in frequency having little or not impact on other narrowband radio systems operating on the same frequency (Qiu et al., 2005); (3) the received UWB signal is rich in multipath diversity introduced by the large umber of propagation paths existing in a UWB channel. For the most important fact that transmitting an ultra-short pulse through a multipath UWB channel leads to a received UWB signal that can be approximated by a linear combination of a few elements from a pre-defined basis, yielding thus a sparse representation of received UWB signal. Next, we briefly describe the CS framework in (Candès et al., 2006) and (Donoho, 2006), and apply this framework into the UWB channel.

2.1. CS overview

Consider the problem of reconstructing an $N\times1$ discrete-time signal vector $x \in \mathbb{R}^N$. It can be shown that if x is sparse, in the sense that x can be represented as a superposition of a small number of vector taken from a dictionary $D=[D_1,\ldots,D_N]$ of tight-frames, which provides a K-sparse representation of x, that is

$$x = \sum_{n=0}^{N\text{-}1} D_n u_n = \sum_{l=1}^{K} D_{n_l} u_{n_l} \tag{1}$$

Where x is a linear combination of K vector chosen from the arbitrary basis D, and $K<<N$; $\{n_l\}$ are the indices of those vectors; $\{u_l\}$ are the weighting coefficients; $\alpha=K/N$ is called sparse-ratio. Alternatively, in this chapter we write the signal vector in matrix notation

$$x = Du \tag{2}$$

Where $u=[u_0, u_1, \cdots, u_{N-1}]^T$ which has K nonzero coefficients, where $K<<N$. In CS, signal x can be represented by K entries of u in place of N entries of x, that reduces dimension of the signal of interest. We need to estimate only K real-parameters not N to reconstruct x from a channel realization. When sparse-ratio α is very small, the compressive gain becomes high.

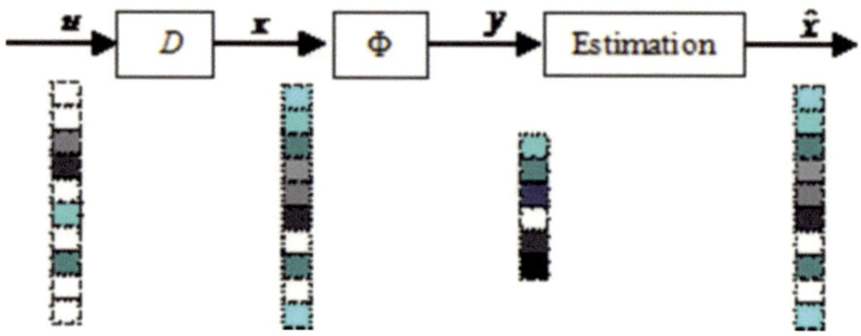

Figure 1. Compressive sensing model

In the viewpoint of the CS theory, a sparse signal can be reconstructed successfully from far fewer data/measurements than what is usually considered necessary (Donoho, 2006). Figure 1 shows the CS theory frame. It gives the whole course of the signal projection and reconstruction. The sampling from x becomes a linear transformation, that is x can be reconstructed from M measurements and $M<<N$. By projecting x onto a random measurement matrix $\Phi \in \mathbb{R}^{M \times N}$, a set of measurements $y \in \mathbb{R}^{M}$ can be obtained as

$$y = \Phi x = \Phi D u \tag{3}$$

where Φ is called measurement matrix, which is incoherent with D; and y is the signal we received in receiver, who has M entries, each becomes a measurement of x. Instead of using the N-sample x to find the weighting coefficients u, M-sample measurement vector y can be used. Accordingly, u can be estimated as

$$\hat{u} = \min \|u\|_1 \text{ s.t. } y = \Phi D u \tag{4}$$

Where l_p -norm is defined as $\|u\|_p = \left(\sum_{n-1}^{N} |u_n|^p \right)^{1/p}$. Note that, the advantage of estimating u from the vector y instead of x is that the former having much fewer samples corresponds to a much lower sampling rate at the receiver. If the dictionary D and measurement matrix Φ are acquired, and they satisfy $M=CK\log N<<N$, signal x can be recovered from measurements y using reconstruction algorithms with overwhelmingly high probability, even we don't know the sparse pattern of the unknown signal u (Candès & Tao, 2006). $C \geq 1$ is then called the oversampling factor.

In short, sampling and processing signals in the CS framework can be concluded just like this: First, we must design tight-frames D according to the character of signal of interest. That is to design a overcomplete dictionary to get the sparse representation of x; after the first stage, one should design a MxN sensing matrix Φ, through which measurement y can be achieved. Finally, x can be recovered with y, D and Φ employing reconstruction algorithm. In next section, we will present how this concept can be used for UWB channel estimation.

2.2. CS for UWB channel estimation

The CS theory explained in (2)–(4) can be applied to UWB channel estimation for the fact that the Gaussian pulse response of the UWB channel is sparse. We show the simulation results in section 5. Suppose that $r \in \mathbb{R}^N$ is the discrete-time representation of the received signal given as

$$r = Ph + n \tag{5}$$

Where $P \in \mathbb{R}^{N\times N}$ is a scalar matrix representing the time-shifted pulses, h= [α_1, α_2, , α_3]T are the channel gain coefficients, and n are the AWGN terms. Since the UWB channel structure is sparse, h has only Knonzero coefficients. Similar to (3), the received signal r can be projected onto a random measurement matrix $\Phi \in \mathbb{R}^{M\times N}$ so as to obtain $y \in \mathbb{R}^M$ as

$$y = \Phi Ph + \Phi n = Ah + v \tag{6}$$

Due to the presence of the noise term v, the channel h can be estimated as

$$\hat{h} = \min\|h\|_1 \ \text{ s.t. } \ \|Ah - y\|_2 \le \varepsilon \tag{7}$$

Where ε is related t o the noise term as $\varepsilon \ge \|v\|_2$. Considering (7), the channel estimation performance depends on the sparsity of h(i.e., the value of K), as well as the number of observations M. It is therefore necessary to understand the discrete-time equivalent structure of h and the effects of standardized channel models.

3. UWB channel estimator based on CS

While CS research has focused primarily on signal reconstruction and approximation, the CS framework can be extended to a much broader rang of statistical inference tasks, well suited for applications in wireless UWB communications. UWB channel estimation is one of those applications which will be used extensively in this section. Next we will investigate the effect of the IEEE 80.15.4a UWB channel models (Molisch et al., 2006) on the channel estimation performance from a practical implementation point of view. Then a new sparse channel estimation method is proposed by improving the random measurement method based on CS for discrete time signals in (Paredes et al., 2007). According to the amplification of channel noises as well as measurement signals, we designed a new channel estimation method with FIR filtering matrix.

3.1. UWB channel

In the following, we initially present the discrete-time equivalent channel h followed by the UWB channel models. In order to obtain h, the general channel impulse response (CIR) should be presented first. Accordingly, the continuous-time channel $h(t)$ can be modeled as

$$h(t) = \sum_{l=0}^{L-1} \alpha_l \delta(t - \tau_l) \tag{8}$$

Where α_l is the l-th multipath gain coefficient, τ_l is the delay of the l-th multipath component, $\delta(\cdot)$ is the Dirac delta function and L is the number of resolvable multipath.

The continuous-time CIR given in (8) assumes that the multipath may arrive any time. This is referred to as the τ-spaced channel model (Erküçük et al., 2007). Suppose that two consecutive multipath with delays τ_k and τ_{k+1} arrive very close to each other. Further suppose that a pulse of duration T_s is to be transmitted through this channel. If $T_s > |\tau_{k+1} - \tau_k|$, then the pulse at the receiver cannot be resolved individually for each path, and experiences the combined channel response of the kth and $(k+1)$th paths. Let us define an approximate T_s-spaced channel model that combines multipath arriving in the same time bin, $[(n-1)T_s, nT_s]$, $\forall n$. Accordingly, for $[(n-1)T_s, nT_s]$, $\forall n$, the delays $\{\tau_l | 0, 2, \ldots, L\text{-}1\}$ that arrive in the corresponding quantized time bins can be determined, and the associated $\{\alpha_l | 0, 2, \ldots, L\text{-}1\}$ gains can be linearly combined to give the new channel coefficients $\{\alpha_n | 1, 2, \ldots, N\}$. Note that some of the $\{\alpha_n\}$ values may be zero due to no arrival during that time bin, hence, the number of nonzero coefficients K satisfies the condition $K \le L \le N$. The equivalent T_s-spaced channel model can be expressed as

$$h(t) = \sum_{n=1}^{N} \alpha_n \delta(t - nT_s) \tag{9}$$

Where $T_c = NT_s$ is the channel length. Using (9), the discrete-time equivalent channel can be written as

$$h = [\alpha_1, \alpha_2, \ldots, \alpha_N]^T \tag{10}$$

where the channel resolution is T_s. Then the discrete-time equivalent channel vector obtained above can be used in (5)–(7) in the context of CS theory. Next, we consider the UWB channel models to be used with the channel vector h.

The CS based UWB channel estimation studies assume that the UWB channel vectordefined above is sparse. However, this is a vague assumption. In order to classify a channel as sparse, initially the channel environment should be examined. In (IEEE Std 802.15.4a, 2007), members of the IEEE 802.15.4a standardization committee have developed a comprehensive standardized model for UWB propagation channels. Accordingly, they have considered different environments and have conducted measurement campaigns in order to model the UWB channels for each environment. The channel environments that they have parameterized include indoor residential, indoor office, outdoor, industrial environments, agricultural areas and body area networks. The details of the related channel models and their associated parameters can be found in (Molisch et al., 2006). We motivate our study with the selection of a variety of environments either having a line-of-sight (LOS) or a non-LOS (NLOS) transmitter-receiver connection. Accordingly, the CM1 (indoor residential LOS), CM2 (indoor residential NLOS), CM5 (outdoor LOS) and CM8 (industrial NLOS) channel models are widely used in UWB research. We now summarize the characteristics of channel models CM1, CM2, CM5 and CM8 in the following.

CM1: This is by-far the most commonly used channel model in order to assess the system performance. It models an LOS connection in an indoor residential environment. It is the most sparse channel model where few Rake fingers can collect considerable amount of signal energy.

CM2: This is a channel model with an NLOS connection in an indoor residential environment. It complements CM1. It is a sparse channel model but usually contains more multipath compared to CM1.

CM5: This is a channel model with an LOS connection in an outdoor environment. Typically, the multipath arrive in a few clusters.

CM8: This is a channel model with an NLOS connection in an industrial environment. The multipath arrive densely so that the channel does not have a sparse structure.

Using the T_s-spaced channel model in (9) and the parameters for channel models CM1, CM2, CM5 and CM8 in (Molisch et al., 2006), a realization for CM1 channel model is plotted in Figure 4 when the channel resolution is T_s=0.66ns. It can be observed that the typical channel properties listed above can be observed. The impulse response of the UWB channel is sparse.

3.2. Random measurement estimation

According to the models proposed by the IEEE 802.15.4a working group, the impulse response of the UWB channel is modeled as function (8) follow in time domain. Consider the simple communications model of transmitting a pulse $p(t)$ throughout a noiseless UWB communication channel $h(t)$. The received UWB signal can be modeled as

$$g(t) = p(t) * h(t) = \sum_{l=0}^{L-1} p_l \delta(t - \tau_l) \tag{11}$$

In this chapter, we suppose $p(t)$ is a first derivative of the Gaussian pulse with unit power. Then the estimate value of $g(t)$ is $\hat{g}(t)$, which represents a referent template for subsequent correlation detection of UWB signals. The received UWB signal given by (11) has been sampled to define the discrete-time vector g, which is taken as a signal targeted for estimation. It is available to get the estimate of g by sampling directly from $g(t)$. However, the extremely high bandwidth of the received UWB signal requires high-speed A/D converters. Some approaches for UWB receivers are needed to attain the required sampling rates. The random measurement method focuses on this goal by sensing data at the receiver using a $M \times N$ measurement matrix Φ_1, which is obeying the restricted isometry property (RIP) (Bajwa et al., 2007), leading to measurements y_1

$$y_1 = \Phi_1 g = \Phi_1 P h \tag{12}$$

The referent template $\hat{g}(t)$ access to correlation detection can be reconstructed successfully with only M measurements at the receiver, provided that g is sparse in some space. The principle architecture of the random measurement estimation is given in Figure 2.

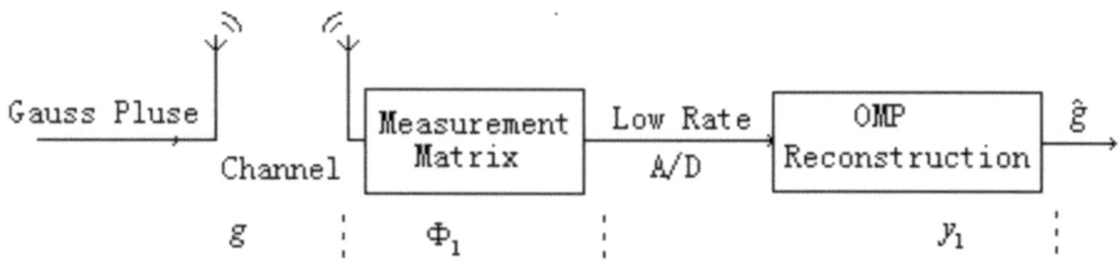

Figure 2. The random measurement estimation

In (Paredes et al, 2007), a over-complete dictionary is designed, in which the signal g has concise representations when expressed. Then a better performance of the channel estimation was guaranteed. The results, however, are based on a premise that there is no channel noise on pilot symbols. It is not true in actual channel. When the noise is introduced, we have

$$g_n(t) = p(t) * h(t) + n(t) = g(t) + n(t) \tag{13}$$

Where $n(t)$ denotes additive white Gaussion noise (AWGN) in UWB channel, follow the $N(0, \delta^2)$ distribution. We restrict our attention to discrete signals, then the measurement process on the signal itself exploiting measurements matrix Φ_1 is described as this

$$y_1 = \Phi_1(g + n) = \Phi_1 Ph + \Phi_1 n_1 \tag{14}$$

Note that, the random measurement method processes the noise from N-dimension to M-dimension via the projection. In terms of the conversion in dimension, the noise power translates into $(N/M)^2\delta^2$ versus δ^2 before projecting. Because of $M \ll N$, that means the sampling rate reduces at the expense of magnification of the channel noise.

3.3. FIR filtering matrix estimation

In this chapter, we propose a new method based on filtering matrix for UWB channel estimation with CS framework. In order to improve the estimation performance, some implements should be taken to suppress the magnification of the noise from measurements matrix. The concrete step is illustrated in Figure 3, which gives the architecture of the proposed method. The processing flows for both the transmitter and the receiver have been adjusted leading to higher accuracy and lower complexity in receiver.

As can been seen from Figure 3 (Yu Huanan & Guo Shuxu, 2010), a UWB signal is transmitted by a UWB pulse generator and through an FIR filter. Then, the received signal is directly sampled through a low-rate A/D conversion after the propagation paths. Finally, the estimation of the impulse response $\hat{h}$ can be reconstructed via OMP algorithm (Pati et al., 1993).

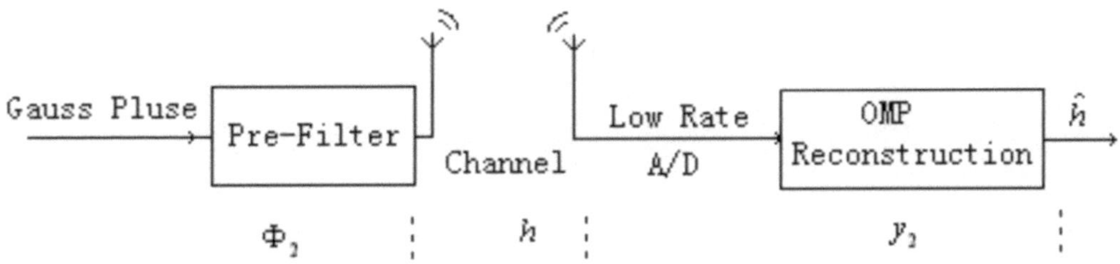

Figure 3. The FIR filtering matrix estimation

According to function (1), the transmitting signal $x(t)$, which is K-sparse over some overcomplete dictionary P,

$$x(t) = \sum_{n=0}^{N-1} P_n(t)h_n = P(t)h \tag{15}$$

Where

$$P(t) = [P_0(t), P_1(t), \ldots, P_{N-1}(t)] \tag{16}$$

$$h = [h_0, h_1, \ldots, h_{N-1}]^T \tag{17}$$

Note that there are only K non-zeros in h. $x(t)$ is then fed into a L-length FIR filter. Suppose that $m(t)$ is the impulse response of the FIR filter, the received signal for the UWB communication is given by

$$y_2(t) = p(t) * m(t) * h(t) + n(t) \tag{18}$$

Since the UWB channel is sparse, the impulse response of the UWB channel $h(t)$ can be viewed as a sparse signal. Let h be the discrete-time representation of $h(t)$, which is set up as the estimation target in this section. In addition the identity matrix is used as an overcomplete dictionary because of the sparsity of h. Define $c(t) = p(t) * m(t)$, then (18) becomes

$$y_2(t) = c(t) * h(t) + n(t) \tag{19}$$

Where $c(t)$ and $h(t)$ are processed using a low-rate A/D, which is M-dimension. The output $y_2(t)$ is then uniformly sampled with sampling period T_s. T_s follows the relation $T_s/T_h = q$, where q is a positive integer, and T_h denotes the time delay between each adjacent channel. M samples are collected so that $M \cdot T_s = \lfloor L \cdot T_h + T_x \rfloor$, which is the duration of $y(t)$. Now we have the down-sampled output signal $y_2(mT_s)$, m=0, 1,…, $M-1$

$$\begin{aligned} y_2(mT_s) &= c(mT_s) * h(mT_s) + n(mT_s) \\ &= \int_0^T c(mT_s - \tau)h(\tau)d\tau + n(mT_s) \\ &= \sum_{l=0}^{L-1} c_i h(mT_s - iT_h) + n(mT_s) \end{aligned} \tag{20}$$

The output $y_2(t)$ is uniformly sampled with sampling period T_s. Now we rewrite (20) in matrix notation

$$y_2 = \Phi_2 h + n \tag{21}$$

where Φ_2 is a quasi-Toeplitz matrix. It has such property: each row of Φ_2 has L non-zero entries and each row is a copy of the row above, shifted by q places. Following (Bajwa et al., 2007), it is illustrated that the quasi-Toeplitz matrix obeys the RIP.

According to (20) and (21), let y_2 be the random projected signal where $\Phi_2=p(t)*m(t)$ is the measurement matrix, and identity matrix is used as over-complete dictionary. The random projected signal y_2 can be acquired with M-dimension low-rate A/D converters, and the OMP algorithm is then applied on y_2 to recover $\hat{h}$. While the convolution process is following (21), associated with the reconstruction results above, the referent template $\hat{g}$ for correlation detection is acquired.

The whole process above accounts to a filtering action on the Gaussian pulse, which we selected as the transmitted pulse waveform. The receiver becomes very simple, with only one M-dimension low-rate A/D to collect measurement samples after the filter and channel. It can be seen that the noise does not go through the projection from N-dimension to M-dimension, thus the noise has not been magnified in the CS framework. Furthermore the proposed method has a better measurement matrix compared with the random measurement method. Hence, a better performance of the estimation accuracy can be achieved.

4. Correlation detection for the UWB signals

The random measurements method is focused on CS reconstruction of noiseless UWB signals, which relies on the assumption that the noiseless composite pulse-multipath waveform is sparse in a pre-designed dictionary. In a more realistic UWB communication scenario, however, the received signal is contaminated with noise and interferences, and the challenges fall in the design of a UWB receiver with the ultimate goal of signal detection.

4.1. Correlation detection

Suppose that the impulse response of the UWB channel $h(t)$ is invariant in each data frame, including N_p pilot symbols and N_s data modulated symbols. The total number of symbols in one burst is N_p+N_s And N_f first derivative of the Gaussian pulse $p(t)$ are repeated over consecutive frames to transmit one pilot or binary symbol. For the sake of damping the effect of AWGN, we average the received signal during a data frame. The maximum excess delay of the dense multi-path channel is given by T_{med}. $p(t)$ is of unit energy and has time duration T_p, and also the duration of a frame is given by T_f. In order to avoid inter-symbol interference (ISI) and intra-symbol interference (Yang & Giannakis, 2004b), it is assumed that $T_f > T_p + T_{med}$.

In the UWB correlation detector, if there exits a module to time precisely, the pilot and data symbols can be exactly separated. When the pilot is canceled, and also the referent template $\hat{g}$ estimated above is employed into the correlation detection of the received signals, the transmitted signal during a data frame for UWB communication is shown as follows

$$s(t) = \sum_{j=0}^{N_s-1} \sum_{n=0}^{N_f-1} b_j p(t - jN_f T_f - nT_f - N_p N_f T_f) \tag{22}$$

here, $b_j \in \{\pm 1\}$ are the j-th information bits. Signal $s(t)$ propagates through an L-path fading channel whose response to $p(t)$ is $\sum_{l=1}^{L} h_l p(t - \tau_l)$ such that the received signal at the receiver is

$$r_b(t) = \sum_{j=0}^{N_s-1} \sum_{n=0}^{N_f-1} \sum_{l=1}^{L} b_j \alpha_l p(t - jN_f T_f - nT_f - N_p N_f T_f - \tau_l) + n(t) \tag{23}$$

Where $n(t)$ is thermal noise with two-sided power spectral density $N_0/2$. The integral term implements the correlation operation between the received UWB signal $r_b(t)$ and the estimate template $\hat{g}(t)$, and then the information bits can be acquired as

$$\hat{b}_j = \mathrm{sign}(\sum_{j=0}^{N_f-1} \int_{jT_f + kT_s}^{(j+1)T_f + kT} r_b(t) \hat{g}(t - jT_f - kT_s) dt \tag{24}$$

It can be seen from (24), since T_s is the sampling period, only the M-dimension low-rate A/D is needed.

4.2. Signal reconstruction algorithms

Then, the Orthogonal Matching Pursuit, the Basis Pursuit De-noising and the Dantzig Selector are used to detect original signal to give the opinions for choosing suitable reconstruction algorithms.

When $M \ll N$, the (8) is an uncertain function, so the search for the most sparse solution becomes an NP-hard problem. The literature (Donoho, 2006) proved that this problem can be inverted to the problem of answering a programming problem. In this chapter we study the three algorithms: the BPDN algorithm and the DS algorithm in ℓ_1-norm and the OMP algorithm in the greedy algorithm.

The BPDN algorithm derived from the Basis Pursuit (BP) algorithm, so it is an optimizing strategy. The BPDN algorithm tries to deal with such problems:

$$\min_x \frac{1}{2}\|y - \Phi x\|_2^2 + \lambda \|x\|_1 \tag{25}$$

It can be rewrite as follow

$$\min_x \|x\|_1 \quad \text{s.t.} \quad \|y - \Phi x\|_2^2 \le \sigma \tag{26}$$

The other is the DS algorithm based on ℓ_1-norm, which can deal with the following problem

$$\min_x \|x\|_1 \quad \text{s.t.} \quad \left\|\Phi^{\mathrm{T}}(y - \Phi x)\right\|_\infty \le \sigma \tag{27}$$

Comparing (26) with (27), we see that the DS algorithm is similar with the BPDN algorithm. The main difference is that the BPDN algorithm relies on punishing residuals using the ℓ_2-norm to realize the optimization, while the DS algorithm relies on minimizing correlation between residuals and all atoms.

OMP is a kind of greedy algorithm, deviating from the ℓ_1-norm shrinkage strategy. One of the most important properties of the algorithm is that it does not choose the same atom twice, so the estimation value satisfies $\|\hat{x}\|_0 = K$, after K iterations.

Literature (Pati et al., 1993) proved that all of the BPDN algorithm, DS algorithm and OMP algorithm can obtain $E\log m$ times of the mean square error about the Oracle estimator, and E is a constant here.

5. Simulations and results

In this section, the performance of the CS based on FIR filter matrix meuof the new method and random measurement estimation are made. Then three experimentations have been designed as follows.

The simulation parameters are set as follows: the transmitted UWB signal pulse $p(t)$ is the first-order derivative of the Gaussian pulse and is normalized to have unit energy. The duration of the time resolution of the channel is T_s=0.66*ns*, T_f =110*ns*. The UWB channel model CM1 (LOS) proposed by IEEE working group are adopted in our simulation. Table 1 shows the principal parameters of the channel models.

Simulation 1: consider the pulse propagating through a noiseless propagation scenarios. We adopt a UWB channel that models an indoor residential environment with line-of-sight IEEE 802.15.4a channel model. Figure 4 shows the impulse response of the UWB channel. Then the first derivative of the Gaussian pulse is selected as the transmitted pulse waveform, according to which the response of the channel is shown in Figure 5. It is just the real value of the referent template for the subsequent correlation detection.

Average cluster arrival rate (ns)	0.0265
Multipath component delay factor (ns)	6.7
Average pulse arrival rate (1/ns)	3.2
Cluster delay factor (ns)	7.8
Standard deviation of the channel gain (dB)	4.0
Standard deviation of the channel coefficient in cluster (dB)	4.2243
Standard deviation of the channel coefficient between clusters (dB)	4.2243

Table 1. Parameters in UWB channel models proposed by IEEE

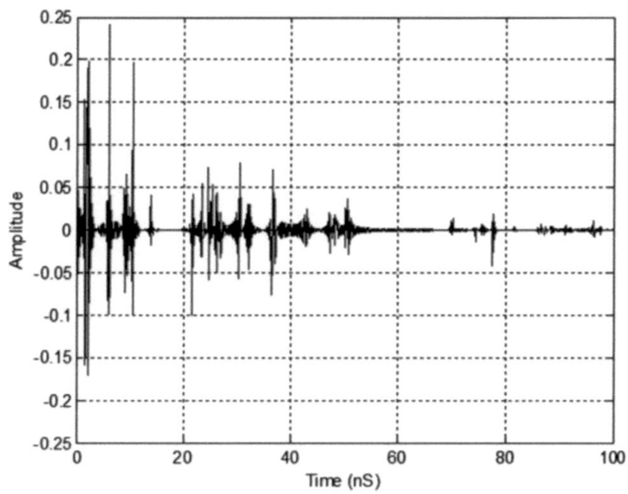

Figure 4. Impulse response of the UWB channel

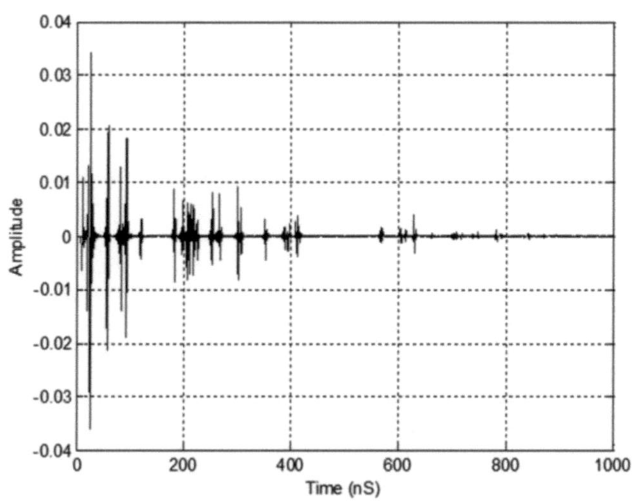

Figure 5. Gaussian pulse response of the UWB channel

Simulation 2: $\hat{g}$ is reconstructed via OMP algorithm in the absence of noise. The simulation parameters are set as follows: N=1000, M=360, K=180. Figure 6 and Figure 7 show the simulation results and reconstruction error respectively for both random measurement method and FIR filtering matrix method.

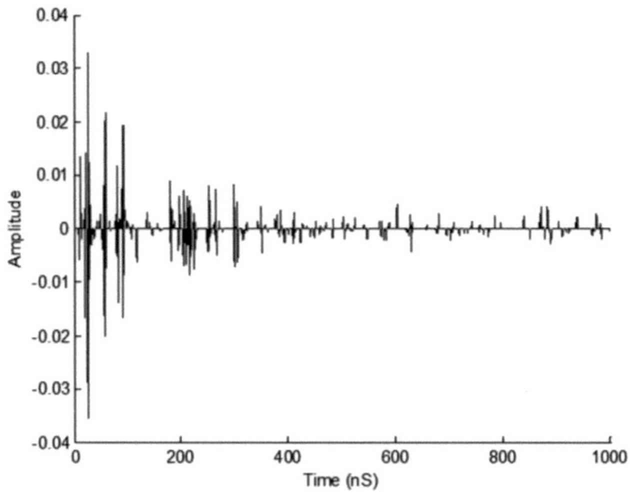

(a) The reconstruction result

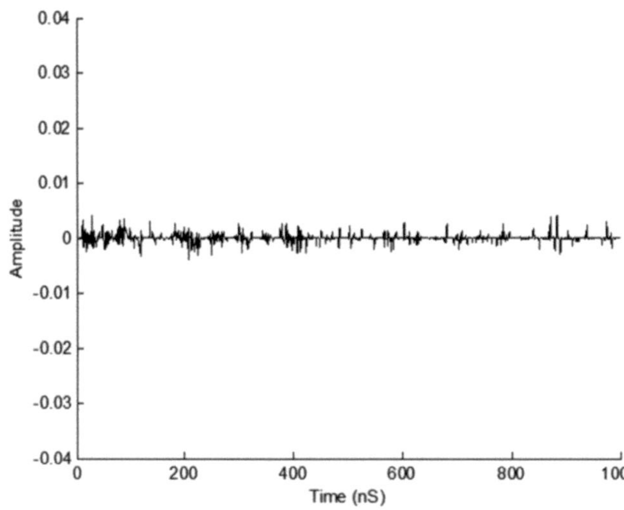

(b) The reconstruction error

Figure 6. The reconstruction performance of the random measurement method

As Figure 6 and Figure 7 shown, both the method proposed in this chapter and the random measurement method can successfully implement channel estimation for UWB communication. Moreover, both the methods sample at a reduced sampling rate, which is only M/N=1/3 of the sampling directly rate. Since a better measurement matrix is used in this chapter, better performance of the estimation accuracy can be achieved. As depicted in the second picture of Figure 7, reconstruction error approximate to zero or negligible values. In the CS theory, more measurements can improve the advancement of the estimation precision at the expense of increasing the sampling rate of the A/D conversion. This work can be advanced by designing better over-complete dictionary or better measurements matrix, which is an ongoing research.

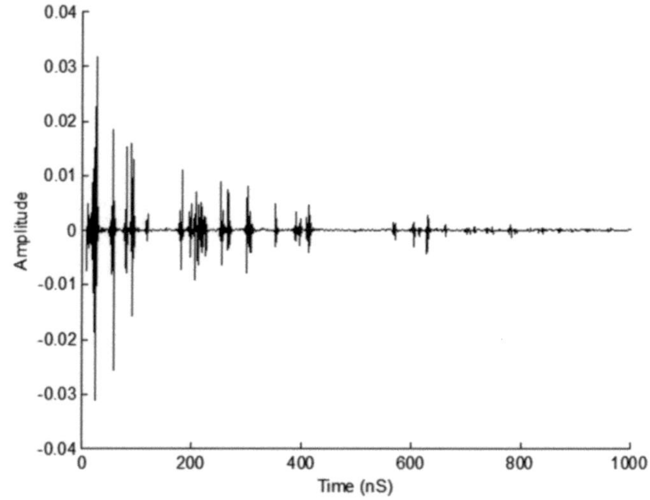

(a) The reconstruction result

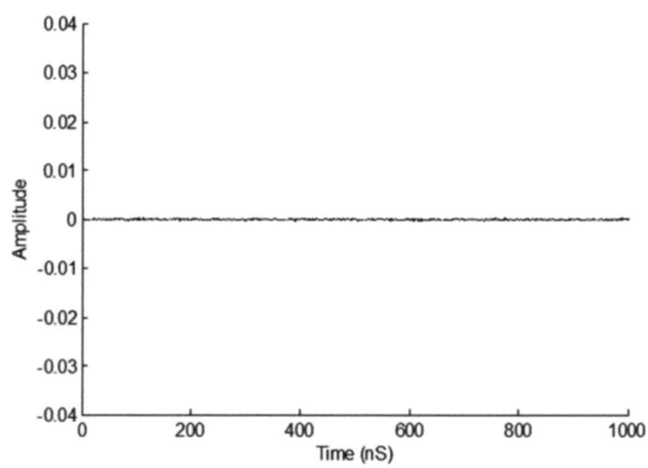

(b) The reconstruction error

Figure 7. The reconstruction performance of the FIR filtering matrix method

Simulation 3: UWB signals have been detected via correlation detection method and the reconstructed referent template has been acquired via three estimation approaches: the random measurement method, FIR filtering matrix method and direct-sampling method. Further, the parameters in (22) are set to N_p=25 and N_s=10 When 1000 information bit is transmitted through the UWB channel, the effect of AWGN is taken into consideration.

Figure 8 illustrates the BER performance on the assumption that the pilot symbols are transmitted in the absence of noise, whereas Figure 9 illustrates the BER performance when the pilot symbols are affected by the AWGN.

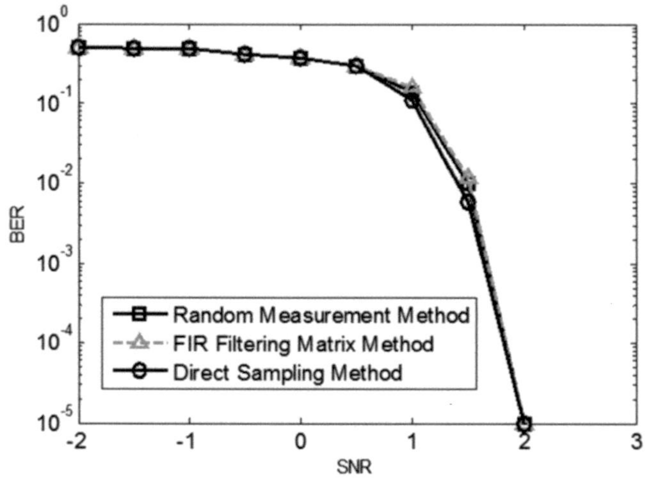

Figure 8. The BER performances of three estimation methods with noiseless UWB signals

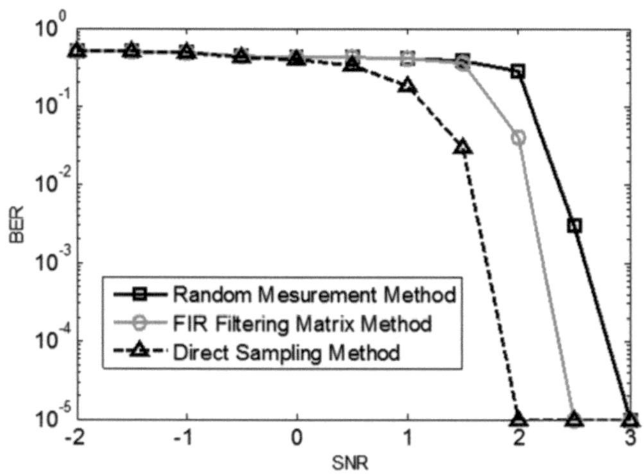

Figure 9. The BER performances of three estimation methods with noiseless UWB signals

In addition, the results of simulation 2 illustrates that FIR filtering matrix method has a better performance in channel estimation than random measurement method under the same simulation conditions. As Figure 8 indicated, when the pilot symbols go through the noiseless channel, the results via direct-sampling estimation are the optimized-template signals, however, it require an A/D converter with much higher sampling rate. Moreover, the BER curves shown above illustrate that both the random measurement method and the method proposed in this chapter can estimate the template signals precisely. When they are compared with the direct-sampling estimation, all the BER curves are close to each other terribly. It is obvious that, the accuracy of the channel estimation has little impact on the BER of the correlation detector. So far as referent template can be reconstructed successfully through the reconstruction method based on the CS theory, the BER curves approach each other.

As Figure 9 demonstrated, when taking into account of the effect of AWGN, the FIR filtering matrix method based on CS has an obvious advantage of the BER performance. While that of the random measurement estimation is the worst one comparing with the others. This performance is expected since the noise has been magnified through measurement matrix. That is, the unsuccessful result of the reconstruction algorithm at low signal-to-noise ratio (SNR) is inevitable. Note that the BER of the correlation detector increases horribly as the referent template has not been estimated efficiently.

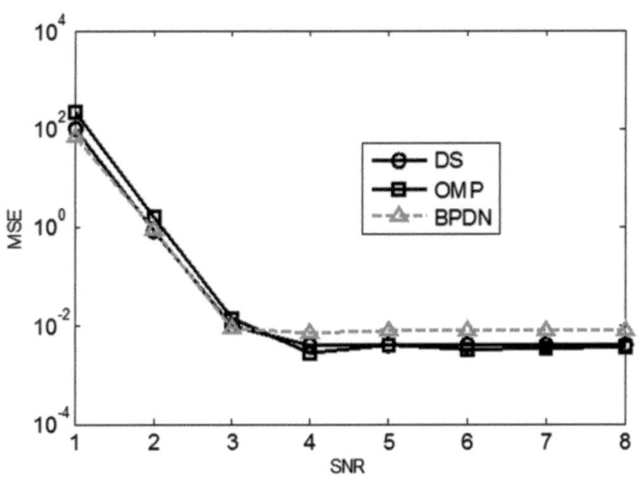

Figure 10. The comparison of the mean square error for the three reconstruction algorithms

Simulation 4: The performance test applies the OMP, the DS and the BPDN to reconstruct the original signals. The main parameters for these methods are set as follows. The maximum number of OMP iterations is set to 100 and the target residual energy is set to 0.3% of the energy for the projected signal, i.e. $\sigma = 3\times10^{-3}$. The target residual energy σ is also used in the BPDN and the DS method. For the BPDN, the relaxation parameter is supposed to be 0.05, i.e. $\varepsilon = 0.05$. For the DS, the tolerance for primal-dual algorithm is $\xi=10^{-3}$ and the max value of primal-dual iterations is set to 50. Thus, we use the mean squared error (MSE) as the performance criterion, so the tests results are achieved by 50 operations for average.

Figure 10 shows the MSE performance of the CS-based channel estimation for the three reconstruction methods. We observe that: (1) the MSE performance of the DS is slightly better than that of the BPDN. (2) In the higher operating SNR, the OMP has strong competitive advantages to the other two methods, however, turn into worse under lower operating SNR. This simulation shows that the FIR filtering matrix method is indeed leading to the improved performance for the CS reconstruction.

6. Conclusion

In this chapter, we proposed a pre-filtering method for UWB channel estimation based on the theory of CS, whose measurement matrix is just a Toeplitz matrix, and the channel estimation accuracy is improved. The method proposed in this paper avoided the magnification to the noise. Thus when the reconstructed signal is used as a referent template at the receiver in the noise realization, a better BER performance can be achieved.

The correlation detector for UWB communication discussed in this paper employs the channel estimates to the conventional correlation detection directly, while the design of the whole system combining the channel estimation and signals detection will be a further research. Moreover, it is the key point of improving the BER performance of the correlation detector to search for a CS reconstruction method, which can successfully recover the referent template under the noise realization and fewer measurements with overwhelming probability. In addition, we analyze the choices of reconstruction algorithms using several simulations. Both the OMP and the BPDN algorithms are compared to the Dantzig selector for different signal noise ratio to give the opinions for choosing suitable reconstruction algorithms.

Admittedly, there are several other theoretical and practical aspects of UWB channel estimation methods based on compressed sensing that need discussing in future. Below, however, we briefly comment on some of these aspects. First, the different types of measurement matrix according to the UWB channels should be in further study. In this paper, we do some attempts to construct the quasi-Toeplitz matrix developing the model of UWB channel estimator. Somewhat similar theoretical arguments can be made to argue the other type of measurement matrix to get better estimation performance. Second, extensive numerical simulations carried out in literatures for a number of CS estimators have established that the performance of CS estimation methods is markedly superior to that of traditional methods based on LS criterion. However, the nontraditional methods based on MUSIC and ESPRIT algorithms are not optimal for estimating sparse channels. This is because it is possible for a channel to have a small number of resolvable paths but still have a very large number of underlying physical paths, especially in the case of diffuse scattering. So the two algorithms can be employed combining with the compressed sensing framework. Third, one expects the representation of real-world multipath channels in certain bases to be only effectively sparse. The channel model and channel parameters are localized with the perfect channel model in this paper. Finally, and perhaps most importantly for the success of the envisioned wireless systems, the CS can be leveraged to design efficient overcomplete dictionary for estimating sparse UWB channels.

Author details

Huanan Yu and Shuxu Guo*
College of Electronic Science and Engineering, Jilin University, Changchun, China

* Corresponding Author

7. References

Bajwa W. U., Haupt J., Aayeed A. M. & Nowak R. (2010). Compressed channel sensing: a new approach to estimating sparse multipath channels. *IEEE Proceedings*, vol. 98, no.6, pp. 1058-1076.

Bajwa W., Haupt J., Raz G. & Nowak R. (2007). Toeplitz structured compressed sensing matrices. *Processing SSP'07, Madison, WI*, August 2007, pp. 294-298.

Berger C. R., Wang Z., Huang J. & Zhou S. (2010). Application of compressive sensing to sparse channel estimation. *IEEE Communications Magazine*, vol. 48, no.11, pp. 164-174.

Candès E. J., Romberg J. & Tao T. (2006). Robust uncertainty principles: exact signal reconstruction from highly incomplete frequency information. *IEEE Transactions Information Theory*, vol. 52, no.2, pp. 489-509.

Candès E. J. & Wakin M. B. (2008). An introduction to compressed sampling. *IEEE Signals Processing Magazine*, vol. 25, no. 2, pp. 21-30.

Candès E. J. & Tao T. (2006). Near optimal signal recovery from random projections: universal encoding strategies. *IEEE Transactions Information Theory*, Vol. 52, no. 12, pp. 5406-5425.

Donoho D. (2006). Compressed sensing. *IEEE Transactions Information Theory*, vol. 52, no.4, pp. 1289-1360.

Erküçük S., Kim D. I. & Dwak K. S. (2007). Effects of channel models and rake receiving process on UWB-IR system performance. *IEEE Processing ICC'07*, June 2007, pp. 4896-4901.

IEEE Std 802.15.4a-2007. (2007). Part 15.4: wireless medium access control (MAC) and physical layer (PHY) specifications for low-rate wireless personal area networks (WPANs).

Liu T. C.-K, Dong X. & Lu W.-S. (2009). Compressed sensing maximum likelihood channel estimation for ultra-wideband impulse radio. *IEEE Processing ICC'09*, June 2009.

Lottici V., Andrea A. D. & Mengali U. (2002). Channel estimation for ultra-wideband communications. *IEEE Journal Selected Areas Communications*, vol..20, no.12,pp. 1638-1645.

Molisch A. F., Cassioli D., Chia-Chin Chong, Emaimi S., Fort A., Kannan B., Karedal J., Kunisch J., Schantz H. G., Siwiak K., & Win M. Z. (2006). A comprehensive standardized model for ultra-wideband propagation channels. *IEEE Transactions Antennas and Propagation*, Vol. 54, no.11, pp. 3151-3166.

Naini F. M., Gribonval R., Jacques L. & Vandergheynst P. (2009). Compressive sampling of pulse trains: spread the spectrum. *IEEE Processing ICASSP'09*, April 2009, pp. 2877-2880.

Paredes J. L., Arce G. R. & Zhongmin Wang. (2007). Ultra-wideband compressed sensing: channel estimation. *IEEE Journal of Selected Topics in Signal Processing*, vol.1, no.3, pp. 383-395.

Pati Y. C., Rezaiifar R. & Krishnaprasad P. S. (1993). Orthogonal matching pursuit: recursive function approximation with applications to wavelet decomposition. *Processing 27th Annual comferenc Signals, Systems and Computers*, November 1993.

Qiu R. C., Liu H. & Shen X. (2005). Ultra-wideband for multiple access communications. *IEEE Communications Magazine*, vol. 43, no. 2, pp. 80-87.

Reed J. H.(2005) An introduction to ultra wideband communication systems. *Ser. Prentice Hall Communications Engineering and emerging technologies series, T. S. Tappaport, Ed. Upper Saddle Tiver, NJ: Prentice-Hall,* 2005.

Wan F., Zhu W. P. & Swamy M. N. S. (2010). Semi-blind most significant tap detection for sparse channel estimation of OFDM systems. *IEEE Transactions Circuits and Systems,* vol. 57, no. 3, pp. 703-713.

Win M. Z. & Scholtz R. A. (1998). Impulse radio: How it works. *IEEE Communication Letters,* Vol. 2,no.2, pp. 36-38.

Witrisal K., Leus G., Janssen G. J. M., Pausini M., Troesch F., Zasowski T. &Romme J. (2009). Noncoherent ultra-wideband systems. *IEEE Signal Processing Magazine,* vol..26, no.4, pp. 48-66.

Yang L. & Giannakis G. B. (2004a). Ultra-wideband communications - an idea whose time has come. *IEEE Signal Processing Magazine,* vol..21, no.6, pp. 26-54.

Yang L. & Giannakis G. B. (2004b). Optimal pilot waveform assisted modulation for ultra-wideband communications. *IEEE Transactions on Wireless Communications,* vol. 3, no. 4, pp. 1236-1249.

Yu Huanan & Guo Shuxu. (2010). Pre-filtering ultra-wideband channel estimation based on compressed sensing. *IEEE Processing CMC'10,* April 2010, pp. 110-114.

Chapter 5

The UWB Channel in Medical Wireless Body Area Networks (WBANs)

Carlos Pomalaza-Ráez and Attaphongse Taparugssanagorn

Additional information is available at the end of the chapter

http://dx.doi.org/10.5772/48634

1. Introduction

UWB is a technology that has several advantages when considered for a Wireless Body Area Network (WBAN). A WBAN is a network with its communications devices in very close proximity to the human body. In medical applications these devices are connected to sensors that can monitor vital signs such as ECG, temperature, and mobility. A WBAN allows for the remote monitoring of a patient's health minimizing the number of cables needed. The monitoring of vital signals usually require a relatively low data-rate which in the case of UWB translates into very small transmitting power, long battery life, and less potential side effects caused by electromagnetic radiation. All of these features are very desirable for devices that are close to the body and meant to be used for extended periods of time.

The human body is a complex structure and human tissues have different electrical properties which affect the propagation of electromagnetic signals. Moreover, as the human body moves, the characteristics of the radio links changes, e.g. the link from the chest to a wrist will change from line-of-sight to non-line-of-sight as a person walks.

To be able to design and develop UWB devices that can interface with WBANs it is then necessary to understand well the characteristics of the radio propagation channel at UWB frequencies and in close proximity to the human body. UWB measurements around a human body have been carried out by several researchers (Fort et al., 2006). There is however a lack of measurements, and subsequent analysis, carried out in real medical environment such as hospitals. The studies described in this chapter focus on scenarios most likely to be found in medical applications and as such they do not assume a large amount of antennas in close proximity to the skin. Among the several issues taken into account are the effects of mobility, and the interaction of the UWB signal with medical implants.

2. Hospital scenarios

Fig. 1 shows a common hospital room scenario. Medical information is collected by sensors on the patient's body. The sensors are interfaced to a WBAN which transmit the information to be displayed on a bedside monitor. This information can also be transmitted to another hospital location for remote monitoring, e.g. a nurse's station. The radio links present in this type of scenario include the ones between sensor nodes (link A1), the links between the sensors and a gateway node (links A2), and the links from wireless devices carried by visitors or healthcare professionals (link A3). Other possible radio links are from the gateway to wireless networks such as 802.11 b/g/n and WiMAX.

Figs. 2 and 3 show real hospital scenarios where measurements described in this chapter were taken (Taparugssanagorn et al. 2010). A regular hospital room is shown in Figure 2. This room's dimensions are: 6.3 m x 7.2 m x 2.5 m. A surgery room, with dimensions 6 m x 4.7 m x 2.5 m, is shown in Figure 3. Both radio links A1 and A2 were measured in each room. Within the hospital several scenarios were considered. Table 1 summarizes the measurements and scenarios in this study. A detailed description of the various experiments and the results can be found in (Taparugssanagorn et al. 2010). To illustrate the experiments and analyses performed the case of a subject standing in hospital room is discussed in more detail in the next section.

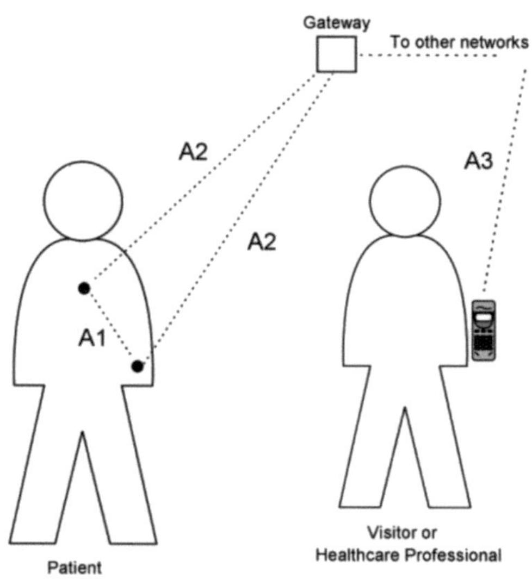

Figure 1. A typical hospital room scenario: A1 is a link between sensor nodes, A2 is a link between a sensor node and a gateway, and A3 is a link to other wireless networks.

To measure the A1 link, the receive (Rx) antenna was located at the centre of the front torso and the transmit (Tx) antenna was placed on the left wrist. These locations are comfortable for most patients and are also convenient places for sensors such as electrodes in the chest areas to generate an electrocardiogram (ECG) and a pulse oximeter on a finger trip to monitor the patient's oxygenation. To measure the A2 link, the Rx antenna was placed on a

2 m high pole located 2 m away from the subject and the Tx antenna was on the left side of the waist.

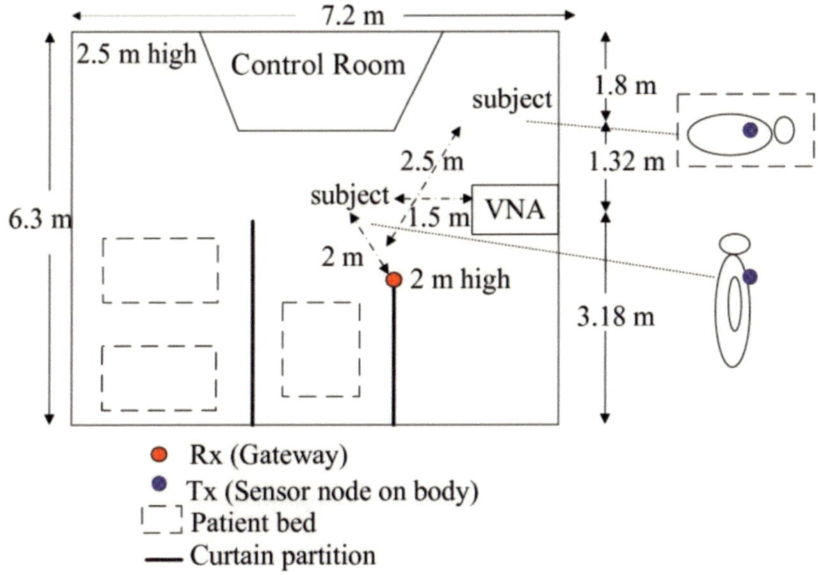

Figure 2. Floor plan of a regular hospital room.

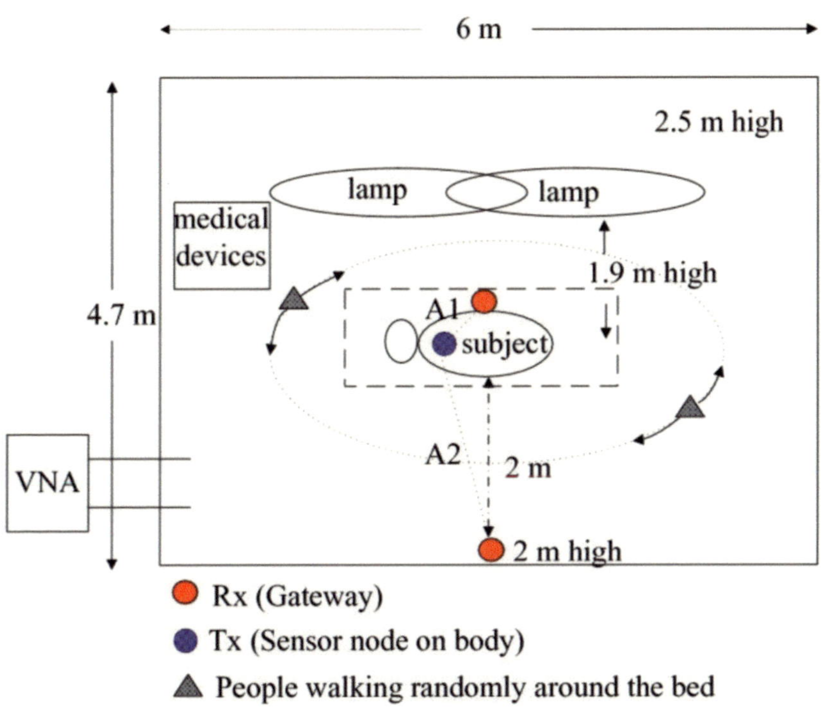

Figure 3. Floor plan a surgery room

Environment	Scenarios	
Regular Room	Subject Standing Links A1 and A2	Subject Lying Links A1 and A2
Surgery Room	Subject Lying 2 people walking around the bed Links A1 and A2	Subject Lying 2 people walking around the bed, one of them is using a mobile phone Links A1 and A2

Table 1. Measurements and scenarios.

3. Channel measurements and models

The measurements were carried out using an Agilent 8270ES vector network analyzer (VNA). The antennas used were SkyCross SMT-3TO10M-A. These antennas are linearly polarized and azimuthally omnidirectional. The cables used were 5 m long SUCOFLEX RF with 7.96 dB loss. The data acquisition system included a computer with LabVIEW™ software. The VNA was operated in a transfer function measurement mode, where Port 1 and Port 2 are the transmitting and receiving ports respectively, as seen in Fig. 4.

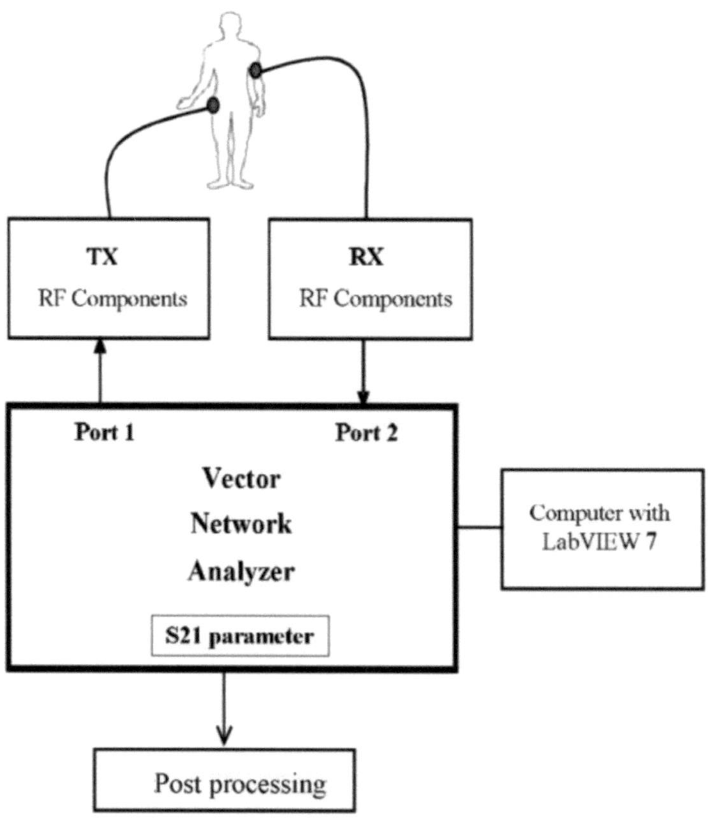

Figure 4. Measurement setup

This setup corresponds to a measurement of the S21 parameter where the device under test (DUT) is the radio channel. The range of the frequency spectrum covered was from 3.1 GHz to 10.6 GHz. For each experiment setup 100 frequency responses were measured. The measurement parameters are summarized in Table 2.

Parameter	Value
Frequency range	3.1 to 10.6 GHz
Bandwidth	6.9 GHz
VNA IF bandwidth	3.0 kHz
Number of samples per sweep	1601
Maximum detectable delay	231 ns
Sweep time	800 ms
Average noise floor	- 120 dBm
Transmit power	0 dBm
Tx and Rx cables' loss	7.96 dB

Table 2. Measurement parameters

The measured transfer function frequency values were converted to the time domain (channel impulse response) using an inverse Fast Fourier Transform. A Hamming window was used to reduce sidelobes.

3.1. Channel impulse response

Fig. 5 shows the average of the channel impulse response, corresponding to link A1, when the subject is standing in the hospital room shown in Fig. 2.

The effect of the human body and the environment can be clearly differentiated. These results are significantly different than the ones obtained in an empty hospital room (Hentilä et al, 2005). In Fig. 5 the first region of the IR shows a fast decay of the energy during the first 5-6 ns due to the effects of the human body. The decay of the second region in the response is slower and contains the diffuse multipath components and a few subclusters caused by the reflections coming from the room. In this particular case the first of such subclusters, arriving at around 8 ns, is due to a measuring equipment (VNA) which was located 1.3 m in front of the subject when is standing.

For each particular hospital scenario listed in Table 1 the measurements obtained share the general characteristics shown in Fig. 5.

Fig. 6 corresponds to the case when the subject is lying down on bed in a hospital room. Fig. 7 corresponds to the case when the subject is lying down on a bed in a surgery room and two other people are randomly walking around the bed. The Least Squares (LS) fitted lines shown in these figures are used to model the variability of the amplitudes as described in Section 3.2.

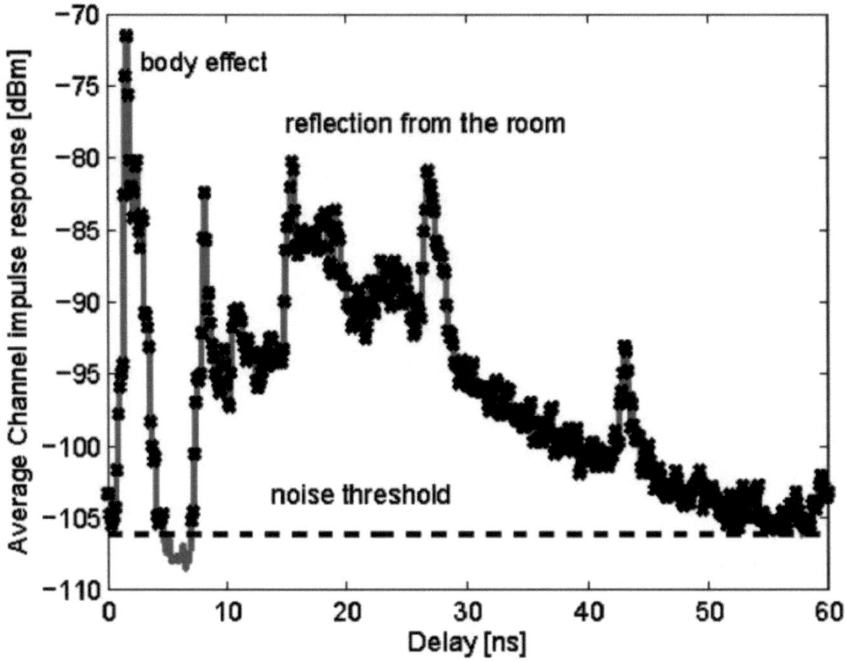

Figure 5. Average channel impulse response of the radio link A1. The subject is standing in a regular hospital room.

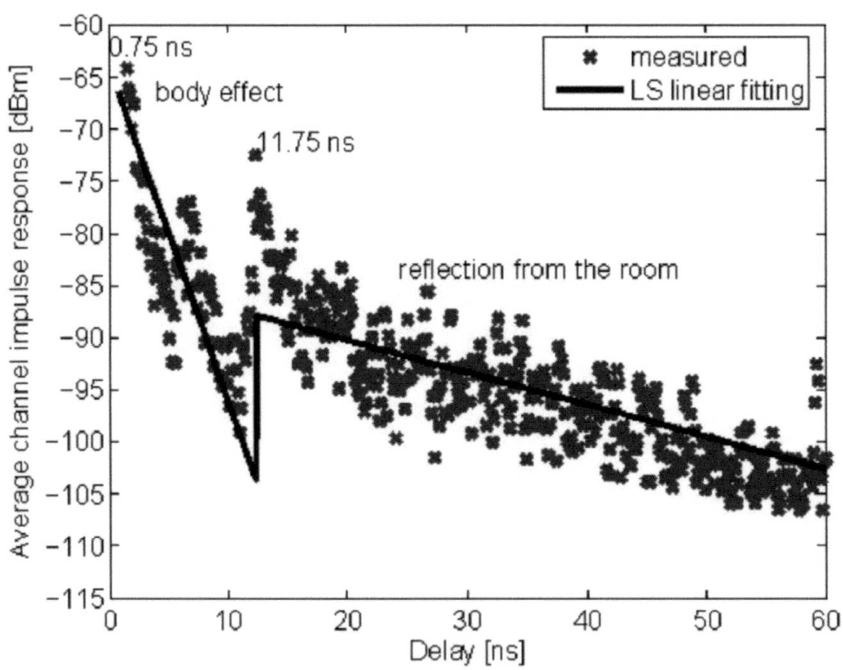

Figure 6. Average channel impulse response of the radio link A1. The subject is lying down on a bed in a regular hospital room.

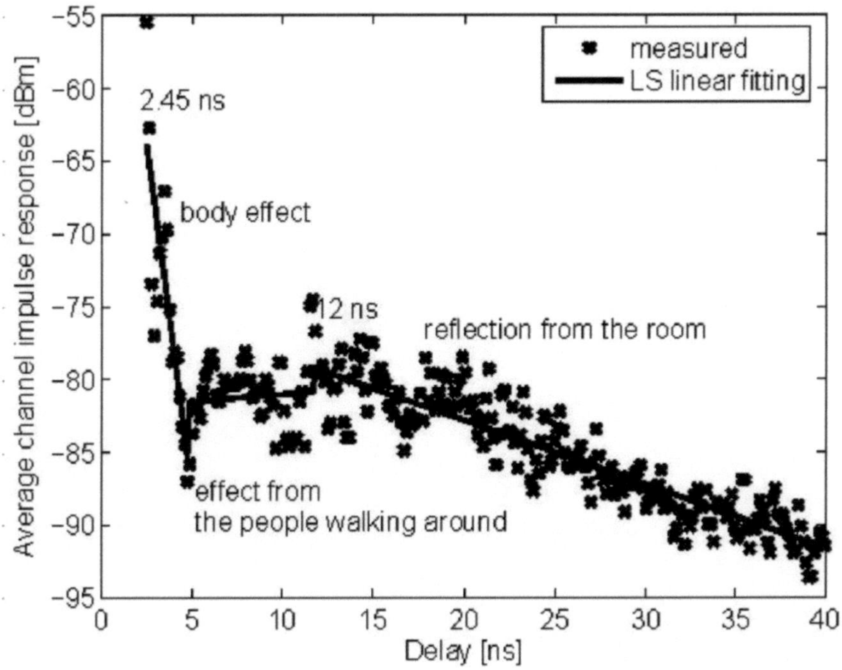

Figure 7. Average channel impulse response of the radio link A1. The subject is lying down on a bed in a surgery room and two other people are randomly walking around the bed.

3.2. Channel models

Once a set of measurements have been obtained they are used to estimate the parameters of a common mathematical representation of communications channels, a tapped delay line.

$$h(\tau) = \sum_{l=0}^{L-1} a_l \delta(\tau - \tau_l)\exp(j\phi_l) \quad (1)$$

In equation (1) L is the number of paths. For the ith path, a_l is the signal amplitude, τ_l is the arrival time, and ϕ_l is the phase. For the case shown in Fig. 5 (when the subject is standing in a regular hospital room) it is more appropriate to consider two regions for the modeling of the signal clusters, each with it is own set of distributions for the characterization of the amplitudes decays and intearrival times.

3.2.1. Exponentially decaying factor

As illustrated in Fig. 8, the values of a_l in equation (1) can be approximated by two exponential decaying functions, one for each region. Using a least squares (LS) method these functions (when expressed in dB) are best fitted with a Rician factor γ_0 and an exponential decaying factor Γ (equation (2)).

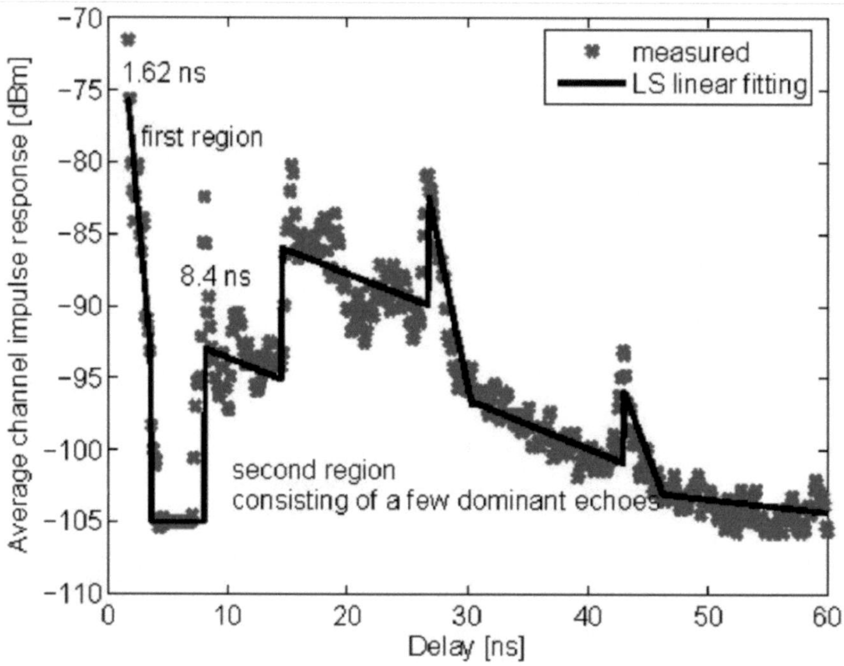

Figure 8. Least Squares fitting of the average channel response of the radio link A1. The subject is standing in a regular hospital room.

$$10\log_{10}|a_l| = \begin{cases} 0, & l = 0 \\ \gamma_{01} + 10\log_{10}\left(\exp\left(\frac{-t_l}{\Gamma_1}\right)\right), & 1 \le l \le l_1 \\ \sum_{m=1}^{M}\left(\gamma_{02m} + 10\log_{10}\left(\exp\left(\frac{-t_l}{\Gamma_{2m}}\right)\right)\right), & l_2 \le l \le L-1 \end{cases} \quad (2)$$

where $\gamma_{01}, \gamma_{02m}, \Gamma_1$, and Γ_{2m} are the corresponding parameters for each region. $l_1$and l_2 are the index for the last path of the first region and the first path of the second region respectively.

For the measurements shown in Fig. 5 the values of the function's parameters for the first region are $\gamma_{01} = -61$ dB and $\Gamma_1 = 1.11$. For the second region the values are $\gamma_{02m} = \{-91, -82, 19, -87, -6, -99\}$ dB and $\Gamma_{2m} = \{30.30, 31.25, 2.44, 29.41, 4.55, 108.70\}$.

3.2.2. *Amplitude variation and path arrival times*

Amplitude variations for each region have been modelled as a log-normal distribution with zero-mean and standards deviations $\sigma_1 = 2.45$ and $\sigma_{2m} = \{2.07, 2.21, 1.62, 1.44, 1.20, 0.91\}$.

Path arrival times $(t_l - t_{l-1})$, i.e. the time difference between consecutive arrival paths, have been modelled, using the LS method, with the following exponential distributions.

$$p(t_l|t_{l-1}) = \begin{cases} \lambda_1 \exp(-\lambda_1(t_l - t_{l-1})), & 1 \leq l \leq l_1 \\ \lambda_2 \exp(-\lambda_2(t_l - t_{l-1})), & l_2 \leq l \leq L-1 \end{cases} \tag{3}$$

The path arrivals can then be modelled as a Poisson process.

$$p(L) = \frac{\mu_L \exp(\mu_L)}{L!} \tag{4}$$

For the case discussed in this section $\lambda_1 = 0.269$ and $\lambda_2 = 0.163$ and the value of μ_L is 324.

4. Effect of body motion

UWB signals are very sensitive to absorption by objects that have a high percentage of water, such as human bodies and plants. It is then expected that WBANs using UWB technology will have particular characteristics when the human body is in motion. To study the effects of motion, measurement while a subject is walking (as shown in Fig. 9) have been performed (Taparugssanagorn et al. 2009).

A real-time measurement and processing of the channel fluctuations due to a body in motion is not possible with the equipment described in Section 3. Instead, a pseudo-dynamic method was used, where each position of the walking cycle shown in Fig. 9 was kept still for the whole time it takes to take 100 snapshots. The average of the magnitude of the channel impulse response is shown in Fig. 10.

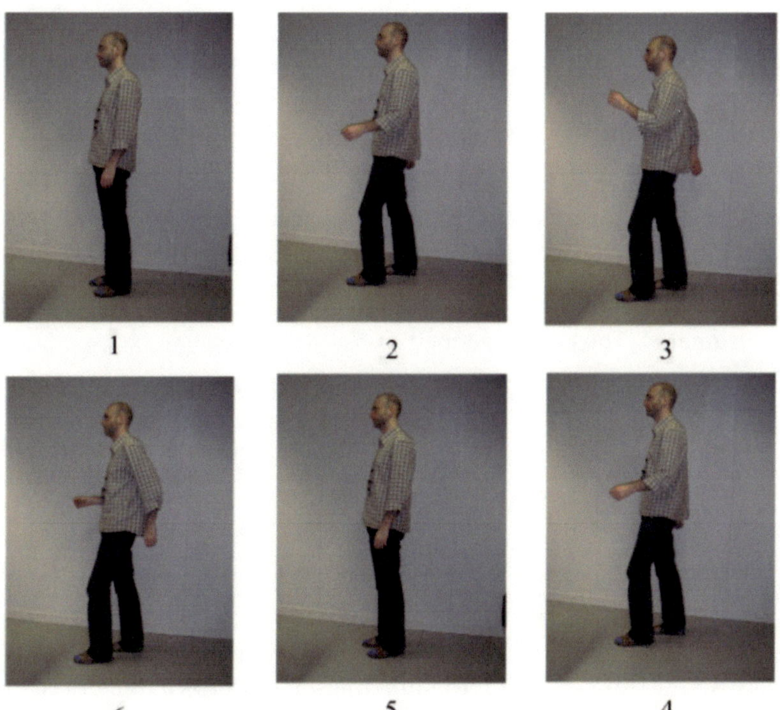

Figure 9. Positions within a walking cycle

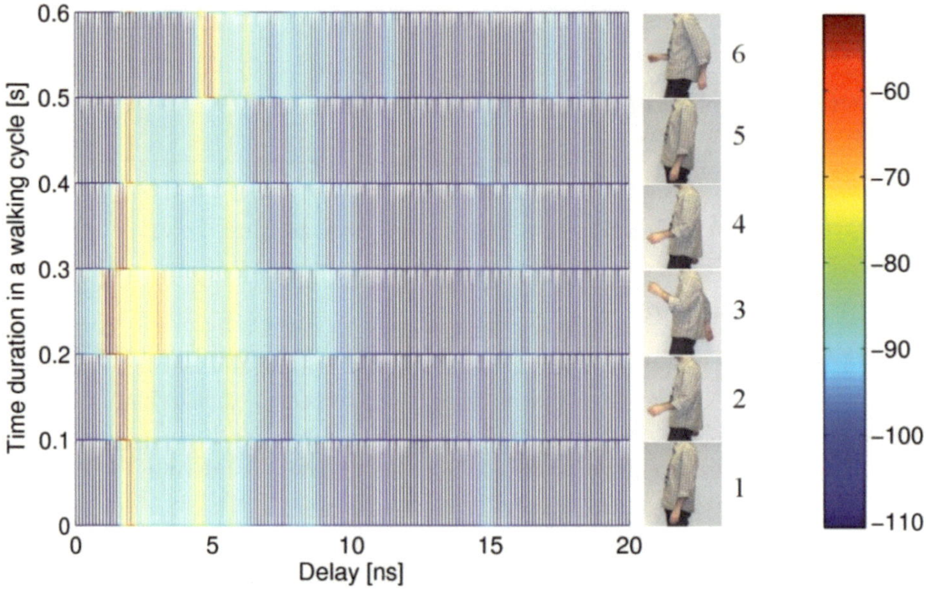

Figure 10. Magnitude of the channel impulse response for each position of the walking cycle.

These experimental results indicate that the arm movements have a significant impact on the radio link A1 (Tx antenna on the left wrist and Rx antenna in the center of the front torso). For instance, when the hand moves to position three the strongest path arrives earlier than in the other positions due to the shorter distance between the antennas. There are also more significant paths due to the interaction of the electromagnetic waves with of the arm and the shoulder. The shadowing of the signal due to blocking by the body is evident in position six, where the left hand moves to the lowermost location.

Fig. 11 provides an alternative view of the channel impulse response where the delay of the most prominent peak is clearly shown (Taparugssanagorn et al. 2011).

To evaluate the delay dispersion within the channel the root mean square (RMS) delay spread τ_{RMS} is estimated. The τ_{RMS} is defined as,

$$\tau_{RMS} = \sqrt{\frac{\sum_{i=0}^{L-1}(\tau_i-\tau_m)^2|h(\tau_i)|^2}{\sum_{i=0}^{L-1}|h(\tau_i)|^2}} \tag{5}$$

where τ_m is the mean excess delay defined as,

$$\tau_m = \frac{\sum_{i=0}^{L-1}\tau_i|h(\tau_i)|^2}{\sum_{i=0}^{L-1}|h(\tau_i)|^2} \tag{6}$$

$h(\tau)$ is the channel impulse response, L is the number of paths and τ is the delay. For the case discussed here the estimates for mean and the standard deviation of τ_{RMS} are 0.1371 ns and 0.0670 ns respectively. Also the probability distribution function that best fits the variations of the amplitude is the Weibull distribution.

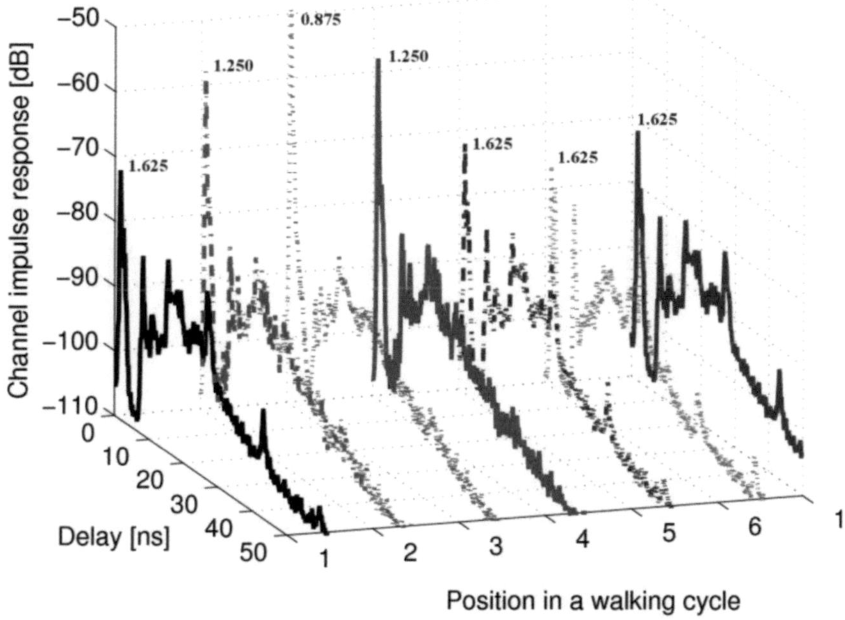

Figure 11. Channel impulse response (link A1) for each position of the walking cycle.

The results presented in this and in the previous section highlight the importance of properly understand and model the UWB channel when designing physical and Medium Access Protocols (MAC) to be used in medical applications (Viittala et al., 2009).

5. UWB radar in medical applications

The potential use of UWB technology goes beyond transmitting information, collected by sensors, to a control station. The nature of the UWB signal is such that it can be used as in common radar applications, e.g. to detect and estimate dynamic parameters of an object. Fig. 12 shows the channel impulse responses for the case of subjects with and without an aortic valve implant (Taparugssanagorn et al. 2009). The Rx antenna was located at the middle of the front torso and the Tx antenna close to the heart, 10 cm away from the Rx antenna. P200 BroadSpec™ UWB antennas were used for this experiment.

It apparent that the responses are different, i.e. the one corresponding to the subject with an aortic implant has lower peaks. A possible explanation for the difference in the responses is the scattering caused by the metallic (titanium alloy) valve. Subsequent simulation studies carried out using a 3D immersive visualization environment has confirmed this type of results (Yang et al., 2011). Further investigations could lead to the use of the response to infer the nature of the implant behaviour.

The use of UWB signals to directly monitor vital signs is currently a very active research area. Thus for example, the estimation of the breathing rate and the heart beat frequency has been studied in (Lazaro et al., 2010). Using a mathematical model of the human body as related to its effect on the propagation of the UWB signals the feasibility of medical diagnosis using UWB radar technology has been assessed in (Pancera et al., 2011).

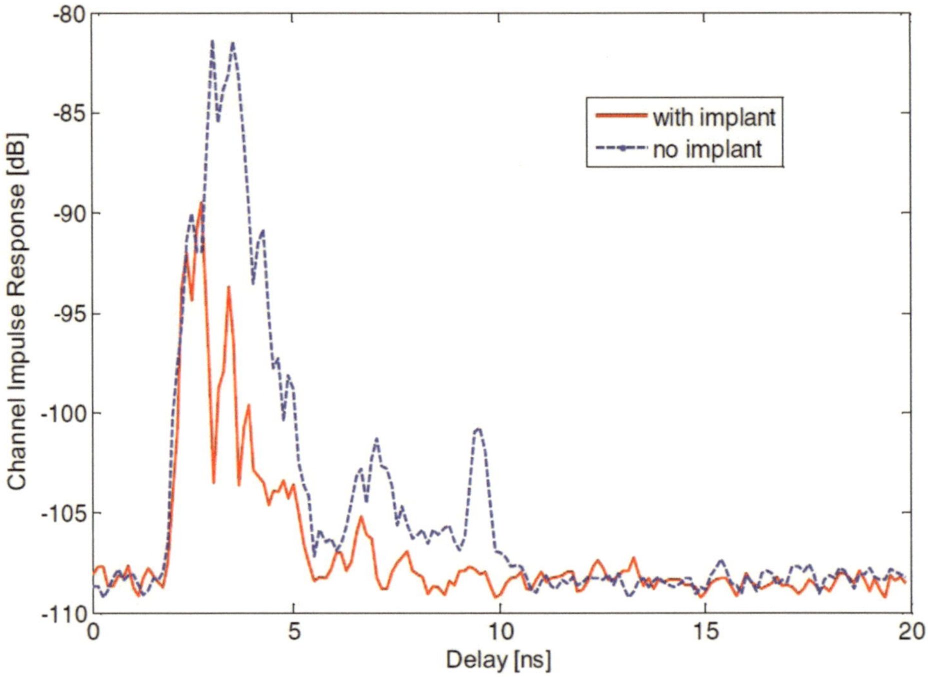

Figure 12. Average channel impulse response for subjects with and without aortic valve implant.

In summary UWB technology can be used not only to transmit information collected by sensors such as ECG electrodes and pulse oximeters but also to actively monitor vital signals and the behaviour of artificial implants.

6. Standards

For WBANs to become widely adopted it is important to have standards. The IEEE 802.15 group of standards focuses on short range communications, low complexity, and low power consumption making them suitable for use in WBANs. Two standards from this group have specifically addressed the use of UWB technology with medical applications in mind. This section describes UWB features of the IEEE 802.15.4 standard of the recently approved IEEE 802.15.6 standard.

The IEEE 802.15 task group 6 (TG6) developed a UWB channel model as part of the process of developing the IEEE 802.15.6 standard (Yazdandoost & Sayrafian-Pour, 2009). A comparison of the IEEE 805.15.6 channel model and the one described in this chapter can be found in (Viittala et al., 2009).

6.1. IEEE 802.15.4

The IEEE 802.15.4 standard and the industrial consortium supporting it, the ZigBee alliance, are widely used in wireless sensor networks (WSNs) applications. The IEEE 802.15.4 standard provides alternative physical layers for devices with precision ranging and extended range (IEEE Std 802.15.4, 2011). The UWB physical layer option of this standard provides for features that are desirable in medical applications such as very low power. The data rates supported are 110 kb/s, 851 kb/s, 1.70 Mb/s, 6.81 Mb/s, and 27.24 Mb/s. Whereas this standard has desired features to be used in medical applications it does not support the levels of safety, quality of service, and security features wanted in many of those applications. Thus, the remainder of this section deals with the IEEE 802.15.6 standard which has features specifically designed to support medical applications.

6.2. IEEE 802.15.6

The final version of this standard has been recently released (IEEE Std 802.15.6, 2012). It specifically deals with wireless communications in the vicinity of, or inside, a human body. It uses existing industrial scientific medical (ISM) bands and other bands. It allows devices to operate on very low transmit power and thus minimizes the specific absorption rate (SAR) into the body as well as increases the battery life. It also supports data rates up to 10 Mbs, quality of service (QoS) and it provides for strong security. The standard takes into account the use of portable antennas in the presence of a human body.

The default mode should support impulse radio UWB (IR-UWB) with a mandatory uncoded data rate of 487.5 kbs. It should also support, as optional PHY, wideband frequency modulation UWB (FM-UWB) with a mandatory uncoded data rate of 250 kbs.

The standard provides specifications for the physical layer (PHY) and the medium access control (MAC) sublayer. Three PHYs are supported by the IEEE 802.15.6 as illustrated in Fig. 13.

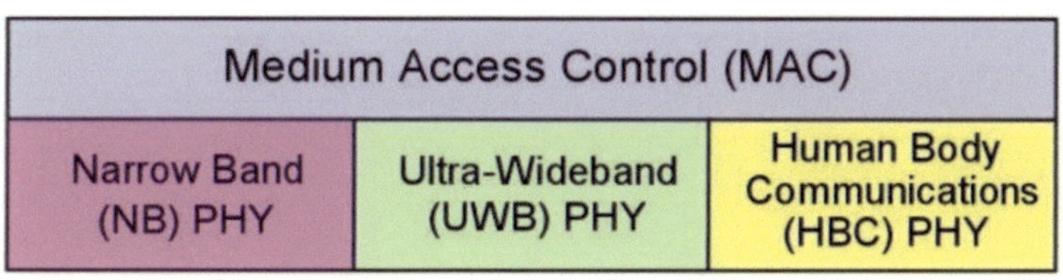

Figure 13. IEEE 802.15.6 MAC and PHY layers

The UWB PHY layer constructs the PHY layer protocol data unit (PPDU) by concatenating the synchronization header (SHR), physical layer header (PHR), and the physical layer service data unit (PSDU). The SHR has two parts. The first part is a preamble, intended for timing synchronization, packet detection, and frequency offset recovery. The second part is the start-of-frame delimiter (SFD). Kasami sequences of length 63 are used to build the preamble. The usage of preamble sequences improves coexistence of WBANs and interference mitigation as different WBANs use different preamble sequences. The PPUD is illustrated in Fig. 14. The PPUD bits are converted into RF signals for transmission in the wireless medium.

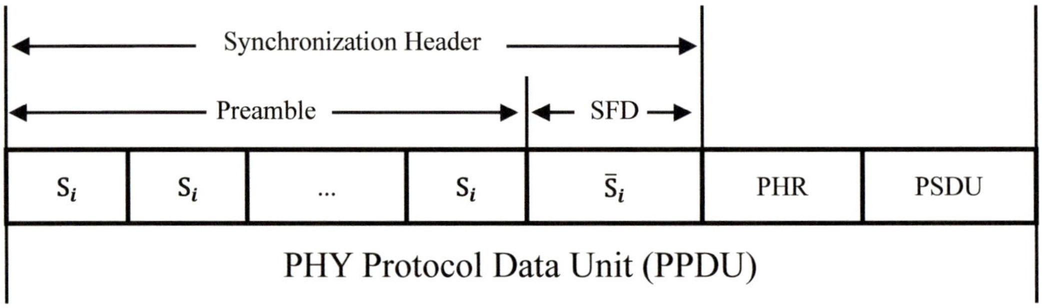

Figure 14. IEEE 802.15.6 UWB physical layer protocol data unit (PPDU)

The PHR contains information about the data rate of the PSDU. The coded bit rates supported are shown in Table 3.

UWB - PHY	Data rate 0 (kb/s)	Data rate 1 (kb/s)	Data rate 2 (kb/s)	Data rate 3 (kb/s)	Data rate 4 (kb/s)	Data rate 5 (kb/s)	Data rate 6 (kb/s)	Data rate 7 (kb/s)
On-Off	394.8	789.7	1579	3159	6318	12636		
DBPSK/DQPSK	487	975	1950	3900	7800	15600	557	1114
FM	202.5							

Table 3. IEEE 802.15.6 UWB-PHY coded bit rates.

According to the IEEE 802.15.6 standard all nodes and hubs are organized into logical sets called body area networks (BANs) as illustrated in Fig. 15. There is one and only one hub in a BAN. The number of nodes in a BAN ranges from zero to nMaxBANSize=64.

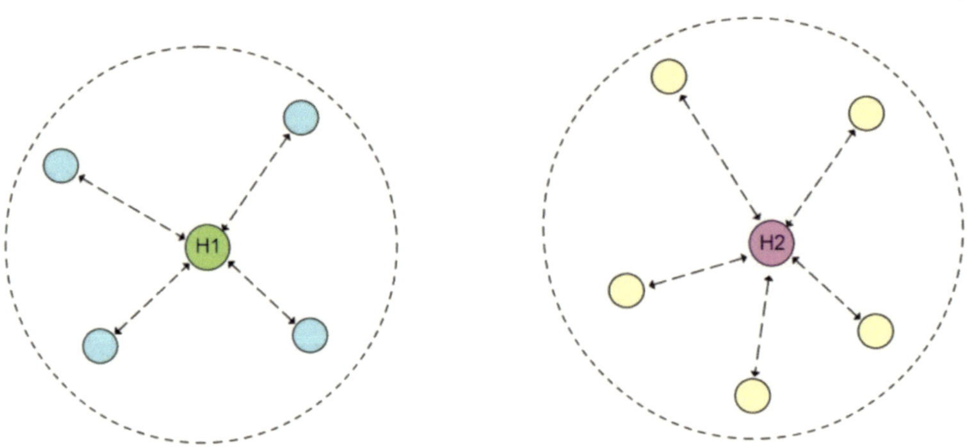

Figure 15. Network Topology

To provide or support time reference allocations a hub establishes a time base that divides the time into beacon periods (superframes). A hub transmits a beacon in each superframe, except in inactive superframes, or does not transmit a beacon in any superframe. A hub can operate in one of the following modes:

- Beacon mode with superframes
- Non-beacon mode with superframes
- Non-beacon mode without superframes

Fig. 16 shows the superframe structure when the hub operates in the beacon mode with superframes. A node can obtain, and initiate frame transactions, in the EAP1, RAP1, EAP2, RAP2, and CAP periods in any active superframe using CSMA/CA or slotted Aloha based random access protocols.

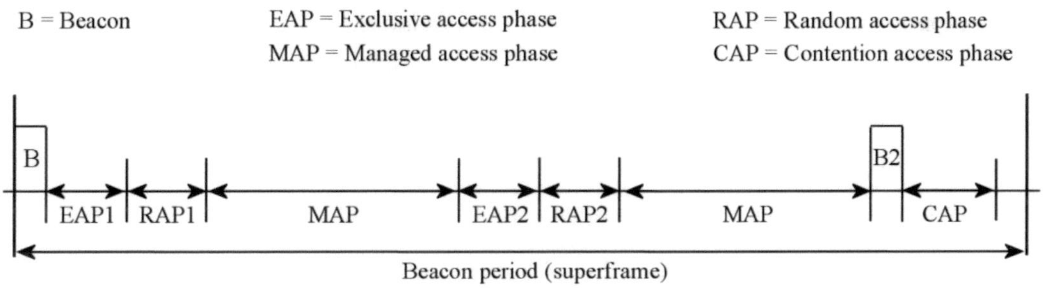

Figure 16. Layout of access phases in a superframe period for beacon mode

The EAP1 and EAP2 periods are used for the highest priority traffic, i.e. emergency information. The RAP1, RAP2, and CAP period are used for regular traffic. In a MAP period a hub can arrange scheduled uplink, downlink, and bilink allocation intervals. It can also provide unscheduled bilink allocation intervals.

In non-beacon mode with superframes the entire superframe period is a MAP phase. In the non-beacon mode without superframe boundaries the hub provides polled allocations whose length is specified in terms of the number of frames granted for transmission (type-II polled allocation).

According to this standard all nodes and hubs can choose the following three security levels:

- Level 0 – unsecured communications. The messages are transmitted in unsecured frames. There are no measures for data authentication and integrity validation, confidentiality and privacy protection, and replay defense.
- Level 1 – authentication only. Messages are transmitted in secured authenticated but not encrypted frames. Confidentiality and privacy is not supported.
- Level 2 – authentication and encryption. Messages are transmitted in secured authenticated and encrypted frames. Confidentiality, privacy protection, and replay defense are supported.

Security starts with a negotiation of the desired security suite between a node and a hub. Once the security selection is negotiated the two communicating parties activate a pre-shared or generate a new shared master key (MK).

7. Future research directions

The UWB channel has been measured and modelled extensively in recent years. Experimental WBANs using this technology have been developed and studied. Now that standards are in place the expectations, for the near future, is to have actual deployment of UWB BANs in medical environments.

Commercial applications of UWB have been limited to situations where precise localization is needed. Once medical applications are deployed the excellent ranging characteristics of the UWB signals could also be used to localize patients and medical equipment.

More experimental work is needed to learn the capabilities of UWB to directly monitor human organ functions as well as the workings of medical implants. In addition to experimental work, there is the need to develop accurate mathematical models that can be used in simulation studies as well as in 3D immersion systems.

Assuming UWB WBANs are widely deployed long term future applications is their use to extend their range and data delivery capabilities by having the BANs work in a cooperative fashion (Kaveh et al., 2011).

8. Summary

This chapter describes features of the UWB channel that should be taken into account when it is being considered for medical applications, in particular in hospital scenarios. These scenarios include cases where the human body is in motion. Using actual measurements mathematical models of the channel have been proposed.

It is also possible to use UWB technology to measure the workings of medical implants or body activities. This chapter presents the case of the response of an artificial aortic valve to UWB waves and its potential use to evaluate the working status of the valve.

Finally there is a brief discussion of engineering standards applicable to the use of UWB technology in the medical field.

Author details

Carlos Pomalaza-Ráez and Attaphongse Taparugssanagorn
Centre for Wireless Communications, University of Oulu, Finland

9. References

Fort A.; Ryckaert, J.; Desset, C.; De Doncker, P.; Wambacq, P.; & Van Biesen, L. (2006). Ultra-Wideband Channel Model for Communication Around the Human Body, *IEEE Journal on Selected Areas in Communications*, vol. 24, No. 4, (April 2006), pp. 927-933, ISSN: 0733-8716.

IEEE 804.15.4 Standard (2011). Low-Rate Wireless Personal Area Networks (LR-WPANs).

IEEE 804.15.6 Standard (2012). Wireless Body Area Networks.

Ghaboosi K.; Pahlavan K.; & Pomalaza-Ráez C. (2011). A Cooperative Medical Traffic Delivery Mechanism for Multi-hop Body Area Networks, *Proceedings of the 2011 IEEE 22nd International Symposium on Personal Indoor and Mobile Radio Communications (PIMRC)*, ISBN: 978-1-4577-1348-4, Toronto, Canada, September 11-14, 2011.

Hentilä L.; Taparugssanagorn A.; Viittala H.; & Hämäläinen M. (2005). Measurement and modelling of an UWB channel at a hospital, *Proceedings of the International Conference on UWB ICUWB)*, ISBN: 0-7803-9397-X, Zurich, Switzerland, September 5-8, 2005.

Lazaro, A.; Girbau D. ; & Villarino R. (2010). Analysis of Vital Signs Monitoring Using an IR-UWB Radar, Progress in Electromagnetics Research, vol. 100, pp. 265-284, ISSN: 1070-4698.

Pancera, E.; Xuyang Li; Jalilvand, M.; Zwick, T.; & Wiesbeck, W. (2011). UWB medical diagnostic: in-body transmission modeling and applications, *Proceedings of the 5th European Conference on Antennas and Propagation* (EUCAP), ISBN: 978-1-4577-0250-1, Rome, Italy, April 11-15, 2011.

Taparugssanagorn A.; Pomalaza-Ráez C.; Isola A.; Tesi R.; Hämäläinen M.; & Iinatti J. (2010). Channel Modelling for Wireless Body Area Networks in Medical Applications, *Proceedings of the Third International Symposium on Medical Information & Communication Technology* (ISMICT), Montreal, Canada, February 24-26, 2009

Taparugssanagorn A.; Pomalaza-Ráez C.; Tesi R.; Hämäläinen M.; & Iinatti J. (2010). UWB Channel Modelling for Wireless Body Area Networks in a Hospital, *International Journal of Ultra Wideband Communications and Systems* (IJUWBCS) , vol. 1, No. 4, pp. 226-236, 2010, ISSN: 1758-7298.

Taparugssanagorn A.; Zhen B.; Tesi R.; Hämäläinen M.; Iinatti J; & Kohno R. (2011). A UWB WBAN channel model based on a pseudo-dynamic measurement, Annals of Telecommunications, vol. 66, No. 3-4, pp. 177-185, DOI: 10.1007/s12243-010-0228-5.

Viittala H.; Hämäläinen M.; Iinatti J.; & Taparugssanagorn A. (2009). Different experimental WBAN channel models and IEEE802.15.6 models: comparison and effects, *Proceedings of the 2nd International Symposium on Applied Sciences in Biomedical and Communication Technologies* (ISABEL 2009), E-ISBN: 978-1-4244-4641-4, Bratislava, Slovakia, November 24-27, 2009.

Yang W.; Sayrafian-Pour K.; Hagedorn J.; Terrill J.; Yazdandoost K.; Taparugssanagorn A.; Hämäläinen M.; & Iinatti J. (2011). Impact of an Aortic Valve Implant on Body Surface UWB Propagation: A Preliminary Study, *Proceeding of the Fifth International Symposium on Medical Information and Communication Technology* (ISMICT), ISBN: 9-781424-494446, Montreux, Switzerland, March 27-30, 2011.

Yazdandoost K. & Sayrafian-Pour K. (2009). Channel Model for Body Area Network (BAN), *Report to the IEEE*, P802.15. ID: IEEE 802.15-08-0780-02-0006, April 2009.

Chapter 6

Cooperative Communication over Multi-Scale and Multi-Lag Wireless Channels

H. Lu, T. Xu and H. Nikookar

Additional information is available at the end of the chapter

http://dx.doi.org/10.5772/48719

1. Introduction

The development of wireless communication applications in the last few years is unprecedented. Wireless communication has evolved in various ways. The next generation of wireless systems should service more users while supporting mobility and high data rates. These requirements necessitate efficient use of available resources to provide acceptable service quality.

In the wireless channel, fading can be coped with by using diversity techniques or by transmitting the signal over several independently fading channels and combining different signal at the receiver before demodulation and detection. Spatial diversity techniques are known to increase the system reliability without sacrificing time and bandwidth efficiency. However, due to the limitation of the diversity order and correlated channel, multiple antenna diversity is not always practically feasible.

Spatial diversity has been studied intensively in the context of Multiple-Input-Multiple-Output (MIMO) systems [1]. It has been shown that utilizing MIMO systems can significantly improve the system throughput and reliability [2]. However, MIMO gains hinge on the independence of the paths between transmit and receive antennas, for which one must guarantee antenna element separation several times the wavelength, a requirement difficult to meet with the small-size terminals. To overcome this problem, and to benefit from the performance enhanced by MIMO systems, cooperative diversity schemes for the relay transmission have been introduced in [3-5].

Cooperative diversity [6] is an alternative way to achieve spatial diversity when the multiple antenna structure is not an option. Cooperative communications offer diversity based on the fact that other users in the cooperative network are able to overhear the transmitted signal and forward the information to the destination through different paths. Cooperative communications have been receiving a lot of attention recently as an attractive way to combat frequency-selectivity of wireless channel, as they consume neither time nor frequency resources. Furthermore, cooperative communications are promising techniques to increase the transmission reliability, since they can achieve spatial diversity by using the relay nodes as virtual antennas, and mitigating fading effects. By adopting relay to forward information, we can increase the capacity, lower the bit-error rate, and increase the achievable transmission range.

There are mainly two relaying protocols in cooperative communications: Amplify-and-Forward (AF) and Decode-and-Forward (DF). In AF, the received signal is amplified and retransmitted to the destination. The advantage of this protocol is its simplicity and low cost implementation. However, the noise is also amplified at the relay. In the DF, the relay attempts to decode the received signals. If successful, it re-encodes the information and retransmits it. Therefore, DF relaying usually enjoys a better transmission performance than the AF relaying. The time-consuming decoding tradeoff for a better cooperative transmission, and finding the optimum hybrid cooperative schemes, that include both DF and AF for different situations, is an important issue for the cooperative networks design.

In this chapter, we investigate the performance and relay selection issue in cooperative wideband communication systems. Wideband communication systems are defined as having a fractional bandwidth—the ratio of single-sided bandwidth to center frequency—that exceeds 0.2 [7]. Wideband channels are of interest in a variety of wireless communication scenarios including underwater acoustic systems and wideband terrestrial radio frequency systems such as spread-spectrum or ultra wideband. Due to the nature of wideband propagation, such channels exhibit some fundamental differences relative to so-called narrowband channels. In the wideband systems, the effects of mobility in the multipath mobile environment are not well described by frequency-domain spreading, but rather by time-domain scale spreading. More specifically, in narrowband channels, the transmitted signal experiences multiple propagation paths each with a possibly distinct Doppler frequency shift, and thus these channels are also known as multi-Doppler shift, multi-lag channels. For wideband channels, on the other hand, each propagation path experiences a distinct Doppler scale, hence the term, multi-scale, multi-lag channel. For both types of wideband and narrowband time-varying channels, so-called canonical channel models have been proposed [8-11], limiting the number of channel coefficients required to represent the channel.

In particular, there has been significant success in the application of canonical models to narrowband time-varying channels [8]. For wideband time-varying channels a canonical model has been proposed in [9-11], which is also dubbed as the scale-lag canonical model. This model has been adopted for direct sequence spread spectrum (DSSS) communication systems [11] to develop a scale-lag RAKE receiver to collect the diversity inherent in the multi-scale multi-lag channel. In addition, this model has spurred the use of wavelet signaling due to the fact that when the wavelets are "matched" to the scale-lag model, the receiver structure is greatly simplified – the signals corresponding to different scale-lag branches of the model are orthogonal when a single wavelet pulse is transmitted. The single pulse case is examined in paper [10]. Multi-scale multi-lag wavelet signaling is possible as well [12], [13], although inter-scale and inter-delay interference results. In paper [12], multiple receiver designs to combat such interference are provided exploiting the banded nature of the resulting interference.

Note that scale-spreading arises from the same fundamental mechanism that causes Doppler spreading. This scale-lag diversity is better described by the wavelet transform than the conventional time-frequency representation for the narrowband linear time-varying (LTV) system, and is so called the wideband LTV representation [10, 11]. Wideband LTV representation has been proven and verified for many applications in terms of high data rate wireless communications [14-17], high-speed underwater acoustic communications [18-20], vehicle-to-vehicle (V2V) wideband communication systems [21, 22], and radar/sonar systems [23]. In general, the transmit waveform could be designed to optimally enable the scale-lag diversity in the wideband LTV system.

Doppler scaling and multipath spread in the wideband system implementations are usually treated as distortions rather than potential diversity sources, and always compensated after estimation. In this chapter, Doppler scaling and multipath spread are utilized to obtain a joint scale-lag diversity with the discrete multi-scale and multi-lag wireless channel model by properly designing signaling and reception schemes using the discrete wavelet transform. The wavelet technique used in the wideband system is well motivated since wideband processing is intimately related to the wavelet theory [24-26]. The wideband LTV representation has proven useful in many applications as noted above. However, no cooperative wavelet implementations have been exploited to provide further increased performance for wideband systems.

In this chapter, we will design a cooperative wavelet communication scheme to exploit the joint scale-lag diversity in the wideband LTV system. Furthermore, we propose the analytical Bit Error Rate (BER) expression for the cooperative wideband system, and provide a dynamic optimal selection strategy for relay selection to gain from multi-relay, multi-scaling, and multi-lag diversity, and maximize the whole system transmission performance.

The rest of chapter is organized as follows. In section 2, an overview of the multi-scale and multi-lag diversity in wideband system is provided. In section 3, we investigate the general hybrid cooperative scheme that includes both DF and AF relays, and review the SNR thresholding scheme as well as dynamic optimal combination strategy for the hybrid DF-AF cooperation to achieve the optimal system BER performance. In section 4, we construct the cooperative wavelet wideband transmission strategy, and derive the analytical BER expression for the cooperative wavelet communications in the multi-relay, multi-scale and multi-lag channel. In Section 5, we represent the dynamic optimal selection strategy for the relay selection. Simulations results are provided in Section 6 and are compared to the analytical formulas. The relay selection is also illustrated in this section. Finally, Section 7 concludes the chapter.

2. Wideband multi-scale and multi-lag representation

Multi-scale and multi-lag representation is suitable for the wideband systems to satisfy either of the two conditions, i.e., absolute condition or relative condition. First is the absolute condition, which requires the signal fractional bandwidth (ratio of bandwidth to center frequency) to be larger than 0.2. Second one is the relative condition, i.e., the motion velocity v, the propagation speed c and the signal time bandwidth (TB) product should satisfy $2v/c \gg 1/(TW)$, where T stands for the transmitted signal duration and W denotes the transmitted signal bandwidth. Therefore, the multi-scale and multi-lag system can be defined as a system that operates at high fractional bandwidths or large TB products or when the v/c ratio is large.

For example, an ultra wideband (UWB) system transmits signals with high fractional bandwidths (> 0.2) or large TB products (10^5- 10^6) to improve resolution capacity and increase noise immunity [27]. Or an underwater acoustic environment with fast moving objects could result in a large ratio due to the relatively low speed of sound [28]. In these situations, multi-scale and multi-lag representation is needed to account for the Doppler scale effects, but not Doppler shift.

Under the wideband background mentioned above, we consider a multi-scale and multi-lag system, that is, a signal, $x(t)$, transmitted over a wideband propagation medium is received as

$$y(t) = \int_{A_l}^{A_u} \int_0^{T_d} h(\tau, a) \sqrt{a} x(a(t-\tau)) d\tau da + n(t), \tag{1}$$

where $a \approx 1 + 2v/c$ is the Doppler scale, results in a time compression or expansion of the waveform caused by a relative velocity v between transmitter and scatterer. When the Doppler scale is such that $a > 1$, then the scatterer is approaching the transmitter and the transmitted signal is compressed with respect to time; in contrast, when $0 < a < 1$, the received signal is dilated and the scatterer is moving away from the transmitter. τ is the propagation delay due to reflections of $x(t)$ by scatterers in the medium. Channel gain $h(\tau, a)$ can be modeled as a stochastic process, when the system is randomly varying [10]. Due to physical restrictions on the system, we can assume that $h(\tau, a)$ is effectively nonzero only when $0 < A_l \leq a \leq A_u$ and $0 \leq \tau \leq T_d$, where $A_u - A_l$ is the Doppler scale spread and T_d is the multipath delay spread. The noise process, $n(t)$ is modeled as a white Gaussian random process.

Note that regardless the noise term, Eq. (1) is in the form of an inverse wavelet transform with $x(t)$ acting as the wavelet. Therefore, according to the wavelet theory, we sample the multi-scale and multi-lag plane in a dyadic lattice as shown in Fig. 1 [10, 29].

Without loss the generality, we consider BPSK modulation, the information-bearing symbol of the transmitted signal is $b_0 = \pm 1$. From the multi-scale and multi-lag channel defined in Fig. 1, the overall baseband signal at the receiver can be rewritten as:

$$y(t)=b_0\sum_{m=M_0}^{M_1}\sum_{l=1}^{L(m)}h(m,l)\sqrt{2^m}\,x\left(2^m t-\frac{l}{W}\right)+n(t), \tag{2}$$

where $L(m)$ denotes the number of the multilag for corresponding scaling index m, as shown to be the number of cross points on each row in Fig. 1. M_0 and M_1 are lower and upper bounds of *m*, respectively. In fact, the multilag resolution in a wideband channel is $1/(aW)$ if the signal is scaled by a. When the number of scatterers contributing to the discrete channel gain $h(m,l)$ is exceedingly large, the random variables $h(m,l)$ can be assumed Gaussian and therefore independent.

Consequently, the inverse discrete wavelet transform description in Eq. (2) effectively decomposes the wideband channel into

$$M=\sum_{m=M_0}^{M_1}\left(L(m)\right), \tag{3}$$

orthogonal, flat-fading channels. This results in a potential joint scale-lag diversity order M that can be exploited to increase system performance.

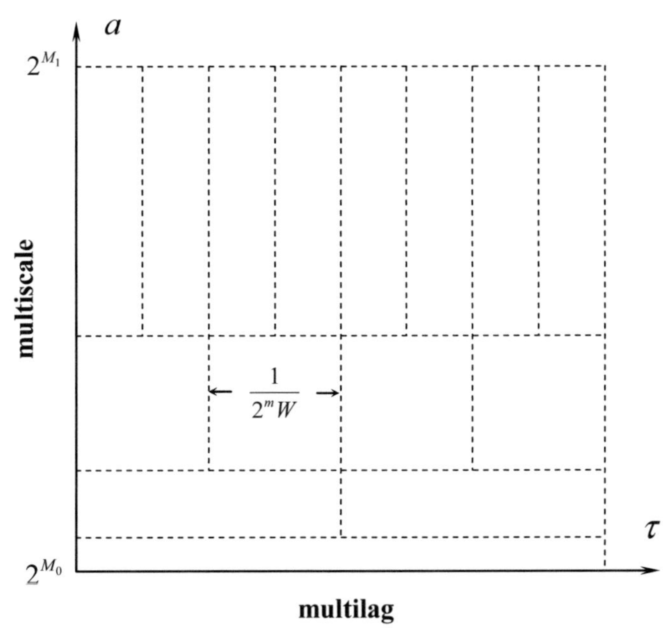

Figure 1. Dyadic sampling in the multi-scale and multi-lag plane, the dyadic scale is $a=2^m$ and for the given $M_0 \le m \le M_1$, m, M_0, M_1 are all integer. The multi-lag resolution is $1/\left(2^m W\right)$, given the signal bandwidth W.

In order for a scale-lag RAKE receiver to collect the aforementioned diversity components, transmitted signal should be designed as a wavelet-based waveform. Wideband multi-scale and multi-lag channel performs the inverse discrete wavelet transform on the transmitted signal $x_{m,l}(t)$. At the receiver side, for the diversity component corresponding to the m-th scale and l-th lag, the detection statistic

$$\lambda_{m,l} = \left\langle y(t), \sqrt{2^m} x\left(2^m t - \frac{l}{W}\right)\right\rangle = \int_{-\infty}^{\infty} y(t) \sqrt{2^m} x^*\left(2^m t - \frac{l}{W}\right) dt \tag{4}$$

is the correlator output of the received signal $y(t)$ and the basic waveform $\sqrt{2^m} x\left(2^m t - l/W\right)$. Therefore, the detection statistic $\lambda_{m,l}$ can be obtained by the dyadic scale-lag samples of the discrete wavelet transform of $y(t)$ associated with the wavelet function $x(t)$, which forms a scale-lag RAKE receiver. Then, the channel gain is combined coherently to obtain the estimate of the transmitted information symbol b_0 as

$$\hat{b}_0 = \text{sign}\left\{ \text{Re}\left(\sum_{m=M_0}^{M_1} \sum_{l=1}^{L(m)} h^*(m,l) \lambda_{m,l} \right) \right\}. \tag{5}$$

We note that this coherent detection of the scale-lag RAKE receiver corresponds to a Maximum Ratio Combination (MRC).

Wideband LTV multi-scale and multi-lag channels are of interest in a variety of wireless communication scenarios including wideband terrestrial radio frequency systems such as spread-spectrum systems or Ultra Wideband (UWB) systems and underwater acoustic systems. Due to the nature of wideband propagation, such wideband multi-scale and multi-lag channels exhibit some fundamental differences compared to the so-called narrowband channels. In particular, these multi-scale, multi-lag channel descriptions offer improved modeling of LTV wideband channels over multi-Doppler-shift, multi-lag models [10-12]. Orthogonal Frequency Division Multiplexing (OFDM) technology has been introduced and examined for wideband LTV channels. Approaches include splitting the wideband LTV channel into parallel narrowband LTV channels [30] or assuming a simplified model which reduces the wideband LTV channel to a narrowband LTV channel with a carrier frequency offset [31].

Receivers for single-scaled wavelet-based pulses for wideband multi-scale, multi-lag channels are presented in [10, 11], and a similar waveform is adopted in spread-spectrum systems [32] over wideband channels modeled by wavelet transforms; while [33] considers equalizers for block transmissions in wideband multi-scale, multi-lag channels. In order to achieve better realistic channel matching, single-scaled rational wavelet modulation was designed in [34]. The above mentioned schemes all employ single-scale modulation and thus do not maximize the spectral efficiency. In order to exploit the frequency diversity, a new form of Orthogonal Wavelet Division Multiplexing (OWDM) has been previously examined in [35] for additive white Gaussian noise channels.

However, no cooperative schemes for multi-scale, multi-lag channels have been exploited to provide further increased performance for wideband systems. In this chapter, we will design a cooperative wavelet communication scheme to exploit the joint scale-lag diversity in the wideband LTV system. Furthermore, we propose the analytical BER expression for the cooperative wideband system, and provide a dynamic optimal selection strategy for relay selection to gain from multi-relay, multi-scaling, and multi-lag diversity, and maximize the whole system transmission performance.

3. Cooperative DF and AF relay communication

3.1. Cooperative DF/AF system model

In the cooperative communications system, DF relaying performs better than AF relaying, due to reducing the effects of noise and interference at the fully decoding relay. However, in some case, DF relaying entails the possibility of forwarding erroneously detected signals to the destination as well; causing error propagation that can diminish the performance of the system. The mutual information between the source and the destination is limited by the mutual information of the weakest link between the source–relay and the combined channel from the source-destination and relay-destination.

Since the reliable decoding is not always available, which also means DF protocol is not always suitable for all relaying situations. The tradeoff between the time-consuming decoding, and a better cooperative transmission, finding the appropriate hybrid cooperative schemes, which include both DF and AF for specific situations, is a critical issue for the cooperative relaying networks design.

In this Section, we review the cooperative strategy with the combination of the DF and AF relay as shown in Fig. 2, where we transmit data from source node S to destination node D through R relays, without the direct link between S and D. This relay structure is called 2-hop relay system, i.e., first hop from source node to relay, and second hop from relay to destination. The channel fading for different links are assumed to be identical and statistically independent, quasi-statistic, i.e., channels are constant within several symbol durations. This is a reasonable assumption as the relays are usually spatially well separated and in a slow changing environment. We assume that the channels are well known at the corresponding receiver sides, and a one bit feedback channel from destination to relay is used for removing the unsuitable AF relays. All the Additive White Gaussian Noise (AWGN) terms have equal variance N_0. Relays are re-ordered according to the descending order of the Signal-to-Noise Ratio (SNR) between S and Q, i.e., $\mathrm{SNR}_{SQ_1} > \cdots > \mathrm{SNR}_{SQ_R}$, where SNR_{SQ_r} denotes the r-th largest SNR between S and Q.

In this model, relays can determine whether the received signals are decoded correctly or not, just simply by comparing the SNR to the threshold, which will be elaborated in Section 3.2. Therefore, the relays with SNR above the threshold will be chosen to decode and forward the data to the destination, as shown with the white hexagons in Fig. 2. The white circle is the removed AF relay according to the dynamic optimal combination strategy which will be described in Section 3.3. The rest of the relays follow the AF protocol, as shown with the white hexagons in Fig. 2 [36].

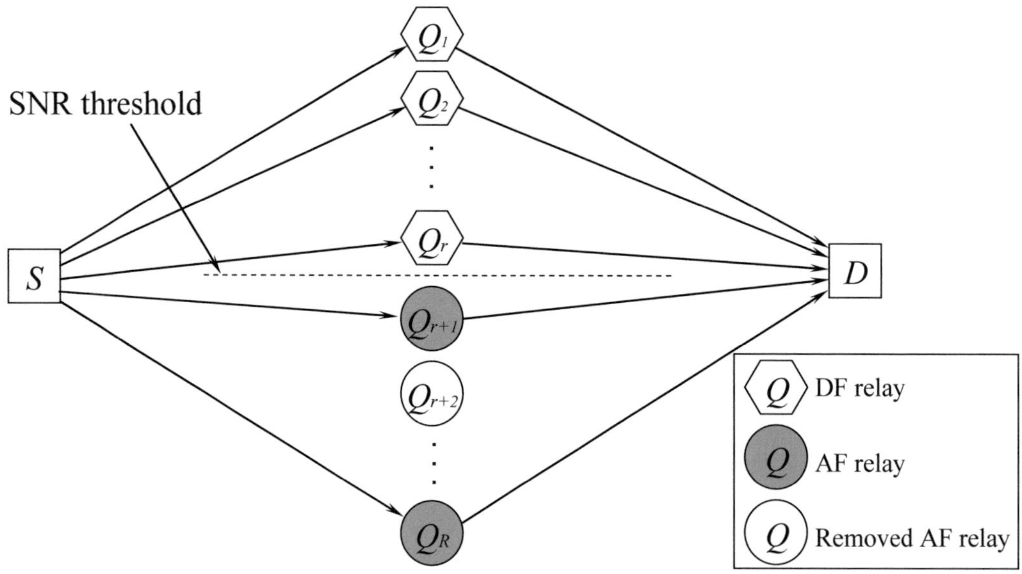

Figure 2. Cooperation communications with dynamic optimal combination of DF/AF relays (*S*: Source, *D*: Destination, Q_r: *r*-th Relay)

The received SNR at the destination in the hybrid cooperative network can be denoted as

$$\gamma_h = \sum_{Q_i \in \mathrm{DF}} \frac{E_Q h_{Q_i,D}}{N_0} + \sum_{Q_j \in \mathrm{AF}} \frac{\frac{E_S h_{S,Q_j}}{N_0} \frac{E_Q h_{Q_j,D}}{N_0}}{\frac{E_S h_{S,Q_j}}{N_0} + \frac{E_Q h_{Q_j,D}}{N_0} + 1}, \tag{6}$$

where $h_{Q_i,D}$, h_{S,Q_j} and $h_{Q_j,D}$ denote the power gains of the channel from the *i*-th relay to the destination in DF protocol, source node to the *j*-th relay in AF protocol and *j*-th relay to the destination in AF protocol, respectively. E_S and E_Q in Eq. (6) are the average transmission energy at the source node and at the relays, respectively. By choosing the amplification factor A_{Q_j} in the AF protocol as

$$A_{Q_j}^2 = \frac{E_S}{E_S h_{S,Q_j} + N_0}, \tag{7}$$

and forcing the E_Q in DF equal to E_S, it will be convenient to maintain constant average transmission energy at relays, equal to the original transmitted energy at the source node.

The receiver at the destination collects the data from DF and AF relays with a MRC. Because of the amplification in the intermediate stage in the AF protocol, the overall channel gain of the AF protocol should include the source to relay, relay to destination channels gains and amplification factor. The decision variable *u* at the MRC output is given by

$$u=\sum_{Q_i\in \mathrm{DF}}\frac{\left(H_{Q_i,D}\right)^{*}Y_{Q_i}}{\left(H_{Q_i,D}\right)^{*}H_{Q_i,D}}+\sum_{Q_j\in \mathrm{AF}}\frac{\left(H_{S,Q_j}A_{Q_j}H_{Q_j,D}\right)^{*}Y_{Q_j}}{\left(H_{S,Q_j}A_{Q_j}H_{Q_j,D}\right)^{*}\left(H_{S,Q_j}A_{Q_j}H_{Q_j,D}\right)}, \tag{8}$$

where Y_{Q_i} and Y_{Q_j} are the received signal from DF *i*-th relay and AF *j*-th relays, respectively, and $(\cdot)^{*}$ denotes the conjugate operation. $H_{Q_i,D}$, H_{S,Q_j} and $H_{Q_j,D}$ are frequency response of the channel power gains, respectively.

In the hybrid DF/AF cooperative network, DF plays a dominant role in the whole system. However, switching to AF scheme for the relay nodes with SNR below the threshold often improves the total transmission performance, and accordingly AF plays a positive compensating role.

3.2. SNR thresholding scheme for DF relays Cooperative DF/AF system model

In general, mutual information I is the upper bound of the target rate B bit/s/Hz, i.e., the spectral efficiency attempted by the transmitting terminal. Normally, $B \le I$, and the case $B > I$ is known as the outage event. Meanwhile, channel capacity, C, is also regarded as the maximum achievable spectral efficiency, i.e., $B \le C$.

Conventionally, the maximum average mutual information of the direct transmission between source and destination, i.e., I_D, achieved by independent and identically distributed (i.i.d) zero-mean, circularly symmetric complex Gaussian inputs, is given by

$$I_D=\log_2\left(1+\mathrm{SNR}\,h_{S,D}\right), \tag{9}$$

as a function of the power gain over source and destination, $h_{S,D}$. According to the inequality $B \le I$, we can derive the SNR threshold for the full decoding as

$$\mathrm{SNR}\ge\frac{2^B-1}{h_{S,D}}. \tag{10}$$

Then, we suppose all of the X relays adopt the DF cooperative transmission without direct transmission. The maximum average mutual information for DF cooperation I_{DF_co} is shown [3] to be

$$I_{DF\_co}=\frac{1}{X}\min\left\{\log_2\left(1+\sum\nolimits_{r=1}^{R}\mathrm{SNR}\,h_{S,Q_r}\right),\log_2\left(1+\sum\nolimits_{r=1}^{R}\mathrm{SNR}\,h_{Q_r,D}\right)\right\}, \tag{11}$$

which is a function of the channel power gains. Here, R denotes the number of the relays.

For the *r*-th DF link, requiring both the relay and destination to decode perfectly, the maximum average mutual information I_{DF_li} can be shown as

$$I_{DF\_li}=\min\left\{\log_2\left(1+\mathrm{SNR}\,h_{S,Q_r}\right),\log_2\left(1+\mathrm{SNR}\,h_{Q_r,D}\right)\right\}. \tag{12}$$

The first term in Eq. (12) represents the maximum rate at which the relay can reliably decode the source message, while the second term in Eq. (12) represents the maximum rate at which the destination can reliably decode the message forwarded from relay. We note that such mutual information forms are typical of relay channel with full decoding at the relay [37]. The SNR threshold of this DF link for target rate B is given by $I_{DF\_li} \geq B$ which is derived as

$$\mathrm{SNR} \geq \frac{2^{B}-1}{\min\left(h_{S,Q_r}, h_{Q_r,D}\right)}. \tag{13}$$

In the proposed hybrid DF/AF cooperative transmission, we only consider that a relay can fully decode the signal transmitted over the source-relay link, but not the whole DF link, thus, the SNR threshold for the full decoding at the r-th relay reaches its lower bound as

$$\gamma_{th} \geq \frac{2^{B}-1}{h_{S,Q_r}}. \tag{14}$$

For the DF protocol, let R denotes the number of the total relays, M denotes the set of participating relays, whose SNRs are above the SNR threshold, and the reliably decoding is available. The achievable channel capacity, C_{DF}, with SNR threshold is calculated as

$$C_{DF} = \sum_{M} \frac{1}{R} \mathbf{E}\left(\log_2\left(1 + y \middle| M\right)\right) \Pr(M), \tag{15}$$

where $\mathrm{E}(\cdot)$ denotes the expectation operator, $y|M = (R-K)\gamma_{S,D} + \sum_{Q\in M}\gamma_{Q,D}$ denotes the instantaneous received SNR at the destination given set M with K participating relays, where $\gamma_{n,m}$ denotes the instantaneous received SNR at node m, which is directly transmitted from n to m. Since $y|M$ is the weighted sum of independent exponential random variables [38], the probability density function (PDF) of $y|M$ can be obtained using its moment generating function (MGF) and partial fraction technique for evaluation of the inverse Laplace transform, see Eq. (8d) and Eq. (8e) in paper [38]. $\Pr(M)$ in Eq. (15) is the probability of a particular set of participating relays which are obtained as

$$\Pr(M) = \prod_{Q\in M} \exp\left(-\frac{R\gamma_{th}}{\Gamma_{S,Q\in M}}\right) \prod_{Q\notin M}\left(1\text{-}\exp\left(-\frac{R\gamma_{th}}{\Gamma_{S,Q\notin M}}\right)\right), \tag{16}$$

where $\Gamma_{u,v}$ denotes the average SNR over the link between nodes u and v.

Combining Eq. (11), Eq. (15) and Eq. (16) with the inequality $I_{DF\_co} \leq C_{DF}$, since the maximum average mutual information, I, is upper bound by the achievable channel capacity, C, we can calculate the upper bound of SNR threshold γ_{th} for fully decoding in the DF protocol.

Now, we can obtain the upper bound and the lower bound of the SNR threshold γ_{th} for hybrid DF/AF cooperation. However, compared to the upper bound, the lower bound as shown in the Eq. (14) is more crucial for improving the transmission performance. This is because the DF protocol plays a dominant role in the hybrid cooperation strategy, and accordingly we want to find the lower bound which provides as much as possible DF relays. We will elaborate this issue later. Fully decoding check can also be guaranteed by employing the error detection code, such as cyclic redundancy check. However, it will increase the system complexity [39].

3.3. Error probability for hybrid DF/AF cooperative transmission

In the maximum ratio combining the transmitted signal from R cooperative relays nodes, which underwent independent identically distributed Rayleigh fading, and forwarded to the destination node are combined. In this case the SNR per bit per relay link γ_r has an exponential probability density function (PDF) with average SNR per bit $\overline{\gamma}$:

$$p_{\gamma_r}(\gamma_r) = \frac{1}{\overline{\gamma}} e^{-\gamma_r/\overline{\gamma}}. \tag{17}$$

Since the fading on the R paths is identical and mutually statistically independent, the SNR per bit of the combined SNR γ_c will have a Chi-square distribution with $2R$ degrees of freedom. The PDF $p_{\gamma_c}(\gamma_c)$ is

$$p_{\gamma_c}(\gamma_c) = \frac{1}{(R-1)!\overline{\gamma}_c^{\,R}} \gamma_c^{\,R-1} e^{-\gamma_c/\overline{\gamma}_c}, \tag{18}$$

where $\overline{\gamma}_c$ is the average SNR per channel, then by integrating the conditional error probability over $\overline{\gamma}_c$, the average probability of error P_e can be obtained as

$$P_e = \int_0^\infty \hat{Q}\left(\sqrt{2g\gamma_c}\right) p_{\gamma_c}(\gamma_c) d\gamma_c, \tag{19}$$

where g = 1 for coherent Binary Phase Shift Keying (BPSK), g = ½ for coherent orthogonal BFSK, g = 0.715 for coherent BFSK with minimum correlation, and $\hat{Q}(\cdot)$ is the Gaussian Q-function, i.e., $\hat{Q}(x) = 1/\sqrt{2\pi}\int_x^\infty \exp(-t^2/2)dt$. For the BPSK case, the average probability of error can be found in the closed form by successive integration by parts, i.e.,

$$P_e = \left(\frac{1-\mu}{2}\right)^R \sum_{k=0}^{R-1} \binom{R-1+k}{k} \left(\frac{1+\mu}{2}\right)^R, \tag{20}$$

where

$$\mu = \sqrt{\frac{\overline{\gamma}_c}{1+\overline{\gamma}_c}}. \tag{21}$$

In the hybrid DF/AF cooperative network with two hops in each AF relay, the average SNR per channel $\overline{\gamma}_c$ can be approximately derived as

$$\overline{\gamma}_c = \frac{\gamma_h}{K + 2 \times J}, \tag{22}$$

where K and J are the numbers of the DF relays and AF relays, respectively. γ_h can be obtained from Eq. (6) In the DF protocol, due to the reliable detection, we can only consider the last hops, or the channels between the relay nodes and destination node.

As the average probability of error P_e is a precise indication for the transmission performance, we consequently propose a dynamic optimal combination strategy for the hybrid DF/AF cooperative transmission. In this algorithm the proper AF relays are selected to make P_e reach maximum.

First of all, like aforementioned procedure, relays are reordered according to the descending order of the SNR between source and relays, as shown in the Fig. 2. According to the proposed SNR threshold, we pick up the DF relays having SNR greater than threshold. Then, we proceed with the AF relay selection scheme, where the inappropriate AF relays are removed. The whole dynamic optimal combination strategy for the hybrid DF/AF cooperation is shown in the flow chart of Fig. 3.

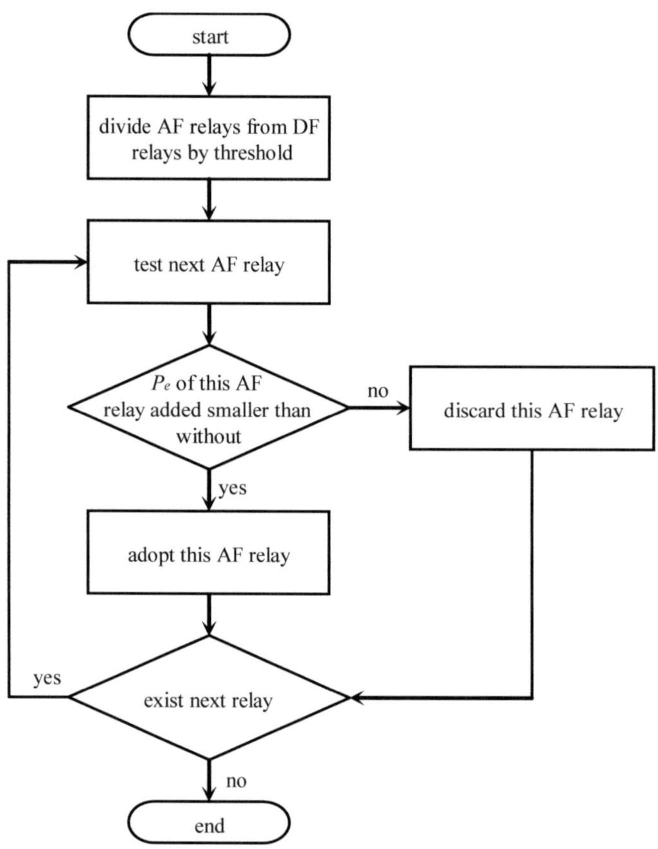

Figure 3. Flow chart of the dynamic optimal combination strategy for the hybrid DF/AF cooperation

4. Cooperative wavelet communication scheme

In this section, by taking advantage of the MRC property of the above mentioned multi-scale and multi-lag wideband channel and wavelet transceiver model, we consider a wideband cooperative wavelet communication scheme as shown in Fig. 4 [40], where we transmit data from source node S to destination node D through R DF relays, without the direct link between S and D.

We only consider and illustrate DF relay case; it is because only DF relays strictly fulfill the MRC property. The hybrid DF/AF scheme can approximately fulfill the MRC property and with some errors. This can be shown by the simulation results as well. If error requirement is not very demanding, the relay selection strategy for DF relay case can easily extended for the hybrid DF/AF case.

Different relays operate at different frequency bands and all relay links undergo multi-scale and multi-lag wideband channel. We assume that the channels are well known at the corresponding receiver sides. All the AWGN terms have equal variance N_0. Relays are re-ordered according to the descending order of the (Signal to Noise Ratio) SNR between S and Q, i.e., $\mathrm{SNR}_{SQ_1} > \cdots > \mathrm{SNR}_{SQ_R}$, where SNR_{SQ_r} denotes the r-th largest SNR between S and Q.

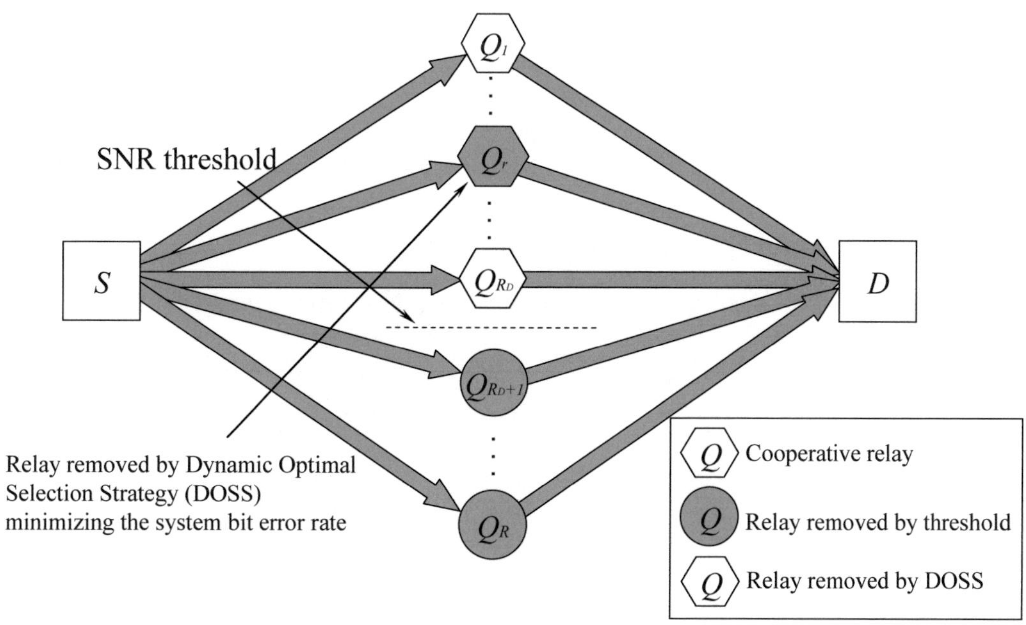

Figure 4. Cooperative wavelet communication scheme with dynamic optimal selection of DF relays in wideband multi-scale and multi-lag channel (S: Source, D: Destination, Q_r: r-th Relay).

In this model, relays can determine whether the received signals are decoded correctly or not, just simply compare its received SNR to the threshold. The SNR threshold for the full decoding at the r-th relay reaches its lower bound as

$$\gamma_{th} \geq \frac{2^B - 1}{h_{S,Q_r}}, \tag{23}$$

where B is the target rate of link between source node and relay, and h_{S,Q_r} denotes the power gain of the channel from source node to the r-th relay. Therefore, the relays with SNR below the threshold will be removed first, as shown with the gray circles in Fig. 4. The other R_D relays shown with hexagons are DF relays. According to the dynamic optimal selection strategy, which will be proposed in the next section, we select proper DF relays for cooperation. A one bit feedback channel from destination to relay is used for removing the unsuitable DF relays.

Haar wavelet signaling is adopted in the cooperative wideband system to transfer the multi-scale and multi-lag channel into the total M_D flat-fading channels

$$M_D = \sum_{r=1}^{R_D} \sum_{m=M_{Q_r,0}}^{M_{Q_r,1}} \left(L(m)\right), \tag{24}$$

where $L(m)$ denotes the number of the multilag for corresponding scaling index m, for the Doppler scale index m with spread $M_{Q_{r,}1} - M_{Q_{r,}0}$, at r-th cooperative link. For capturing the multi-scale and multi-lag diversity in the wideband channel, other wavelets, such as Daubechielkhs wavelets, Symlets, etc., have the same capability, since they all possess orthogonality in both scale and lag domain. Rational orthogonal wavelets can be adopted for the scale factor of $a_0{}^m$, $1 \leq a_0 \leq 2$, which is more suitable for the practical scenario [34]. However, wavelet selection problem is beyond the scope of this thesis. In this chapter, we focus on the multi-relay, multi-scale, and multi-lag diversity issue of the cooperative wideband system.

5. Dynamic optimal relay selection strategy

In the maximum ratio combining, the transmitted signal from R_D cooperative relays nodes over all multi-scale and multi-lag channel, which underwent independent identically distributed (i.i.d.) complex Gaussian fading, are forwarded to the destination node and combined. In this case, the average probability of error can be found in the closed form as

$$P_e = \left(\frac{1-\mu}{2}\right)^{M_D} \sum_{k=0}^{M_D - 1} \binom{M_D - 1 + k}{k} \left(\frac{1+\mu}{2}\right)^{M_D}, \tag{25}$$

where

$$\mu = \sqrt{\frac{\overline{\gamma}_c}{1+\overline{\gamma}_c}}, \tag{26}$$

In the proposed DF cooperative wideband network, because of the fully decoding at the relays, we only consider the link between relays and destination. Therefore, the average SNR per channel $\overline{\gamma}_c$ can be derived as

$$\overline{\gamma}_c = \left(\prod_{Q_r \in DF} \prod_{m=M_{Q_r,0}}^{M_{Q_r,1}} \prod_{l=0}^{L(m)} \frac{E_Q h_{Q_r,D}(m,l)}{N_o} \right)^{1/M_D}, \quad (27)$$

where $h_{Q_r,D}(m,l)$ denotes the power gains corresponding to the m-th scale and l-th lag, of the channel from the r-th relay to the destination in DF protocol. Combining Eq. (25), (26) and (27), we derive the analytical expression of the BER performance for the proposed DF cooperative wideband network.

As the average probability of error P_e is a precise indication, we can use it to predict the comprehensive transmission performance, only given the channel gains and SNRs at the destination. Consequently, we propose a dynamic optimal selection strategy for the cooperative multi-scale and multi-lag communications. In this algorithm the proper relays are selected to make P_e reach the minimum. First of all, relays are ordered according to the descending order of the SNR between source and relays, as shown in the Fig. 4. According to the proposed SNR threshold, we pick up the DF relays whose SNR is above the threshold.

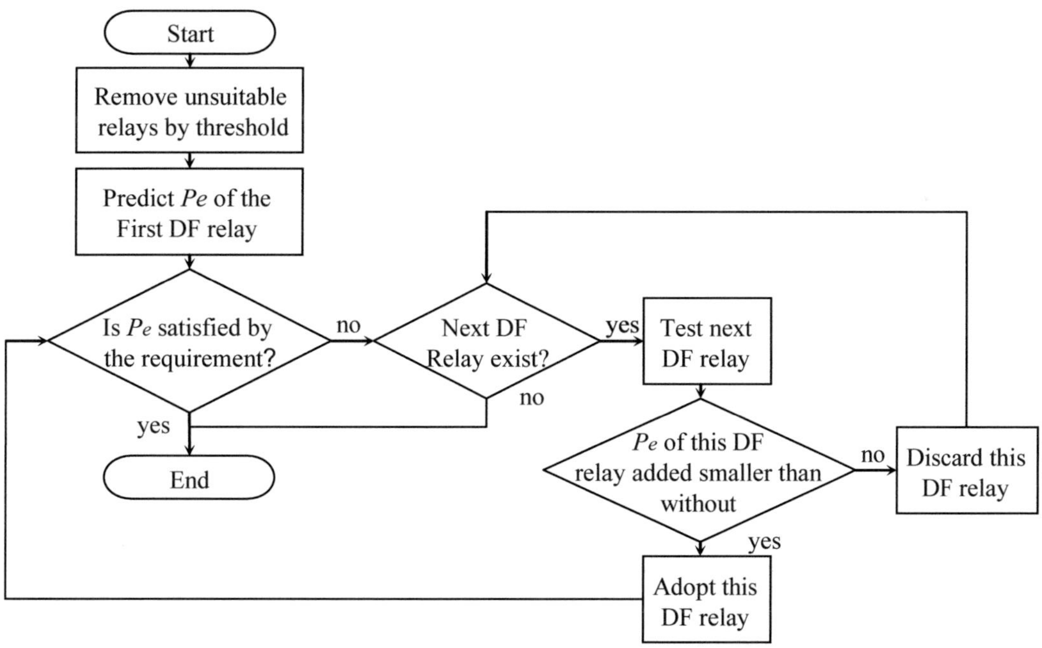

Figure 5. Flow chart of the dynamic optimal selection strategy for the cooperative wideband communications

Then, we proceed with the relay selection to maximize the entire BER performance and try to satisfy the P_e requirement, where the inappropriate DF relays are removed. The whole dynamic optimal selection strategy for the cooperative wideband communication is shown in the flow chart of Fig. 5.

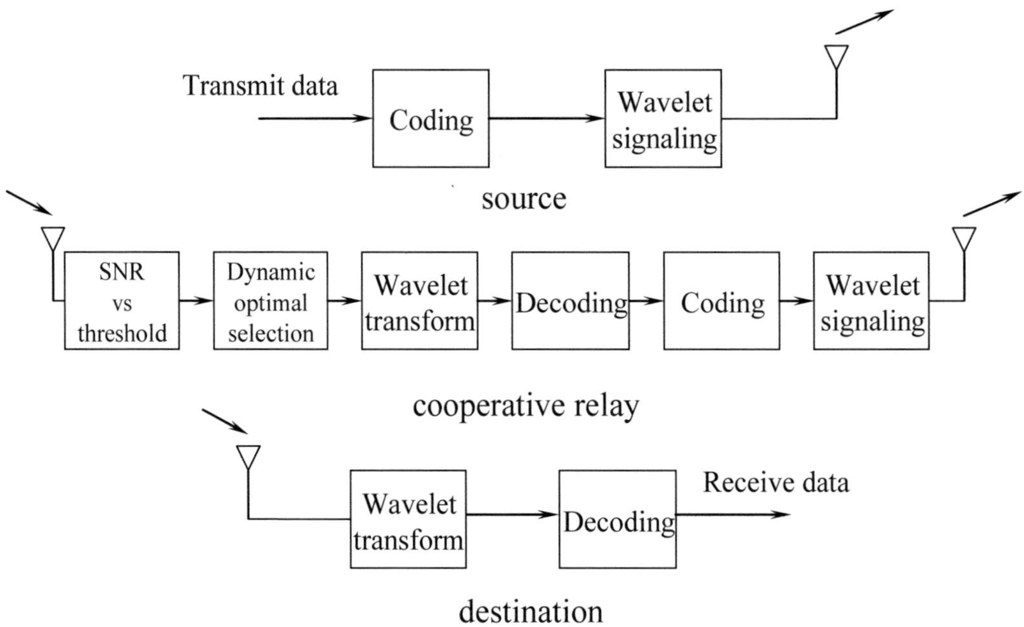

Figure 6. Relay selection in the cooperative wavelet wideband wireless transmission strategy (top: source, middle: cooperative relay, bottom: destination)

The wavelet signaling and transceiver design are shown in the Fig. 6. Before the transmission, Haar wavelet signaling is adopted to capture the multi-scale and multi-lag diversity in the wideband channel. In the relaying section, we first remove un-decodable relays by SNR threshold. Then, those relays which undergo the deep fading between relay-destination links will be removed by using the dynamic optimal selection strategy, in order to meet the P_e requirement. After recoding again, Haar wavelet signaling is applied again on the signal. At the destination node, after inverse wavelet transformation, the resulting signals are used for the combination and detection.

6. Simulation results and analysis

Test Case 1 (BER performance based Relay selection for cooperative communications):

In this example, first, we simulated BPSK modulation, Rayleigh channel, flat fading, without OFDM, and supposed the SNR threshold for correct decoding is $4E_b/N_0$, then we assumed $h_{Q_i,D} = h_{S,Q_j} = h_{Q_j,D} = 1$, for all branches, to verify proposed analytical BER expression. The resulting average BER were plotted against the transmit SNR defined as SNR = E_b/N_0. As shown in the Fig. 7, the theoretical curves of multi-DF cooperation derived from our analytical closed-form BER expression clearly agree with the Monte Carlo simulated curves, while the theoretical curves of 2-AF and 3-AF cooperation match the simulation result only at the low SNR region.

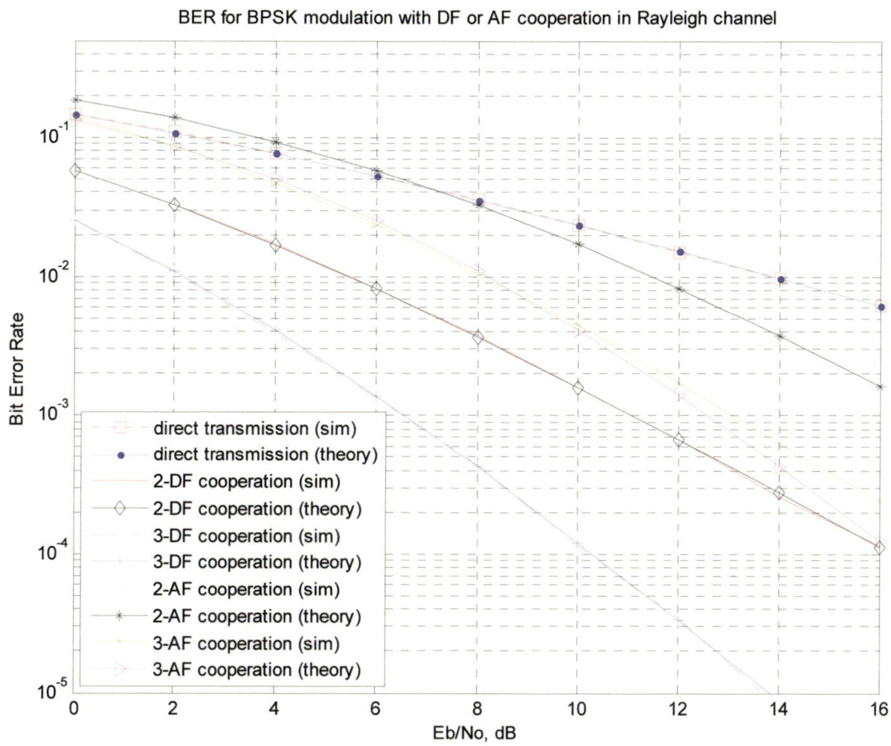

Figure 7. BER performance for DF or AF cooperation.

Fig. 8 shows the BER performance for hybrid DF-AF cooperation. For the DF-dominant hybrid cooperation, the theoretical curves exhibit a good match with the Monte Carlo simulation results curves. The slight gap between theoretical and simulation BER results for the hybrid case of 1-DF + multi-AF can be explained by the AF relay fading which was considered as a double Gaussian channel, a product of two complex Gaussian channel [49]. Obviously, the distribution of combined SNR (i.e., γ_c) will no longer follow the chi-square distribution giving rise to this slight difference.

Comparing 2-DF to 2-AF in Fig. 7, or 2-DF plus 1-AF to 1-DF plus 2-AF in Fig. 8, or other hybrid DF-AF protocols with the same *R*, we can see that the fully decoded DF protocols always show a better BER performance than AF protocols. Therefore, DF protocols with a reliable decoding play a more important role in hybrid cooperative networks than AF protocols. Meanwhile, we can see from the figure that, changing to the AF scheme for the relay nodes with SNR below the threshold also improves the BER performance, as well as the diversity gain of the whole network. In fact, this is a better way than just discarding these relay nodes.

Test Case 2 (Relay selection for cooperative communications over multi-scale and multi-lag wireless channels):

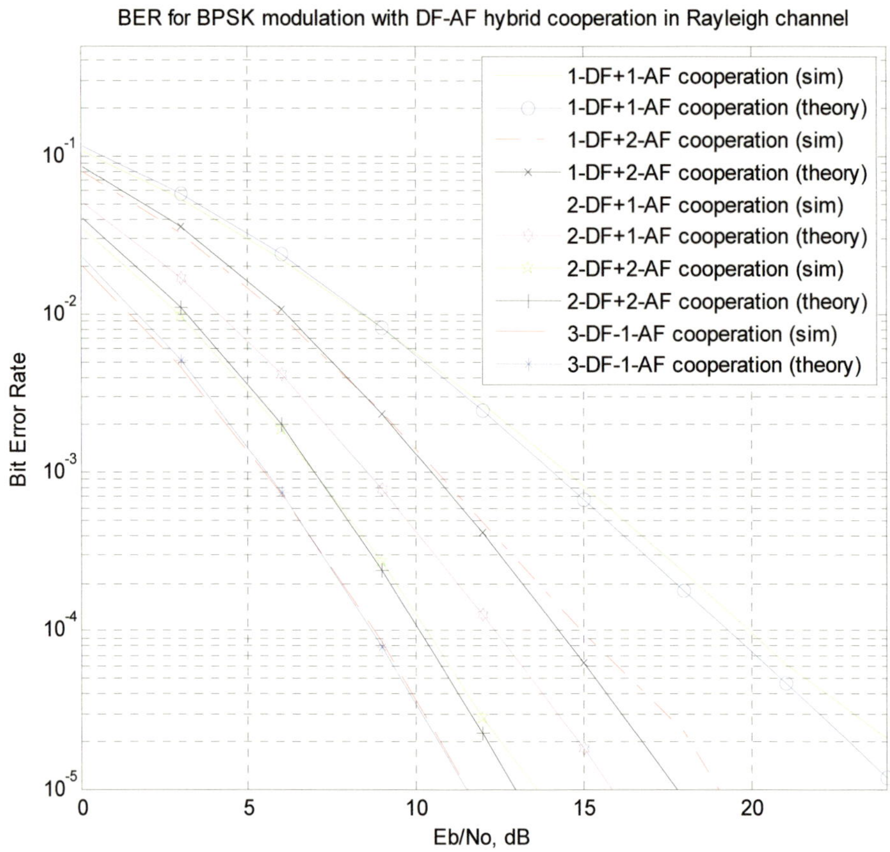

Figure 8. BER performance for hybrid DF+AF cooperation.

In this case, we use the simulation results to verify our theoretical claims on the analytical BER expression and illustrate the dynamic optimal selection strategy.

In the first example, simulation results justify the proposed analytical BER expression for cooperative wavelet communications with multi-scale and multi-lag wireless channel, i.e., the combination of Eq. (25), (26) and (27), which can be used to predict the transmission performance and enable the dynamic optimal selection strategy as shown in the Fig. 5. BPSK is adopted as the modulation scheme, and the 2-decomposition level Haar wavelet transform is adopted as a RAKE receiver to capture the multi-scale and multi-lag diversity components, and transfer the multi-relay, multi-scale and multi-lag channel into the orthogonal flat-fading channels. Therefore, we consider 2-relay three orthogonal channels in this simulation. Relay 1 has 1-scale and 2-lag diversity components, the power gains are $h_1 = 4$ and $h_2 = 1$. Relay 2 has 1-scale and 1-lag diversity component, the power gain is $h_3 = 1$. The resulting average BER are plotted against the transmit SNR defined as SNR = E_b/N_o. As shown in the Fig. 9, the theoretical curves of different diversities derived from our analytical closed-form BER expression clearly agree with the Monte Carlo simulated curves.

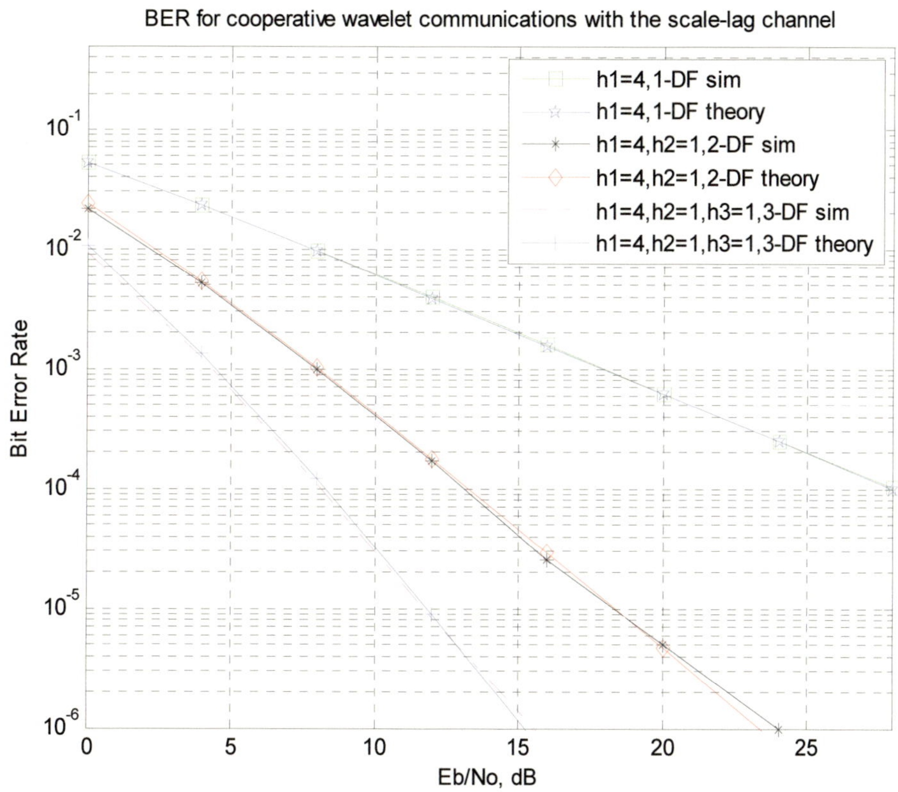

Figure 9. BER performance for cooperative wavelet wideband communications

In the second example, we illustrate how to exploit the proposed analytical BER expression together with dynamic optimal selection strategy to select relays for the cooperative wideband communications. We suppose the target P_e at SNR E_b/N_o = 10dB is 10^{-4}. In the original state, we suppose that we already have 1-Relay with 1-scale and 2-lag diversity components, with power gains $h_1 = 4$ and $h_2 = 1$. The BER performance is shown by the triangle marked curve in Fig. 10. The P_e requirement is not met by the original state, so we expect to cooperate with more relays, to gain from more diversity components. For the test case 1, we test and combine with a deep fading relay with only one scale-lag diversity, power gain $h_3 = 0.04$. Analytical BER expression predicts that adding this deep fading relay deteriorates the BER performance. Therefore, we discard this relay. For the test case 2, we test and combine with a relay with one scale-lag diversity, power gain $h_3 = 4$, which improves the BER performance, and satisfies the P_e requirement. Therefore, we adopt this relay.

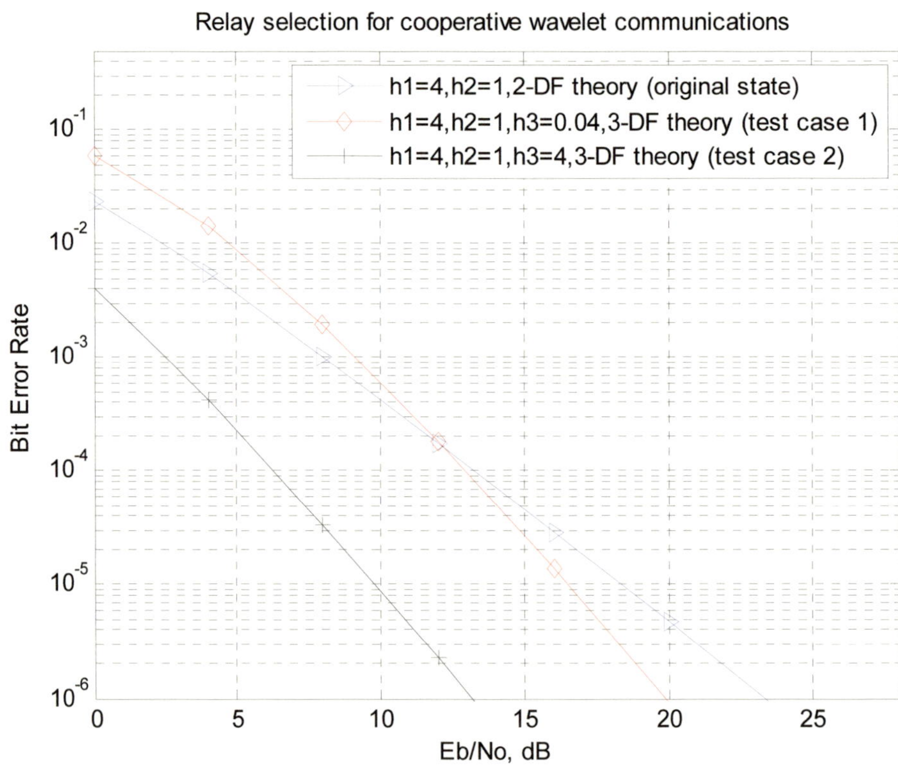

Figure 10. Relay selection for cooperative 2-decomposition level wavelet wideband communication

7. Conclusions

Wideband scale-lag channels can be found in many applications, including ultra-wideband communications and underwater acoustic communications. Signaling and reception schemes using the wavelet theory enable the multi-scale and multi-lag diversity in the wideband system. In this chapter, we designed a cooperative wavelet system to capture the joint cooperative-scale-lag diversity. We proposed the analytical BER expression for the cooperative wavelet wideband communication. The agreement between the analytical curves and numerical simulated results shows that the derived analytical BER expression is suitable for the performance prediction of cooperation wavelet wideband communication. The compact and closed-form BER expression can easily provide an insight into the results as well as a heuristic help for the design of future cooperative wavelet wideband systems. For the suggested cooperative wavelet protocol, we also presented a dynamic optimal selection strategy for the optimal relay selection, which maximizes the whole system transmission performance.

Author details

H. Lu and H. Nikookar
International Research Centre for Telecommunications and Radar (IRCTR),
Dept. EEMCS, Delft University of Technology, Delft,
The Netherlands

T. Xu
Circuits and Systems Group (CAS),
Dept. EEMCS, Delft University of Technology, Delft,
The Netherlands

8. References

[1] Barbarossa S., *Multiantenna Wireless Communication Systems*, Norwood, MA: Artech House, 2005.

[2] Foschini G. J. & Gans M., On the limits of wireless communication in a fading environment when using multiple antennas, *Wireless Personal Commun.*, vol. 6, Mar. 1998, pp. 311-335.

[3] Laneman J. N., Cooperative diversity in wireless networks: algorithm and architectures, *Ph.D. dissertation*, MIT. Cambridge, MA, Sep. 2002.

[4] Sendonaris A., Erkip E., & Aazhang B., User cooperation diversity—Part I: System description, *IEEE Trans. Commun.*, vol. 51, no. 11, Nov. 2003, pp. 1927–1938.

[5] Sendonaris A., Erkip E. & Aazhang B., User cooperation diversity—Part II: Implementation aspects and performance analysis, *IEEE Trans. Commun.*, vol. 51, no. 11, Nov. 2003, pp. 1939–1948.

[6] Liu K. J. R., Sadek A. K., Su W. & Kwasinski A., *Cooperative Communications and Networks*, Cambridge University Press, 2009

[7] Federal Communications Commission, Revision of part 15 of the commission's rules regarding ultra-wideband transmission systems, First report and order, ET Docket 98-153, FCC 02-48, Feb. 14, 2002, pp. 1–118.

[8] Sayeed A. M. & Aazhang B., Joint multipath-Doppler diversity in mobile wireless communications. *IEEE T. Commun.*, vol. 47, no. 1, 1999, pp. 123–132.

[9] Rickard S., Balan R., Poor V. & Verdu S., Canonical time-frequency, time-scale, and frequency-scale representations of time-varying channels. *J. Comm. Infor. Syst.*, vol. 5, no. 5, 2005, pp. 1–30.

[10] Jiang Y. & Papandreou-Suppappola A., Discrete time-scale characterization of wideband time-varying systems. *IEEE T. Signal. Proces.*, vol. 54, no. 4, 2006, pp. 1364–1375.

[11] Margetts A. R., Schniter P., and Swami A., Joint scale-lag diversity in wideband mobile direct sequence spread spectrum systems. *IEEE T. Wirel. Commun.*, vol. 6, no. 12, 2007, pp. 4308–4319.

[12] Xu T., Leus G. & Mitra U., Othogonal wavelet division multiplexing for wideband time-varying channels. *Proc. of IEEE ICASSP*, Prague (Czech Republic), May 2011, pp. 3556-3559.

[13] Leus G., Xu T. & Mitra U., Block Transmission over Multi-Scale Multi-Lag Wireless Channels. *Proc. of the Asilomar Conference on Signals, Systems, and Computers*, Pacific Grove, California, USA, Nov. 2010.

[14] Rickard S., Time-frequency and time-scale representations of doubly spread channels, *Ph.D. dissertation*, Applied and Computational Mathematics Dept., Princeton Univ., Princeton, NJ, Nov. 2003

[15] Bircan A., Tekinay S. & Akansu A., Time-frequency and time-scale representation of wireless communication channels, *Proc. of IEEE Int. Symp. Time-Frequency/Time-Scale Analysis*, Pittsburgh, PA, Oct. 1998, pp. 373–376

[16] Capoglu I. R., Li Y. & Swami A., Effect of doppler spread in OFDM based UWB systems, *IEEE Transactions on Wireless Communictations*, vol. 4, no. 5, Sep. 2005, pp. 2559–2567.

[17] Zhang H., Fan H. H. & Lindsey A., A wavelet packet based model for time-varying wireless communication channels, *Proc. of IEEE Workshop Signal Processing AdvancesWireless Communications*, Mar. 2001, pp. 50–53.

[18] Johnson M., Freitag L. & Stojanovic M., Improved Doppler tracking and correction for underwater acoustic communications, *Proc. of IEEE Int. Conf. Acoustic, Speech, Signal Processing*, vol. 1, Germany, Apr. 1997, pp. 575–578

[19] Sharif B. S., Neasham J., Hinton O. R. & Adams A. E., A computationally efficient Doppler compensation system for underwater acoustic communications, *IEEE J. Oceanic Eng.*, vol. 25, no. 1, Jan. 2000, pp. 52–61.

[20] Mitra U. & Leus G., Equalizers for multi-scale / multi-lag wireless channels, *Proc. of IEEE Global Telecommunications Conference, 2010.* pp. 1–5.

[21] Acosta G., Tokuda K. & Ingram M. A., Measured joint Doppler-delay power profiles for vehicle-to-vehicle communications at 2.4 GHz, *Proc. of IEEE Global Telecommunications Conference, 2004.* pp. 3813–3817.

[22] Acosta G. & Ingram M. A., Six time- and frequency-selective empirical channel models for vehicular wireless LANs, *Proc. of 1st IEEE International Symposium on Wireless Vehicular Communications*, September 2007.

[23] Blahut R. E., Miller W. Jr. & Wilcox C. H., Eds., *Radar and Sonar, Part I, IMA Volumes in Mathematics and its Applications*. New York: Springer-Verlag, 1991.
[24] Weiss L. G., Wavelets and wideband correlation processing, *IEEE Signal Process. Mag.*, vol. 11, no. 1, Jan. 1994, pp. 13–32.
[25] Young R. K., *Wavelet Theory and Its Applications*. Norwell, MA: Kluwer, 1993.
[26] Tsatsanis M. K. & Giannakis G. B., Time-varying system identification and model validation using wavelets, *IIEEE Trans. Signal Process.*, vol. 41, no. 12, Dec. 1993, pp. 3512–3523.
[27] Iverson D. E., Coherent processing of ultra-wide-band radar signals, *Proc. Inst. Elect. Eng. Radar, Sonar, Navigation*, vol. 141, Jun. 1994, pp. 171–179.
[28] Sharif B. S., Neasham J., Hinton O. R. & Adams A. E., A computationally efficient Doppler compensation system for underwater acoustic communications, *IEEE J. Oceanic Eng.*, vol. 25, no. 1, Jan. 2000, pp. 52–61.
[29] Wornell G.W., Emerging applications of multirate signal processing and wavelets in digital communications, *Proc. of IEEE*, vol. 84, Apr. 1996, pp. 586–603.
[30] Leus G. & van Walree P., Multiband OFDM for covert acoustic communications, *IEEE J. Sel. Area. Comm.*, vol. 26, no. 9, 2008, pp. 1662–1673.
[31] Li B., Zhou S., Stojanovic M., Freitag L. & Willett P., Multicarrier communication over underwater acoustic channels with nonuniform doppler shifts, *IEEE J. Oceanic. Eng.*, vol. 33, no. 2, 2008, pp. 198–209.
[32] Martone M., Wavelet-based separating kernels for array processing of cellular ds/cdma signals in fast fading, *IEEE T. Commun.*, vol. 48, no. 6, 2000, pp. 979–995.
[33] Mitra U. & Leus G., Equalizers for multi-scale / multi-lag wireless channels, *Proc. of IEEE Global Telecommunications Conference, 2010.* pp. 1–5.
[34] Yu L. & White L. B., Optimum receiver design for broadband Doppler compensation in multipath/doppler channels with rational orthogonal wavelet signaling, *IEEE T. Signal. Proces.*, vol. 55, no. 8, 2007, pp. 4091–4103.
[35] Linfoot S. L., Ibrahim M. K. & Al-Akaidi M. M., Orthogonal wavelet division multiplex: An alternative to ofdm, *IEEE T Consum. Electr.*, vol. 53, no. 2, 2007, pp. 278–284.
[36] Lu H., Nikookar H. & Lian X., Performance evaluation of hybrid DF-AF OFDM cooperation in Rayleigh Channel, *Proc. of European Wireless Technology Conference*, Sep. 2010, Paris, France.
[37] Cover T. M. & El Gamal, A. A. Capacity theorems for the relay channel, *IEEE Trans. Inform. Theory*, vol. 25, no. 5, Sep. 1979, pp. 572-584.
[38] Farhadi G. & Beaulieu N. C., On the Ergodic Capacity of Wireless Relaying System over Rayleigh Fading Channels, *IEEE Trans. Wireless Communications*, vol. 7, no. 11, Nov. 2008, pp. 4462-4467.
[39] Lin S. & Constello D. J., *Error Control Coding: Fundamentals and Applications*. Englewood Cliffs, NJ: Prentice–Hall, 1983.

[40] Lu H., Xu T., Lakshmanan M. and Nikookar H., Cooperative wavelet communication for multi-relay, multi-scale and multi-lag wireless channels, *Proc. of IEEE Vehicular Technology Conference (VTC)*, Budapest, Hungary, May, 2011, pp. 1-5.

Chapter 7

Ultra-Wideband Antenna and Design

Xian Ling Liang

Additional information is available at the end of the chapter

http://dx.doi.org/10.5772/47805

1. Introduction

Ultra-wideband (UWB) antennas are gaining prominence and becoming very attractive in modern and future wireless communication systems, mainly due to two factors. Firstly, people increasingly high demand for the wireless transmission rate and UWB properties such as high data rate, low power consumption and low cost, which give a huge boost to the UWB antennas' research and development in industry and academia since the Federal Communications Commission (FCC) officially released the regulation for UWB technology in 2002. Secondly, now the wireless portable device need antenna operated in different frequencies for various wireless transmission functions, and operation bands and functions are increasing more and more, which may result in challenges in antenna design, such as antenna space limitation, multi antennas interference, and etc. One UWB antenna can be used to replace multi narrow-band antennas, which may effectively reduce the antenna number.

The bandwidth is the antenna operating frequency band within which the antenna performances, such as input impedance, radiation pattern, gain, efficiency, and etc., are desired. The most commonly used definitions for the antenna bandwidth are the fractional bandwidth (for narrow or wideband definition) and the bandwidth ratio (for ultra-wideband definition).

The fractional bandwidth is defined as

$$BW = \frac{f_h - f_l}{f_c} \times 100\% \tag{1}$$

The bandwidth ratio is defined as

$$BW = \frac{f_h}{f_l} : 1 \tag{2}$$

f_l – the lower frequency of the operation band.
f_h – the higher frequency of the operation band.
f_c – the center frequency of the operation band.

2. History of UWB antennas

In 1898, Oliver Lodge [1] firstly introduced the concept of UWB antenna design, such as spherical dipoles, square plate dipoles, triangular or "bow-tie" dipoles, and biconical dipoles. Fig.1 depicts Lodge's biconical antennas which are unmistakenly used in transmit-receive links. After that, a number of types of UWB antennas were developed in the following several years [2-7]. *i.e.*, one of interest rediscoveries of the biconical antenna and conical monopole is done by P. S. Carter in 1939 [2] (See Fig.2). Carter improved upon Lodge's original design by incorporating a tapered feed, one of the key steps towards the design of broadband antennas.

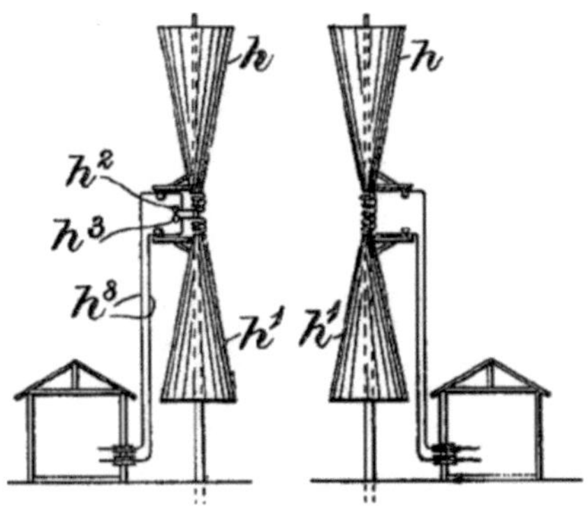

Figure 1. Lodge's biconical antennas (1898).

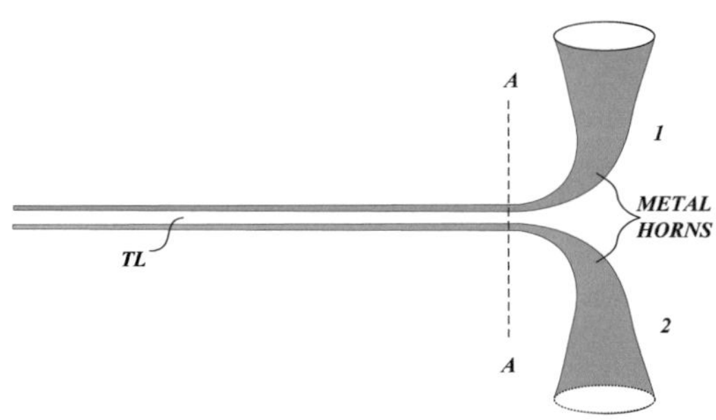

Figure 2. Carter's improved match biconical (1939).

In 1940, S. A. Schelkunoff [3] proposed elaborate conical waveguides and feed structures in conjunction with a spherical dipole (see Fig.3). Unfortunately, his design of the spherical dipole antenna was not very useful. Almost at that time, the most well-known UWB antenna was the coaxial horn proposed by N. E. Lindenblad [4]. In order to make the antenna more broadband, Lindenblad took the design of a sleeve dipole and introduced a continued impedance change, as shown in Fig.4.

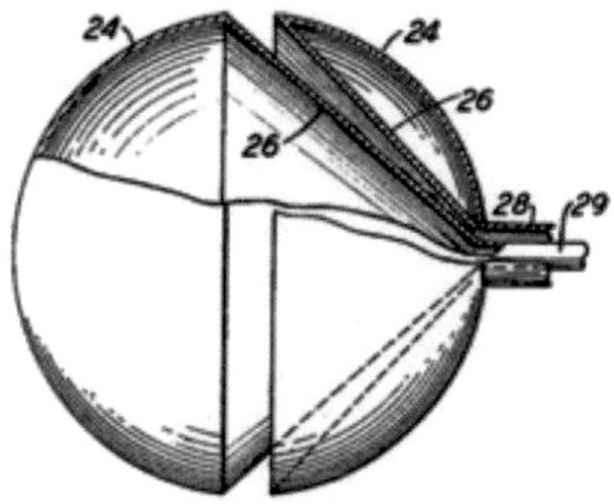

Figure 3. Schelkunoff's spherical dipole (1940)

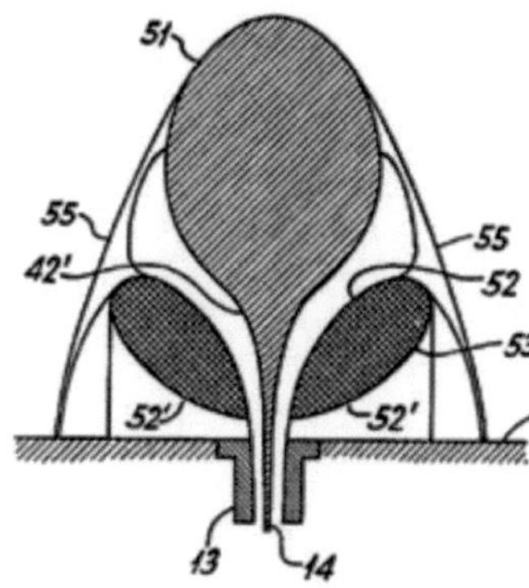

Figure 4. Lindenblad's coaxial horn (1941).

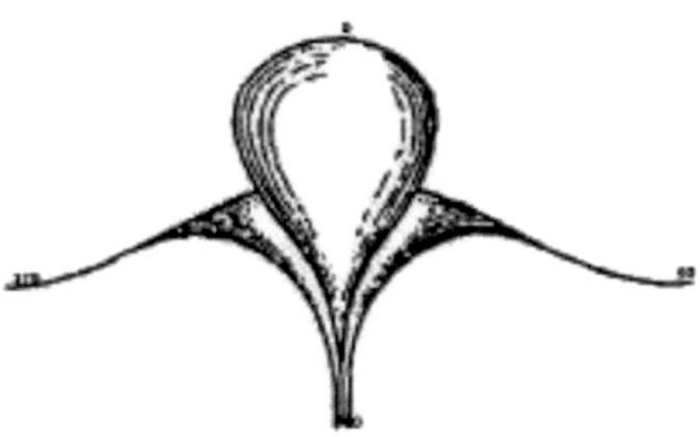

Figure 5. Volcano smoke antennas (1940).

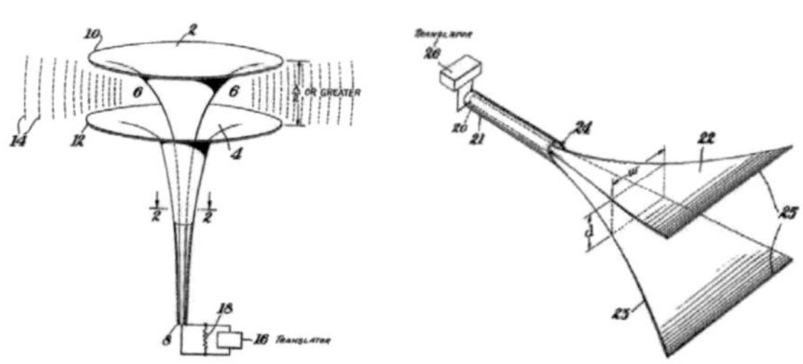

(a) Omni-directional radiation (b) Directional radiation

Figure 6. Brillouin's coaxial horn (1940).

In 1940, J. C. Kraus [5] also developed an antenna similar to the Lindenblad's coaxial horn and named it volcano smoke antenna (See Fig.5), which played a significant role as the cornerstone of television development. Investigations carried out on this antenna showed that this bulbous monopole-like structure yields an impedance bandwidth ratio of 5:1. During that period, coaxial transitions became one of the design techniques for other antenna researchers and designers. In 1948, L. N. Brillouin [6] developed omni-directional and directional coaxial horns, as shown in Fig.6. But these two antennas are difficult to manufacture and use because of their widely structure. Thus, some aspects such as manufacturing cost and complexity of procedures become the important considerations in the design of broadband antennas. The well-known "bow-tie" antenna reveals those benefits, which was originally proposed by Lodge and later rediscovered by G. H. Brown and O. M. Woodward. In 1947, R. W. Masters [7] proposed a similar type of antenna, the inverted triangular dipole, which was later referred to as the "diamond antenna". More recent, other UWB antennas were also developed. W. Stohr [8] introduceed the ellipsoidal monopole and dipole antennas in1968, as shown in Fig. 8. P. J. Gibson proposed the Vivaldi antenna [9] as an amalgamation of slot and Beverage antenna, collectively called tapered slot antenna, in 1979.

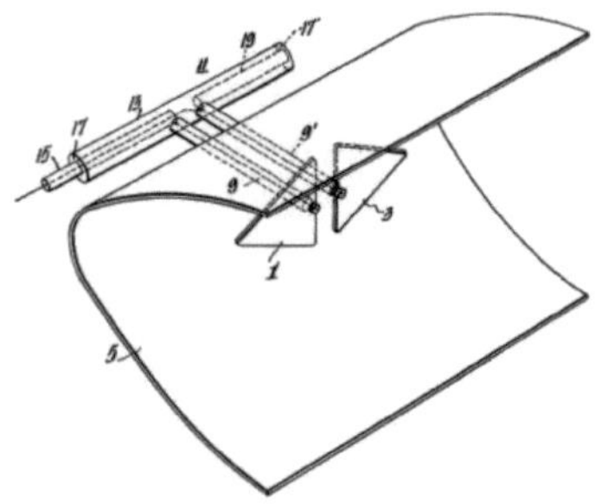

Figure 7. Master's diamond dipole (1947).

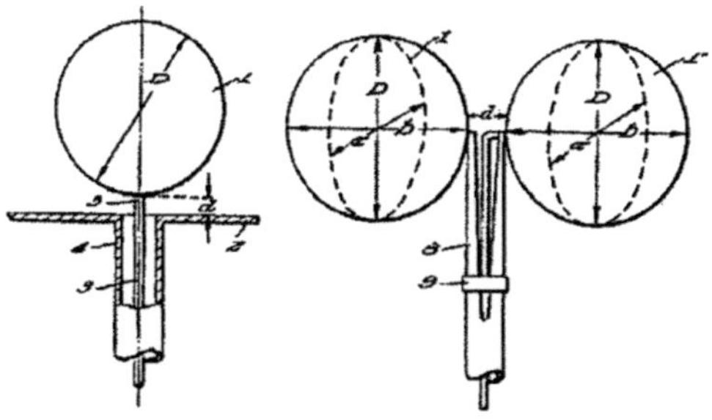

(a) (b)

Figure 8. Stohr's ellipsoidal monople and dipole (1968).

The conventional UWB antennas have been wide used in the broadcast communication applications, but they are not suitable for some high frequency applications in modern and further due to their solid structure and un-integration. In the following sections, some new types of UWB antennas will be introduced for high frequency applications..

3. Omni-directional UWB antenna and design

Along with the wireless system miniaturization and operation frequency increasing, some novel types of onmi-directional UWB antennas have been developed in the last decade. Mainly consisting of two types, the UWB planar monopole antenna and the UWB printed monopole antenna, both types are basically developed from the principles of conventional UWB antennas, such as the biconical antenna, the cone-disc antenna, the cage antenna, and etc. Based on several techniques in terms of bandwidth enhancement, omni-directional radiation improvement and size reduction, they can provide almost the same bandwidth and radiation performances as the conventional UWB antennas but with much smaller volumes.

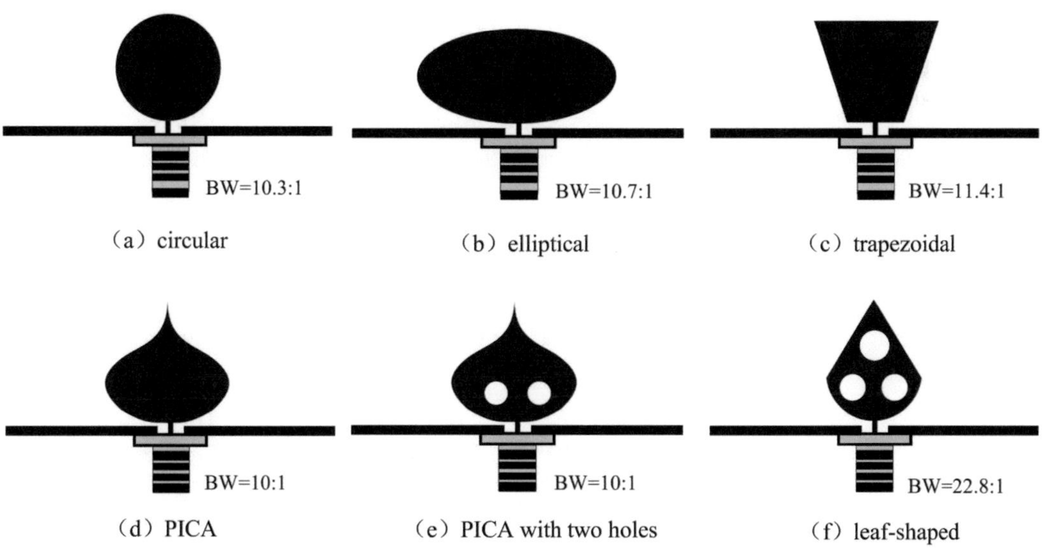

Figure 9. Various geometries of planar monopole antennas [11-14].

3.1. UWB planar monopoles

The planar monopole antenna was firstly reported in 1976 by G. Dubost and S. Zisler [10]. It can be realized by replacing a conventional wire monopole with a planar monopole, where the planar monopole is located above a ground plane and commonly fed using a coaxial probe. Up to now, many planar monopole antennas have been introduced due to their wideband performance. Several representative structures are shown in Fig.9, and these antennas achieve the impedance bandwidth ratio from 2:1 to more than 10:1. *i.e.*, Agrawall *et al.* [11] carried out a bandwidth comparison of several planar monopoles with various geometries, such as circular, elliptical, rectangular, and trapezoidal monopoles. The results show that the circular and elliptical monopoles exhibit much wider bandwidth performance than those of others, and both can obtain the impedance bandwidth ratio exceeding of 10:1 (Circular monopole: 1.17~12 GHz, Elliptical monopole: 1.21~13 GHz). Evans *et al.* [12] proposed a trapezoidal planar monopole antenna above the ground plane, also achieved a bandwidth ratio exceeding of 11:1. Besides the regular structures, Suh *et al.* [13] proposed an interesting structure, the planar inverted cone antenna (PICA), which can provide an impedance bandwidth ratio of more than 10:1 and the pattern bandwidth ratio of about 4:1. To improve the pattern bandwidth ratio, two circular holes are added in the PICA, as shown in Fig.9(e). This alteration improves the radiation pattern performance dramatically without impairing the impedance performance, where the radiation pattern of the two-circular-hole PICA antenna provides a good omnidirectional performance over a bandwidth ratio up to 7:1 and has a very low cross polarization, 20 dB or less. Later, Bai *et al.* [14] presented a modified PICA, where a leaf-shaped metal plate with three circular holes is vertically mounted on the ground plate and is covered by a dielectric plate instead of the conventional metal plate. It achieves the impedance bandwidth ratio better than 20:1, covering the frequency range from 1.3 to 29.7GHz, as shown in Fig.9(f).

Among various planar monopole antennas, the square planar monopole is the simplest in geometry, and its radiation pattern is usually less degraded within the impedance bandwidth. These favourable features attract many studies, mainly on the bandwidth enhancement since the square planar monopole only owns an impedance bandwidth ratio of 2:1. From the antenna geometry, the feed gap, the feed point location and the shape of the monopole's bottom, all may affect the impedance matching. Thus, several techniques such as notching, bevelling, double feed, trident-shaped feed, and etc., were proposed to expand the bandwidth of the square monopole antenna, as shown in Fig.10 *i.e.*, Su *et al.* [15] proposed a method of cutting a pair of notches at the two lower corners of the square planar monopole. With suitable dimensions of the notches chosen, the impedance bandwidth can be greatly enhanced to be about 3 times that of a corresponding simple square planar monopole antenna (2~12.7 GHz compared to 2~4.5 GHz). Antonino-Daviu *et al.* [16] proposed a method of double feed with aims to intense the vertical current distribution and suppress the horizontal distribution in the square planar monopole, which contributes to improvement in the impedance bandwidth and polarization properties. This method achieves an impedance bandwidth ratio of 9.5:1. However, in this double-feed design, an additional feeding network under the ground plane is required to excite the antenna at two separate feeding positions. To avoid using the external feeding network, Wong *et al.* [17] proposed a trident-shaped feeding strip structure. With the use of a trident-shaped feeding strip, the square planar antenna's impedance bandwidth can be enhanced to be larger than 3.5 times that of a simple feeding strip (1.4~11.4 GHz compared to 1.5 ~3.3 GHz). Moreover, hybrid techniques are also proposed, *i.e.*, Thomas *et al.* [18] used a sleeved transmission line as a transformer together with the bevelling technique, the impedance bandwidth ratio reaches about 12:1 (0.5 ~ 6 GHz). Ammann *et al.* [19] adopted a method of combining the bevelling and the shorting techniques, the square planar monopole antenna's bandwidth ratio can be expanded to more than 13:1 (0.8 ~10.5 GHz).

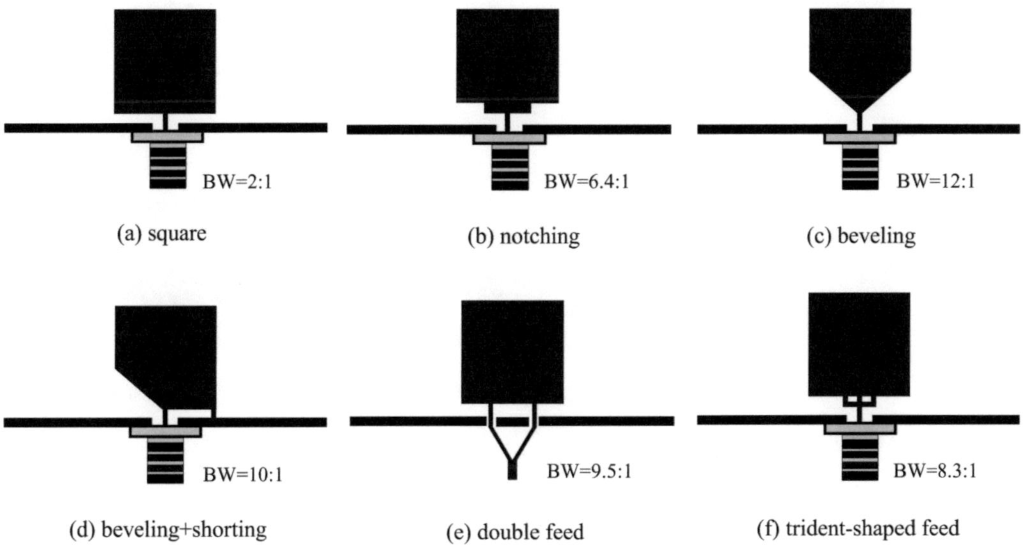

Figure 10. Various wideband techniques for planar square monopole antennas [15-19].

3.2. UWB printed monopoles

The aforementioned planar monopole antennas achieve an ultra-wideband performance based on various techniques, but they all need a perpendicular ground plane, resulting in increasing of the antenna volume and inconvenience for integration with monolithic microwave integrated circuits (MMICs). For the portal wireless device applications, the printed UWB monopole antennas are more popular due to their easier integration than the planar UWB monopole antennas.

The printed UWB monopole antenna commonly consists of a monopole patch and a ground plane. Both of them are printed on the same or opposite side of a substrate, and a microstrip or CPW feedline is used to excite the monopole patch. Since Choi *et al.* [20, 21] introduced this type of antenna with the wideband characteristics in 2004, various printed monopole antennas were studied in the following several years, mainly on the geometries of the monopole and the ground plane.

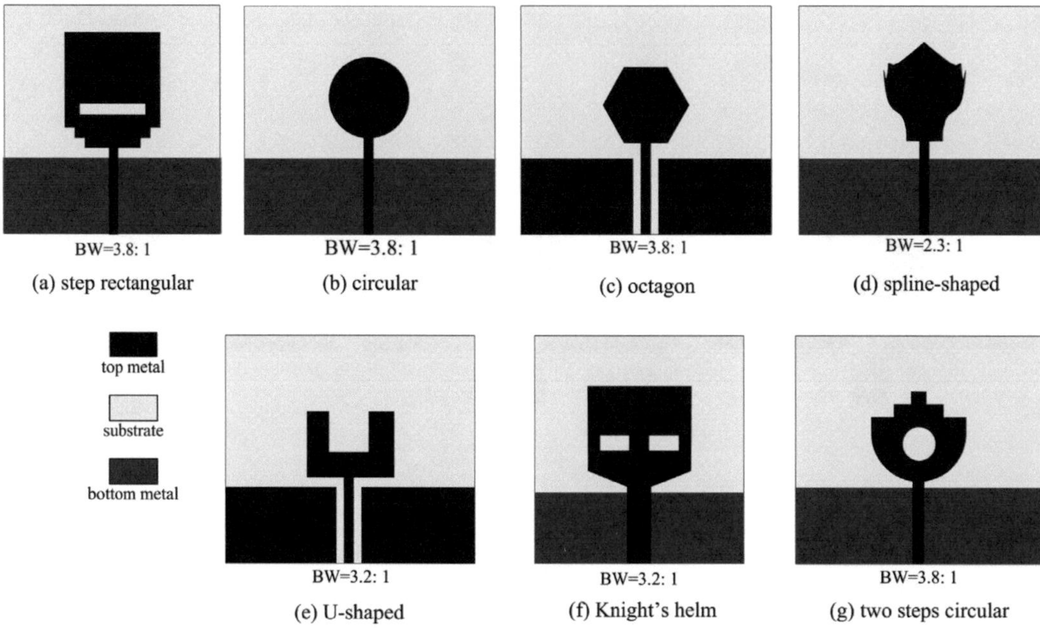

Figure 11. Various monopole structures [21-27].

For geometry of the monopole patch, Fig.11 presents several representative structures. These antennas achieve the impedance bandwidth ratios from 2.3:1 to 3.8:1. Among various geometries of the monopole patches, the printed circular monopole antenna is one of the simplest [22], which achieves the impedance bandwidth ratio of 3.8:1 (2.69~10.16 GHz) with satisfactory omnidirectional radiation properties. Other monopoles such as octagon monopole [23], spline-shaped monopole [24], U-shaped monopole [25], knight's helm shape monopole [26] and two steps circular monopole [27], as shown in Fig.11, were also proposed and studied,. *i.e.*, Ooi *et al.* [23] introduced the two-layer octagon monopole antenna based on the low-temperature co-fired ceramic (LTCC) technique, also obtaining an impedance bandwidth ratio of 3.8:1 (3.76~14.42 GHz). Lizzi *et al.* [24] proposed the spline-shaped monopole UWB antenna able to support multiple mobile wireless standards, covering DCS, PCS, UMTS, and ISM bands, with the bandwidth ratio of 2.3:1 (1.7~2.5 GHz).

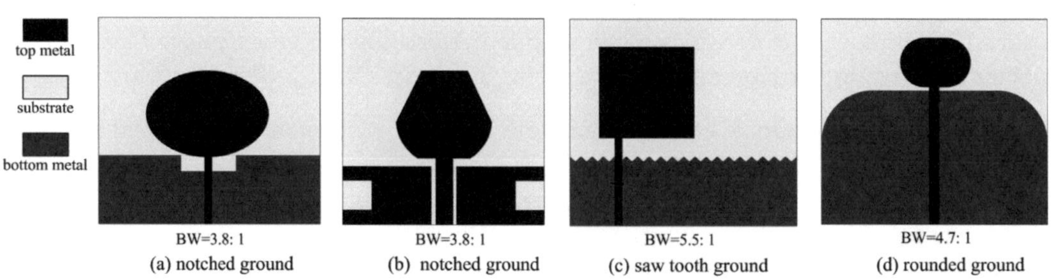

Figure 12. Various ground structures [28-31].

For geometry of the ground plane, several representatives are also shown in Fig.12, and obtain the impedance bandwidth ratios from 3.8:1 to more than 10:1. *i.e.*, Huang *et al.* [28, 29] introduced an impedance matching technique by cutting a notch at the ground plane, and the impedance bandwidth can be enhanced by suitable size and position of notch chosen. Azim *et al.* [30] proposed to improve the impedance bandwidth by cutting triangular shaped slots on the top edge of the ground plane. The printed square monopole antenna with symmetrical saw-tooth ground plane obtains the impedance bandwidth ratio of 5.5:1 (2.9~16GHz). Considering high concentration of currents in the corners of the patch or ground, Melo *et al.* [31] studied a rounded monopole patch with a rounded truncated ground plane. It provides an impedance bandwidth ratio of larger than 4.7:1 (2.55 ~12 GHz).

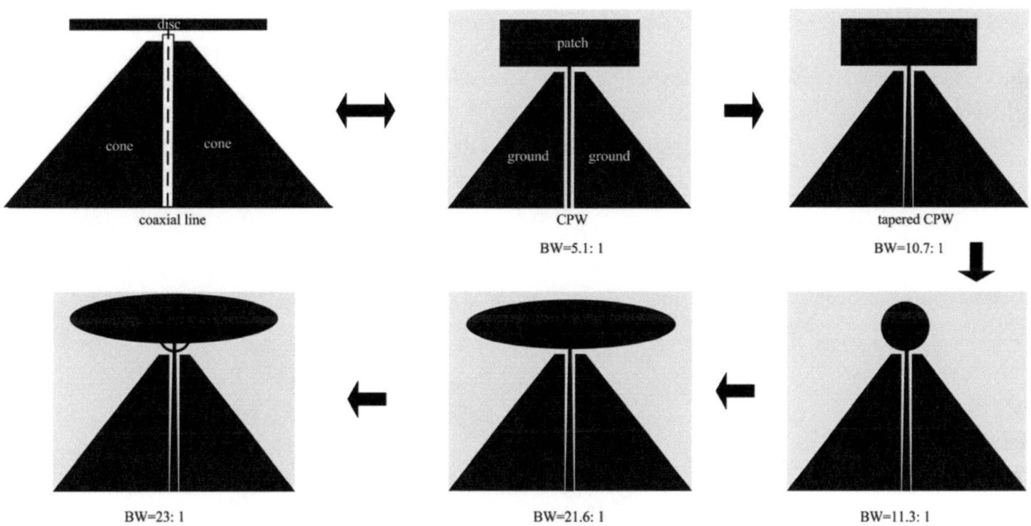

Figure 13. Various printed monopole antennas with trapeziform ground [32-36].

One of interesting UWB printed monopole antenna designs is a trapeziform ground plane with a rectangular patch monopole aroused from the discone antenna, where the rectangular patch is used to replace the disc, the trapeziform ground plane is used to replace the cone, and the CPW is used to replace the coaxial feed, as shown in Fig.13 [32]. It is found that the printed rectangular antenna with a trapeziform ground plane achieves an impedance bandwidth ratio of 5.1:1, which is similar to that of a discone antenna. To enhance the bandwidth further, the input impedance is investigated by comparing bandwidths for various characteristics impedance of CPW feedline. The impedance bandwidth ratio expands to 12:1 when the characteristic impedance of CPW feedline is about 100Ω, which means the impedance bandwidth is enhanced by a factor of about 2.3. In order to match 50Ω SMA or N-type connectors, a linearly tapered central strip line is used as an impedance transformer, and an impedance bandwidth ratio of 10.7:1 (0.76~11 GHz) is obtained. Moreover, various printed monopoles and feed structures are also studied to enhance the bandwidth further [33-36].

In fact, geometries of the monopole and the ground plane not only affect the antenna impedance bandwidth but the antenna radiation pattern and phase center over a wide bandwidth, which is an important phenomenon, especially for pulsed devices that need minimum signal distortion. Thus, the relation between antenna structure and its radiation are also studied. *i.e.*, Fortino *et al.* [37] introduced a CPW-fed printed triangular monopole antenna with a specific ground plane, such as the double folded structure, where the ground plane is optimized to obtain more constant radiation patterns versus frequency. Wu *et al.* [38] also pointed out the ground plane may affect the antenna's H-plane radiation pattern as its width could be comparable with the wavelength at the higher operating frequency. Thus, a compact ground plane with a trident-shaped feed structure is proposed to significantly improve the H-plane radiation pattern in a very wide impedance bandwidth, as shown in Fig.14.

Moreover, the wireless portable devices become smaller and smaller, which contributes the printed UWB antenna design on miniaturization. One of creative miniaturization techniques is proposed by Sun *et al.* [39], where a 40% size reduction is realized by simply exploiting its structural symmetry, as shown in Fig.15. It is found that the miniaturized bevelled planar monopole antenna exhibits a much wider impedance bandwidth, higher cross-polar radiation, and slightly lower gain at higher frequencies as compared with its un-miniaturized counterpart.

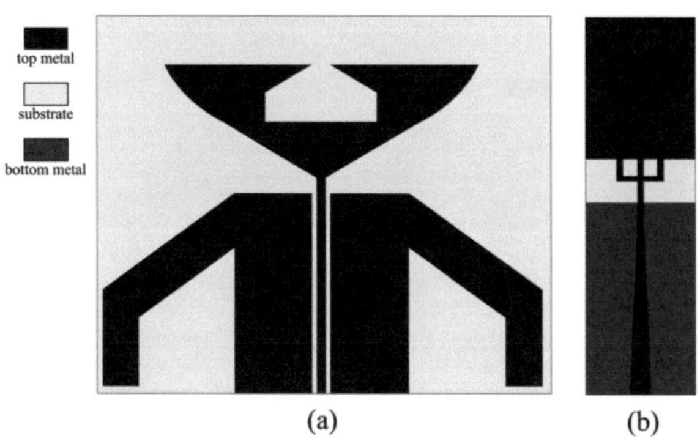

Figure 14. Radiation improvement techniques [37, 38].

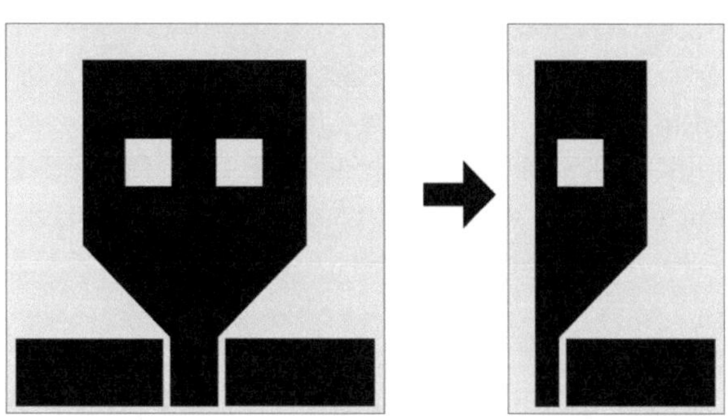

Figure 15. Miniaturization techniques [39].

4. Directional UWB antenna and design

Comparing to the omni-directional UWB antenna, the directional UWB antenna with a much higher gain is also required to meet various applications. In this section, several types of directional UWB antenna design such as the UWB printed wide-slot antennas, the UWB DRAs, and the DRAs with radiation reconfiguration, etc., are detail introduced.

4.1. UWB printed wide-slot antenna

The printed wide-slot antenna is another type of the most suitable candidates for UWB applications. This type of antenna is commonly consists of a wide-slot and a tuning stub connected with a microstrip or CPW feedline. Up to now, many wide-slot antennas, including different wide-slots or tuning stubs, have been extensively studied on the antenna operation bandwidth. Among various shapes of slot, the rectangular wide-slot is the simplest structure. Based on the rectangular wide-slot, several shapes of the tuning stub are studied, where the representative geometries are shown in Fig.16 and the bandwidth ratios vary from 1.8:1 to 3.6:1. *i.e.*, Jang *et al.* [40, 41] presented two rectangular wide-slot antennas fed by the microstrip line with a cross-shaped stub and a Π-shaped stub, respectively. The antenna bandwidth greatly depends on the length of the horizontal and vertical feed lines as well as on the offset position of the feedline. The two slot antennas achieve the impedance bandwidth ratios of 2.8:1 (1.7~ 4.9 GHz) and 3.5:1 (1.7~ 6.0 GHz), respectively. Later, Yao *et al.* [42] proposed a fan-shaped microstrip stub together with a strip, which contributes to a little wider with the bandwidth ratio of 3.6:1 (0.5~ 5.7 GHz). A rectangular slot with a rectangular tuning stub was also studied [43], but with the bandwidth ratio of 1.8:1.

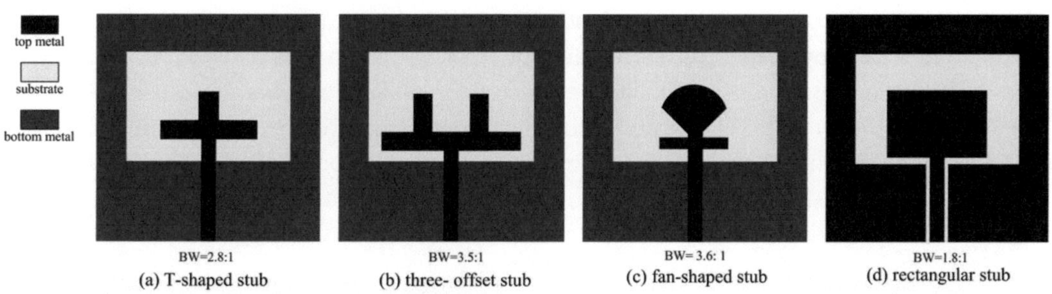

Figure 16. Various shapes of tuning stubs [40-43].

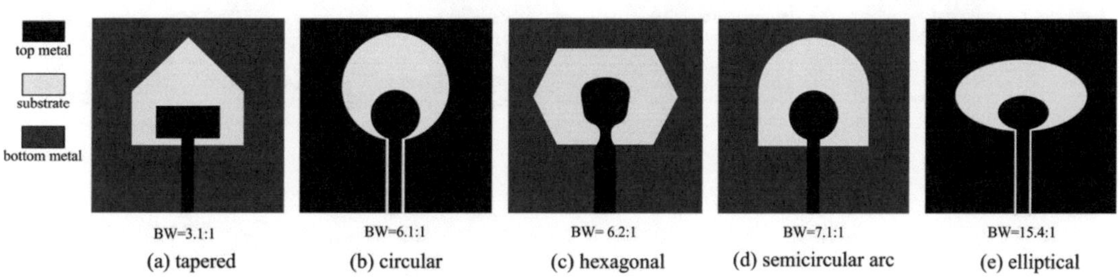

Figure 17. Various shapes of wide-slots [44-48].

It is also noted that the slot shape plays more important on affecting the antenna bandwidth compared to the tuning stub shape. Fig.17 gives several shapes of wide-slots, such as the tapered slot, the circular slot, the hexagonal slot, and etc. These antenna can provide the impedance than ratios from 3.1:1 to 15.4:1, which are much wider bandwidth those of rectangular wide-slot antennas. *i.e.*, Azim *et al.* [44] presented a tapered-shape slot excited by a rectangular tuning stub, which achieves an impedance bandwidth ratio of 3.1:1 (3~11.2 GHz) with a stable gain and radiation over the bandwidth. Denidni *et al.* [45] proposed a circular slot fed by a circular patch through a CPW feedline. This configuration offers a much larger bandwidth ratio of 6.1:1 (2.3~13.9 GHz). Furthermore, Angelopoulos *et al.* [46] investigated the elliptical slot with an elliptical tuning stub achieving an impedance bandwidth ratio of 15.4:1 (1.3~20 GHz), which is the widest bandwidth of the printed slot antennas in open literature.

Different from above regular shapes of the slot or tuning stub, several special geometries of printed slot antennas, such as dual annular slot, semi-elliptic slot, and etc., were also introduced for UWB applications, as shown in Fig.18. These antennas achieve the impedance bandwidth ratios from 3.7:1 to 7:1. For instance, Ma *et al.* [49] introduced a tapered-slot-fed annular slot antenna. The tapered-slot feeding structure serves as an impedance transformer and guides the wave propagating from the slot line to the radiating slot without causing pernicious reflection. The radiating slot is curved to distribute part of the energy to the reverse side of the feeding aperture. This antenna achieves an impedance bandwidth ratio of 3.7:1 (2.95~11 GHz). Gopikrishna *et al.* [50] introduced a semi-elliptic slot antenna. The antenna features a CPW signal strip terminated with a semi-elliptic stub and a modified ground plane to achieve a wide bandwidth ratio of 7:1 (2.85~20 GHz). Pourahmadazar *et al.* [51] studied a special square slot antenna for circular polarization, which is composed of a square ground plane embedded with two unequal-size inverted-L strips around two opposite corners of the square slot. The antenna owns an impedance bandwidth ratio of 4.7:1 (2.67~13 GHz) and a circular polarization bandwidth ratio of 1.5:1 (4.9~6.9 GHz). Sim *et al.* [52] proposed a compact microstrip-fed narrow slot antenna design for UWB applications. By properly loading a notch to the open-ended T-shaped slot and extending a small section to the microstrip feedline, an impedance bandwidth ratio of 3.7:1 (3.1 ~11.45 GHz) is obtained.

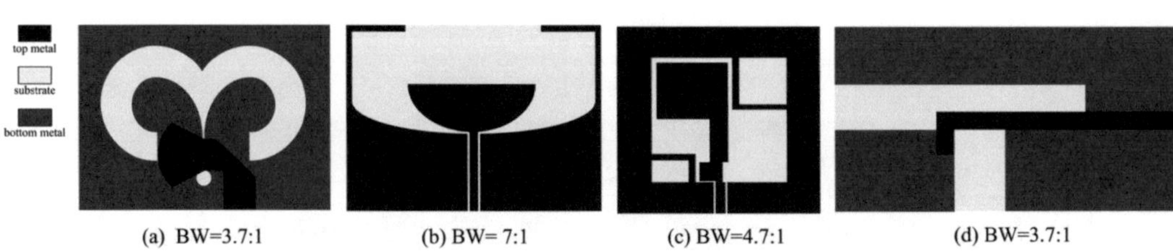

Figure 18. Special geometries of wide-slot antennas [49-52].

Commonly, the operation bandwidth depends on the requirement of the wireless communication system which needs various bandwidths. From above mentioned slot antennas, it is known that the printed wide-slot antenna may achieve a wide range bandwidth based on design of various special slots or stubs. Unfortunately, it maybe wastes a lot of time for antenna designers to find a suitable slot antenna structure according to a required operation bandwidth. So the investigation on the relationship between the slot structure and the bandwidth becomes very useful. For this purpose, reference [53] presented an interesting deep study on printed binomial-curved slot antennas, where the slot and the tuning stub both formed by a binomial curve function, thus various bandwidths can be obtained based on the structure with different binomial curves.

The CPW-fed printed binomial-curved slot antenna is shown in Fig.19. It consists of a wide slot, a tuning stub, and a CPW feedline, all printed on a single-layer metallic substrate of thickness h and relative permittivity ε_r, and with dimension of $L \times W$. The slot's outline size is denoted as $l \times w$, where the coordinate of point **A** on the up-right of the slot is fixed to $(w/2, l)$ and the edge is formed by a binomial curve function, expressed as follows:

$$y = f_1(x) = l \cdot (2x / w)^N, \qquad 0 \leq x \leq w/2 \tag{3}$$

where N is the order of the binomial curve function. The slot is excited by a CPW feedline, where the signal strip width is w_f, and the gap spacing between the signal strip and the coplanar ground plane is g. To achieve an efficient excitation and a wide impedance matching, the signal strip is terminated to a tuning stub with the same shape as the slot but with a smaller size, which has an offset d away from the bottom edge of slot. The stub-to-slot's outline size-ratio is denoted as τ. Therefore, the coordinate of point **B** on the upper-right of the tuning stub is denoted as $(\tau w/2, \tau l+d)$, and the binomial curve function for the edge of the tuning stub can be rewritten as follows:

$$y = f_2(x) = d + \tau l \cdot (2x / \tau w)^N, \qquad 0 \leq x \leq \tau w/2 \tag{4}$$

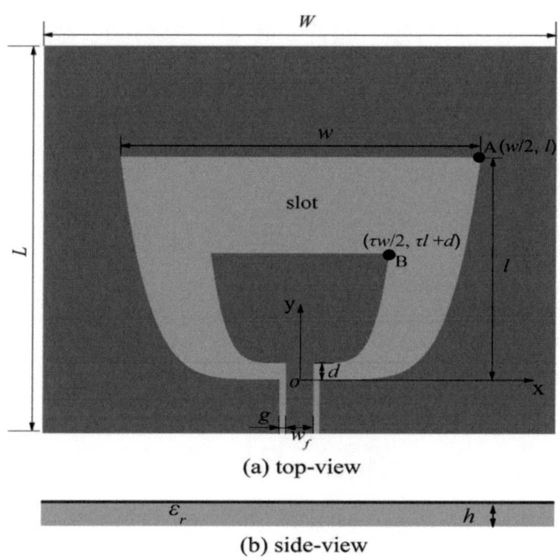

Figure 19. Binomial slot antennas.

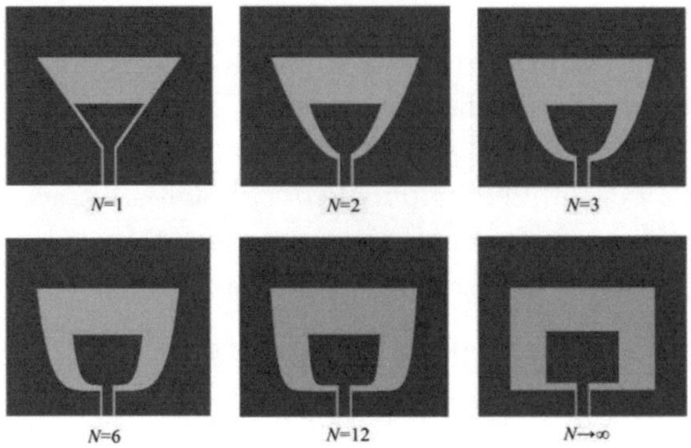

Figure 20. Various binomial slot antennas.

Several shapes for different N are given in Fig.20. As N equals to 1, both the slot and the tuning stub are the triangular shape. As N increases, the bottom widths of the slot and the tuning stub expand gradually, and their shapes look like bowls. As N approaches infinity, both the slot and the tuning stub are transformed to the rectangular shape. From Fig.21, it is found that the operation bandwidth range varies from 10~20 % to 20~40 %, 25~60 %, 60~90 %, 70~110% and 85~110%, as the order N increases from 1 to 2, 3, 6, 12 and ∞, respectively, meaning that the larger order N is selected, the wider bandwidth may be obtained. It is also known that a proffered wideband slot antenna can be easily designed based on suitable selection of parameters, such as w/l, N, according to the operation bandwidth in applications, which is convenient for the wideband antenna design.

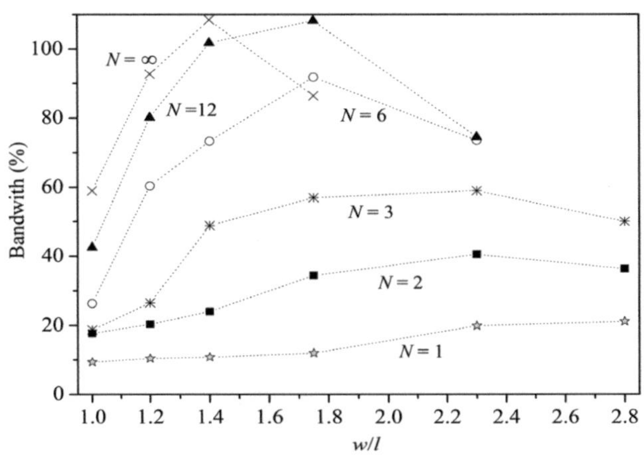

Figure 21. Various bandwidths for orders N.

4.2. UWB dielectric resonator antenna

Dielectric resonator antenna is a new type of directional UWB antenna. It owns a much smaller size and higher efficiency than the UWB printed monopole and wide-slot antennas. Recently, many studies have been proposed to expand the DRA's bandwidth and promote it into the UWB antenna. One of effective methods to expand the DRA's bandwidth is the hybrid technique, which combine the DRA and the monopole antenna. Both antennas provide the similar radiation patterns but with different operation frequency bands. Several representative UWB hybrid DRAs are shown in Fig.22 *i.e.*, Lapierre *et al.* [54] firstly proposed a hybrid antenna design by combining the properties of a quarter wavelength monopole antenna with an annular DRA. This design can be used to retrofit existing monopole antennas: by introducing an appropriate DRA, the original narrow-band monopole can be transformed to achieve an UWB performance, with bandwidth ratio of 2.6:1 (6.5~16.8 GHz). To enhance the antenna bandwidth further, Ruan *et al.* [55] proposed a double annular ring DR with different permittivity, which achieves an impedance bandwidth ratio of 3.7:1 (1.8~6.9 GHz). Later, Jazi *et al.* [56] proposed a skirt monopole antenna used to excite an inverted conical-ring-shape dielectric resonator. In this design, three different methods of impedance matching, dielectric, and ground plane shaping procedures have been applied to increase the antenna bandwidth. The results show that by shaping the dielectric structure and impedance matching method, the input impedance and location of the higher part of the frequency bandwidth can be controlled by exciting higher order mode ($TM_{012+\delta}$) of the same family with dominant resonant mode inside the DR ($TM_{01\delta}$). The lower part of the input impedance bandwidth can be adjusted using the ground plane shaping and matching method at the feed. This antenna achieves the bandwidth ratio of 3.8:1 (1.8~6.9 GHz).

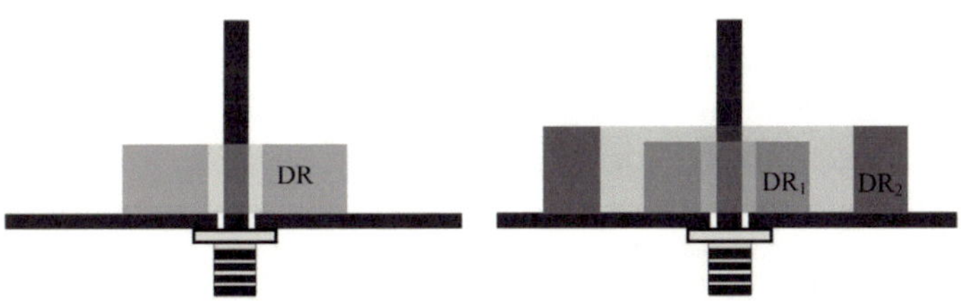

(a) Annular ring DR+monopole (b) Conical-ring DR+monopole (c) Double annular rings DR+monopole

Figure 22. UWB hybrid DRAs [54-56].

(a) L-shaped (b) U-shaped (c) Z-shaped

Figure 23. Photos of various UWB DRAs [58-60].

Apart from the UWB hybrid DRA design, recently, Liang *et al.* [57-61] proposed a patch feed technique and DRAs with various alphabet structures, such as cross-T-shaped [58], L-shaped [57], U-shaped [59], and Z-shaped [60], as shown in Fig.23, where the DRA bandwidth ratios of 2.1:1~9.4:1 were obtained. Fig.24 presents the trapezoidal patch-fed L-shaped DRA design process. To explain the wideband operation of the patch-fed L-shaped DRA, three reference antennas are used as references. Fig.24 (a) is a rectangular DR on a ground plane; Fig.24 (b) is a rectangular DR on a single-sided copper-clad substrate, and Fig.24 (c) is an L-shaped DR on a single-sided copper-clad substrate. The three antennas are excited by the probe feed mechanism and their optimized numerical results in terms of bandwidth are compared with patch-fed L-shaped DRA, as shown in Fig.24. It is observed that the impedance bandwidth can be expanded by using an inserted intermediate substrate, an L-shaped DR, and an inverted-trapezoidal patch feed mechanism. Table 1 lists several proposed UWB DRAs, including the antenna geometry, the feed mechanism and the bandwidth.

The above mentioned UWB antennas could provide the monopole-like radiation or mushroom-like radiation in an ultra-wideband. While some portal UWB wireless devices need both the monopole-like radiation and the mushroom-like radiation since their position are not fixed in communication. For this purpose, Liang *et al.* [62] proposed a UWB DRA with the configurable radiation pattern design, where a rectangular DR excited by dual bevel-rectangular patches. Two bevel-rectangular metal patches are attached on the opposite sides of the DR for excitation and both connect to the 50-ohm microstrip lines, as shown in Fig.25. The bevel-rectangular patch-fed DRA has been proposed to achieve an UWB operation. A reconfigurable radiation pattern performance in terms of the monopole-like radiation and the mushroom-like radiation is obtained through the in-phase feed and the out-phase feed of two input ports, respectively. For the in-phase feed, the antenna performs the monopole-like radiation pattern in the entire operation band. While for the out-phase feed, the antenna performs the mushroom-like radiation in the same operation band.

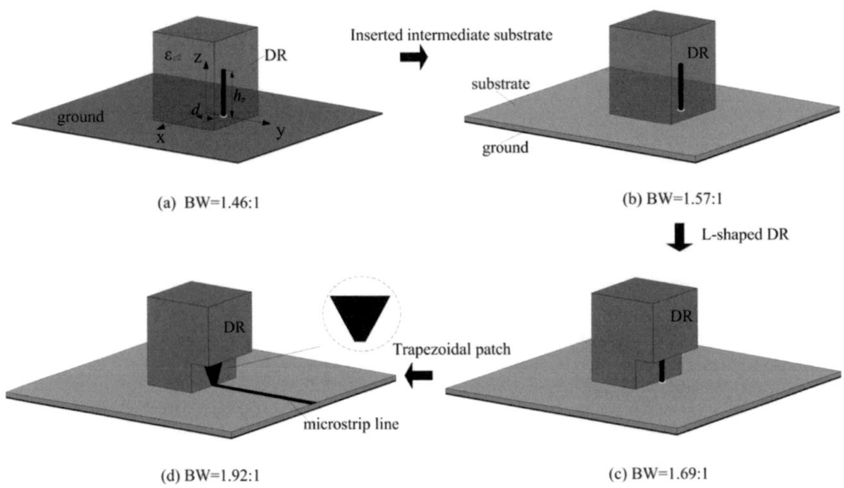

Figure 24. Wideband DRA design.

No	Antenna Geometries	Feed mechanisms	εDR	VSWR ≤ 2	
				Frequency range	Bandwidth ratio
1	Annular ring DR+monopole [54]	probe	10	6.5~16.8 GHz	2.6:1
2	Double annular-ring DR+monopole [55]	probe	4&36	3~11.2 GHz	3.7:1
3	Conical-ring DR + skirt monopole [56]	probe	10	1.8~6.9 GHz	3.8:1
4	L-shaped DR [57]	Trapezoidal patch	9.8	3.87~8.17 GHz	2.1:1
5	Cross-T-shaped DR [58]	Trapezoidal patch	9.8	3.56~7.57 GHz	2.1:1
6	U-shaped DR [59]	Triangle patch	9.8	3.1~7.6 GHz	2.4:1
7	Z-shaped DR [60]	Beveled rectangular patch	9.8	2.5~10.3 GHz	4.1:1
8	Circular DR [61]	Crescent patch	35	1.6~15 GHz	9.4:1
9	Rectangular DR[62]	Bevel-rectangular patch	9.8	3.9~12.2 GHz	3.1:1

Table 1. Various bandwidths of UWB DRAs.

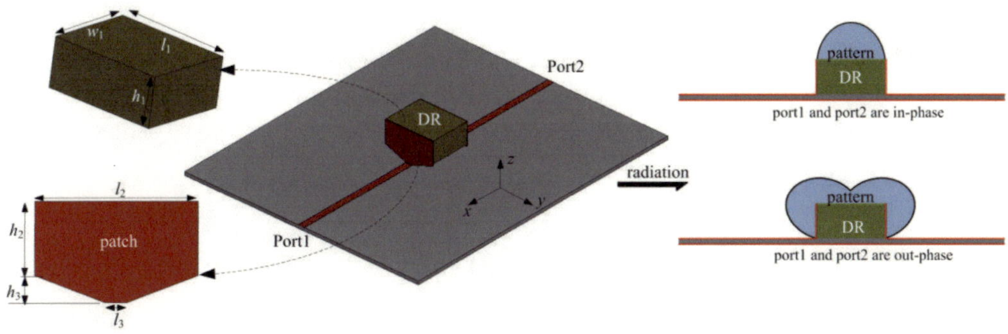

Figure 25. UWB DRAs with a radiation reconfiguration.

5. Band-notched UWB antenna and design

The Federal Communication Commission released the frequency band 3.1~10.6 GHz for the UWB system in 2002. But along with the UWB operating bandwidth, there exist some narrowband wireless services, which occupy some of the frequency bands in the UWB band. The most well-known among them is wireless local area network (WLAN) IEEE802.11a and HIPERLAN/2 WLAN operating in 5.15~5.35 GHz and 5.725~5.825 GHz bands. Apart from WLAN, in some European and Asian countries, world interoperability for microwave access

(WiMAX) service from 3.3 to 3.6 GHz also shares spectrum with the UWB. This may cause interference between the UWB system and other exist communication systems. To address this problem, one way is to use filters to notch out the interfering bands. However, the use of an additional filter will result in increasing the complexity of the UWB system and also the insertion loss, weight and size for the UWB trans-receivers. Therefore, various UWB antennas with notched functions have been researched to overcome this electromagnetic interference. This section concludes the existing band-notched techniques, which can be classified into the following categories: embedding slot, parasitic stub, bandstop transmission line, and hybrid techniques.

5.1. Embedding slot

Among various proposed techniques on the band-notched UWB antenna design. One common and simple way is to etch slots on the radiation patch or ground plane. Up to now, many shapes of embedding slots were studied, and some representatives are shown in Fig.26 *i.e.*, Kim *et al.* [63] proposed a CPW-fed planar UWB antenna with a hexagonal radiating element. By inserting a V-shaped thin slot with a length of $\lambda c/4$ (λc is the wavelength of notched frequency) on the hexagonal radiating element, the narrow frequency band-notched is created, where the fractional bandwidth is approximately 8~10%. Chung *et al.* [64-66] introduced the printed UWB monopole antenna by inserting an inverted U-shaped, Π-shaped or rectangular slot. At the notched frequency, current is concentrated around the edges of the slot and is oppositely directed between the interior and the notched frequency. This leads to the desired high attenuation near the notched frequency. In [67], a band-notched printed monopole antenna is provided by using two modified U-shaped slots on the monopole. The U-shaped slot perturbs the resonant response and also acts as a half-wave resonant structure. At the notched frequency, the desired high attenuation near the notched frequency can be produced. Jiang *et al.* [68, 69] introduced a pair of inverted-L-shaped slots around the microstrip line on the ground plane; a frequency-notched response can also be achieved.

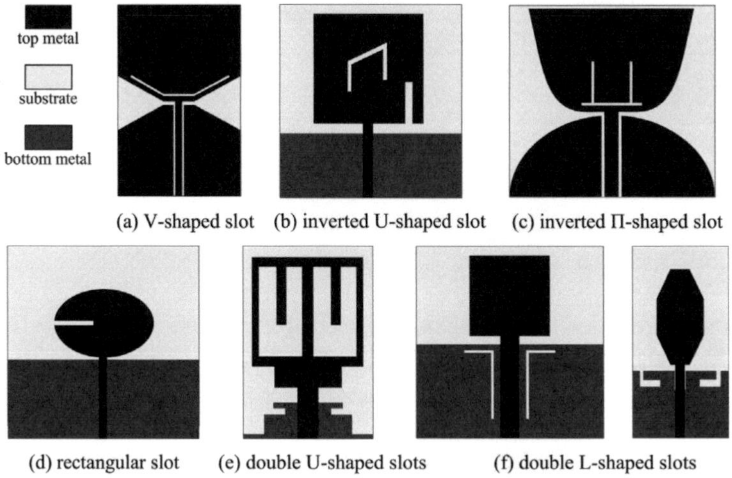

Figure 26. Notched-band designs with various slots on patch or ground [63-69].

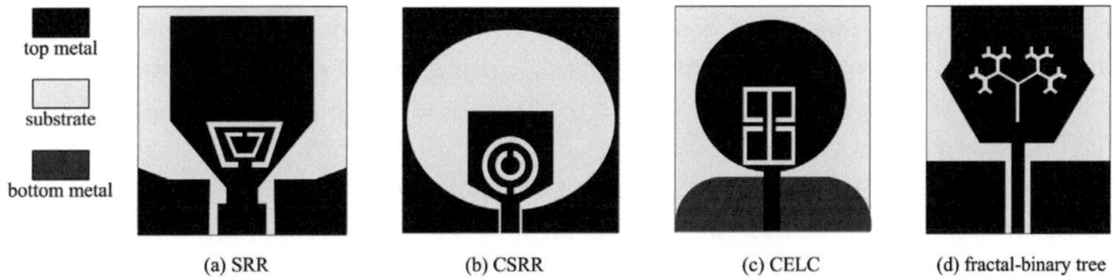

Figure 27. Notched-band designs with various periodic structure slots [70-73].

Since split-ring resonator (SRR), electric-LC (ELC) resonator, complementary split-ring resonator (CSRR) and complementary electric-LC (CELC) resonator are commonly used to design a material with negative permittivity and permeability, all these structures can also be applied in UWB antennas for the notched band design. Several representatives are shown in Fig.27. The SRR is generally composed of two concentric split ring strips. It has a favourable aspect in size since it can be designed as small as one-tenth of the resonance wavelength. In [70], a dual reverse split trapezoid slots, instead of the conventional strip-type SRRs, was proposed and implemented for a bandstop application. In [71], a slot-type CSRR is etched inside the tuning stub of the printed elliptical slot antenna, and implemented for a band-stop application. It was found that an alterable notched band could take place by adjusting the radiuses of the CSRR. In [72], the CELC resonator is etched inside the circular patch of the monopole antenna to achieve the notched frequency band. The CELC could provide a predominantly magnetic response. At the notched frequency, the current flows into the CELC region so that the desired high attenuation near the notched frequency would be produced. In [73], a fractal-binary tree slot embedding technique for the band-notched characteristics design was introduced. By etching a dual band-notched resonance slot using a four-iteration fractal binary tree, two additional filters are applied to the radiating element of the antenna. The fractal, which effect increases the possible length of isolated current paths on the radiating element, has clear and useful properties for band-stop design within small antenna footprints.

5.2. Parasitic stub

Similar to the embedding slot technique in the UWB antenna design, another commonly used technique is a parasitic strip or stub in the aperture area of the antenna or a nearby radiator that forms a resonant structure and leads to a sudden change in the impedance in the notched band. Many parasitic strips or stubs were studied and several representative structures are presented in Fig.28.

For the UWB printed wide-slot antenna design, Liu *et al.* [74] proposed a UWB rectangular slot antenna with a fractal tuning stub to realize the notched function. Chui *et al.* [75] proposed a branch with a length of a quarter of the wavelength adding on the tuning stub to obtain the band-notched property. Cai *et al.* [76] studied a pair of elliptic arc-shaped strips inserted into a suitable aperture region to disturb the field distribution that generates the resonance at the designed notched-band. The total length of this pair of strips is adjusted to about half-wavelength at the desired notched-band, resonance will occur at the strips.

For the printed UWB monopole design, Zhang *et al.* [77] introduced a segmented circular planar monopole antenna with a notched band. Through cutting apart a circular monopole patch with a pair of symmetrical slots, the patch is divided into three segments: the center patch and two side patches. Practically, the side patches function as two parasitic elements and work as bandstop filters. Then, the band-notched property is achieved. Wu *et al.* [78] introduced a square looped resonator and an end-coupled resonator to achieve the gain suppression in the notched band. The square-looped resonator consists of two square loops whose physical length approximates half a wavelength at the notched frequency. Meanwhile, the end-coupled resonator is composed of a strip line with a pair of quarter wavelength folded open stubs. Compared with the band-notched methods using thin slits and plastic strips, this resonator has a small size and a fast rolloff rate well as 10 - 25 dB gain suppression (generally, the gain suppression of thin slits and plastic strips are usually less than 10 dB).

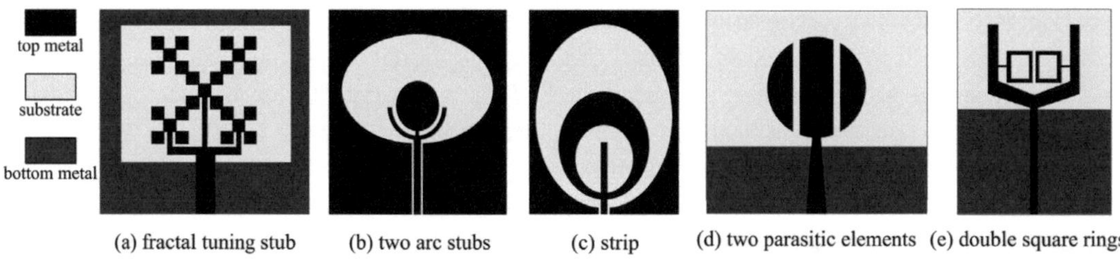

Figure 28. Notched-band designs with various stubs [74-78].

5.3. Bandstop transmission line

The above mentioned notched-band techniques, such as embedding slot or ELC resonator, parasitic stub, will result in affecting the antenna radiation, especially for increasing of the cross-polarization. A transmission line with a bandstop characteristic to feed the UWB antenna can be considered as an integration design of the printed UWB antenna and the filter, which may have little affection to the antenna radiation. Several designs of microstrip feedline with the notched-band function are proposed, as shown in Fig.29 *i.e.*, Zhang *et al.* [79] proposed a U-shaped slot embedded in a microstrip feedline, and a band-notched characteristic was realized. Later, Nouri *et al.* [80] used the defect ground technique to realize the microstrip filter, where a vertical metal strip connected to the rectangular ring is embedded in the shovel-shaped slot that is located under the feedline at the center of ground plane. The notched frequency can be controlled by adjusting the dimensions of the filter structure. Moreover, the electromagnetic band-gap (EBG) structure has a characteristic of preventing wave propagation in special directions or at certain frequencies. In [81], square EBG cells are placed close to the microstrip feedline to obtain the desired notched bands. In [82], the dual band-notched characteristic has been achieved by introducing two open-circuited stubs from two sides of the microstrip feedline. By adjusting the length of two open circuited stubs approximately to one quarter of wavelength, a destructive interference of the current distribution takes place causing the antenna non-radiating at that notched frequency.

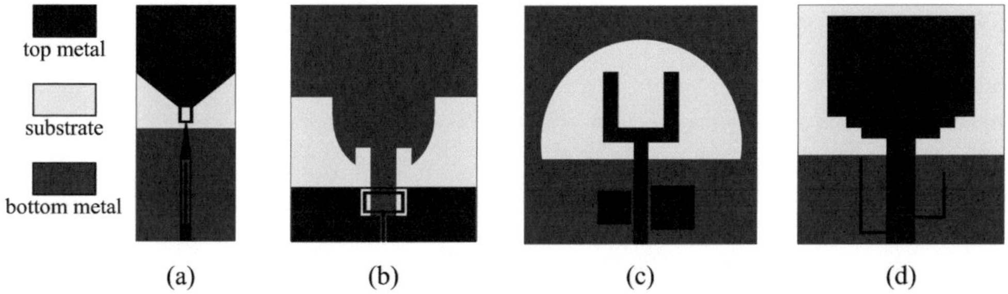

Figure 29. Various bandstop transmission lines [79-82].

5.4. Hybrid techniques

Using one notched-band technique will face two problems. Firstly, it is relatively difficult to create multiple frequency notches with a sharp and narrow stop band. Secondly, multi-notched bands do not have any means to control independently because of the same technique. Therefore, various notched-band techniques have been together used to realize the WiMAX and WLAN bands rejection. The representative hybrid techniques are shown in Fig.30 *i.e.*, Abdollahavand *et al.* [83] studied the hybrid technique by adding parasitic strip and bandstop transmission line. A compound band-notched structure is formed by embedding Γ-shaped stubs in the radiation patch and a modified G-slot defected ground structure in the feeding line, which can provide two filtering frequencies in a certain band and function as a second-order filter. Ye *et al.* [84] studied the hybrid technique of a parasitic strip and a parasitic slit, where the parasitic strip is embedded inside the polygon slot and an isolated slit employed in the beveled T-stub. The desired excellent band-notched UWB operation can be obtained by choosing the sizes of the parasitic strip and slit. Zhou *et al.* [85] presented the hybrid technique of adding parasitic and embedding slot to realize dual notched bands of WiMAX and WLAN. Firstly, the circular patch is cut to an annular ring and a pair of Y-shaped strips is connected to the annular ring, the notched band of WiMAX centered at 3.5 GHz is realized. Then, an inverted V-shaped slot is etched on the patch, a notched band of 5.2 - 5.98 GHz for WLAN band is achieved. Niu *et al.* [86] used the hybrid technique of CCRC resonator and embedding slot, where a CCRC is to realize the 5 GHz WLAN notched-band and an elliptic arc-shaped slot is to realize the WiMAX notched band. Kim *et al.* [87] suggested a triple-band notched hybrid technique, which is based on a geometric combination of a meander shaped stub and the two rectangular complementary split ring resonators (CSRRs) on the feedline, and an inverted U-shaped slot on the center of the patch.

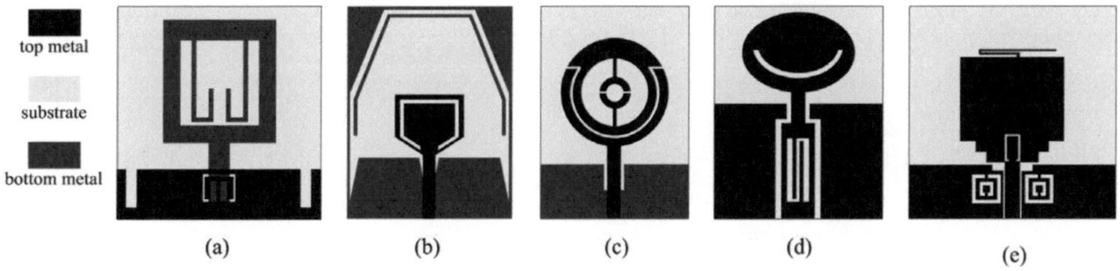

Figure 30. Hybrid notched-band techniques [83-87].

Author details

Xian Ling Liang
Department of Electronic Engineering, University of Shanghai Jiao Tong, P.R. China

6. References

[1] Lodge, Electric telegraphy. U.S. Patent 609,154 (August 16, 1898).
[2] Carter PS. Wideband, short wave antenna and transmission line system. U.S. Patent 2,181, 870 (December 5, 1939).
[3] Schelkunoff SA. Advanced antenna theory. New York: John Wiley and Sons; 1952. p 160.
[4] Lindenblad NE. Wide Band Antenna. U.S. Patent 2,239,724 (April 29, 1941).
[5] Paulsen L, West JB, Perger WF, Kraus J. Recent investigations on the volcano smoke antenna. IEEE Antennas and Propagation International Symposium (Digest), Columbus, OH, Jun. 2003.
[6] Brillouin LN. Broad Band Antenna. U.S. Patent 2,454,766 (November 30, 1948).
[7] Masters RW. Antenna, U.S. Patent 2430353 (November 4, 1947).
[8] Stohr W. Broadband ellipsoidal dipole antenna. U.S. Patent 3,364,491, (January 16, 1968).
[9] Gibson PJ. The Vivaldi aerial, Proc. 9th Europe Microwave Conf., Brighton, U.K. 1979.
[10] Dubost G and Zisler S. Antennas a Large Bande. New York: Masson, 1976.
[11] Agrawall NP, Kumar G, and Ray KP. Wide-band Planar Monopole Antennas. Transactions on Antennas and Propagation 1998 ; 46(2): 294-295.
[12] Evans JA and Ammann MJ. Planar Trapezoidal and pentagonal monopoles with impedance bandwidth in excess of 10:1. IEEE Antennas and Propagation International Symposium (Digest), Orlando, FL; 1999.
[13] Suh SY, Stutaman WL and Davis WA. A new ultrawideband printed monopole antenna: the Planar Inverted Cone Antenna (PICA). IEEE Transactions on Antennas and Propagation 2004; 52(5): 1361-1365.
[14] Bai XF, Zhong SS and Liang XL. Leaf-shaped monopole antenna with extremely wide bandwidth. Microwave and Optical Technology Letters 2006; 48(7): 1247-1250.
[15] Su S, Wong K and Tang C. Ultra-wideband square planar antenna for IEEE 802.16a operating in the 2–11 GHz band. Microwave and Optical Technology Letters, 2004; 42 (6): 463-466.
[16] Antonino-Daviu E, Cabedo-Fabres M, Ferrando-Bataller M and Valero-Nogueira A. Wideband double-fed planar monopole antennas. IEE Electronics Letters 2003; 39(23): 1635-1636.
[17] Wong KL, Wu CH and Su SW. Ultrawide-band square planar metal-plate monopole antenna with a trident-shaped feeding strip. Transactions on Antennas and Propagation 2005; 53(4): 1262-1269.
[18] Thomas KG, Lenin N and Sivaramakrishnan R. Ultrawideband planar disc monopole. IEEE Transactions on Antennas and Propagation 2006; 54 (4): 339-1341.
[19] Ammann MJ and Chen ZN. A wide-band shorted planar monopole with bevel. IEEE Transactions on Antennas and Propagation 2003; 51(4): 901- 903.

[20] Wang W, Zhong SS and Liang XL. A broadband CPW-fed arrow-like printed antenna. IEEE Antennas and Propagation International Symposium (Digest), Monterey, CA, Jul. 2004.
[21] Choi SH, Park JK, Kim SK and Park JY. A new ultra-wideband antenna for UWB applications. Microwave and Optical Technology Letters 2004; 40(5): 399 - 401.
[22] Liang JX, Chiau CC, Chen, XD and Parini CG. Study of a printed circular disc monopole antenna for UWB systems. IEEE Transactions on Antennas and Propagation 2005; 53(11): 3500-3504.
[23] Ooi BL, Zhao G, Leong MS, Chua KM and Lu Albert CW. Wideband LTCC CPW-fed two layered monopole antenna. IEE Electronics Letters 2005; 41(16): 889-890.
[24] Lizzi L, Azaro R, Oliveri G and Massa A., Printed UWB antenna operating for multiple mobile wireless standars. IEEE Antennas and Wireless Propagation Letters 2010; 10: 1429-1432.
[25] Kim JP, Yoon TO, *et al.* Design of an ultra wide-band printed monopole antenna using FDTD and genetic algorithm. IEEE Microwave and Wireless Components Letters 2005; 15(6): 395-397.
[26] Low ZN, Cheong JH and Law CL, Low-cost PCB antenna for UWB applications. IEEE Antennas and Wireless Propagation Letters 2005; 4: 237-239.
[27] Osama A, Sebak AR. A printed monopole antenna with two steps and a circular slot for UWB applications. IEEE Antennas and Wireless Propagation Letters 2008; 7: 411-413.
[28] Huang CY and Hsia WC. Planar elliptical antenna for ultra-wideband communications. IEE Electronics Letters 2005; 41(6): 296-297.
[29] Zhang X, Wu W, Yan ZH, Jiang JB and Song Y. Design of CPW-fed monopole UWB antenna with a novel notched ground. Microwave and Optical Technology Letters 2009; 51(1): 88-91.
[30] Azim R., Islam MT and Misran N, Ground modified double-sided printed compact UWB antenna. IEE Electronics Letters 2011; 47(1): 9-11.
[31] Melo DR, Kawakatsu MN, Nascimento DC and Dmitriev V. A planar monopole UWB antennas with rounded patch and ground plane possessing improved impedance matching. Microwave and Optical Technology Letters 2012; 54(2): 335-338.
[32] Liang XL, Zhong SS and Wang W. Tapered CPW-fed printed monopole antenna. Microwave and Optical Technology Letters 2006; 48(7): 1242-1244.
[33] Liang XL, Zhong SS, Wang W and Yao FW. Printed annular monopole antenna for ultra- wideband applications. IEE Electronics Letters 2006; 41(2): 71-72.
[34] Liang XL, Zhong SS and Wang W. UWB printed circular monopole antenna. Microwave and Optical Technology Letters 2006; 48(8): 1532-1534.
[35] Zhong SS and Liang XL. Compact elliptical monopole antenna with impedance bandwidth in excess of 21:1. IEEE Transactions on Antenna Propagation 2007; 55(11): 3082-3085.
[36] Liu JJ, Zhong SS, Esselle KP. A printed elliptical monopole antenna with modified feeding structure for bandwidth enhancement. IEEE Transactions on Antenna and Propagation 2001; 59(2): 667- 670.
[37] Fortino N, Dauvignac JY, Kossiavas G and Staraj R. Design optimization of UWB printed antenna for ominidirectional pulse radiation. IEEE Transactions on Antenna Propagation 2008; 56(7): 1875-1881.

[38] Wu Q, Jin RH, Geng JP and Ding M. Printed omni-directional UWB monopole antenna with very compact size. IEEE Transactions on Antenna and Propagation 2008; 56(3): 896-899.
[39] Sun M, Zhang YP and Lu YL. Miniaturization of planar monopole antenna for ultrawideband radios. IEEE Transactions on Antenna Propagation 2010; 58(7): 2420- 2425.
[40] Jang YW. Broadband cross-shaped microstrip-fed slot antenna. IEE Electronics Letters 2000; 36(25): 2056-2057.
[41] Jang YW, Experimental study of large bandwidth three-offset microstripline-fed slot antenna. IEEE Microwave and Wireless Components Letters 2011; 11(10): 425-427.
[42] Yao FW, Zhong SS and Liang XL. Wideband slot antenna with a novel microstrip feed. Microwave Optical Technology Letters 2005; 46(3): 275-278
[43] Chen HD, Broadband CPW-fed square slot antennas with a widened tuning stub. IEEE Transactions on Antenna Propagation 2003; 51(8): 1982-1986.
[44] Azim R, Islam MT and Misran N. Compact tapered-shape slot antenna for UWB applications. IEEE Antennas and Wireless Propagation Letters 2011; 10: 1190-1193.
[45] Denidni TA and Habib MA. Broadband printed CPW-fed circular slot antenna. IEE Electronics Letters 2006; 42(3): 135-136.
[46] Evangelos SA, Anastopoulos AZ, *et al.* Circular and elliptical CPW-fed slot and microstrip-fed antennas for ultrawideband applications. IEEE Antennas and Wireless Propagation Letters 2006; 5: 294-297.
[47] Chen D and Cheng C H. A novel ultra-wideband microstrip-line fed wide-slot antenna. Microwave and Optical Technology Letters 2006; 48(4): 776-777.
[48] Ghaderi MR and Mohajeri F. A compact hexagonal wide-slot antenna with microstrip - fed monopole for UWB application. IEEE Antennas and Wireless Propagation Letters 2011; 10: 682-685.
[49] Ma TG and Jeng SK. Planar miniature tapered-slot-fed annular slot antennas for ultra-wideband radios. IEEE Transactions on Antenna Propagation 2005; 53(3): 1194- 1202.
[50] Gopikrishna M, Krishna DD, Anandan CK, *et al.* Design of a compact semi-elliptic monopole slot antenna for UWB systems. IEEE Transactions on Antenna Propagation 2009; 57(6): 1834-1837.
[51] Pourahmadazar J, Ghobadi C, Nourinia J, Felegari N and Shirzad H. Broadband CPW-fed circularly polarized squre slot antenna with inverted-L strips for UWB applications. IEEE Antennas and Wireless Propagation Letters 2011; 10: 369-372.
[52] Sim CYD, Chung WT and Lee CH. Compact slot antenna for UWB applications. IEEE Antennas and Wireless Propagation Letters 2010; 9: 63-66.
[53] Liang XL, Denidni TA, Zhang LN, *et al.* Printed binomial- curved slot antennas for various wideband applications. IEEE Transactions on Microwave Theory and Techniques 2011; 59(4): 1058-1065.
[54] Lapierre M, Antar YMM, Ittipiboon A and Petosa A. Ultra wideband monopole/ dielectric resonator antenna. IEEE Microwave and Wireless Components Letters 2005; 15(1): 7-9.
[55] Ruan YF, Guo YX and Shi XQ. Double annular-ring dielectric resonator antenna for ultra-wideband application. Microwave and Optical Technology Letters 2007; 49(2):362-366.

[56] Jazi MN, Denidni TA. Design and implementation of an ultrawideband hybrid skirt monopole dielectric resonator antenna. IEEE Antennas and Wireless propagation Letters 2008; 7: 493-496.
[57] Liang XL, Denidni TA. Cross-T-shaped dielectric resonator antenna for wideband applications. IEE Electronics Letters 2008; 44(2): 1176-1177.
[58] Liang XL, Denidni TA. Wideband L-shaped dielectric resonator antenna with an inverted-trapezoidal patch feed. IEEE Transaction on Antennas and Propagation 2009; 57(1): 272-274.
[59] Zhang LN, Zhong SS and Liang XL. Wideband U-shaped dielectric resonator antenna fed by triangle patch. Microwave and Optical Technology Letters 2010; 52(11): 2435- 2438.
[60] Denidni TA, Weng ZB andmahmoud NJ. Z-shaped dielectric resonator antenna for ultrawideband application. IEEE Transactions On Antennas and Propagation 2010; 58 (12): 4059-4062.
[61] Weng ZB, Wang XM, Jiao YCand Zhang FS. CPW-fed dielectric resonator antenna for ultra-wideband applications. Microwave and Optical Technology Letters 2010; 52(12): 2709- 2712.
[62] Liang XL, Fu X, Jin RH, Geng JP and Ye S. An ultra-wideband dielectric resonator antenna with pattern reconfiguration, International Workshop on Antenna Technology, Hong Kong, China, March, 2011.
[63] Kim Y and Kwon DH. CPW-fed planar ultra wideband antenna having a frequency band notch function. IEE Electronic Letters 2004; 40(7): 403-405.
[64] Chung K, Kim J and Choi J. Wideband microstrip-fed monopole antenna having frequency band-notch function. IEEE Microwave and Wireless Components Letters 2005; 15(11): 766-768.
[65] Zhao YL, Jiao YC, Zhang G, et al. Compact planar monopole UWB antenna with band-notched characteristic. Microwave and Optical Technology Letters 2008; 50(10):2656-2658.Peng L, Ruan CL and Yin XC. Analysis of the small slot-loaded elliptical patch antenna with a band-notched for UWB applications. Microwave and Optical Letters 2009; 51(4): 973-976.
[66] Ojaroudi M, Ghanbari G, Ojaroudi N. and Ghobadi C., Small square monopole antenna for UWB applications with variable frequency band-notch function. IEEE Antennas and Wireless Propagation Letters 2009; 8: 1061-1064.
[67] Jiang JB, Song Y, Yan ZH, Zhang X andWu W. Band-notched UWB printed antenna with an inverted-L-slotted ground. Microwave and Optical Letters 2009; 51(1): 260-263.
[68] Sim CYD, Chung WT and Lee CH. An octagonal UWB antenna with 5GHz band-notch function. Microwave and Optical Letters 2009; 51(1): 74-78.
[69] Tand T, Lin DB, Liou GH and Horng JH, Miniaturized 5.2GHz notched UWB CPW-fed antenna using dual reverse split trapezoid slots. Microwave and Optical Letters 2009; 50(3): 652-655.
[70] Liu L, Yin YZ, Jie C, Xiong JP and Z Cui. A compact printed antenna using slot-type CSRR for 5.2GHz/5.8GHz band-notched UWB application. Microwave and Optical Letters 2008; 50(12): 3239- 3242.

[71] Movahedinia R and Azarmanesh MN. A novel planar UWB monopole antenna with variable frequency band-notch function based on etched slot-type ELC on the patch. Microwave and Optical Letters 2010; 52(1): 229-232.

[72] Jahromi MN, Falahati A and Edwards RM. Application of fractal binary tree slot to design and construct a dual band-notch CPW-ground-fed ultra-wide band antenna, IET Microwave Antennas and Propagation 2011; 5(12):1424-1430.

[73] Lui WJ, Cheng CH, Cheng Y and Zhu H. Frequency notched ultra-wideband microstrip slot antenna with fractal tuning stub. IEE Electronics Letters 2005; 41(6): 294-296.

[74] Chiu CW and Li CS. A CPW-fed band-notched slot antenna for UWB applications. Microwave and Optical Letters 2009; 51(6): 1587-1592.

[75] Cai YL and Feng ZH. A UWB band-notched antenna with novel branches on the back of substrate. Microwave and Optical Letters 2008; 58(12): 3274-3278.

[76] Zhang K, Li YX and Long YL. Band-notched UWB printed monopole antenna with a novel segmented circular patch. IEEE Antennas and Wireless propagation Letters 2010; 9: 1209-1212.

[77] Wu SJ andTarng JH. Planar band-notch ultra-wideband antenna with square-looped and end-coupled resonator. IET Microwave Antennas and Propagation 2011; 5(10): 1227-1233.

[78] Zhang LN, Zhong SS, Liang XL, Du CZ. Compact omnidirectional band-notch ultra-wideband antenna. IEE Electronic Letters 2009; 45(18): 659-660.

[79] Nouri A and Dadashzadeh GR. A compact UWB band-notched printed monopole antenna with defected ground structure. IEEE Antennas and Wireless propagation Letters 2011; 10: 1178-1181.

[80] Yang Y, Yin YZ, Sun AF and Jing SH. Design of a UWB wide-slot antenna with 5.2-/5.8-GHz dual notched bands using modified electromagnetic band-gap structures. Microwave and Optical Letters 2012; 54(4): 1069-1075.

[81] Panda JR and Kshetrimayum RS. 5.5 GHz dual-band notched UWB printed monopole antenna with two open-circuited stubs in the microstrip feedline. Microwave and Optical Letters 2011; 53(12): 2973- 2978.

[82] Abdollahavand M, Dadashzadeh G and Mostaga M. Compact dual band-notched printed monopole antenna for UWB application. IEEE Antennas and Wireless propagation Letters 2009; 9: 1148-1351.

[83] Ye LH and Chu QX. Improved band-notched UWB slot antenna. IEE Electronic Letters 2009; 45(25): 1283-1285.

[84] Zhou ZL, Li L andHong JS. Compact UWB printed monopole antenna with dual narrow band notches for WiMAX/WLAN bands. IEE Electronic Letters 2011; 47(25): 1111-1112.

[85] Niu SF, Gao GP, Li M, Hu YS and Li BN. Design of a novel elliptical monopole UWB antenna with dual band-notched function. Microwave and Optical Letters 2010; 52(6): 1306-1310.

[86] Kim DO and Kim CY. CPW-fed ultra-wideband antenna with triple-band notch function. IEE Electronic Letters 2010; 46(18): 1246-1248.

Chapter 8

Architecture and Design Procedure of a Generic SWB Antenna with Superb Performances for Tactical Commands and Ubiquitous Communications

D. Tran, N. Haider, S. E. Valavan, I. E. Lager,
A. Szilagyi, O. Yarovyi and L. P. Ligthart

Additional information is available at the end of the chapter

http://dx.doi.org/10.5772/48487

1. Introduction

Ultra-wideband radio technology (UWB-RT) inherited a potential of extremely high rate of data communications, Claude Shannon discovered this in 1948 and derived the later-called-as Shannon-Hartley's channel capacity laws. This famous theoretical law however was not able to substantiate in practice until the development of the sampling oscilloscope by Hewlett-Packard in 1962, which, in accordance with the Nyquist-Shannon sampling theorem, was then capable to reconstruct at-that-time rather large UWB signals (Wilson, 2002).

UWB-RT, thus far, has been around for half a centuries but most research confined only in military applications and systems. The release of the 7.5 GHz of unlicensed spectrum by the US Federal Communications Commission (FCC, 2002) for commercial usages and applications in 2002 sparked a renewed interest in R&D of UWB-RT in industries, universities and governments. Today "ultra-wideband" usually refers UWB-RT where the electronic systems should be able to coexist with other electronic users (FCC, 2004).

UWB-RT and systems are becoming important not only to communications due to its high transmission capacity and speeds, but also gained strong foothold in many applications in the areas of industries, health monitoring, law-enforcement, defense and public security, etc.,.

Today, UWB-RT is not limited to carrier-free signaling but modulated in both analogue and digital domains typically in collision avoidance, medical imaging, security imaging systems, through-wall, ground penetrating and LPI/LPD tactical command radar systems etc.,.

The UWB-RT has established indeed as an inevitable technology in the fabric of our everyday life, however, there remains significant number of challenges for the technology to become ubiquitous, especially in the safety and security issues. Enhanced by the decision on choosing mm-wave-based airport security passenger screening sensors by the Transportation Security Administration, new research directives in public security domain are the search for sensors with higher channel capacity, and screening with higher resolution. To compromise both range and Doppler resolution (Thor, 1962), Sensors which support super-wideband (SWB) signaling could be the solution for the problem at hand. Super-wideband radio technology (SWB-RT) could possibly be a potential approach enables high-resolution sensing in free space and in matter including ground-penetrating radar and through-wall sensing. SWB-RT has unique advantages as compared to narrowband technology, and also comprised all UWB-RT's advanced features but with more channel capacity, higher precision and super resolution in communication, ranging and screening, respectively.

The system's performance and characteristics are heavily dependent on the design of the radiating element. The requirements placed on UWB antennas in terms of *impedance bandwidth, size, phase linearity* and *spectral efficiency* are more demanding than for narrowband antennas (Valderas, 2011).

One of the challenges in the realization of SWB-RT is the development of a suitable antenna that sustains SWB-signaling. To obtain wider bandwidth, several bandwidth enhancement techniques have been studied such as: using log periodic arrays in which the different elements are deduced from an homothetic ratio (Rahim & Gardner, 2004), introducing a capacitive coupling between the radiating element and the ground plane (Rmili & Floc'h, 2008), using microstrip-line feed and notching the ground plane (Tourette et al., 2006), using symmetrical notch in the CPW-feeding (Zhang et al., 2009), asymmetrical feeding by micro-strip line together with reduced ground plane and appropriate gap-patch distance (Karoui et al., 2010), adding T-slots for both patch and feeding strip (Rahayu et al., 2008a), using cross-lot in the truncated circular patch with tapered micro-strip feed line (Kshetrimayum et al., 2008). All these techniques are based on the modification of the surface current distribution to broaden the antenna's impedance bandwidth.

We report here a *generic* SWB antenna architecture (SWBA), whose structure is purposely designed to support the functional section block division design approach (FSD, i.e. dividing the antenna structure into functional sections). The FSD in turn was utilized to accelerate the bandwidth optimization process; the SWBA and FSD together have conclusively enabled the designer to obtain antennas with SWB performances by optimization of just a few most-significant-parameters. It is noted here that in our SWB antenna, hereafter-named prototype 4, modifications were made not only at the feed-section but also the transition section and the radiation-section as well.

Anticipating and combining of all the advantages of its predecessors (prototypes 1, 2, 3), our new SWP-prototype 4 has been designed, fabricated and evaluated. Measurement results revealed that its SWP-performances are superior to other SWP-radiators reported in open-literature.

This chapter is organized as follow:

First, in §2 we discussed briefly the wire-version-antennas that all planar UWB and SWB antennas were derived thereof, the concepts of quasi-electric and quasi-magnetic for planar antennas are typically discussed, also definitions pertaining to qualitatively expressing the SWB antenna's impedance bandwidth were considered. In §3, the SWBA and the FSD are proposed and discussed. In §4, we briefly examine the performance of radiator prototype 1 as a "proof of concept" and provide a methodological procedure for simplifying the multivariate optimization (MVO) process, which were intensively used in the design of other prototypes 2, 3, 4, also parametric investigations and numerical simulations of the proposed SWB radiator are shown in this section. Technical issues related to the practical consideration of the design and fabrication are discussed in §5. Measurements of impedance bandwidth, reception patterns and pulse characteristics in both frequency- and time-domain are reported in §6. Acknowledgements are expressed in §7, and final conclusions are summarized in §8.

2. Fundamentals

In designing of cm-, mw-, and mm-wave antennas and components, comparing with the traditional wire- and waveguide-technologies, the planar technology offered numerous advantages such as planar, light weight, small volume, low profile, low cost, compatible with integrated- or with active-circuits, easy integrate into passive or active phased arrays and communication systems.

The planar technology facilitates the designers much flexibility in creating a myriad of different UWB and SWB antennas, the architectures of these antennas may different but their topology mainly resembles the traditional wire-version of monopole- or dipole-antennas. The nomenclatural names of planar antennas are confusingly taken over from the wire-version with similar topology. Topological similarities may support such borrowed name, nevertheless it is incorrect—as discussed in the next subsections—regarding to electromagnetic-properties point of view. To avoid such inconsistent idealizations new names and definitions, which support both topological and electromagnetic point of views, will be introduced in this section. We briefly start with recalling the traditional wire-version dipole and monopole antennas in §2.1, then the correct nomenclatures—for the planar-version magnetic and electric antennas—are introduced in §2.2, and end with their similarities: quasi-electric and quasi-magnetic antennas in §2.3

2.1. Dipole and monopole antennas

The Dipole antenna is one of the oldest radiators with theoretical expressions for the radiation fields being readily available (Balanis, 1997, p.135). The shapes of their radiation patterns are also well-known [ibis., p.154]. The antenna first used by Hertz in his early RF experiments in the late 19th century, as an example, was a half-wave dipole (Krauss, op. cit.) and the shapes of its 3D-radiation pattern had a similar appearance to a full-doughnut or figure-eight (Balanis op. cit. p.163).

The Monopole antenna is formed by replacing one half of the dipole antenna with the ground plane, when the ground plane is large enough the monopole behaves like the dipole, except that its radiation pattern is just one half of the dipole, its gain approaches twice, while its length is one half of the dipole.

Magnetic dipole and electric dipole are standardized and well documented in [IEEE STD 145-1983, p.11-16]. The terms magnetic antenna and electric antenna were logically defined but occasionally used in literature, the first term used to describe radiators which possess radiation properties resembling those of thin wire loop (Balanis, op. cit., p.217), while the second is for those resembling of thin wire linear antennas.

2.2. Electric and magnetic antennas

Planar UWB and SWB antennas which geometrically resemble its counterpart (wire)-monopole antennas are widely called monopole. However, this topological naming for the planar radiators is incorrect and confused, because the radiation pattern of all the so-called (planar)-monopole antennas have not the shaped of the monopole but of the dipole antenna i.e., having the shape of the full-doughnut.

To avoid ambiguities, formal definitions for planar antennas are hereafter provided.

The definitions planar-magnetic antenna and planar-electric antenna were constructed by means of an analogy to the wire-loop antenna and the wire-electric dipole which are documented in (IEEE std, op. cit.; Schantz et al., 2003, 2004; and Tanyer et al. 2009a), to keep this chapter self-content and avoid cross-reference, we summarize them here:

As a first step, let the base plane *B*, be the plane comprises the antenna's effective radiating/receiving aperture, and let **n** be the unit normal vector to this plane, with reference to Figure 5 , *B* can be assimilated into xOy, while $n = i_z$.

Assume that the field has a transverse electromagnetic (TEM) distribution propagating along the base plane. Then the following cases can be distinguished:

- In the case when the base plane magnetic field $\mathbf{H}(\mathbf{r})$, with $r \in B$, is directed along **n**, the radiator is referred as *magnetic antenna*.
- In the case when the base plane electric field $\mathbf{E}(\mathbf{r})$, with $\mathbf{r} \in B$, is directed along **n**, the radiator is referred as *electric antenna*.

The above definitions are strictly applied to structures that support propagating-and-non-zero TEM- field distributions only, so that the waveguide case is automatically excluded by this TEM regard. We note here that the above definition have not taken in to account the diffraction effects at the edges/vertexes/corners of the metallic/dielectric material that constituent the transmitting/receiving aperture.

2.3. Quasi-electric and quasi-magnetic antennas

Most of the cases, particularly in planar antenna configuration, the topology of the radiating apertures may prevent the above-indicated conditions from being rigorously satisfied. Even in such cases, either one or the other of the two situations may prevail, thus correctly determine the type of the antenna. For instant, a radiator for which the magnetic field strength H(r) or the electric field strength **E**(**r**) is parallel to **n** over most of the effective aperture will be denoted as *quasi-magnetic* antenna, or *quasi-electric* antenna, respectively.

Obviously, planar antennas fed by microstrip-line or co-planar-waveguide can be classified as quasi-electric or quasi-magnetic antennas, respectively.

As will be demonstrated hereby, our prototypes fall in the class of quasi-magnetic antennas, whilst for all patch antennas fed by micro-strip line, as an example, the RAD-NAV antenna (Tran et al., 2010) belongs to the class of quasi-electric antennas.

2.4. Bandwidth definitions

There are several definitions of bandwidth circulated among our antennas and propagation society; those frequently met are octave-, decade-, ratio-, fractional-, percent-, and ratio-bandwidths. The two definitions, that most frequently used, are the percent bandwidth and the ratio bandwidth. They are defined respectively as follows:

$$BW_p \triangleq 100\% \times BW / f_c \tag{1}$$

$$BW \triangleq f_H - f_L \tag{2}$$

$$BW_{UWB} = \begin{cases} BW_{UWB,DARPA} \triangleq BW_p \geq 25\% \\ BW_{UWB,FCC} \quad \triangleq BW_p \geq 20\% \end{cases} \tag{3}$$

$$BW_{R:1} \triangleq BW / f_L, \text{ when } BW_p \geq 100\% \tag{4}$$

Where:
f_H, f_L are the maximum and minimum frequency at -10 dB, respectively.
BW is the nominal bandwidth defined by $BW = f_H - f_L$
f_C is the central frequency defined by $f_C = (f_H + f_L)/2$
B_P is the percent bandwidth and,
$BW_{R:1}$ is the *Ratio bandwidth,* commonly read as *R-over-1 bandwidth,* in which R is the normalized ratio of f_H to f_L defined as $R = f_H/f_L$,($f_L \neq 0$)

The *percent bandwidth* (1) has originally been used to describe the narrow-bandwidth of conventional antennas and microwave-devices. Its usage is quite popular and often considered as a standard in many textbooks, nevertheless, it is mathematically not a solid definition because it possesses a defect when f_L approaching zero. For example, suppose that the nominal bandwidth of antennas #1 is 2GHz (0-2GHz), and antenna #2 is 20GHz (0-

20GHz). It is clearly that the nominal bandwidth *BW* of the second antenna is 10 times wider than the first one; however, formula (1) indicates that both antennas have the same percent bandwidth. Another weak point is the percent bandwidth of formula (1) is always less than or equal to 200% irrespective of how wide the antenna's nominal bandwidth was. Note also that formula (1) is often mistakenly called as *fractional bandwidth,* indeed the formula (1) consolidates its meaning "fractional bandwidth" only when the factor 100% is removed.

Alternatively, the *ratio bandwidth* (2) can also be used for expressing the bandwidth of UWB and SWB antennas and devices. The defect at zero- frequency point still lurks there but the 200%-limit is lifted. The use of the ratio bandwidth is more adequate to envision the wideband characteristics of devices under investigation. We choose for the second formula (2) for describing the bandwidth for the SWB-prototype discussed in this chapter.

How to choose two formulas, although no official consent however, the first formula is often used for cases that the bandwidths are less than 100%, whilst the second is for UWB and SWB antennas/devices.

Traditional communications systems typically used signals having a percent bandwidth of less than 1%, while standard CDMA has an approximately of 2%. Early definition in the radar and communications fields considered signals with percent bandwidth of 25% or greater (measured at the -3 dB points) to be ultra-wideband. The recent FCC regulations (FCC,2004), which will be used as a standard throughout this text, defined UWB devices/signals as having an nominal bandwidth which exceeds 500 MHz or percent bandwidth of over 20%, measured at -10 dB points.

The term SWB has been often used to indicate bandwidth, which is greater than a decade bandwidth. Since the percent bandwidth confused and failed to envision the SWB property adequately as discussed in §2.4, the "ratio bandwidth" is more suitable and often be used for describing bandwidth of 10:1 or larger, we adopt this convention throughout this report.

The proposed antenna possesses

3. UWB/SWB antennas, a review

3.1. Planar dipole antennas

Common and most-used planar dipole antennas are shown in **Figure 1**. They are well-known for their enormous impedance bandwidth (**Figure 2**), and have been used in many applications. However, their radiated/reception patterns, as shown in **Figure 1**, are not stable/usable in the whole of their claimed bandwidth, so the naming "*UWB/SWB*" may be argued, and the needs for true UWB/SWB antennas which consolidate their name in not only impedance-bandwidth, but also in other criteria as important as radiation patterns, gain, phase, group delay, etc.,. The pattern distortions have been studied and clearly pointed out by many several researchers, excellent research-works on this topic are reported, we cataloged here only some of the most pronounced works, to name a few, Marsey(2007), Biscontini(2006), Hayed(2007), Chavka(2006), Ruengwaree(2007), Garbaruk(2008), Welch

(2002). Close inspection the patterns displayed in fig. 1 (and in all referenced work above) we discovered a **remarkable** general property that pattern dispersion became lesser as the antenna's sharp corners becoming more rounded. So our generic radiator

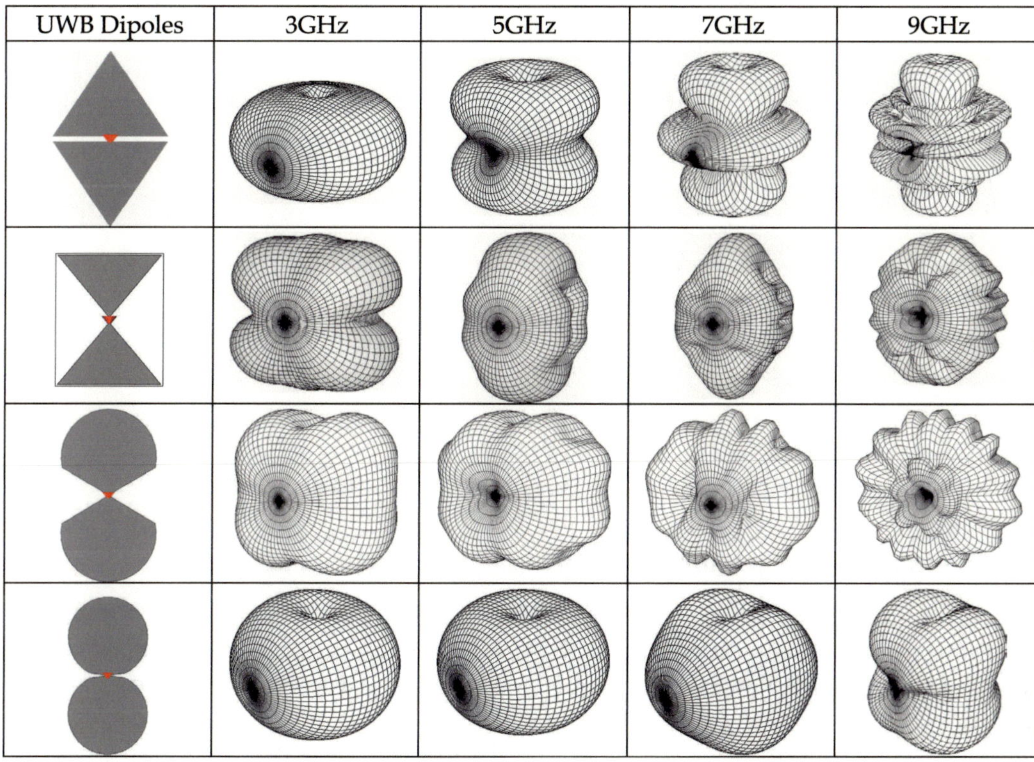

Figure 1. Distortion of radiation patterns of common planar UWB/SWB radiators

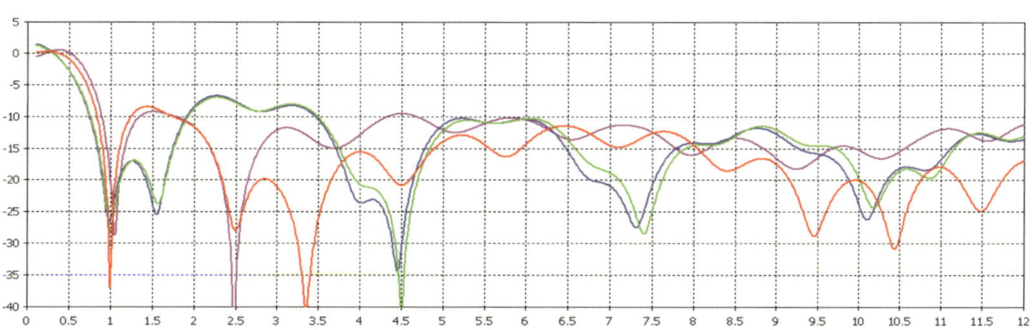

Figure 2. Typical extremely large impedance bandwidth of planar dipole antennas

3.2. Planar monopole antennas

There are countless numbers of UWB/SWB monopole antennas have been developed in the last 20 years, the variety in shapes and architectures vary enormous. **Figure 3** represents the most important monopoles which have been already designed, patented and published in both open and close literature.

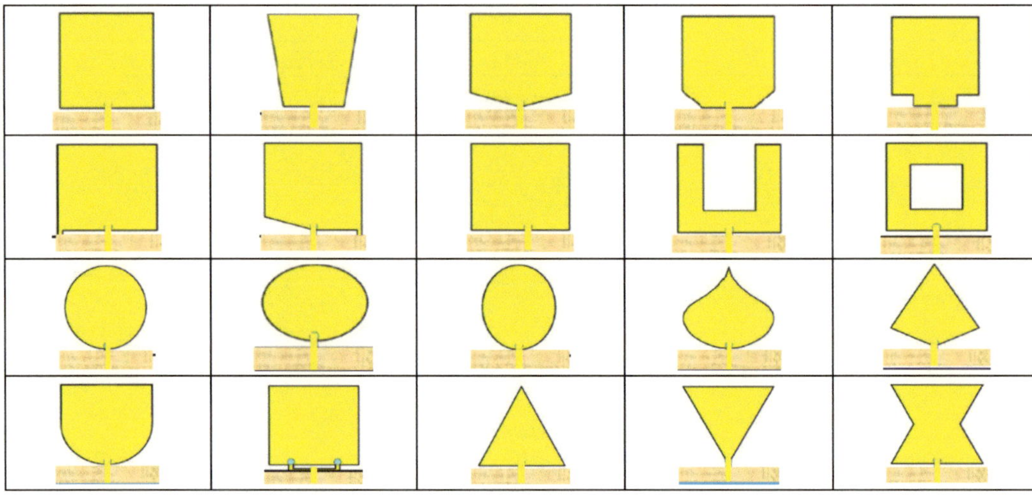

Figure 3. Monopole planar antennas(Courtesy of Dr. S.W Su, Department of EE, NSYSU)

3.3. Antenna miniaturization

The bandwidth and geometrical dimensions have been remarkably reduced, best achievements are timely listed in Table 1, for the used miniaturized concepts and techniques, the readers please consult the references in the listed table.

Antenna	Feeding	Dimension~	Bandwidth	Created by
Slot	Coax-CPW	130x130 mm²	0.5-7 GHz	Yeo, et al., 2004
Monopole	CPW	70x70 mm2	1.5-3 GHz	Chiou et,al, 2003
Pulson 200	Microstrip	40x80 mm²	2.5-6 GHz	Schantz, Time Domain. 2003
Elliptical slot	Coax-CPW	55x50 mm2	3-12 GHz	Ying et al., 2004
Square monopole	Microstrip	50 x50 mm²	3-11 GHz	Battler, M. et all,2006
Monopole	Microstrip	30x40 mm2	2-12 GHz	Xiao et.al., 2009
Monopole	CPW	30x40 mm2	3-12 GHz	Shastry et al., 2009
Elliptical slot	differential	27x46 mm2	3-12 GHz	Power et al., 2004
Vivaldi	Microstrip	35x35 mm²	1-12 GHz	Abbosh et al., 2007
Monopole	Microstrip	30x35 mm²	3-12 GHz	Choi et.al., 2004
Monopole	Microstrip	30x33mm2	3-11 GHz	Kimouche et al., 2009
Monopole	Microstrip	30x32 mm2	2.9-13.2 GHz	Choi et.al., 2009
Monopole	Microstrip	30x30 mm2	3-11 GHz	Rahayu et al., 2008b
Monopole	Microstrip	25x25 mm²	3.2-12 GHz	Cho et al. 2006
Generic	CPW	15x15 mm²	5-150 GHz	This work

Table 1. Most pronounced UWB/SWB monopole antennas

4. Antenna topology, architecture and the FSD methodology

Since the release of the license-free band and the regulation of the emission spectra by the FCC in 2002, a myriad of UWB antennas have been created and invented by both industry and academia, most of them are limited to the FCC-band, this 7.5GHz bandwidth corresponds to a moderate short pulse in order of nanoseconds, these short pulses are good enough for high capacity communication, accurate ranging and imaging but not enough for the more stringent needs of precise localization, high resolution screening, sensitive sensing. To satisfy such stringent requirements, challenges are placed on the design of sensors that support signaling of extreme short pulse in the order of hundreds of picoseconds or less. Sensors in the terahertz region support such short pulse and unarguably provide sharpest images, nonetheless the detection range is too short and the sensors are very costly. Note that in the terahertz region, a radiator with only 5% is capable to support , for example, a Gaussian pulse of 20 ps (assumed unity time bandwidth product), while in the RF-region one must have a SWB radiator of over 11:1 (or 167%, by a lowest frequency of 5GHz) for signaling such a short pulse.

There existed broadside and end-fire UWB antennas with different topologies, which comprised of many configurations are available in open literature. The pattern stability of several antenna's topologies and architectures had been thoroughly investigated and reported by (Massey, 2007, p.163-196). It seemed that there was no broadside antenna architecture could exhibit stable patterns within a bandwidth wider than 10 GHz, and most of them are UWB-radiators with ratio bandwidth much less than 10:1.

We propose here an SWB antenna architecture which possessed not only SWB bandwidth lager than 10:1 but also exhibited a much stable patterns in its SWB bandwidth than all those which have been studied and reviewed by (Massey, op. cit.). The SWB prototype 4, and all other prototypes reported in this chapter had been designed, fabricated and evaluated at our IRCTR.

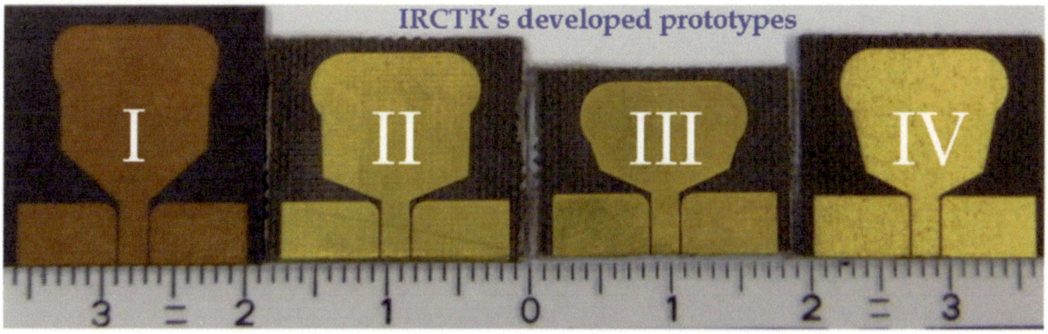

Figure 4. Prototypes developed at IRCTR, resulted design dimensions are listed in table 2.

The SWB radiator reports in this chapter, was indeed a revolutionary improved version from IRCTR's previous developed prototypes. All the developed prototypes, shown together in Figure 4, shared the same topology and architecture as depicted in Figure 5a.

The kernel of this topology is its simplicity, and the essence of the antenna's architecture is the logicalness of dividing the antenna into functional section blocks that enables simplifying the MVO process into a sequence of single-variable optimization (SVO) one.

Simplicity: the topology of the proposed antenna is simple in design with just a copper pattern on top of dielectric layer, the employment of dielectric layer is just for the purpose of structural rigidity of the prototype. In fact, without the dielectric layer, the propose antenna performs much better in terms of matching, and having more perfect symmetric and stable patterns and lower cross-polarization.

Compactness: Thank to the functional section block design (FSD, to be discussed in next section), we are able to miniaturize the generic antenna in an area as small as 15x15mm^2, Table 2 shows comparative indication of miniaturization effectiveness of different proposed architectures.

The merits of the topology, architecture and logical functional blocks, and optimization process will be discussed in §4.1, 4.2 and 4.3, respectively.

The original antenna topology and architecture of prototype 1 leaved many flexible possibilities for adjusting parameters or scaling dimensions to meet new requirements or applications without much entangled in complicated MVO process. These possibilities have been exploited to double the antenna's bandwidth of prototype 1 (Tran et al, 2007) from 2:1 to 4:1 by (Tanyer et al., 2009a), and further broaden to 9:1 (Tanyer et al., 2009b), and scaled down to the FCC-band for IR-applications (Tanyer et al., 2010).

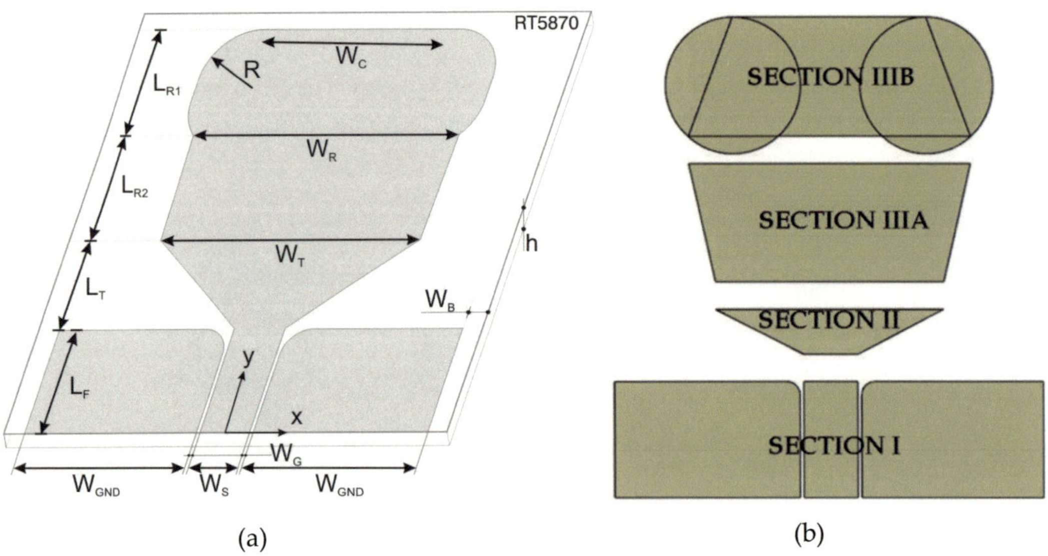

Figure 5. Proposed configuration: a) Architecture and parameters, b) logical functional sections.

The prototype 4 is a proof of concept in proving viability of the design of radiators, which are capable of supporting such extreme short-pulse in RF/mmw-region. The proposed antenna's topology, architecture and the FSD approach, as will be discussed in subsequence sections, were objectively aimed at two main goals:

- Choosing an antenna topology whose shape is kept simple regarding ease of fabrication, prevent diffractions, easy to scale up/down to meet designer's requirements in size and performance (§4.1).
- Creating an architecture that providing a basis platform for implementing the "FSD approach", and deploying the conceptual "thumb-rules" for effective simplifying the MVO process in to SVO, a much simpler one (§4.2).

4.1. Antenna topology

The starting point in the design of the SWB antenna reports in this section is mainly credited to the original radiator (Tran et al., op. cit.), whose topology is planar, as sketched in Fig.5a, with structural topology comprised of following stack-ups:

- Single dielectric layer to provide structural rigidity.
- CPW feeding structure on top of the structure.
- An antenna patch is directly connected to the CPW-feeding, so that they together formed a single planar pattern run on top of the structure.

The CPW feeding structure has been chosen because of it well-behaved properties: such as negligible radiation, low loss, the effective dielectric is constant over a sustained wide frequency range, where the later property is more suitable for super-wide-band feeding a SWB-radiator than the micro-strip line (Simons, 2001).

4.2. Antenna architecture and the FSD approach

In antenna design if dimensions are unconstrained and by a proper design, antenna will behave as a high pass filter, and if its dimensions are physically large enough then all frequencies will pass through. Research of published papers over UWB and SWB antennas reveals that most of the UWB and SWB antennas had a considerable larger size, mostly lager than $\lambda_L/2$ of the lowest frequency f_L, and on broadening the bandwidth, a first option, was to resort to stochastic optimization methods, nevertheless, these methods are known for carrying extremely high computational load.

A more feasible alternative approach was provided by the critical analysis of the relationship between the geometrical parameters and the physics of the problem at hand.

The FSD approach was intentionally created in such a way that the overall dimensions are constrained, kept small and the process of optimization can also be simplified.

The logical architecture together with its parameters and FSD are sketched in Figure 5a&b.

The FSD approach follows the bottom-up strategy, i.e.:

- Starting from the *feed section*—its CPW feed supports the required impedance bandwidth for SWB signaling.

- Then to the *internal transition section*—this section is intentionally inserted between the feed and the patch for the purpose of impedance matching, its shape was logically chosen so that it able to serve the design-properties: resonance shifting, impedance matching, and also enable parameter to serve as independent optimized parameter.
- In addition to the *radiating section*, we divided it in to two sub-sections, which are the *patch section* and the *external transition*. The patch section creates an extra degree of freedom to ease the optimization process, and the radiating section provides parameter for shaping radiation patterns. Two round areas had been added to the top corners of the patch for the purpose of 1) diminishing of diffraction at the antenna's top vertices (deformation of sharp corner in to circular edge) and, 2) controlling and retaining the shaped of the radiation patterns at high frequency band of the spectrum, and, 3) subduing the number of parameters to accelerate the optimization process. These four sections are orderly numerated as I, II, IIIA, and IIIB in Figure 5b.

In summary, the FSD approach assumed the following steps:

- Keep the antenna's overall dimension small and fixed;
- Start orderly from section I, II, IIIA and IIIB (this bottom-up strategy prefers matching impedance bandwidth in prior of pattern bandwidth);
- Separate and understand the role of each section;
- Identify the parameters associate with that section, select the parameter that predominantly influences the function of that section.
- Isolate the effect of that parameter so that an optimization only on that parameter can be undergone, without affecting too much the performance of other sections.

Based on the FSD approach the prototype 1 was first designed and evaluated (Tran et al. 2007). Its designed parameters are used as *start values* for the optimization process of all the later prototypes. Four prototypes (1, 2, 3, 4 shown in fig.4) were successful designed and evaluated, the prototype 4 proved to be a radiator which is superior to the others in that both of its impedance bandwidth and radiation pattern are SWB-sustainable.

4.3. Generic architecture

With the proposed antenna topology and FSD-architectures, we obtained a generic configuration that can be used as a configurative template for finding the perfect antenna configuration (shape + dimension) which fulfills the constraints place on sizes, bandwidth, and dispersion (pattern, gain flatness, linear phase).

Figure 6 shown typically some of the many possible UWB/SWB antenna configurations, which are all generated by varying just some of the design parameters of the same original generic architecture, proposed in Figure 5. Note that all of these antennas are UWB/SWB, and although they are subsets of the generic configuration, but only one third of them worth to be called UWB/SWB antennas when their usable UWB/SWB patterns are considered. It is worthy to note that the antennas (2, 3, 4, 5, 6, 7, 9, 10, 11, 13) generated by our generic architecture have been intended/published/patented by other researchers several years ago.

Our generic architecture generalized, showed the connections between them, and proved that they are just particular cases of the introduced architecture.

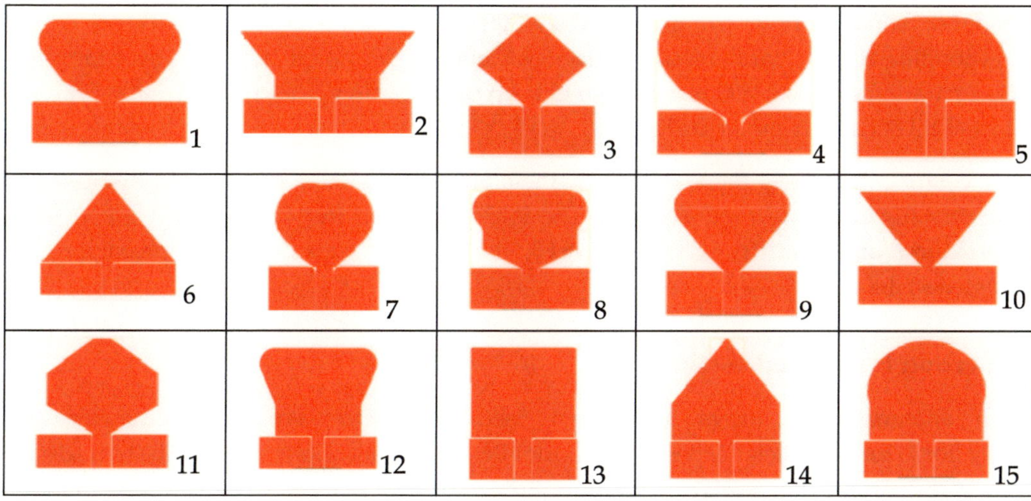

Figure 6. Proposed configuration: a) Architecture and parameters, b) logical functional sections.

4.4. Parameter identification and optimizations

Prototype 1: elaborated by (Tran et al. op. cit.), formed a basis for the designs of the later prototypes. Its 10GHz impedance bandwidth and designed parameters are plotted in fig.9 and listed in table 2, respectively. For the detailed works and the measurement results, the reader should refer to (Tran et al., op. cit). The other prototypes all took the designed parameters of prototype 1 as start-values, and used the FSD to identify the significant parameter for its SVO process of bandwidth broadening.

Prototype 2: is obtained by taking the start values of the prototype 1, and using FSD to identify the LT as its significant parameter for SVO process. The resulting designed parameters and the impedance bandwidth are listed in 2nd column of table 2, and Fig.9, respectively. The SVO demonstrated the solidity of the FSD approach that the bandwidth can be doubled (from 10GH to 20 GHz) by optimization of just a single parameter LT. For details of the elaborated work, please refer to (Tanyer et al., 2009a, op cit.)

Prototype 3: by keeping LT fixed, the prototype 3 is obtained by optimizing the lower part of the radiation section which comprised of two parameters LR2 and WT, the first make the total length of the radiator shorter (thus affects the higher frequency), whist the second parameter provide an better match (i.e., lower reflection coefficient). By doing two SVO sequentially (first with parameters LR2, then WT) we again doubled the nominal BW (form 20GHz to 45GHz), in term of ratio bandwidth it is of BR = 9:1 as shown in Fig. 10. The resulting parameters are listed in column 3rd table 2. Further details of measurements and other properties of this prototype 3, are detailed in (Tanyer et al., 2009b, op. cit).

Prototype 4: is obtained by the combined optimization of the FSD-identified parameters WT and WC, By keeping the optimized parameters of the previous optimization steps (LT of prototype 3, all others of prototype 1) fixed, and doing 2 SVO sequences, we obtain the SWB impedance bandwidth of BR greater than 30:1, the result is plotted in Figure 9, the corresponding design parameters are listed in table 2. We noted here that by controlling radii-separation distance WC we are able to keep the radiation pattern of this prototype less distorted till 50GHz as shown in Figure 8.

The created parameters, their functions, their effects and their usages are discussed in greater details in the next sub-sections. The FSD is detailed in §4.5-4.7, the resulted SWB-performances are given in §4.8 and in §4.9, the optimizations of all the prototypes are discussed in details.

4.5. Section I: The feed section

Planar antennas and arrays have been used for micro-wave and millimeter-wave applications for decades, especially in mobile communications where system design requires low profile, lightweight, and high directivity. The two most used feeding methods are micro-strip line (MS) and coplanar waveguide (CPW), they both carried signal excellently in narrow-band and UWB antennas and devices. Many planar antenna arrays have been designed by using MS, however, until recently, only a few works so far have used CPW to feed the array. The CPW has gained increasing popularity in recent years, since it has several advantages over the MS, such as low radiation losses, less dispersion, easier integration with solid-state active devices, and the possibility of connecting series and shunt elements, and suitable for SMD-technology, also for SWB- antennas/devices CPW feeding provides better match and performs better than the MS line (Simons, 2001).

In search for the SWB radiator, both radiator and the feed must be super wide band. Since the SWB-signal first must able to pass through the feeding-line before reaching the antenna, obviously that the feed must be considered first in advance of other sections, we conduct the work with bottom-up approach, i.e., the feed is consider first, because if the feeding mechanism fail to be SWB, then there is no SWB radiator exists no matter how good the radiator will be. The coplanar waveguide is the first choice for feeding the signal to the radiator, because the CPW's effective dielectric is constant, (this property is a key feature in wide band matching the antenna), over a wider BW than micro-strip line, another advantage is, in contrast with MS line, one of the parameter pair (WS, WG) can be varied in size and shape, whilst the other is correspondingly changes to keep the characteristic impedance stays unchanged, furthermore CPW is low-loss, and the signal width can be chosen width enough to support characteristic impedance from 30Ohm and higher (Simons, 2001, p.52),

The CPW would be a better choice for SWB-feeding because of it considered features, summarized as follows:

- **SWB behavior**: the effective dielectric constant is almost independent of frequency (Simon op cit.), this feature is a priori condition for SWB feeding and matching.

- **Dimensional flexibility**: the width of the signal line, and hence the corresponding gaps, can be freely designed to accordingly support the physical dimension of the transition region and the antenna.
- **Dielectric support**: The dielectric thickness exerts negligible weight on the economy of the CPW-impedance

4.6. Section II: The tapered transition

The tapered transition has been inserted between the CPW-feed and the radiating patch, this section responsible for a smooth transition between the feed and the antenna, and because the current distribution is denser in this region than the others, this property indicated that this section must have strongest influence in matching the impedance bandwidth. This section has two parameters (LT, WT), which are described in details in §4.6.1 and §4.6.2 below.

4.6.1. The internal Transition Length LT

The length LT of the tapered transition (section II) is responsible for the smooth transition between the feed and antenna, and proved to be the most sensitive parameter in the design of our prototypes. Anticipation from the theory and design of micro-strip antenna (MSA), it is well-known that the length of the MSA determines the resonance frequency (by lengthen or shorten this parameter, one can shift the resonance frequency to lower or higher band, respectively).

The antenna's resonance is affected by its length, this length is composed by LT + LR1 + LR2, when this composed length is changed, and the resonance will presumably change accordingly.

It is observed, from the results plotted in Fig.9, that when LT is longer the resonance will shift to lower frequency (as shown by prototype 1, LT = 3.64, Fig.9), and when LT is shorter the radiator's resonance will shift to higher frequency (prototype 2, LT= 1.64, Fig.9)

4.6.2. The internal Transition Width WT

The width WT of the taper transition section (section II), is also a "share-parameter" with lower radiating section (section IIIA); this width provides, as similar role as the width in microstrip patch antenna, a fine-tune mechanism for impedance matching as its nominal value varies. This enhanced matching mechanism are numerically demonstrated with the reflection coefficients of the prototypes 3 and 4 as plotted in Fig. 10, in which they shown a lower reflection coefficient, i.e., a better match.

4.7. Section III: The radiation section

This section comprise of six parameters LR1, LR2, WT, WR, WC, and RC, in which WT is the share parameter described in §4.6.2, we divided this section into sub-sections IIA and IIB as sectioned in Fig. 5b. The parameter set of the internal sub-section (IIIA) and the external parameter sub-section (IIIB) are {WT, WR, LR2} and {WR, LR1, WC, RC}, respectively.

4.7.1. The internal radiating matching section

Two parameters, which identified to be key player for this subsection, are {WT, LR2} (the WR is not touched because it is share parameter of these two sub-sections), by first optimize the width parameter WT, and by keeping this optimized parameter fixed, and continuing to optimize the other parameter LR2. By this token, (Tanyer et al., 2009b, op cit.) obtained a huge enhancement in ratio bandwidth reported as 9:1. The design parameters are listed in table.1, the result is plotted in Fig.9. More detailed works and measurement results, the reader should refer to (Tanyer et al., op cit.).

4.7.2. The external radiating transition section

This sub-section was often neglected by the designers due to the fact that the current distribution is weak along the edges of this section. However, we observed that it plays an important role in maintaining the shape of radiation pattern in a wide range of frequencies, as will be numerically proved in §4.8

The sub-section IIIB consists of a set of parameters {WR, LR1, WC, RC}; WR is a share parameter so we keep it intact. From fig.1 it is seen that WR1 is suppressed and covered by varying the radii-distance WC, we can also single LR1 out because it contribution to the length of the radiator can already be economized by LT and LR2, so WC is the only parameter left that we may use to fine-tune the radiator for both SWB performance and radiation pattern characteristics. Prototype 4 utilized this philosophy by varying LR2 (instead of LR1) and WC to obtain the super wideband performance plotted in Fig.7. It is observed that variation of WC had no significant impact on reflection coefficient (current distribution along the antenna circle edges are rather week compared with those close to the tapering transition region). Nevertheless, by properly controlling WC we are able to maintain the usable shape of the radiation patterns up to 50 GHz as shown in Fig.8. So, WC is clearly to be the parameter to control the interference of the edge/corner scattering and diffraction of the radiator. In transforming the vertex-diffraction to edge-diffraction, we advocate the use of circular shape; nevertheless, other researchers suggested the shapes (elliptical, football cape, etc.). To answer the question which shape would serve best, we need a further in depth study about all possible pattern sensitive shapes before providing a final conclusive appraisal.

4.8. The prototype 4 and its super wide band performances

Fig. 7 shows the simulated result of the magnitude of the reflection coefficient of our SWB prototype 4. We computed and shown here only up to 150GHz. By close inspection of the reflection coefficient, the reader could observe that the prototype 4 shows a trend and exhibits the behavior of an all-high-pass antenna; its impedance bandwidth could be much wider than shown here.

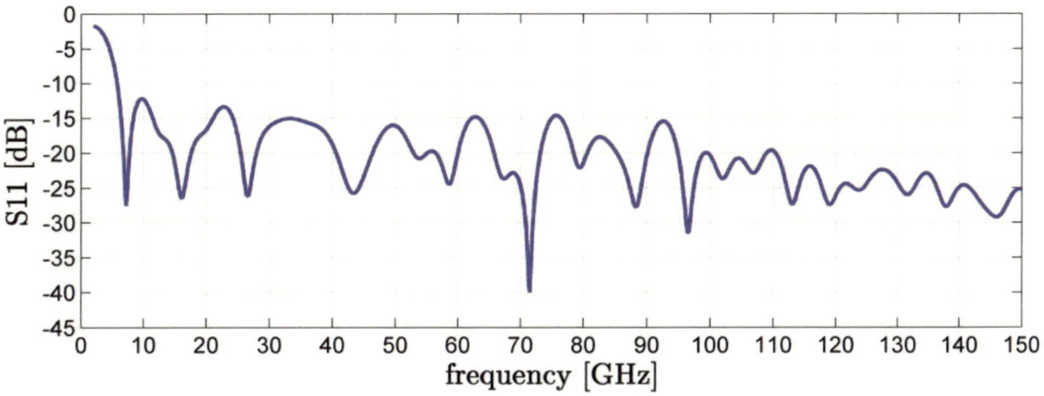

Figure 7. Impedance bandwidth performance of the proposed prototype 4.

The numerical results were computed with the following practical assumptions:

- The relative dielectric constant and the loss are assumed constant over the computational bandwidth.
- The commercial Duroid RT5870 high frequency laminated material we used, in deed could practically support up to 77 GHz only (Huang, 2008, p.64); beyond this frequency we may have to look for other dielectric material.

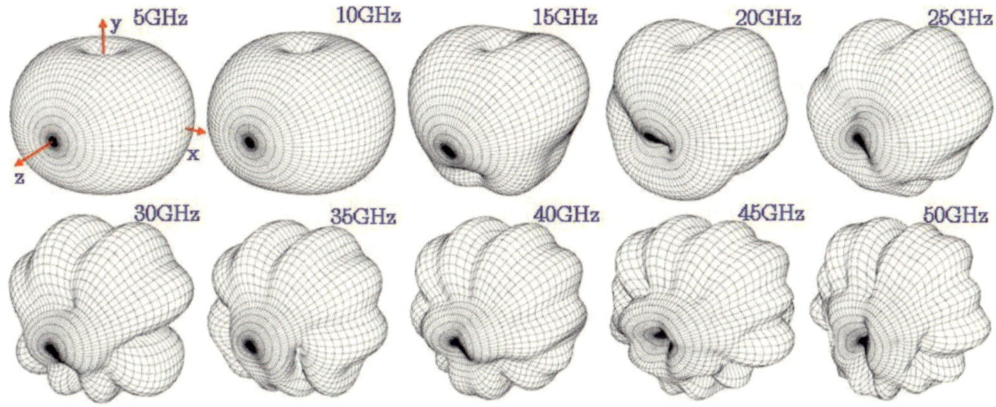

Figure 8. Pattern performance of prototype 4. Complete spectra of 3D-farfield co-polar radiation patterns from 5 to 50 GHz with increment step of 5GHz.

Fig.8 shows the simulated 3D radiation patterns of prototypes 4; the plot indicated that the radiator exhibits a super wide band pattern characteristics, the usable spherical patterns are sustained in a bandwidth wider than 10:1. The SWB properties displayed in fig.7 and fig.8 shown that prototype 4 is a true SWB radiator in both impedance and patterns aspects, and its SWB-behavior is superior to antennas reviewed by (Massey, 2007, pp.163-196). We emphasize here that this radiator is termed as quasi-magnetic antenna, because it is clearly seen that this antenna possessed radiation patterns similar to that of a dipole, therefore calling it monopole reflects wrongly the EM-characteristics that it possesses and exhibits.

4.9. Optimization process and development of prototypes

Although the proposed topology and architecture is simple, however with a total of 14 parameters it would be an impossible task for the multivariate optimization process.

This section reports in details of how to delimit the variables and how to reduce the number of variables for simplifying the MVO process to a SVO one. In addition, some pragmatic rules are also given for identifying the key parameters, and for weighting the priority of those parameters in the sequentially SVO processes.

To keep the optimization process controllable en less entangled in multivariate-optimization process, efforts have been done in the following steps:

Step 1. Topology and architecture: this step is important in that it had to form a basis for the FSD, and had to keep the antenna topology as simple as possible, but not simplest as (Al Sharkawy et al., 2004), Although all efforts has been carried out to ensure a minimum amount of the created parameters, the structure (fig.1a) still have a considerable set of parameters { ε_r, h, t; L_F, W_S, W_{GND}, W_G; L_T, W_T; L_{R2}, W_R; L_{R1}, W_C, R }.

Step 2. The FSD: dividing the radiator into sections depending on their main function. Inspection of the radiator's current distribution and the radiator's topology shown in fig. 1a revealed that the radiator could presumably be divided into functional sections as depicted in fig. 1b. The analyzing and optimizing process are conducted following the bottom-up approach that always started from the feed (section I) and ended at the last radiation-section (section III.B). The analysis, optimization and development of the prototypes all should start and avoid as much as possible the *share-parameters* between sections (W_T , W_R); if impasse is met then, as a thumb-rule, the section on top has priority on taking the share parameter.

Step 3. Excluding of fixed parameters: The numbers of parameters of the radiator have been reduced by singling out the non-optimizable parameters. The first three material-parameters { ε_r, h, t}, because their nominal values were already fixed by the manufacturers, are not quite suitable for optimization process as continuous-parameters, so the parameters of the feed section (section I) are kept fixed and excluded in the optimization process.

Step 4. Setting boundaries: in order to accelerate the optimization process and avoiding the problem of unbounded optimization, we put geometrical restrictions on the total width (2WGND + 2WG + WS) and length (LF + LT + LR2 + LR1) of the antennas fixed to $\lambda b/2$, where λb is wavelength at the design-frequency, and force all the internal parameters and their combination to be constrained inside this antenna's boundary $\lambda b/2$.

Step 5. Reduction of parameters: in this step, we did further reduction of the number of parameters involved. Exploiting the fact that the feed-section's parameters have no significant added values to the total performance once its optimum values are founded, and the following prior measures have been set, 1) for impedance-matching it is fixedly set to 50 Ω and, 2) for field-matching the impedance-parameter-pair (WS, WG) has been chosen such that it is wide enough to support the currents to separately flow along the edges of the signal line WS, and the feed length LF must be long enough to support the transformation from coax's TEM to quasi-TEM of the CPW line, so the feed's parameters can ruled out for optimization, and set to be fixed with values as listed in table.1.

Step 6. Start values: This step initiated the start values for the set of parameter listed in step 1. The initialization of the start values was elaborately detailed in (Tran et al., 2007). Prototype 1 (Fig.4) with the start values as listed in table.1, obtained a BW of 10GHz (fig.9). Detailed discussions and simulated, measured results can be found in (Tran et al., op. cit.).

Step 7. Simplified process: The 7th step is the simplification of the optimization process by breaking the MVO process down into series of SVO one. It was observed that the current intensity is mainly distributed along the edges of the transition region, this suggested that : a) small variation of the transition parameter in this section (LT, WT) would have a strong influence on the flow of the signal current and hence play a large impact on the impedance bandwidth, and exploiting the fact, as similarly in microstrip antenna, that b) the length of the radiator defines the shifting-property of the resonance, whilst c) the width of the radiator effectuates the matching-property, and d) another independent-property is that they can be separately optimized.

So by keeping WT, the share-parameter between sections II and III.A (fig.5b), fixed and took just a single-variable-optimization (LT), the impedance bandwidth was double to 20GHz as obtained by prototype 2, which is plotted in fig.9. The parameters of prototype 2 are listed in table.2, more details of the electromagnetic properties and measurements regarding the performances of this prototype can be found in (Tanyer et al., op. cit.).

Step 8. BW enhancement: Prototype 3 (fig.4) was obtained by analyzing and optimization the lower part of radiation section IIIA (fig.1b). This section has three parameters (WT, LR2, WR); first, following the rules discussed in step 2 we left out the parameter WR since it is the share-parameter between section IIIA and IIIB; next, in order to avoid multivariate-optimization, anticipating the shifting-property and matching-properties discussed as (b) and (c), respectively, in step 7; then two SVO sweepings were carried out, first WT (property c) and then LR1(property b); the order of WT or LR1 can be chosen freely according to independent-property (d, in step 7). The design parameters of prototype 3 are listed in table.1 in which the parameters of the previous optimized prototype are kept fixed, only the two parameters (WT and LR1) belonged to section IIIA are investigated and optimized.

Step 9. The design of prototype 4 is aimed at two SWB-compliances: 1) SWB impedance bandwidth, and 2) SWB radiation pattern. The radiation section consists of the following set of parameters (WT; LR2, WR; LR1, WC, R), the length parameters of this set can be keep fixed, because LR2 and LR1 are resonance-shifting parameters, and the antenna architecture allows us to use other length of the antenna to control the resonance, this was already done by LT of the lower section, so these two parameters can be singled out of the optimization process; the share parameter WR can also be neglected because it matching-property is covered by the set {WC, R}, so by keeping R fixed the parameter set of section III was left with only two parameters left {WC, WT}. The procedure followed: fist, keeping the parameters of prototype 1 with the optimized

LT = 1.64mm, then two SVOs were carried out first for WT and then WC. The reason that WT was chosen first is twofold, 1) WT is more sensitive on matching because the current distribution is denser at the lower part, 2) WC is the share parameter that mainly located in section IIIA where the matching effect is week, and it is purposely inserted to control the radiation patterns instead of wideband matching the radiator. The optimized parameters of prototype 4 are listed in table.2. The impedance bandwidth of the prototype 4 is plotted in fig.9, and fig.7; also, its SWB radiation patterns were in fig.8.

Prototypes 1, 2, 3, 4 have been design, fabricated, measured and evaluated, photographs of them are shown in fig 4. At first sign, they seemingly looked different, however they all shared the same topology and architecture as depicted in fig.5a, only one or two parameters is slightly changed to obtained difference desired performances. For comparison, their correspond impedance bandwidths are plotted together in fig.9. The impedance bandwidth enhancement is improved from 10 GHz, to 20 GHz, 40 GHz and beyond 150 GHz as shown in fig.9 and fig.7

4.10. Comparison of the prototypes

The design parameters of prototype 4 and all other prototypes are listed in table.2, so that, the reader, could independently recheck, or reproduce them without much difficulty.

These are results of the FSD methodology and SVO steps described in previous sub-section (§4.9). Note that the SVO should be carried out orderly by A, B and C. (bold and capitalized in table.2), A is dedicated as the first to be optimized, keep that optimized parameter fixed, and goes on with B, then continue with C. For example, the prototype 4, (A) first fixing the taper's height LR2 to 4.335mm, then (B) optimizing the taper width WT, and then (C) adjust the WC for the radiation-characteristics. The optimized results showed an SWB impedance bandwidth of at least over 150GHz. In fact the result of prototype 4 (with parameters listed in column 4 of Table.2) shown the downtrend of reflection coefficient for increasing frequency (Fig.7), we expect that prototype 4 will well-behave beyond 150GHz as well.

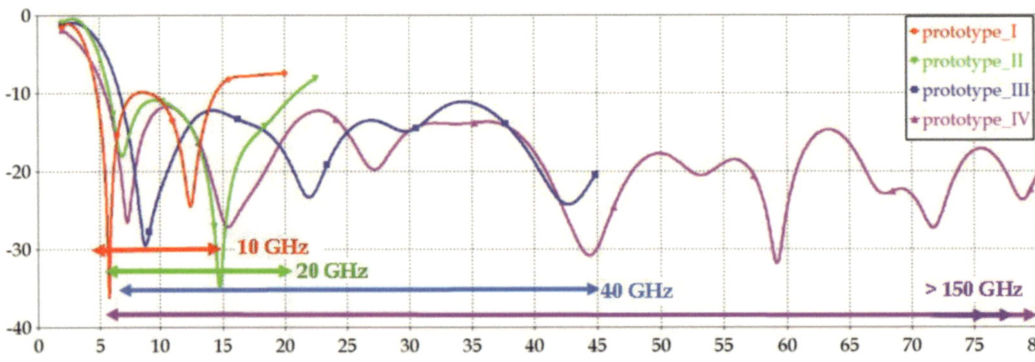

Figure 9. Impedance bandwidth of the developed prototypes. Ordinate: magnitude of the reflection coefficient [dB]; Abscissa: frequency [GHz].

IRCTR developed prototypes

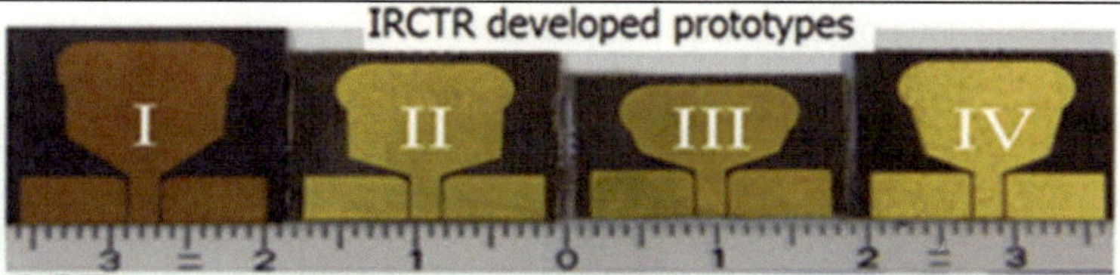

Parameter (mm)	I	II	III	IV
L_F	4.25	4.25	4.25	4.25
R_F	0.5	0.5	0.5	0.5
W_{GND}	6.86	6.86	6.86	6.86
W_S	2	2	2	2
W_G	0.14	0.14	0.14	0.14
L_T	3.64	1.64	1.64	1.64
W_T	10.2	10.2	9.2 A	8.5 B
L_{R1}	4.33	4.33	4.33	4.33
L_{R2}	4.335	4.335	1.835 B	4.335 A
W_R	10.5	10.5	10.5	10.5
W_C	7	7	7	7.2 C
R	2.5	2.5	2.5	2.5
BW (GHz)	4-14	5-25	6.5-45	5-150

Table 2. Parameters of the prototypes; the alphabetical order A, B, C indicates the priority-order of parameters in the SVO process.

5. Design and fabrication

5.1. Design

All prototypes depicted in Fig.4, with their design dimensions listed in table.2, have been fabricated on Duroid RT 5880 high frequency laminate with substrate height h=0.787mm, copper cladding thickness t=17μm, relative dielectric constant ε_r=2.2, electric and magnetic loss tangents are given by tan δ_E=0.0027 and tan δ_H=0, respectively. The foremost reason of choosing this material is that it could relatively afford SWB frequency range up to 77 GHz (Huang et al., 2008, p.64). Other reasons are assessments related to temperature, moisture, corrosion and stability, which were investigated in details by (Brown et al., 1980).

5.2. Feed elongation

Since the dimension of the SMA connector's flange is considerably large in comparison with the antenna dimension (see Fig.10a), this comparable size exerts a huge impact on the antenna's electromagnetic-properties in particularly to the transmission, scattering and radiation mechanism. In order to reduce this obstruction and to measure the antenna's scattering parameters and radiation patterns adequately, it is necessary to elongate the antenna as show in Fig. 10b. To back up the advocating of this elongation, we exploited the facts that the co planar waveguide has negligible radiation, low-loss and constant effective dielectric constant in rather wide range of application from DC to above 50GHz. we decided to elongate the CPW feed L_F to 40mm, and carried out numerical simulations of the same SWB radiators with short and long feed. The magnitudes of the reflection coefficient are compared and plotted in Fig.10c. As expected, the numerical results exposed a negligible differences as theoretically has predicted (Simons, 2001, p.240). Note that these theoretical properties (negligible radiation and low-loss) were also experimentally consolidated by (Tanyer et al, op cit.).

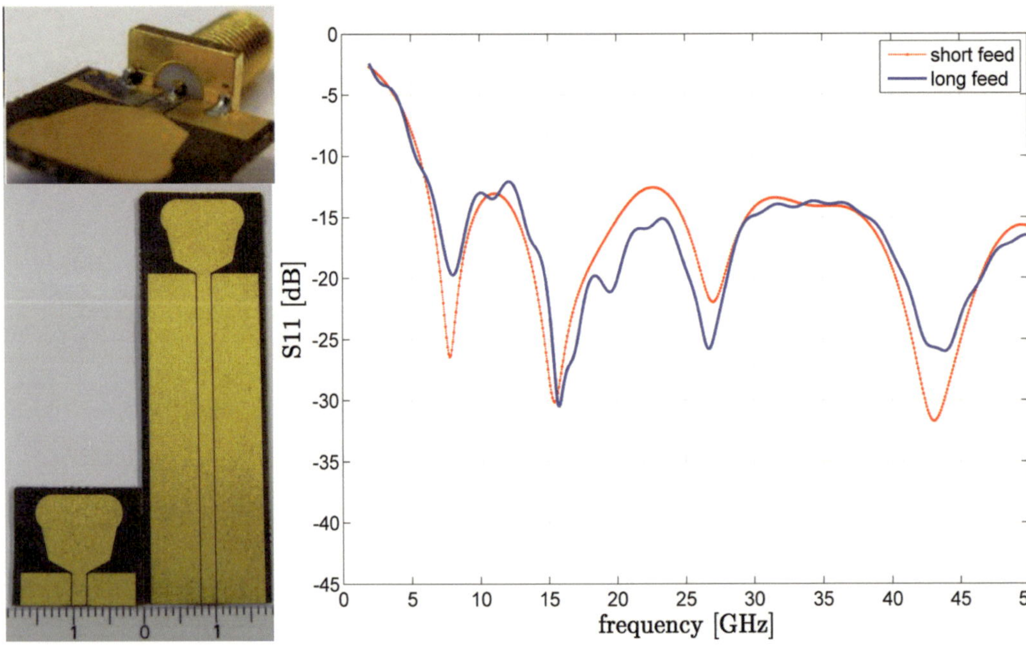

Figure 10. Conceptual demonstration for advocating of CPW feed elongation, a) radiator with SMA connector, b) radiators with short and elongated feed, b) simulated reflection coefficient magnitudes of antenna with short and long feed.

6. Measurements

6.1. Reflection coefficient

The prototype 4 is measured with the Agilent E8364B PNA vector network analyzer, the electronic calibration kit N4693A 2-port ECal-module was used for full-range calibration of the PNA (50GHz).

The reflection coefficient magnitude of prototype 4 is measured and shown in fig. 11, the measurement agreed well with predicted value. Small deviation as frequency higher than 26

GHz, this defect is inherently caused by the failure of the 3.5mm SMA-connector, whose HF-range is cataloged as 18GHz max. The result indicated that the prototype 4 is a SWB-radiator because its measured ratio-bandwidth B_R is certainly greater than 10:1.

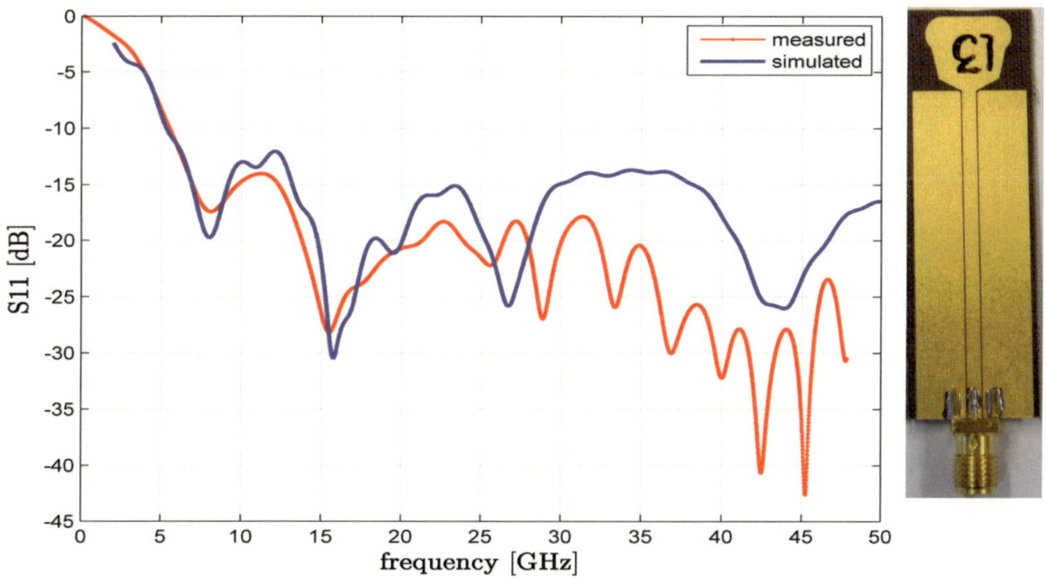

Figure 11. SWB-performance: simulated and measured results.

6.2. Far-field radiation patterns

The far field radiation patterns are measured in the Delft University Chamber for Antenna Test (DUCAT); the anechoic chamber DUCAT (Fig.12a) is fully screened, its walls, floor and ceiling are shielded with quality copper plate of 0.4 mm thick. All these aimed to create a Faraday cage of internal dimension of 6 x 3.5 x 3.5 m3, which will prevent any external signal from entering the chamber and interfering with the measurements. The shielding of the chamber is for frequencies above 2 GHz up to 18 GHz at least 120 dB all around (Ligthart, 2006). All sides are covered with Pressey PFT-18 and PFT-6 absorbers for the small walls and long walls, respectively. It is found that one side reflects less than -36 dB. All these measures were taken together in order to provide sufficient shielding from other radiation coming from high power marine radars in the nearby areas.

TX: Single polarization standard horn is used as transmitter, which can rotate in yaw-y-direction to provide V, H polarizations and all possible slant polarizations. The choice of the single polarization horn above the dual polarization one as calibrator is two-folds: 1) keeps the unwanted cross-polarization to the lowest possible level, 2) and also voids the phase center interference and keeps the phase center deviation to the lowest level.

RX: Prototype 4 is put as antenna under test (AUT) on the roll-z-rotatable column (Fig.8b). For the measurements of polarimetric components (VV, HV, VH, HH, the first letter denotes transmission's polarization state, the second is for the reception), two measurement setups are configured, the 1st is the vertical reception setup (VRS, Fig.12c) for VV, VH and the 2nd is the horizontal reception setup (HRS, Fig.12d) for HH, HV. Combination of the two setups and the TX's two polarizations provide full polarimetric patterns of the AUT.

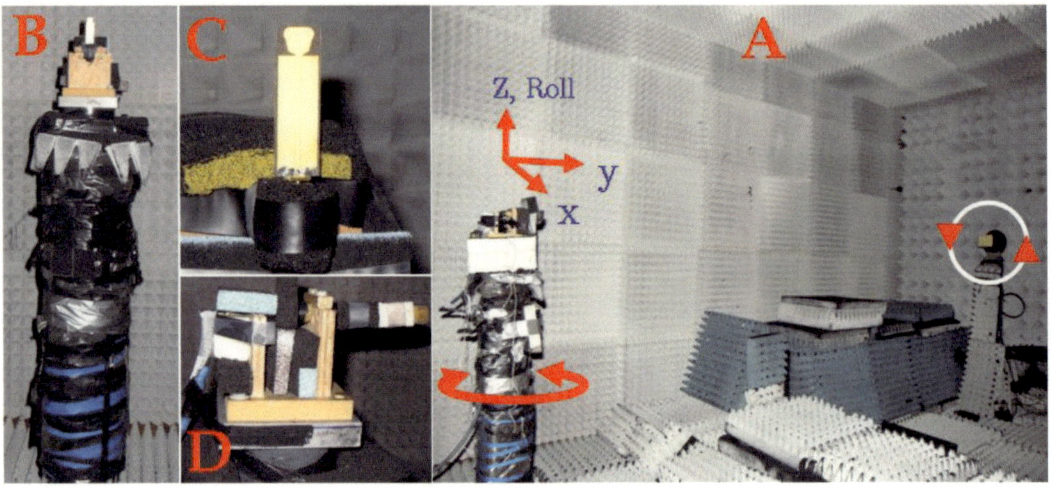

Figure 12. Patterns measurement set up: a) anechoic chamber DUCAT, b) AUT on the rotatable column, c) Vertical configuration and d) Horizontal configuration.

Calibration: the HF-ranges of the Sucoflex-cable, T-adapters and connectors used in this measurement set up all cataloged as 18GHz max, owing to this limitation, we calibrated the PNA with Agilent N4691B cal-kit (1-26GHz).

6.2.1. Co-polar VV radiation patterns

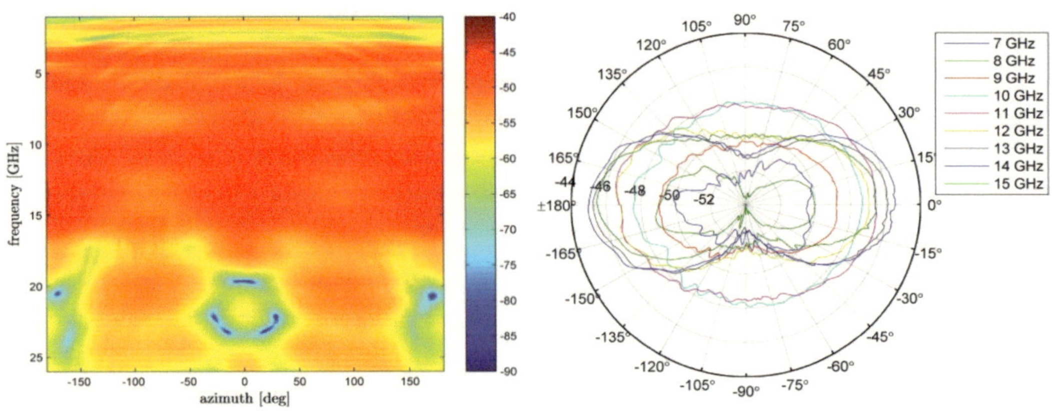

(a) (b)

Figure 13. Full-band VV co-polar measured patterns; RX: VRS; TX: zenithally oriented; a) 2D continuous patterns; b) 1D polar patterns with frequency paramaterization

6.2.2. Cx-polar HV radiation patterns

The HV cx-polar patterns are obtained with the VRS configuration in which TX-polarization in Fig.12a is 900-rotated. Plotted in Fig.14 are the HV cx-polar patterns. As expected, perfect symmetrical and repeatable patterns can be observed in full-calibrated range (1-26GHz).

Fig.14b showed the measured HV cx-polar patterns for the in-band range (7-15GHz, 1GHz increment). The patterns consolidated the repeatable symmetrical receiving/transmitting mechanism of the prototype 4. Also observed is that all EIRP are less than -65dBm, this revealed that a greater than -20dBm XPD is obtained. Note, in the yoz–plane, theoretically no cx-polar components are expected as all cross polar components cancel each other in the 0^0-180^0 and -90^0-90^0 direction. In a real case scenario, some cx-polar components are observed, their level being, nonetheless, extremely low (~ -90dBm)

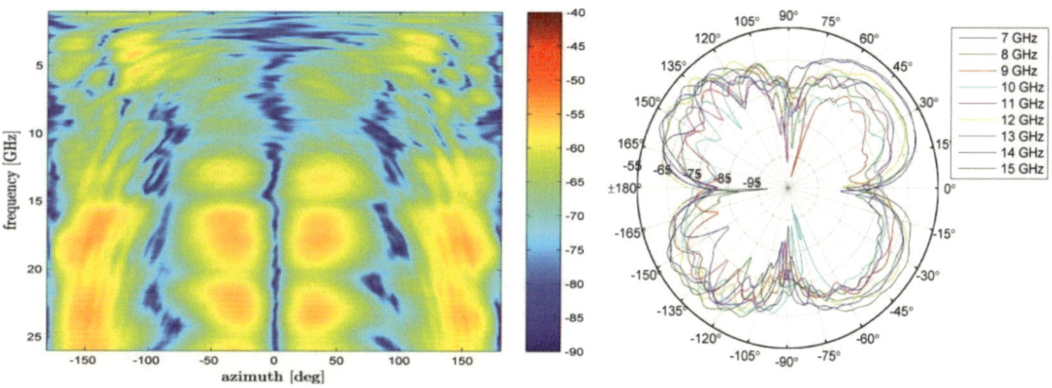

(a) (b)

Figure 14. HV cx-polar measured patterns; RX: VRS; TX: azimuthally oriented; a) 2D full-band patterns; b) Frequency parameterized 1D polar patterns

6.2.3. Co-polarized HH and Cx-polarized VH rradiation patterns

The co-polar (HH) and cx-polar (VH) radiation patterns can be acquired by the HRS with two polarization states of the TX, respectively. However, due to the mounting of the antenna (shown in Fig.12d) it was not possible to measure the backside of the antenna, thus only half of the co-polar and cx-polar patterns were measured. Owing to the frequency limitations of used components (cables, adapters, connectors, absorbents), the DUCAT anechoic chamber specifications (Ligthart, 2006, op. cit.) and the desired band the in-band range is chosen from 7-15GHz.

Fig.15a showed the measured co-polar HH in-band radiation patterns. The patterns are symmetrical and repeatable with all EIRP less than -42dBm. The measured in-band cx-polar patterns for the VH configuration are plotted in Fig.15b, all peak powers have the EIRP in the order of -60dBm. The XPD of between HH and VH of the HRS displays the same discrimination dynamic as that of VV and HV of the VRS.

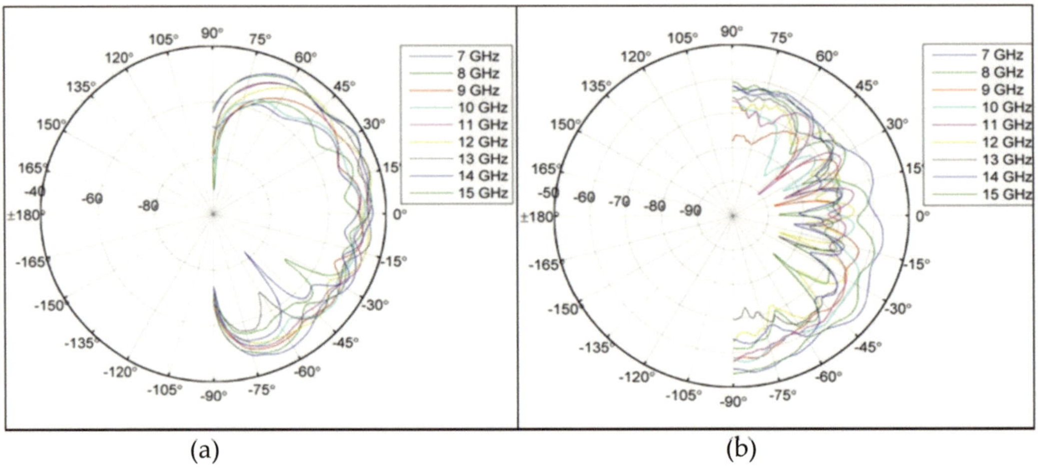

Figure 15. In-band measured patterns; a) HH co-polarized with RX: HRS and TX: azimuthally oriented; b) VH Cx-polarized with RX: VRS and TX: azimuthally oriented.

6.3. Time domain measurements

Fig.16 shows the time domain set up for measurement and evaluation of: 1) pulse deformation, 2) the omni-radiation characteristics of the AUT. The same prototype 4 are used for TX (left) and RX (right), they stand on a horizontal foam bar which situated 1.20m above the floor.

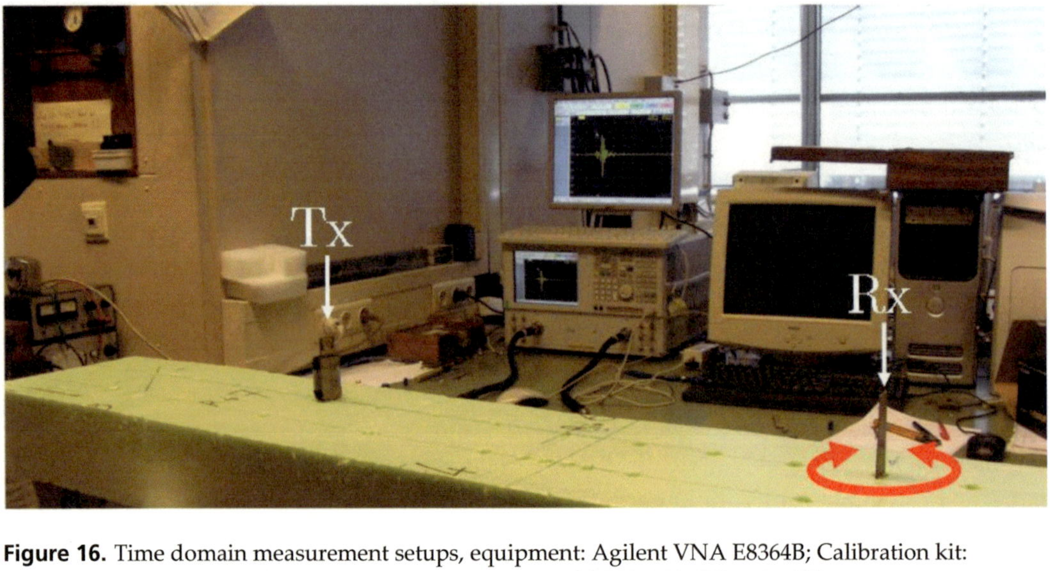

Figure 16. Time domain measurement setups, equipment: Agilent VNA E8364B; Calibration kit: Agilent N4691B, calibrated method: 2-port 3.5 mm, TRL (SOLT), 300 KHz – 26 GHz

6.3.1. Pulse measurements

Pulse spreading and deformation: Fig.17a shows the time-synchronization between the calculated transmit pulse (CTS) and the measured receive pulse (MRP) (for comparison, the CTS has been normalized, time-shifted and compared with the MRP), qualitative inspection shows that the synchronization-timing between transmitted and received pulses is very good, there is no pulse spreading took place, these measured features proved that the device is suitable for accurate ranging/sensing-applications, the small deviation at the beginning of the received pulse is due to RF-leakage (Agilent, AN1287-12, p.38), and at the end of the received pulse are from environments and late-time returns (Agilent, ibid., p.38), Note that the measurements are carried out in true EM-polluted environment as shows in fig.16, and no gating applied.

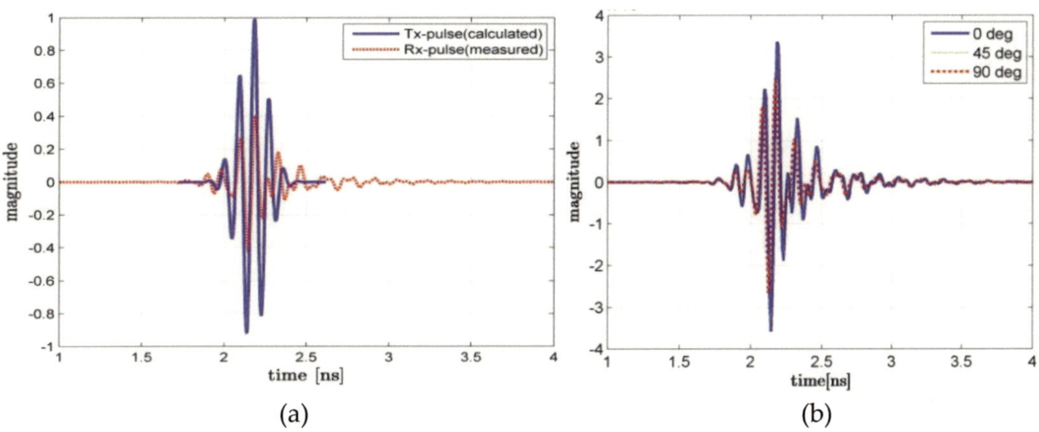

Figure 17. Co-polar transmission results of VRS-configuration; a) face-to-face: calculated vs. measured; b) oblique facing: measured results with RX 0-, 45- and 90-degree rotated.

Omni-radiation characteristics: To correctly evaluate the omni-directional property of the AUT, both quantitative characteristics (spatial) and qualitative characteristics (temporal) are carried out, the spatial-properties of prototype 4 are already tested and evaluated in frequency-domain (as depicted in fig.13), and only the temporal-characteristic is left to be evaluated. To evaluated temporal-omni-radiation characteristics, three principal cuts are sufficiently represent the temporal-omni-radiation characteristics of the AUT in the time domain. Due to the editorial limitation, we report here only the most representative case (omni-directional in the azimuthal plane, i.e. co-polar VRS, which represents the most of all realistic reception scenarios). Fig.17b shows three MRPs of the measurement configuration pictured in Fig.16 with RX 0^0, 45^0, and 90^0 rotated. The results show a perfectly identical in timing, there is no time–deviation or spreading detected between the three cases. Furthermore, although the radiator is planar, it still exhibits a remarkable azimuth-independent property of 3D-symmetric radiators (for the 90^0 configuration, the projection of the receiving aperture vanished, however the prototype still able to receive 90% power as compare to the face-to-face case), this TD-measured results pertained the omni-directional property of the radiator, and this is also in agreement with, and as well consolidate the validity of the measured results carried out in the FD.

6.3.2. *Transmission amplitude dispersion*

To evaluate the amplitude spectral dispersion of the prototype 4, the measured time-domain transmission scattering coefficients of the three co-polar configurations (0^0, 45^0, and 90^0 configurations displayed in fig.16) were Fourier-transformed in to frequency domain. The measured magnitudes are plotted in fig.18a, the measured results show a smooth and flat amplitude distribution in the designated band, and all are lower than -42dBm.

6.3.3. *Transmission phase delay and group delay*

The measured phase responses of the transmission parameter for the three co-polar configurations are plotted in Fig.18b. In narrowband technology, the phase delay defined as $\tau P =- \theta/\omega$, is a metric for judging the quality of the transmission is the phase delay between the input and output signals of the system at a given frequency. In wideband technology, however, group delay is a more precise and useful measure of phase linearity of the phase response (Chen, 2007). The transmission group delays for the three above-mentioned configurations are plotted in Fig.18c. The plots show an excellent and negligible group delays in the order of sub-nanosecond, this is no surprise because the phase responses of the prototype are almost linear (fig.18b), thus the group delay, which is defined as the slope of the phase with respect to frequency $\tau G =- d\theta/d\omega$, resulted accordingly. Note: although the group delay (fig.18c) is mathematically defined as a constituent directly related to the phase, but it was impossible to visually observe directly from the phase plot (fig.18b), but well from the magnitude plot (fig.18a).

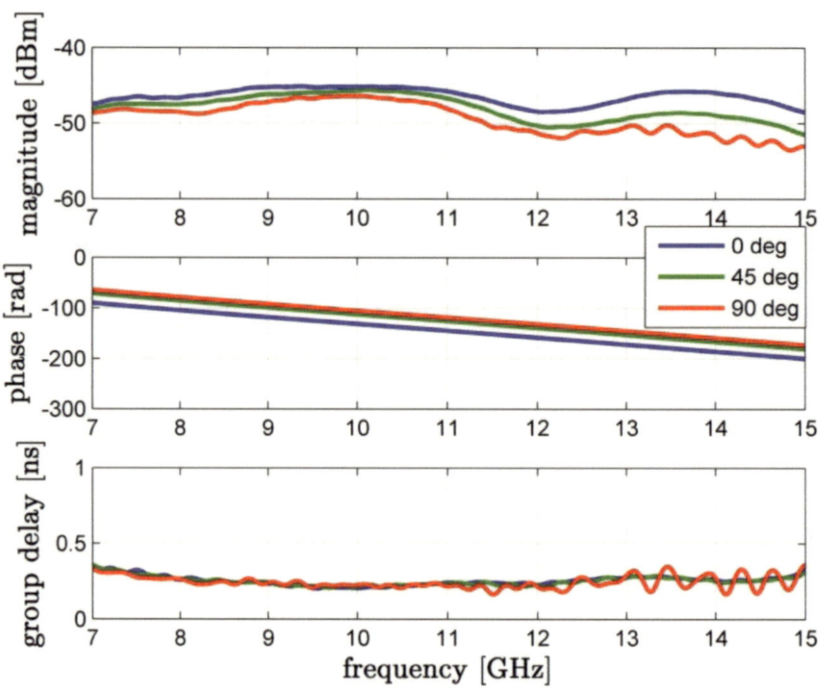

Figure 18. Measured in-band transmission coefficients a) magnitude, b) phase, c) group delay.

7. Conclusions

The intent message of this report focusses on the concept, the design methodology and the pragmatic simplification of MVO process in to a SVO one.

Distinct concepts and definitions are defused and corrected. An SWB-antenna topology with simplest structure is proposed. The single layer topology paved the way for the creating of the obtained SWB antenna architecture. The antenna architecture supported, in turn, the FSD. The introduced design methodology and conceptual concept are consolidated by the developed prototypes.

The antenna architecture provides powerful isolated-parameters to control the antenna characteristics, such as resonance-shifting, resonance matching, bandwidth broadening, diffraction reduction, and SWB pattern maintaining.

The FSD approach is introduced to obtain the required performance, whilst keeping the overall dimension of the radiator fixed, the separated sections provide engineering insights, and can be designed or optimized almost independently.

Parameter order and SVO methodology are elaborated in details, the priority and role of separable parameters are identified, and so, instead of multivariable-optimization, the optimization process can be accelerated by carry out sequence of SVOs. The proposed design, optimization procedure can possibly be used as a gauging-process for designing or optimizing similar SWB structures.

Although the prototype 4 comprised a simplest structure and shape, however superior SWB impedance bandwidth is obtained and stable SWB-patterns are uniquely preserved.

This structure, although, can be modified to obtain huge frequency bandwidth, but cannot be one-size-fit-all for gain-size requirement. However, the architecture is flexible enough for scaling up/down its dimensions to match customer's gain-size requirement.

SWB prototype is designed, fabricated and evaluated for the super wideband impasse, and could possibly be used as an alternative radiator for the sub-millimeter-wave regime.

Performances of the prototype are tested and evaluated. Good agreements between numerical predictions and measurements are obtained.

Due to editorial limits, we exclusively report here only the design methodology and conceptual approach; detailed mathematical formulation and numerical aspects related to this SWB prototype will be published in another occurrence.

Author details

D. Tran, N. Haider, S. E. Valavan, O. Yarovyi and L. P. Ligthart
International Research Centre for Telecommunication and Radar(IRCTR), Netherlands

I. E. Lager
Laboratory of Applied Mathematics, Delft University of Technology, Netherlands

A. Szilagyi
International Research Centre for Telecommunication and Radar(IRCTR), Netherlands
Military Equipment and Technologies Research Agency, Romania

Acknowledgement

The research reported in this work was effectuated with in the frame of the "Wise Band Sparse Element Array Antennas" WiSE-project, a scientific undergone financed by the Dutch Technology Foundation (Stichting Technische Wetenschapen – STW). This support is hereby gratefully acknowledged. This technological-transfer is made possible to the "Sensor

Technology Applied in Reconfigurable Systems for Sustainable Security" STARS-project thanks to the successful collaboration between two national institutions IRCTR (Netherlands), and METRA (Romania), with Thales, TNO, ASTRON and ESA as technological supervisory bodies.

8. References

Abbosh, A.M. et al. (2007), Design of a high fidelity UWB planar antenna operating in a high dielectric medium, Proceed. of the IEEE Ant. and Prop. Soc. Int. Symp. 9-15 Jun. 2007, Honolulu, USA, pp.241-244.

Balanis, C.A.(1997). Antenna theory analysis and design, Wiley Inc, 2nd ed. New York, 1997.

Bataller, M.F. et al.(2006), An overview of planar monopole antennas for UWB application, Proc Eucap 2006, Nov. Nice , France

Biscontini, B. et al.(2006), A Novel Multilayered Planar Antenna for Ultra Wide Band (UWB) Intelligent Antenna Systems, Proc. of the 22nd Ann. Rev. of Prog. in Appl. Comp. Electromagnetics ACES, March 12-16, 2006, Miami, FL; pp. 198 - 204.

Brown, D., Fiscus, T.E, Meierbachtol, C.J(1980). Results of a study using RT 5880 material for a missile radome, In: Symp. on Electromagnetic Windows, 15th, Atlanta, GA, June 18-20, Proc. (A82-2645, 11-32), Georgia Inst. of Technology, p.7-12.

Chavka, G. et al.(2006), "Radiation of ultra-wideband signals by pulse antenna" in proceeding of the Ultra-wideband and Ultra-short Impulse Signals, 18-22 September, 2006, Sevastopol, Ukraine.

Chiou, J.Y. et al.(2003), A Broad-Band CPW-Fed Strip-Loaded Square Slot Antenna, IEEE Transactions on Antennas and Propagation, Vol. 51, No. 4, April 2003, pp.719-721.

Cho, Y.J. et al.(2006), A Miniature UWB Planar Monopole Antenna With 5-GHz Band-Rejection Filter and the Time-Domain Characteristics, IEEE Trans. AP. Vol.54, No. 5, May 2006.pp-1453-1460.

Choi, S.H. et al.(2004), A new ultra-wideband antenna for UWB applications, MOTL, V.40, No.5, Mar. 2004

Choi, S.T. et al.(2009), Small printed CPW-fed triangular monopole antenna for ultra-wideband applications, IEEE Micro & Opt. Tech. Letters, Vol.51, No.5, May 2009,pp.180-182.

FCC (2002). Federal Communications Commission, FCC 02-48, ET-Docket 98-153, "*First Report and Order*", Apr. 2002.

FCC (2004), Federal Communications Commission, FCC 04-285, ET-Docket 98-153, "*Second Report and Order and Second Memorandum Opinion and Order*", Dec. 2004.

Garbaruk, M.(2008)"Design and experimental investigations of UWB microstrip and stripline antennas", Proceeding of the 17th International Conference on Microwaves, Radar and Wireless Communications, MIKON 2008.

Hayes, P.R., et al.(2007), "UWB Antenna Surrogate Design", in: Ultra-Wideband Short-Pulse Electromagnetics 8, Baum, C.E., Ed. 2007.

Huang, Y. & Boyle, K.(2008). Antennas from Theory to Practice, John Willey and Sons, Singapore, 2008, pp.64.

IEEE STD 145-1983,(1983), IEEE Standard Definitions of Terms for Antennas, New York, IEEE Press, 1983, pp.11-16.

Karoui, M.S., et al.(2010), Bandwidth Enhancement of the Square Rectangular Patch Antenna for Biotelemetry applications, International journal of information systems and telecommunication engineering, v.1, 2010, iss.1, pp.12-18.

Kraus, J. D. (1985), Antennas since Hertz and Marconi, IEEE Trans. AP-33, No. 2, Feb.1985, pp.131-137.

Kshetrimayum, R.S., Pillalamarri, R.(2008). Novel UWB printed monopole antenna with triangular tapered feed lines, IEICE Electronics express, vol.5, No. 8, pp. 242-247.

Ligthart, L.P.(2006). Antennas and propagation measurement techniques for UWB radio, Wireless personal communications, 37(3-4), pp.329-360.

Massey, P.(2007), Planar Dipole-like for Consumer Products, In: Ultra-wideband Antennas and Prop. for Comm.s, Radar and Imaging, Allen,B.(ed) Wiley 2007, pp.163-196.

Powell, J.(2004), Antenna Design for Ultra Wideband Radio, MSc. Thesis, Massachuset Inst. Of Tech. 2004,

Powell, J. (2004), Differential and Single Ended Elliptical Antennas for 3.1-10.6 GHz Ultra-wideband communications, . Proc. of the IEEE Ant. and Prop. Soc. Int. Symp. 20-25 Jun. 2004, Cambridge, MA, USA, Vol.3, pp.2935-2938.

Rahayu, Y. et al.(2008a), A Small Novel Ultra Wideband Antenna with Slotted Ground Plane, Proc. Of the Int. Conf. on Comp. and Communications. Eng. 2008, May 13-15, 2008 Kuala Lumpur, Malaysia.

Rahayu, Y., et al. (2008b). Slotted ultra wideband antenna for bandwidth enhancement, 2008, Loughborough Antennas & Propagation Conference 17-18, Mar. 2008, Loughborough, UK.

Rahim, M.K.A. & Gardner, P. (2004). The design of nine element quasi microstrip log periodic antenna, RF and microwave conference, RFM2004, 5-6 Oct. 2004, Selangor, Malaysia, pp.132–135.

Rmili, H. & Floc'h, J. M. (2008), Design and analysis of wideband double-sided printed spiral dipole antenna with capacitive coupling, Microwave and optical technology letters, Vol. 50, No.5, pp. 312-1317.

Ruengwaree, A. (2008), "Design of UWB Radar Sensors" Dissertation, 2008, Kassel univ. press GmbH, Kassel, Germany.

Shastry, P.N. et al.(2009), Planar UWB Conical Skirt Tapered Monopole Antenna; Proceed. of the 9th IEEE Wireless and Micr. Tech. Conf., WAMICON'09. 20-21 April 2009.

Schantz, H.G.(2003), Apparatus for establishing signal coupling between a signal line and an antenna structure, US pat. 6512488B2, Jan28, 2003

Schantz, H.G and Barnes, M. (2003). UWB magnetic antenna, IEEE International Symposium on Antennas and Propagation Digest, 3, 2003, pp.604-607.

Schantz, H.G. (2004). A brief history of UWB antennas, IEEE Aerospace and Electronic Systems Magazine, vol .19(4), pp.22-26, 2004.

Simons, R.N. (2001). Coplanar Waveguide Circuits, Components, and Systems, John Wiley & Sons, New York, 2001.

Tanyer-Tigrek, F.M, Tran, D., Lager, I.E., Ligthart, L.P.(2009a). CPW-fed Quasi–Magnetic Printed Antenna for Ultra-Wideband Application, IEEE Antennas and Propagation Magazine, Vol.51, No.2, April 2009, 1, pp.61-70.

Tanyer-Tigrek, F.M, Tran, D., Lager, I.E., Ligthart, L.P.(2009b). Over 150% bandwidth, quasi-magnetic printed antennas, Antennas and Prop. Soc. Int. Symp. APS2009.

Tanyer-Tigrek, F. M.(2010). Printed antenna elements with attested ultra wideband array capability, dissertation, ISBN-978-90-9024664-2, 2010, p.113-135.

Thor, R.C. (1962). A Large Time-Bandwidth Product Pulse-Compression Technique, *IRE Transaction on military electronics*, Vol. MIl-6, No.2, Apr. 1962, pp.169-173.

Tourette, S., Fortino, N., Kossiavas, G.(2006). Compact UWB printed antennas for low frequency applications matched to different transmission lines, Microwave and optical technology letters, Vol. 49, No.6, pp.1282-1287.

Tran, D., et al. (2007). A Novel CPW-fed Optimized UWB printed Antenna, Proc of the 10th European conference on wireless technology, pp.40-43, Munich, Germany, Oct.8-10, 2007.

Tran, D.P., Coman, C.I., Tanyer-Tigrek, F.M., Szilagyi, A., Simeoni, M., Lager, I.E., Ligthart, L.P. & van Genderen, P.(2010). The relativity of bandwidth – the pursuit of truly ultra wideband radiators, In: Antennas for Ubiquitous Radio Services in a Wireless Information Society, Lager, I.E (Ed.), pp.55-74, IOS Press, 2010, Amsterdam.

Welch, T.B,et al. (2002); "The effects of the human body on UWB signal propagation in an indoor environment", Selected Areas in Communications, IEEE Journal on; 2002

Wilson, J. M (2002). Ultra-wideband a disruptive RF technology?, Intel Research and Development, Version 1.3, Sept. 10, 2002.

Valderas, D., et al., " Ultrawideband Antennas: Design and Applications", Imperial College Press, Covent Garden, London WC2H 9HE, 2011.

Xiao, J.X. et al. (2009), A novel UWB critical slot antenna with band-notched characteristics, J. of Electromagn. Wav. and Appl., Vol. 23, 2009, pp.1377–1384.

Yeo, Y. et al.(2004), Wideband slot antennas for wireless communications, IEE Proc. Microw. Antennas Propag., Vol. 151, No. 4, Aug. 2004, pp.351-355.

Ying, C. et al., An LTCC planar ultra-wideband antenna, Microw. and Opt. Tech. Letters, Vol.21, No,3, Aug.5, 2004, pp.220-222.

Zhang, et al.,(2009), Design of CPW-Fed monopole UWB antenna with a novel notched ground, Microwave and optical technology letters, Vol. 51, No.1, pp. 88-91.

Chapter 9

Active Integrated Antenna Design for UWB Applications

Danial Nezhad Malayeri

Additional information is available at the end of the chapter

http://dx.doi.org/10.5772/50185

1. Introduction

UWB technology is advancing rapidly because of its potential to have high data rates and very low radiation power. In recent years UWB technology has been used in the areas of, sensing, radar and military communications [1]. In February 2002, the Federal Communications Commission (FCC) of the United States issued a report that UWB could be used for wireless data Communications [2]. Since then, a huge surge of research interests have occurred and this technology has been considered as one of the most reliable wireless technologies for various applications that leads to new innovations and greater quality of wireless personal area network services industry.

Reliable wireless connection between computers, portable devices and consumer electronics in short distances and data storage and transfer between these devices are new subjects of scientific and industrial competitions which require data rates much more than now a day accessible ones.

In this chapter, at the second part, some advantages of ultra-wideband technology and its progress trend will be reviewed. In the third part, antennas structures and parameters, especially wideband antennas will be studied for use in the UWB systems. To describe the performance of an antenna, various parameters must be defined. There are several important and practical parameters such as frequency bandwidth, radiation pattern, directivity, gain and input impedance which will be explained briefly. The performance of a UWB antenna is required to be stable and uniform over the ultra wide operational bandwidth. In the other word, antenna radiation pattern, gain and input impedance should be stable across the entire band. Also antenna needs to be small enough to be compatible with the other UWB system elements, especially in portable devices. In addition, basic antenna parameters, such as gain and return loss, must have little variations across the operational band. Various methods have been employed to enhancement antenna bandwidth. In this part, frequency independent antennas will be studied for instance.

In part 4, planner spiral antenna characteristics and features will be reviewed as a frequency independent antenna. Since without optimization, spiral antenna has some limitations for UWB applications, these limitations will be improved by using some optimization techniques. One of new methods is using active circuit in antenna structure. So in the fifth part, improving history of active antenna technology will be reviewed. Integration of active circuit into passive antenna gives a lot of advantages such as increasing the effective length of short antennas, increasing bandwidth, improving noise factor, impedance matching and sensitivity of receiver antennas and some applications such as utilizing active antenna arrays in mobile communications and beam control, solving channel capacity limitation problems by increasing data rate and improving smart antenna technologies [3] and many other advantages. Overall active antenna structure and different types and applications will be discussed in this part.

A review of distributed amplifiers characteristics will be done in the sixth part as the active part of active antenna structure. Here the aim is to design a UWB distributed amplifier with uniform and acceptable parameters such as Gain and VSWR in the 3.1- 10.6 GHz band. Calculation of the optimum load resistance and the number of amplifier stages, and then design, optimization and analysis of the circuit must be done for active circuit design completion. Adding antenna element to the active circuit and combined circuit analysis will be explained in this part too. Finally a brief analysis of design and simulation results of UWB active antennas will be shown in the seventh part and it will be favorable that active antenna parameters such as VSWR and Gain are appreciably optimized rather than passive antenna.

2. UWB Technology

UWB systems historically have been based on impulse radio signals; therefore they can communicate at very high data rates by sending pulses rather than using narrow band frequency carriers. The pulses normally have an ultra-wide frequency spectrum caused by short pulse durations which are about nanoseconds. The concept of impulse radios initially was introduced by Marconi, in the 1900s [1], but since 1960s, impulse radio technologies started being developed for radar and military applications.

In February, 2002, the FCC allocated a bandwidth of 7.5GHz, i.e. from 3.1GHz to 10.6GHz for UWB applications [1]. It was the largest spectrum allocation for unlicensed applications that the FCC has ever permitted. According to the FCC's report, any signal that contains at least 500MHz spectrum can be used in UWB systems. It means that UWB applies to any technology that uses 500MHz spectrum and complies with all other requirements for UWB.

Shannon-Nyquist criterion in Equation (1) shows the relation between channel capacity, bandwidth and signal-to-noise ratio (SNR), when channel is assumed to be ideal band-limited with Additive White Gaussian Noise (AWGN):

$$C = BW \log 2 \left(1 + SNR\right) \quad (1)$$

Where C is the maximum data rate and BW is the channel bandwidth. Equation (1) indicates that by increasing the SNR (which is directly related to transmission power) or bandwidth, transmitting data rate can be increased. Because of power limitations, increasing the SNR is not a general solution [2, 3, 4 and 5]. Therefore to increase channel capacity and achieve high data rate, a large frequency bandwidth is needed. Considering Shannon-Nyquist Equation indicates that channel capacity can be increased more rapidly by enhancing the channel bandwidth than the SNR. Thus, the wider frequency range can lead to the greater channel capacity. This is more applicable for WPAN which works over short distances and SNR is more satisfactory there.

2.1. UWB benefits

UWB has many satisfactory advantages which make it an interesting technology for wireless systems. It is probably the most promising technology for new wireless systems because of some advantages such as low complexity, low power consumption, low cost, high data rate and short-distance wireless connectivity. From circuit point of view, accurate power transfer between transmitter and receiver is the major challenge of UWB system design to obtain a flat received power with minimum ripple.

Here some other benefits are reviewed:

- Shannon-Nyquist theorem shows that channel capacity is proportionally related to bandwidth. According to the ultra-wide frequency bandwidth of UWB systems, they can achieve grate capacity in distances below 10 meters [1].
- UWB systems use very low power transmission levels across an ultra-wide frequency spectrum that lead to reducing the effect of power upon each frequency element below the acceptable noise level [1]. This is illustrated in figure (1),

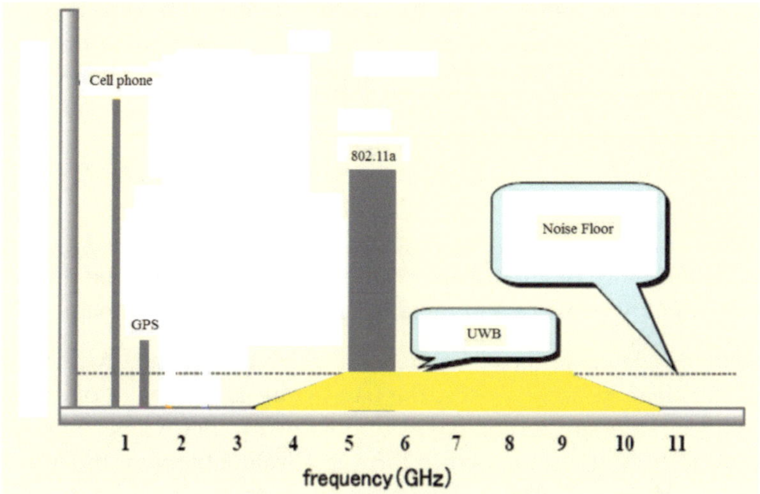

Figure 1. Ultra wideband communications spread transmitting energy

- Because of low energy density of the UWB signal, it is a noise-like signal and therefore its undesirable detection is unlikely. Since it is a noise-like signal which has a particular shape, it can be detected in related receiver. In contrast, real noise has no shape, thus interference cannot distort the pulse shape completely and it can still be recovered to restore primary signal. Hence UWB communications are very secure and reliable means.
- Baseband nature of the UWB signal which is based on impulse radios causes low cost and low complexity of operation systems. Because it does not require system components such as mixers, filters, amplifiers and local oscillators which are necessary for modulation and demodulation units.

Some of the other UWB benefits and advantages are briefly listed below:

- High data transfer rate
- Channel capacity improvement
- Lower power consumption
- Lower cost
- Coexistence possibility with 802.11/b/g
- Accurate position and distance metering
- Improved measurement accuracy of target detection in radar
- Identification of target class and type

And so many other interesting advantages which cannot be explained here. For more information see references [4 and 5].

3. Antenna

A passive antenna is an electrical conductor or array of them which radiates (transmits/ receives) electromagnetic waves. Most antennas are resonant devices, which operate over a relatively narrow frequency band. For any wireless system, antenna is an essential part which must concentrate the radiation energy in some directions at certain frequencies. Thus the antenna is also considered as a directional device too. Depending on application type, antenna must take various shapes to meet the required conditions. Therefore, antenna may be a specific length of a wire, an aperture, a patch and so on. A good designed antenna can improve system performance by complying system requirements.

3.1. Antenna parameters

To discuss the performance of an antenna, various parameters must be defined. Frequency bandwidth, gain, input impedance and radiation pattern are some of them.

Frequency bandwidth; is the range of frequencies which the antenna performance conforms specified characteristics. In other words, the bandwidth is the range of frequencies which the antenna characteristics are acceptable in compare with their values in center frequency. Frequency bandwidth can be expressed in form of absolute bandwidth (Abw) or fractional bandwidth (Fbw). Abw and Fbw can be calculated as given in Equations (2) and (3), respectively:

$$Abw = f_H / f_L \tag{2}$$

$$Fbw = Abw / f_c \tag{3}$$

$$f_c = (f_H + f_L) / 2$$

Where f_H and f_L express the high and low frequencies of the bandwidth respectively and f_c shows the center frequency. Although sometimes the bandwidth is expressed as the ratio of the high to low frequencies of operational bandwidth for broadband antennas:

$$BW = f_H / f_L \tag{4}$$

Radiation Pattern; is the representation of the radiation properties of the antenna as a function of space coordinates. Usually radiation Pattern is determined in the farfield region to avoid effects of the distance on the spatial distribution of the radiated power [1].

The radiation pattern can be expressed in two or three-dimensional spatial distribution and it is usually in normalized form with respect to the maximum values. Radiation properties of an antenna can be described by three types of radiation patterns:

- Isotropic - An ideal lossless antenna with equal radiation in all directions.
- Directional - An antenna with the radiation pattern in some directions significantly greater than the others.
- Omni-directional - An antenna which have a non-directional radiation pattern in one plane and a directional pattern in other orthogonal planes.

Gain and Directivity; directivity is calculated as the ratio of the radiation intensity in a given direction over an isotropic source radiation intensity, to describe the directional radiation properties of an antenna. The directivity is expressed by D letter and can be calculated by equation (5):

$$D = \frac{U}{U_0} \tag{5}$$

$$U_0 = \frac{P_{rad}}{4\pi}$$

Where U0 is the radiation intensity for an isotropic source, U is radiation intensity of antenna and P_{rad} is radiated power.

Antenna gain is related to the directivity and radiation efficiency and it can be calculated by equation (6):

$$G = e_{rad} D \tag{6}$$

Which G is antenna gain and e_{rad} is radiation efficiency.

3.2. UWB antenna characteristics

Although UWB antenna is an important part of conventional wireless communication systems, but designing a UWB antenna needs to consider important notes that some of them are listed below:

- Because of the ultra-wide frequency bandwidth of these systems and to comply with the FCC report, Abw and Fbw of a UWB antenna should not be less than 500MHz and 0.2 respectively
- UWB antenna parameters such as antenna radiation pattern, gain and input impedance must be uniform and stable over the entire operational band.
- Radiation pattern properties are different depending on the practical conditions which the UWB antenna should meet them.
- In many cases such as portable devices, the UWB antenna should be small enough to be compatible with the overall system. In some other applications this antenna must be compatible with printed circuit board (PCB) structures.
- UWB antenna should comply with the FCC power emission mask or other world regulatory bodies.
- UWB antenna must have minimum effects on UWB pulse waveform.

3.3. Frequency independent antennas

Ultra wide operating bandwidth is the main difference and advantage of a UWB antenna. To achieve wideband characteristics for different antennas, various methods can be used. Frequency independent antennas are one important group of antennas which display wideband features. Some principles and conditions of this group are discussed here.

Frequency independent antennas can display almost uniform input impedance and radiation pattern and other radiation properties over a wide frequency bandwidth. Spiral antenna is one practical example of frequency independent antennas. These antennas were called travelling wave antennas by Johnson Wang at first.

Victor Rumsey in the 1950s and Yasuto Mushiake in the 1940s introduced some principles which explain how frequency independent characteristics can be achieved.

The first principle which is introduced by Rumsey suggests that to achieving frequency independent properties of an antenna, its shape must be specified only by angles [6]. One example for this type of antennas is spiral antenna with no limitation of its length. Infinite biconical antennas are other examples whose shapes are completely described by angles.

Self-complementarity is the second principle of frequency independent characteristics which was introduced by Yasuto Mushiake. This principle suggests that if an antenna is complement of itself, frequency independent behavior is achieved. In such an antenna, impedance is constant and equal to $\eta/2$ or 188.5 Ohms [1].

Although these antennas are theatrically frequency independent, but they are some limitations which cause limited bandwidth of this type of antennas and needs some optimization to achieve unlimited bandwidth.

The first problem is the unlimited dimensions of antenna requirements according the Rumsey's principle. It is impossible to have an infinite length for example in a spiral antenna and the antenna size will be a practical challenge. Truncating each of dimensions of antenna can cause limitations in frequency bandwidth.

Second problem is that spiral antenna active zone depends on the signal frequency. For a UWB signal, they are many frequency components and each frequency component is radiating from different part. In other word the smaller parts radiate higher frequencies and the larger scale parts emit lower frequencies of antenna and this may cause dispersive and signal distortion. Therefore these antennas can be cause problems for systems which cannot tolerate dispersion. Signal detection and recovery features are needed for systems which use this type of antennas.

4. Spiral antenna

New generation communication systems requirements have essential role to identifying the type of antenna which is used with them. High data rate and wider bandwidth requirements for data\video transfer and growth in number of users lead to increase operating frequencies of these systems to microwave and millimeter-wave frequencies. In these frequencies performance of standard antennas such as monopole and dipole antennas is considerably weak and it causes to choose planner antennas as a way of surmount this problem. Planner antennas have many benefits such as lower manufacturing costs, considerably smaller size and less weight in compare with the other antennas. This causes to increase their applications in for example mobile phones and communication stations.

In the other hand, their planner structure makes them desirable for use in large arrays and suitable for integrating with electronic circuits such as amplifiers and phase shifters which are the main parts of designing radar, satellite communications and etc. planner structure gives them the ability of using in some applications such as antennas printed in the airplane body which can resolve limitations of antenna size [7]. These features and many other benefits of this type of antennas caused them to become a good subject of new designs of high performance planner antennas in many different applications and raised a new type of antennas with the name of Active Integrated Antennas (AIA) which will discuss more in the next parts.

According to previous discussion about frequency independent antennas such as spiral antenna, they are classical wideband antennas which can display uniform impedance and pattern characteristics over a frequency range wider than 10:1. Thus spiral antenna seems to be a good choice for use in UWB active antenna design. So in the following part, spiral antenna and its parameters are discussed in detail.

4.1. Features

Planner spiral antennas are one of frequency independent antennas with wide bandwidth and good pattern efficiency in compare with other antenna types. Theatrically a spiral antenna with infinite number of turns and dimensions which confirms the frequency independent principles can exhibit infinite pattern efficiency and bandwidth. But practical infinite arm length is impossible and some limitations must be applied [6, 8]. This type of antennas has widespread usage because of their small substrate size and few pulse dispersions in communicating processes. The primarily used single arm spiral antenna is illustrated in the figure (2). Desirable radiation characteristics can be achieved by changing circular radius r, number of turns N and the width of them W [8].

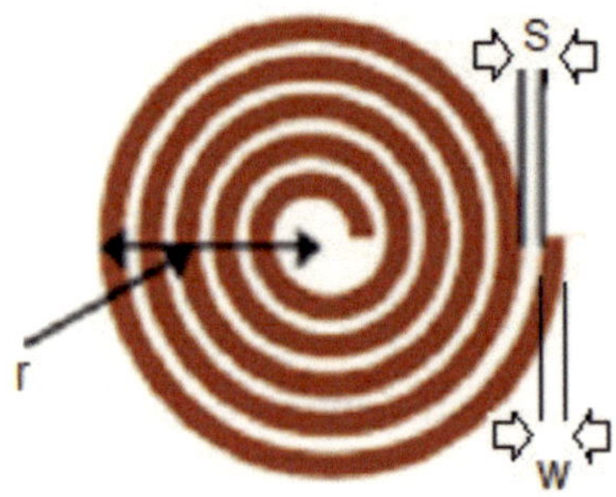

Figure 2. Single arm spiral antenna

If W = S, i.e. the metal and the air parts of the antenna are equal, the spiral antenna is self-complementary.

4.2. Active zone

In spiral antenna radiation is done from that part of antenna which its circumference is equal to or greater than 2λ, where λ is the wave length. This region is called active zone of antenna. Thus low frequency limit of antenna is related to exterior radius of antenna as expressed by (7) relation and the high frequency limitation is related to its interior radius. In the other word, circumference of the radiation zone determines the radiation frequency.

$$f = c / \pi r \rightarrow f_{low} = c / \pi r_2 \quad \text{and} \quad f_{high} = c / \pi r_1 \tag{7}$$

Where c is the light speed, r_1 is the interior radius and r_2 is the exterior radius of the antenna. In fact, relation shows that higher frequencies are emitted from smaller circles and the lower frequencies will radiate from bigger circles of the antenna. However in practice, because of antennas end reflections and source effects, high and low frequencies will be a little different from the calculated values of relation (7) [8].

According to the previous discussion, active zone calculation is one of important and effective parameters to design a spiral antenna. Radius of radiation circle is defined by relation (7). In fact this region is a part of antenna in which maximum power of the device is radiated from and it can be considered as only radiating zone of antenna.

In this area, current amplitude inducted by radiation is considerably higher than the currents in other parts of structure. The region between source point and the active zone which has low power emission is called transmission line zone, because it just transfers power from source to load or antenna. Although the final antenna radiation pattern is not considerably affected by this area, it is effective in the value of input impedance and therefore must be considered to have an optimum design. The area between active zone and the end of antenna arm is out of radiation circuit. In fact, whole power is emitted before reach to this area, therefore it has no effect on the radiation pattern and input impedance [8]. So it can be predicted that by increasing frequency, active region will move along spiral antenna radius such a way that electrical dimensions remain constant in different frequencies. To conforming Rumsey's principle, antenna dimensions must be infinite and such antenna can be considered as frequency independent antenna. But by eliminating antenna structure in both ends because of practical manufacturing limitations, antenna performance will be considerably dependent on active zone properties.

To define active zone, at first the lowest frequency of bandwidth must be selected to calculating the exterior radius of active zone. As is discussed, by increasing frequency, active zone will move to interior parts and smaller radiuses. Then by choosing the highest operating frequency of bandwidth, its related active zone and interior structure boundary can be defined. However usually interior radius is not limited more than requirements of implementing power source in center.

Defining active zone has some benefits; because of predefined manufacturing limitations, antenna structure must be limited in both directions and missing a vast width of frequency band. Therfore active zone definition helps designers to design antenna with better performance in the practical frequencies. Another advantage of defining active zone is to understanding and calculating other effective parameters in antenna performance optimization [8].

5. Active antenna

The idea of using active antennas was introduced in about 1928 by using a small antenna with electron tube in radio receivers. In 1960's and 1970's, active antennas were studied more seriously due to the invention of high frequency transistors [9, 10]. Because of progresses in technology of microwave integrated circuit (MIC), active antennas became an interesting subject of researches at that time [11, 12 and 13]. Integration of active device into passive antenna gives a lot of advantages such as increasing the effective length of short antennas, increasing the bandwidth, improving the noise factor, impedance matching and sensitivity of receiver antennas and some applications such as utilizing active antenna arrays in mobile communication and beam control, using to solve channel capacity limitation problems by increasing data rate and improving smart antenna technologies [3] and many other advantages.

5.1. Active antenna structure

Radiation element or passive antenna is a device that converts received signals from a transmission line into electromagnetic waves and radiates them into free space in a transmitting antenna and vice versa in a receiving one. According to IEEE Standard antenna is a means for radiating or receiving radio waves [1]. Figure (3) shows the conventional receiving configuration of passive and active antennas.

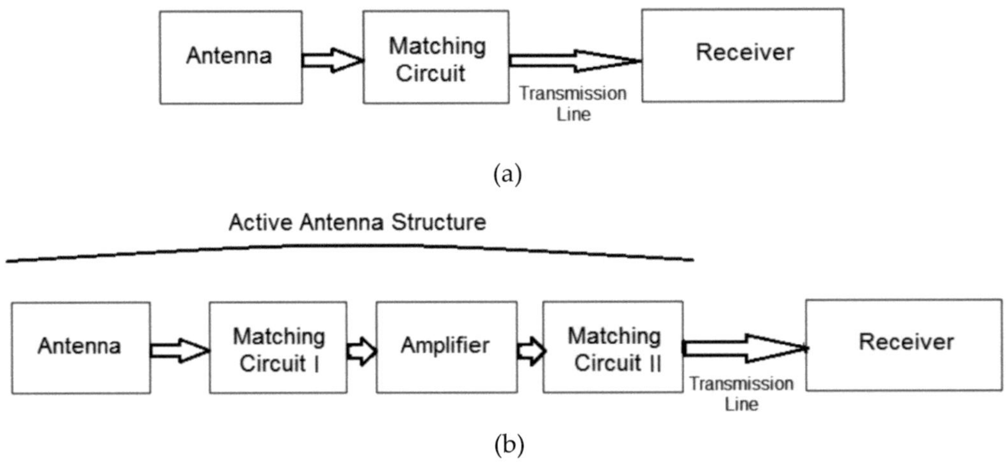

Figure 3. a). Receiving system structure for passive antenna; b). Receiving system structure for active antenna

In 1977, Lindenmeier and Meinke suggested that if the cable length between the antenna and the amplifier is about 1-5 m, the antenna system will be considered as the passive one as shown in Figure (4a). Consequently when the radiating element is closely connected (Integrated) to the active circuit or amplifier, the structure is considered as active antenna [14]. It is important to note that the distance between radiation element and active circuit is related to the operating frequency and electrical length of cable.

As mentioned, the term "active antenna" means that the active device is coupled with the passive antenna to improve antenna performance, while the term "active integrated antenna" expresses more distinctive that the passive antenna element is integrated on the same substrate with the active circuit [9]. From the microwave theory standpoint, an active integrated antenna (AIA) can be regarded as an active circuit which its output or input ports are in free space instead of a conventional 50 Ω interface. In this case, the antenna provides certain functions such as resonating, filtering, and duplexing circuit behaviors. But from the antenna theory sight, the AIA is an antenna which exhibits radio signal generating and processing capabilities such as mixing and amplification [15].

In these systems whole systems is working with antenna and controls antenna as well as load parameters. By connecting antenna and circuit in such a way, transmission line losses are reduced considerably. It will be more important when the frequency is growing up [16]. This is significantly different and more effective than the systems in which radiating element and circuit are designed separately and then connected by a strip line or another type of transmission line. This is important to note that when antenna is consisting of a nonreciprocal circuit, it means that AIA system is non-reciprocal unlike passive antenna alone [7].

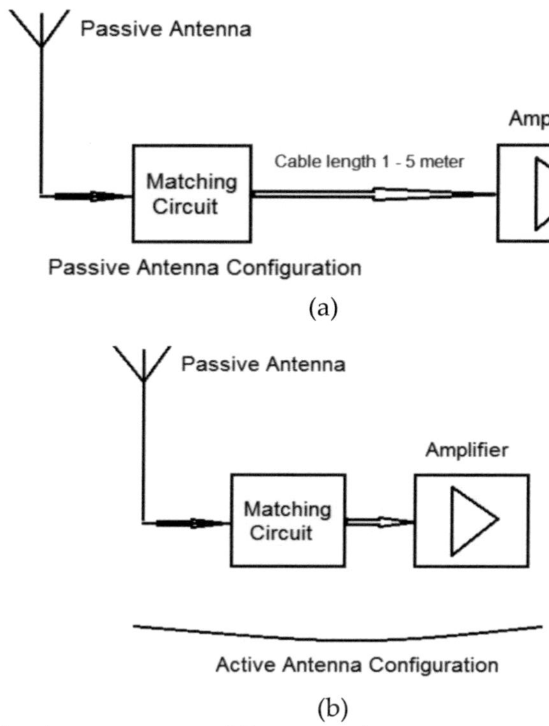

Figure 4. Passive and active antenna assembling comparison

5.2. Applications

The intelligent design of the antenna and integration with active circuit leads to innovative microwave and millimeter-wave application systems and considerable achievements in compactness, low cost, small profile, low power consumption, and multiple applications. This technology caused new designs in both areas of military and industrial applications such as wireless and radar communications, low cost sensors and transceivers [16].

The AIA concept has been extensively employed in the areas of power combining and quasi-optical power combining, beam steering and switching and retro directive arrays.

It also provides an effective solution to several fundamental problems at millimeter-wave frequencies including high transmission-line losses, limited source power, reduced antenna efficiency, and lack of high-performance phase shifters [15].

AIA is also used to design high-efficiency microwave power amplifiers recently. In other word, the antenna element is used as a part of circuit to terminate the harmonics at the amplifier output in addition to its traditional role of radiating electromagnetic waves.

Retro directive arrays are applicable in a wide range of applications such as self-steering antennas, radar transponders, search and rescue and wireless identification systems which are outcome of their omnidirectional coverage and the high level of gain. In these arrays any incident signal will reflect back toward the transmitter without prior knowledge of its location [15].

Transponders are circuits which can be activated by an external explorer system transmitting signal in predefined frequency. In this case, transponder will transmit a response signal to the interrogator. These small low-cost microwave transponders are used for noncontact identifications such as entry systems, toll collection, and inventory control.

Transceivers and millimeter-wave vehicle radars are some other applications which are used respectively for wireless local area networks (WLANs) and for intelligent cruise control.

Additionally AIA can be an ideal choose for designing compact transceivers and transponders for wireless applications. In this case the whole RF subsystem, including active circuit and antenna can be built on a single substrate [15].

Active antennas are categorized depending on the function of active circuit integrated with Them. The main functions of the devices in active antenna structures are generating and amplifying RF signals and frequency conversion. Based on previous discussion, the active antenna functions can be categorized into three types comprised of oscillator type, amplifier type and frequency type. This base unites can prepare possibility of more complicated functions by integrating with antenna such as transponders [16].

Some other benefits of using active antenna in microwave and millimeter wave frequencies are as below [17];

- Bandwidth increment and impedance matching improvement
- Improvement in sensitivity of receiver antennas and reduction of return loss and antenna dimensions
- Possibility of using active antenna arrays in microwave and millimeter wave signal generating
- Possibility of using active circuits in large antenna arrays which can eliminate the need to complicated RF circuits for phase shifters and advanced control electronics
- Development of using active antenna arrays in mobile communications
- Advancing beam steering techniques and their applications in smart antenna concept
- Possibility of using them to resolve channel capacity limitations by increasing data rates in future

As discussed in the previous part, for UWB applications, there are some limitations in using spiral antenna for broadband applications and it must be optimized to exhibit desirable characteristics. Using active antenna technology is one of effective optimization methods which leads to reduction of return losses and improvement in bandwidth, gain and input impedance [8].

To design a UWB active antenna with desirable parameters, study and design of an appropriate active circuit is an important step. Therfore distributed amplifier structure and parameters are reviewed in the next part as the active part of active antenna.

6. Distributed amplifier

Power amplifiers are essential parts of each transmitting system. They are used to amplifying signals to transmit from one point to another. Nonlinear effects of high power signals are the main differences between power amplifiers and the others.

In power amplifier design depending on application, higher efficiency, power and gain can be the main subjects of optimizations and linearity and noise figure are less considered.

Power amplifiers are categorized to different classes depending on active element bias and input/output signal forms.

Travelling wave structures are new methods to design of wideband amplifiers which have vast applications in wideband communications such as wideband travelling wave amplifiers, matrix amplifiers, travelling wave oscillators, mixers and power amplifiers . Concept and basic of travelling wave structures was initially originated by Percival in 1937. In 1940, this method was used to design of wideband vacuum tube amplifiers. But using GaAs MESFET in distributed amplifiers was studied at first by Moser in 1967 and Jutzi in 1969. They designed a distributed amplifier using lumped element technology and showed the ability of these circuits to achieve high gains in a wide frequency band [18].

6.1. Features

Because of the ultra-wide operating band of these amplifiers, they are receiving much attention. In a general amplifier, using parallel transistors lead to increasing the gain which is caused by summation of trans-conductances. But increasing input and output capacitors cause decrease in cutoff frequency. So as is shown in figure (5a), it does not solve the problem, because the multiplication of gain and bandwidth almost remains constant. In a distributed amplifier, low or high cutoff frequencies will be modified by summation of transistors trans-condoctances and realization of additional LC transmission lines in the input/output sides. The result is illustrated in figure (5b) [18].

7. Antenna design and simulation

One of the most important and commonly used parameters in antenna design is the radiation pattern in the space around antenna. By defining antenna radiation pattern[1], radiation power in each direction and the direction which maximum power is emitted will be defined.

Since selected passive antenna is spiral antenna in this work, its radiation pattern is half-space. So if interior and exterior radiuses be selected according to the active zone and frequency range, E-plane and H-plane radiation patterns will be uniform in all frequency bandwidth and have few variations. Changing design parameters has no effect on antenna polarization and it has always circular polarization. All simulations of this chapter are done in ADS (Advanced Design System) software [19].

[1] Radiation Pattern

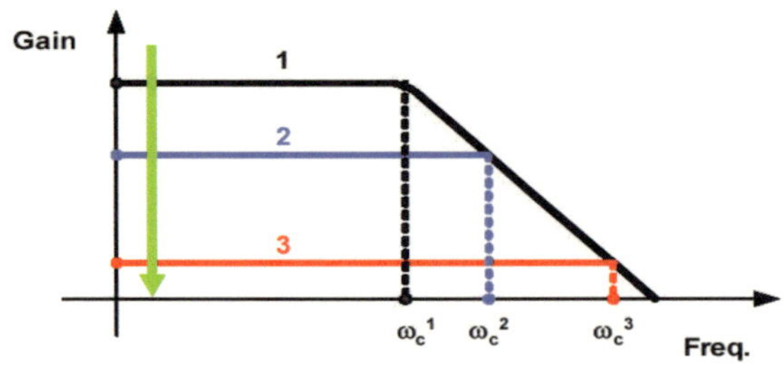

(a)

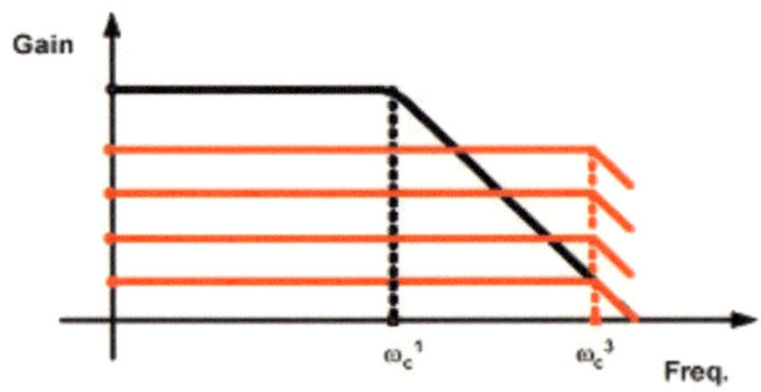

(b)

Figure 5. a).Trade-off between gain and bandwidth; b).high gain and wide bandwidth by distributed amplifier

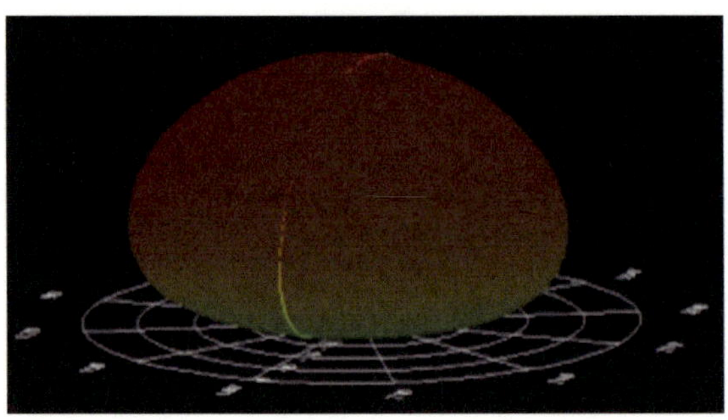

Figure 6. Spiral antenna radiation pattern in 3.36GHz

Other parameters which are effective in defining antenna performance are S_{11}[2], Z_{11}[3], VSWR[4] and Gain. Each of these parameters are simulated and illustrated for spiral antenna. Antenna dimensions are chosen similar to spiral antenna in reference [9], but substrate is different and a distributed power amplifier is used as the active part. To analysis of antenna dimensions and identify the active zone, relations (8) and (9) are used and radiation radius of high and low frequencies are calculated as below:

$$f_{low} = c / \pi r_2 \rightarrow 3 * 10^9 = 3 * 10^8 / \pi\, r_2 \rightarrow r_{out} = 31.8\ \mathrm{mm} \tag{8}$$

---

[2] return loss

[3] Input Impedance

[4] Voltage Standing Wave Ratio

$$f_{high} = c / \pi r_1 \rightarrow 11 * 10^9 = 3 * 10^8 / \pi\, r_1 \rightarrow r_{in} = 8.7\ mm \qquad (9)$$

r_{out} is exterior active zone radius for low frequencies and r_{in} is interior active zone radius for high frequencies. These dimensions show that the radius of r=40 mm is a good value. Simulation results are calculated and depicted in figure (7) and (8).

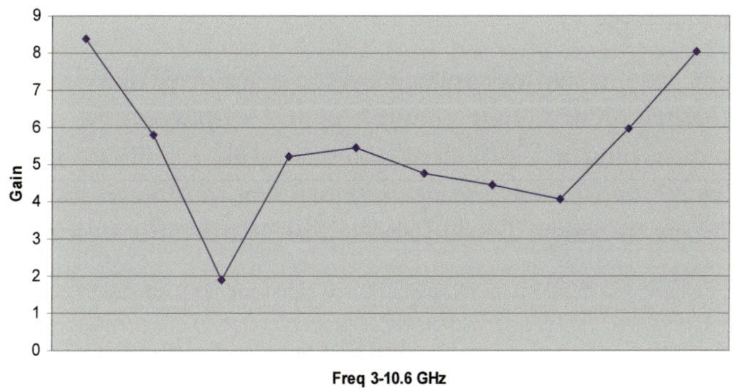

Figure 7. Antenna gain with w = 4.54m, s = 4mm and r = 40mm

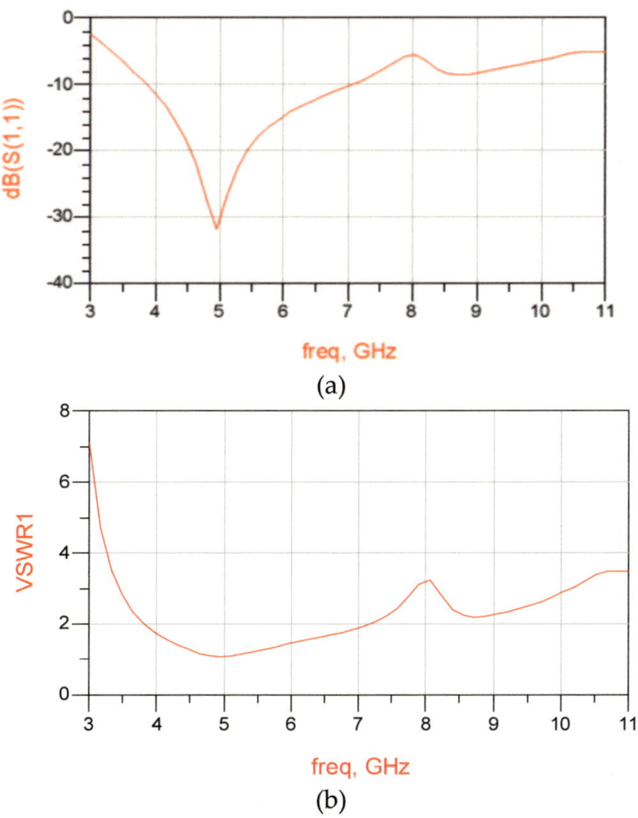

Figure 8. Spiral antenna parameters in the UWB Band; a) S_{11} and b) VSWR

Generally, in wireless communications, the antenna is required to provide a return loss less than -10dB over its frequency bandwidth [1]. As it is obvious, although spiral antenna parameters are good, to have desirable characteristics in the UWB band, optimization is necessary. In this work optimization is done by using a distributed amplifier and designing an active antenna [19].

7.1. Active circuit design

Active antennas are categorized depending on the active circuit behavior integrated with. Main functions of active antennas are generating and amplifying RF signals and frequency conversion. Here, designing a UWB distributed amplifier with uniform gain and return losses on the entire 3.1 to 10.6 GHz frequency band in the linear and nonlinear operation modes is the aim. Steps to design distributed amplifier are briefly described below;

Step 1. Selecting active element according to the project requirements

The first step is selecting a suitable active element which its linear and nonlinear models and parameters are accessible.

Step 2. Defining the optimum load resistance

After choosing suitable transistor, the optimum load resistance must be calculated to achieve maximum power in output

Step 3. Defining optimum number of amplifier stages

Optimum number of amplifier stages can be calculated using equation (10).

Step 4. Calculating 1dB point to define boundaries between linear and nonlinear regions of amplifier operation

For calculation of the optimum load resistance point, load power against load resistance curve is simulated as illustrated in figure (9). The load which gives maximum output power is the point [12]. Here Optimum Load Resistance is equal to 100 Ohms.

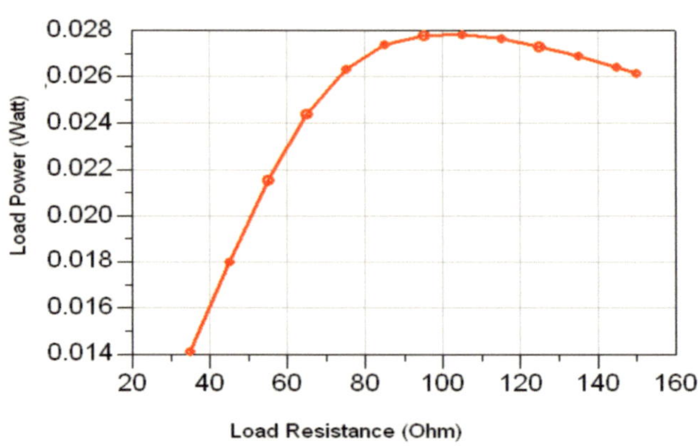

Figure 9. Load power against load resistance

After definition of optimum load and source resistance, now optimum amplifier stages can be calculated as bellow;

$$n_{opt} = \frac{Ln(\alpha_d) - Ln(\alpha_g)}{\alpha_d - \alpha_g} \tag{10}$$

Where α_d and α_g are drain and gate lines attenuation respectively. Calculation must be done in the highest frequency of bandwidth which is 11GHz because the optimum number of amplifier stages n_{opt} is calculated to have less reduction in gain and lower effects of attenuation constant on transistors in higher frequencies,.

Hear n_{opt} was equal to 3. Then the 1dB point was calculated to define linear and nonlinear operation modes of the amplifier. As is shown in figure (10), m2 is 1dB point at 7GHz. In the other word, for input powers below -16 dBm, amplifier will work in the linear mode and for higher input powers it works in the nonlinear operation mode.

Various methods can be used for linear and nonlinear analysis of such a circuit. Some of them are discussed in reference [19]. To more studies, see references [20 to 26].

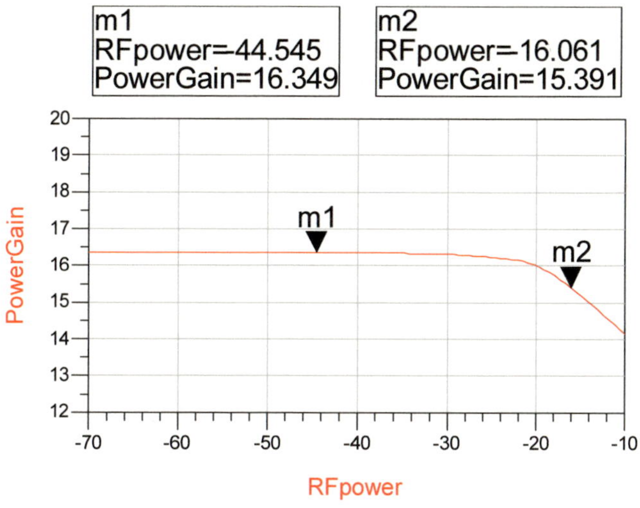

Figure 10. 1dB point calculation in 7GHz. RFpower is input power

7.2. Simulation

Distributed amplifier shown in figure (5) is simulated in the linear region of its operation with 50Ω load and matching network. Distributed amplifier structure is shown in figure (11) [11].

UWB distributed amplifier parameters are shown in figure (12). These results show a rather uniform gain and good return losses over this band. The results of this work and other experiences show that optimizing the integrated circuit of antenna and active circuit is more useful than optimizing each of them separately and then matching them by a matching network [11].

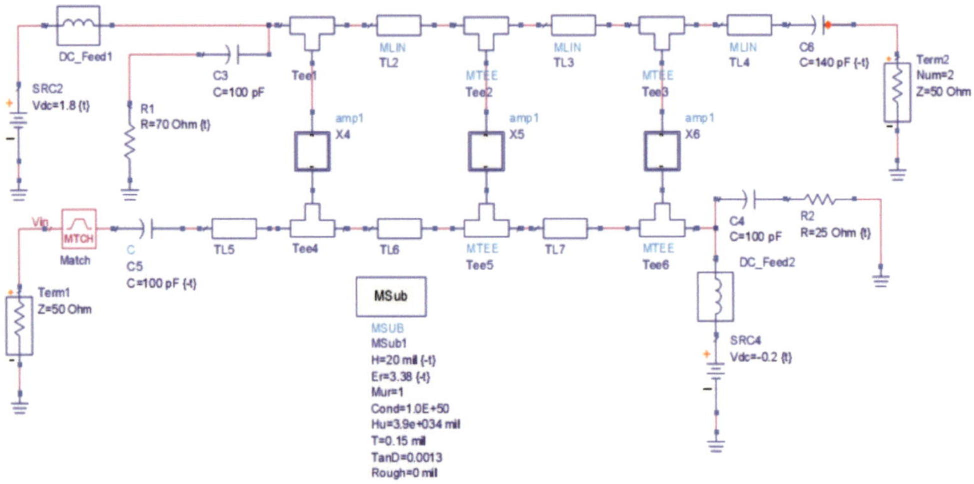

Figure 11. Designed distributed amplifier structure.

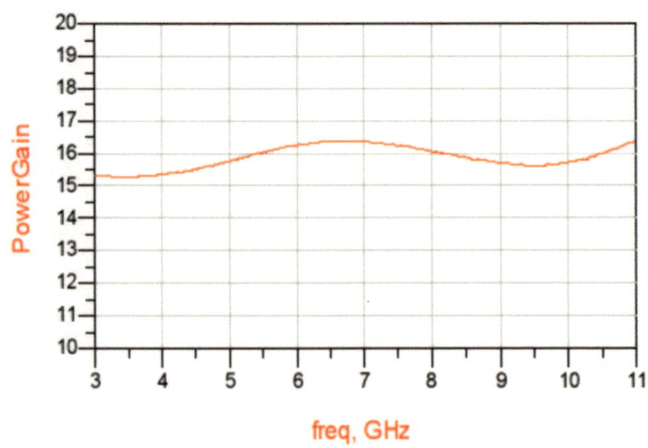

(a)

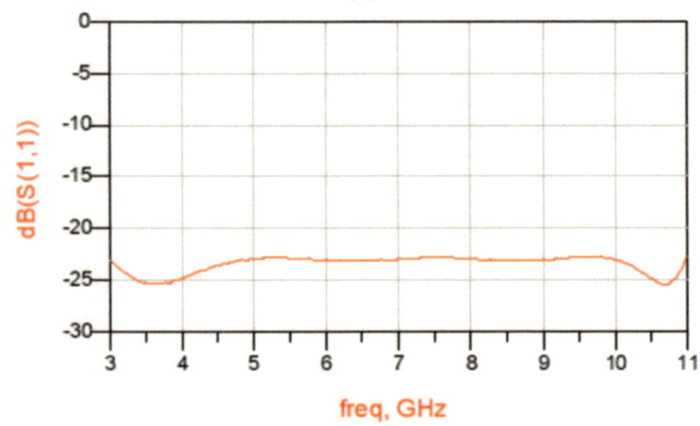

(b)

Figure 12. UWB distributed amplifier parameters; a) Gain. b) S_{11}

7.3. Active antenna simulation results

After design and simulation of passive spiral antenna and UWB distributed amplifier separately, spiral antenna is added to the active circuit as a load and specifications of this combined circuit is analyzed. For more accurate results, simulation is done by electromagnetic simulator "momentum" of ADS software. Simulated parameters of active antenna in linear mode are shown in figure (13).

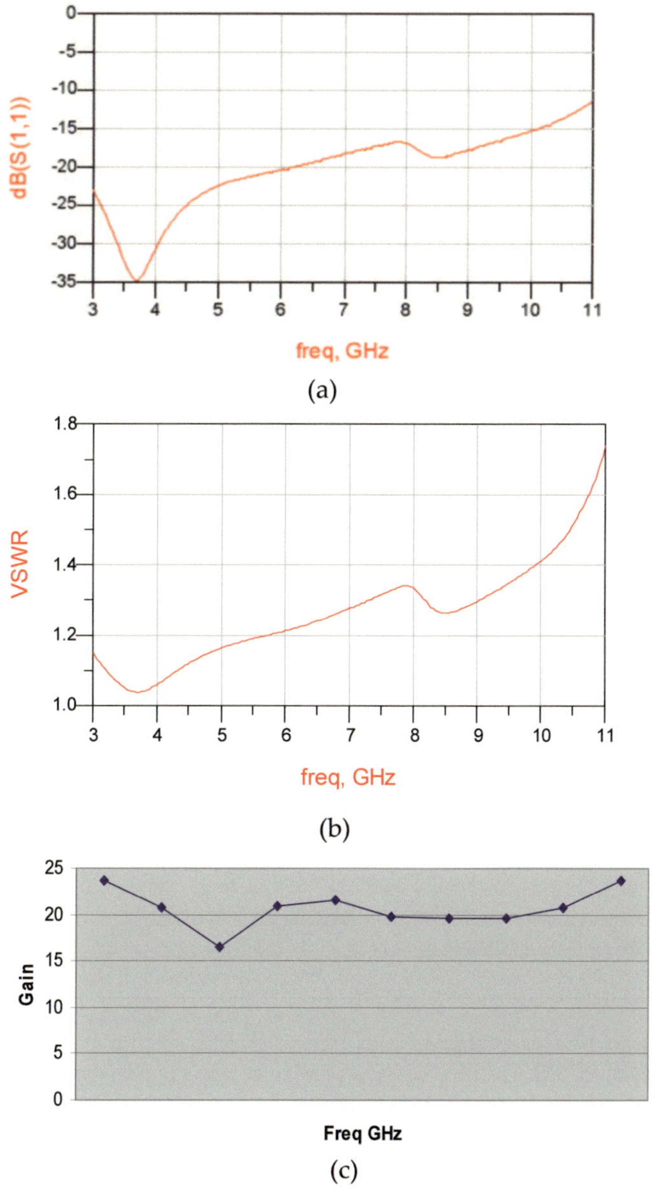

Figure 13. Linear active antenna parameters; a) S11, b) VSWR and c) Gain

In the figure (14), active and passive antenna simulation results are shown. Comparing these results shows that active antenna parameters are considerably optimized rather than passive one. These results were predictable when the features of active antenna were introducing and now it is approved. It is important to know that the final circuit gain is not equal to the summation of passive antenna and active circuit gains. Total gain is calculated by replacing passive antenna as a load to the active circuit and calculating circuit gain [19].

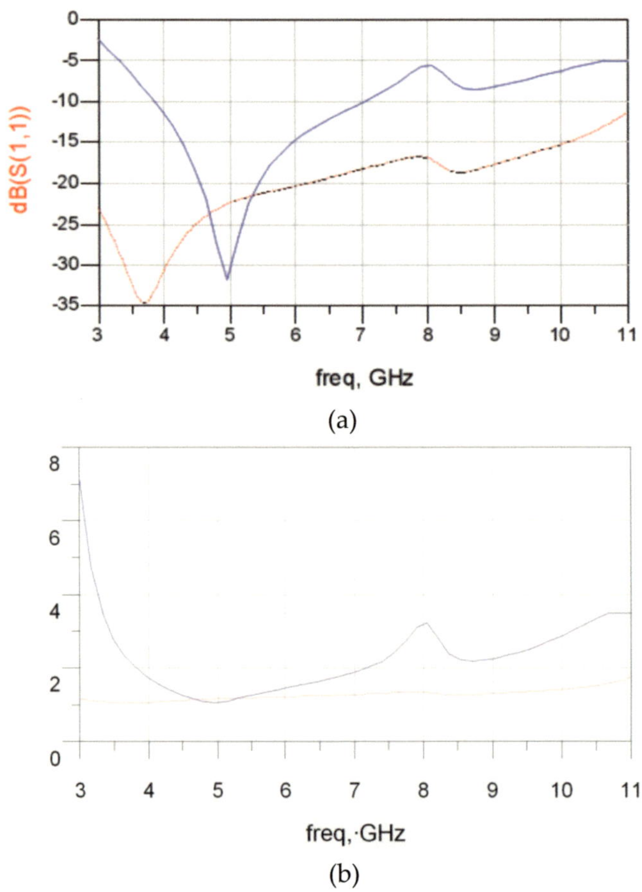

Figure 14. a) S_{11} and b) VSWR. The blue curve is passive antenna parameter and the red one is active antenna parameter. Active and passive antenna parameters comparison shows that linear active antenna parameters are optimized in compare with passive one.

In a similar work, M. Jalali et all. [27] optimized a spiral antenna using active integrated antenna technology for linear operation mode. Results are shown in figure (15).

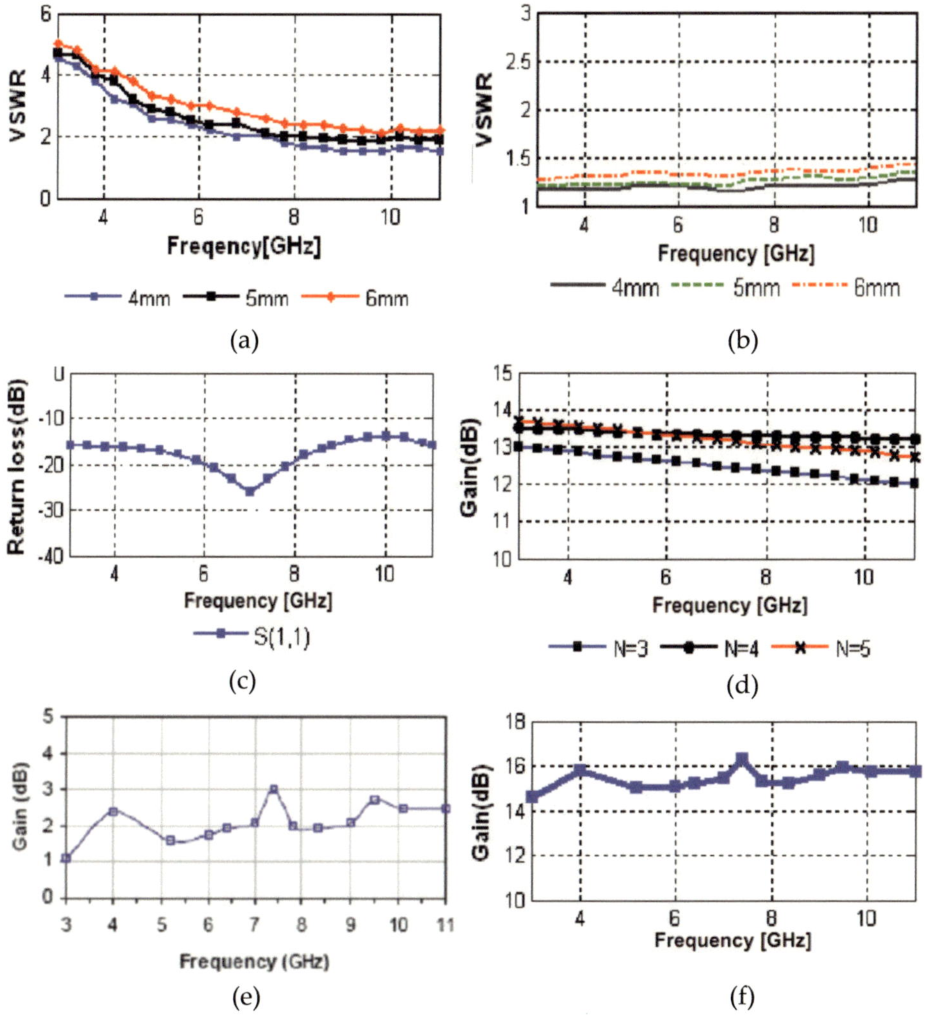

Figure 15. Passive antenna parameters and optimized parameters of active antenna [27]; a) VSWR of passive spiral antenna in w=4, 5 and 6 mm; b) VSWR of optimized active antenna in w=4, 5 and 6 mm; c) Return loss of optimized active antenna; d) Uwb distributed amplifier gain for N=3,4 and 5 (N is the turns of spiral antenna); e) Gain of passive spiral antenna; f) Gain of optimized active antenna

By using active circuit specifications calculated in part 7, nonlinear simulation of active antenna is done here. In this case, input power is more than the power calculated for 1dB point and antenna works in nonlinear mode. For nonlinear analysis of antenna, large signal model and parameters of cicuit elements must be used. For special applications and higher output power requirements, power amplifiers and nonlinear operation may be considered. Results are shown in figure (16). For more information see [19].

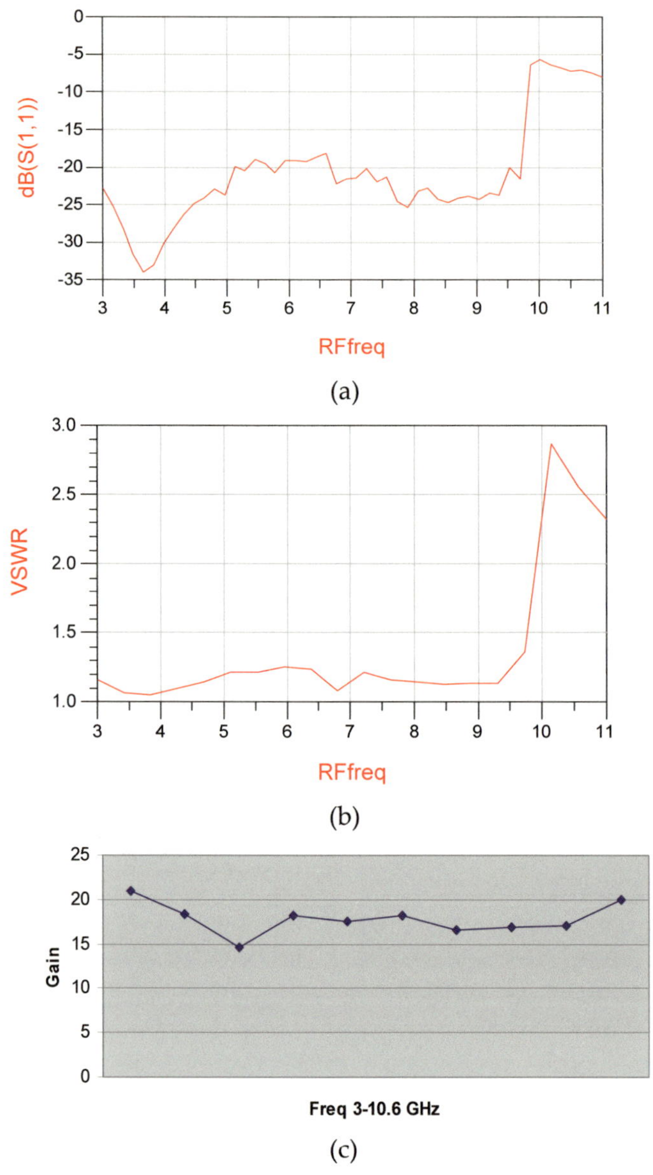

Figure 16. Nonlinear active antenna parameters; a) S11. b) VSWR. c) Gain

8. Conclusion

In the designed UWB active antenna, parameters are almost uniform in the entire band. It is again emphasized that one of very important requirements of UWB systems is having uniform parameters on the all 3.1 to 10.6 GHz frequency band, because all signal frequency elements must amplify uniformly to not distort transmitted signals. Comparing with results in reference [8], gain is increased about 5dB and final amplifier stages are reduced from 4 to 3 and return losses are significantly low. Comparing figures (7 and 13), shows usefulness of adding active circuit to passive antenna to increase gain and make antenna parameters uniform and desirable. This antenna can also amplify narrow band signals which are in this frequency band in addition to amplifying UWB signals with frequency elements in whole band.

Author details

Danial Nezhad Malayeri
South of Kerman Power Distribution Company, Kerman, Iran

Acknowledgement

I thank to Dr Ahmad Hakimi from Shahid Bahonar University of Kerman for his guidance and assistance and Dr Abdipour from Tehran Amirkabir University of Technology. This work was supported by South of Kerman Power Distribution Company and Telecommunication Research Center of Iran.

9. References

[1] J. Liang. Antenna Study and Design for Ultra Wide Band Communication Applications. Department of electronic engineering Queen Mary, University of London, United Kingdom, July 2006
[2] FCC, First Report and Order 02-48, February 2002.
[3] J. Igor Immoreev, "Ultra-wideband Systems. Features and Ways of Development", Ultra Wideband and Ultra Short Impulse Signals, 19-22, Sevastopol, Ukraine, September, 2004.
[4] J. M Wilson, "Ultra-Wideband /a Disruptive RF Technology", Intel research & development, Version 1.3, September 10, 2002.
[5] M. Tanahashi, "Design Technologies of RF devices for UWB Simulation", Ansof High Performance Applications Workshop, TRDA Inc. / Taiyo Yuden, October 27, 2005.
[6] V.H. Rumsey, "Frequency Independent Antenna," IRE National Convention Record, pt.1, pp. 114-118. 1957.
[7] W.R. Deal, V. Radisic, Y. Qian and T. Itoh. "RF Technology for Low Power Wireless Communications ", Wiley & Sons Inc. 2001
[8] M. Pourjalali, "Analysis, design and simulation of active antenna in UWB band ", Tehran Amirkabir University of Technology, MSc. Thesis. 2006
[9] J. Lin, and T. Itoh, "Active Integrated Antennas", IEEE Transactions on Microwave Theory and Techniques, Vol 12. 12th December 1994.
[10] J. R. Copeland, W. J. Robertson, and R G. Verstraete, "Antennafier arrays," IEEE Trans. Antennas Propagate, Vol. AP- 12, pp. 227-233, Mar. 1964.
[11] E. Marzolf, M.H. Drissi, Global Design of an Active Integrated Antenna for Millimeter Wave.

[12] D. E.J. Humphrey, V.F. Fusco and S. Drew, "Active Antenna Array Behavior", IEEE Transactions on Microwave Theory and Techniques, Vol. 43, No. 8, August 1995.

[13] F. Carrez, R. Stolle, J. Vindevoghel, "Integrated Active Microstrip Antenna for Communication and Identification Applications", IEEE. 1997.

[14] A. Ramadan, Active Antennas with High Input Impedance Low Noise and Highly Linear Amplifiers. Universitat der Bundeswehr München, July 2005

[15] Y. Qian, and T. Itoh, "Progress in Active Integrated Antennas and Their Applications", IEEE Transactions on Microwave Theory and Techniques, Vol. 46, No. 11, November 1998

[16] R. Garg, P. Bhartia, I. Bahl, A. Ittipiboon, Microstrip Antenna Design Handbook, Artech House, Inc. 2001.

[17] S.L. Loyka, J.R. Mosig, "Nonlinear Modeling and Simulation of Active Array Antennas", 5th COST 260 MC & WG Meeting, Wroclaw, Poland, 30 June 1999.

[18] S. Asadi, "Design and Simulation of UWB Distributed Power Amplifier ". Amirkabir university of Tehran (Politechniques), MSc. Thesis, 2006

[19] D. Nezhad Malayeri: "Design, Simulation and Nonlinear Analysis of UWB Active Antenna" Shahid Bahonar university of Kerman. MSc. Thesis, 2009

[20] V. Rizzoli, A. Costanzo and P. Spadoni, "Computer-Aided Design of Ultra Wide band Active Antennas by Means of a New Figure of Merit", IEEE Microwave and Wireless Components Letters, Vol. 18, No. 4, April 2008.

[21] C.C. Penalosa, "Numerical Stady-State Analysis of Nonlinear Microwave Circuits with Periodic Excitation", IEEE Transactions on Microwave Theory and Techniques, vol. MTT-31, p. 724-730, Sept. 1983.

[22] N. Soveiko, and M. Nakhla, "Wavelet Harmonic Balance", IEEE Microwave and Wireless Components Letters, December 12, 2002.

[23] Y. Zhang, P. Gardner, H. Ghafouri-Shiraz and P.S. Hall, "FDTD Analysis of Microwave Active Integrated Antenna Including Nonlinear Model of FET Transistor", 3rd International Conference on Microwave and Millimiter wave Technology Proceedings, 2002.

[24] J. Zhang, Y. Wang, and Y. Chen, "Analysis of Active Antenna by using Piecewise Harmonic Balance Technique and Neural Network", International Journal of Infrared and Millimeter Waves, Vol. 18., No. 5, 1997.

[25] R. Goyal and V. Veremey, "Application of Neural Networks to efficient Design of Wireless and RF Circuits and Systems", Xpedion Design Systems, Inc.

[26] V. Rizzoli, A. Neri, D. Masotti and A. Lipparini, "A New Family of Neural Network-Based Bidirectional and Dispersive Behavioral Models for Nonlinear RF/Microwave Subsystems", (InvitedPaper), Wiley & Sons, Inc., John 2002.

[27] M. Jalali, A. Abdipour, A. Tavakoli, and G. Moradi, "Performance Amelioration of an Active Integrated Spiral Antenna System for UWB Applications", Iranian Journal of Electrical and Computer Engineering, Vol. 7, No. 1, Winter-Spring 2008

Chapter 10

Multiple-Input Multiple-Output Antennas for Ultra Wideband Communications

Ali Imran Najam, Yvan Duroc and Smail Tedjini

Additional information is available at the end of the chapter

http://dx.doi.org/10.5772/50181

1. Introduction

UWB is a very promising technology for short-range wireless communications providing the opportunity of high data rate communications. In 2002, the Federal Communication Commission (FCC) regulated the UWB technology utilization for commercial applications in the United States in the frequency range of 3.1–10.6 GHz [1]. Other than the United States, UWB regulations have been issued in Europe, Japan, Korea and Singapore. These regulations did not stipulate the technology type to be used. Later, two distinct techniques were envisaged: the Multi-band Orthogonal Frequency Division Multiplexing (MB-OFDM) and the Impulse-UWB (I-UWB) [2]. The MB-OFDM divides the UWB spectrum in 14 sub-bands, the utilization of the bands is managed for a code time-frequency exploiting the spatial-temporal diversity [3], while the I-UWB transmits pulses of very short duration that occupy the entire allowable frequency band [4]. UWB has vast array of applications in wireless world. The dominant applications include WBAN, WPAN, RFIDs, sensor networks, radars, etc. The relevant IEEE standards for UWB are: 802.15.3a for high data rate and 802.15.4a for low data rate. Digital communication using Multi-Input Multi-Output (MIMO) processing has emerged as a breakthrough for wireless systems of revolutionary importance. All wireless technologies face the challenges of signal fading, multipath, increasing interference and limited spectrum. MIMO technology exploits multipath to provide higher data throughput, and simultaneous increase in range and reliability all without consuming extra radio frequency. Early studies conducted by Foschini and Gans [5] indicated that capacity increases were possible by using MIMO systems. In a rich scattering environment, Telatar showed that the capacity of system consisting of M transmitter and N receiver antennas is $min(M,N)$ times that of a single transmitter receiver system [6]. MIMO systems exploit the antenna diversity (spatial, polarization or pattern diversity) to increase the strength of the transmitted signals and therefore to improve the Signal Noise Ration

(SNR). Spatial multiplexing in MIMO systems helps in increasing data rate. Beamforming is used either to increase data rate or to strengthen the signal. The applications of MIMO include digital television (DTV), wireless local area networks (WLANs), metropolitan area networks (MANs), and mobile communications. IEEE standard related to MIMO technology is 802.11n.

Where one says that UWB is a classical solution to the demand of high data rate communications, there arise the questions while thinking of UWB and MIMO together in the wireless systems. Obviously, it can be answered in terms of the reasons rather motivations behind this combination of two technologies. In effect, it is well known that the main applications of UWB technology are found for WPAN and WBAN (Wireless body area network) in indoor environments where the dense multipath propagation leads to generally detrimental Inter Symbol Interference (ISI). Therefore, to turn this drawback into an advantage, multiple antennas or MIMO techniques can be employed to exploit such rich scattering environments. The more important is that the applications of UWB are limited to the short distance communication due to very low transmission power allowed by the FCC. Hence, using MIMO together with UWB helps in extending the communication range as well as offers higher link reliability. The benefits of UWB-MIMO can be summarized as following [7]-[8]: interference mitigation/suppression, higher data rates, improved link quality, extended coverage, reduced analog hardware requirements, and concurrent localization. Apart from these benefits, there are also challenges for the joint implementation of UWB and MIMO. These challenges include: UWB-MIMO signaling trade-offs, UWB-MIMO channel modeling, the optimization of UWB-MIMO modulation schemes, design of compact and suitable UWB antenna arrays, efficient and cost-effective UWB-MIMO RF circuit design, etc. Among these challenges, the design of compact and suitable UWB diversity antennas has appealed many researchers to work on this topic. The significance of antenna in a wireless communication system cannot be avoided. It also becomes a critical element to be miniaturized along with the other circuit elements. The design of antenna faces a lot of challenges itself in this race. As devices are going to be more compact, therefore the antennas must be positioned within the available space. Moreover, MIMO can be implemented by three ways: beamforming, spatial multiplexing and diversity techniques. Diversity techniques, more specifically antenna diversity techniques (i.e., spatial, polarization and pattern), are adopted for MIMO antenna designs. Hence, the implementation of multi-antenna structures becomes more challenging in the very limited space provided by the small terminals. The limitations to exploit the diversity arise when the antennas are placed in close vicinity. So, it is required to decorrelate their patterns, or in other words, mutual coupling should be minimized. Another challenge is the enhancement of isolation between the access ports of MIMO antennas. In the literature, a relative few MIMO antenna designs have been presented for UWB systems. In this context, this chapter presents an overview of MIMO UWB antennas with the main following objectives: to highlight the additional parameters required to characterize the performance of UWB antennas as well as MIMO antennas, to bring a state of the art in MIMO antennas for UWB systems, and to present a state of art in techniques to be used to reduce mutual coupling and enhance the isolation. This chapter also describes some of our proposed designs and structures of the different types of MIMO antennas for UWB applications exploiting spatial, polarization and pattern diversities, and a solution to enhance the isolation with reduced size of antenna.

2. Antenna theory

The antennas are an essential part of any wireless system. According to the IEEE Standard Definition of terms for Antennas, an antenna is defined as "a means for radiating and receiving radio waves" [9]. In an advanced wireless system, an antenna is usually required to optimize or accentuate the radiation energy in some directions and suppress it in others at certain frequencies. A good design of the antenna can relax system requirements and improve overall system performance. To describe the performance of an antenna, there are several commonly used antenna parameters, including impedance bandwidth, radiation pattern, directivity, gain, efficiency, and polarization [10].

2.1. Specific parameters for UWB antennas

In UWB systems, the previous fundamental and classical parameters must be considered in designing antennas but there are more challenges to monitor them and some additional parameters.

Bandwidth - First of all, what distinguishes a UWB antenna from other antennas is its ultra wide frequency bandwidth. According to the FCC's definition, a suitable UWB antenna should be able to yield an absolute bandwidth of not less than 500 MHz or a fractional bandwidth of at least 0.2. Moreover, UWB antenna must be operable and must have stable impedance matching over the entire 3.1-10.6 GHz frequency range in the case of I-UWB following the FCC defined spectral mask. Sometimes, it is also demanded (e.g., in Europe) that the UWB antennas should provide the band-rejected characteristic to coexist with other narrowband devices and services occupying the same operational band [11].

Radiation Pattern - Directional or omni-directional radiation properties are needed depending on the practical application. Omni-directional patterns are normally desirable in mobile and hand-held systems. For radar systems and other directional systems where high gain is desired, directional radiation characteristics are preferred. High radiation efficiency is usually required for antennas but it is imperative and essential for an ultra wideband antenna because the transmit power spectral density is excessively low. Therefore, any excessive losses incurred by the antenna could potentially compromise the functionality of the system.

Size and Cost - A suitable antenna needs to be small and of light weight enough to be compatible to the application. As we are projecting UWB for the applications that include especially mobile and portable devices, therefore it is highly desirable that the antenna should feature low profile and compatibility for integration with printed circuit board (PCB).

Specific parameters to be required to characterize UWB antennas are now described.

Compliance with Spectral Masks - A good design of UWB antenna should be optimal for the performance of overall system. To avoid the possible inband/outband interference between the UWB systems and existing electronic systems, the antenna should be designed such that the overall device (antenna and RF front end) complies with the mandatory power emission mask given by the FCC or other regulatory bodies. The emission limits will be determined by both the selection of source pulse and design of antennas in UWB systems.

Impulse Response - As the origin of UWB technology stems from time-domain electromagnetics, therefore UWB antenna is required to achieve good time domain characteristics (i.e., good impulse response). The idea is simply to characterize the LTI (Linear Time Invariant) system by its response to an impulsive excitation instead of amplitude and phase and measurements versus frequency (i.e., swept frequency response). For the narrowband case, it is approximated that an antenna has same performance over the entire bandwidth and the basic parameters, such as gain and return loss, have little variation across the operational band. In contrast, I-UWB systems often employ extremely short pulses for data transmission. In other words, enormous bandwidth has been occupied, thus the antenna can't be treated as a "spot filter" any more but a "band-pass filter". In this case, the antenna imposes more significant impacts on the input signal. As a result, a good time domain performance (i.e., minimum pulse distortion in the received waveform) is a primary concern of a suitable UWB antenna because the signal is the carrier of useful information [12]. Therefore, it is indispensable and important to study the antenna's characteristics in time domain.

Group Delay - It is an important parameter that represents the degree of distortion of UWB signal. Group delay is a measure of the slope of the transmission phase response. The linear portion of the phase response is converted to a constant value and deviation from linear phase are transformed into deviations from constant group delay. The variations in group delay cause signal distortion, just as deviations from linear phase cause distortion. It can be given as

$$group\ delay = -\frac{\Delta\varphi}{\Delta\omega} \tag{1}$$

where φ is the total phase shift in radians, and ω is the angular frequency in radians per unit time, equal to $2\pi f$, where f is the frequency. The group delay variations induced by the radiation pattern of the antenna will affect the overall receiver system performance, since it can bring relatively large timing errors. An antenna gain versus frequency without nulls, means a linear phase response, hence a constant group delay.

In summary for the applications of portable devices, general specifications required to design UWB antennas under the FCC regulations can be summarized in Table 1.

Parameter	Value
VSWR bandwidth	3.1-10.6 GHz
Radiation pattern	Omnidirectional
Radiation efficiency	High (> 70%)
Gain	Smooth in the band of interest
Phase	Linear ; nearly constant group delay
Physical profile	Small, compact, planar

Table 1. Characteristics of UWB antenna for portable devices.

2.2. Specific parameters for MIMO antennas

Like the case of UWB, there are also some additional parameters other than the fundamental parameters to be taken into account while designing MIMO antennas.

Mutual Coupling and Isolation - In MIMO applications, the signals transmitted by multiple antenna elements are generally supposed to be independent or uncorrelated. But in reality, the current induced on one antenna produces a voltage at the terminals of nearby elements, termed as mutual coupling [13]. It means there is always mutual coupling present between nearby antenna elements. However, for MIMO applications, the mutual coupling should be minimized to as low value as possible. In a contradictory way, it should be noted that it is also studied that mutual coupling can help to reduce the correlation between the different channel coefficients in nearby placed antenna elements scenario, thus escalating the capacity [14]. This is an important issue for the antenna community. In a general way the coupling has an adverse effect and mutual coupling has to minimize [15].

The port-to-port isolation is defined as the transmission of power between two of the input ports of the multiport antenna under test. It is characterized by $|S_{21}|$ parameter. It must be noted that isolation is a positive quantity and is given as

$$Isolation = -10log_{10}|S_{21}|^2 \tag{2}$$

In MIMO systems, to maximize the energy radiated by an antenna, it should be ensured that negligible amount of transmitted energy is lost into the ports of other antennas terminated by the matched impedances. In other words, MIMO systems require the $|S_{21}|$ to be minimized to as low value as possible as isolation is directly related to the antenna efficiency. A lot of research has been done on the reduction of mutual coupling and the enhancement of the isolation. It is worth mentioning that the mutual coupling is characterized most of the times by the isolation in the literature. However, in [16], it is stated that isolation is not the exact representation of mutual coupling, as it is possible that there is very good isolation but it is not necessary for mutual coupling to be low in this case. Hence, to evaluate the mutual coupling, it is better to observe the surface current distributions on the non-excited radiating element, when nearby radiating element, is excited. Although the ports may be isolated, there is a possibility of having large induced currents in the neighboring antenna, which, in turn, affects the radiation pattern of the antenna considered. It is good to mention that we will take both mutual coupling and isolation into account differently unlike majority of the authors (Table 2) who have given S_{21} parameter as mutual coupling. However, we will always present surface current distributions as well. Further, the techniques to reduce mutual coupling and to enhance the isolation are discussed in detail in section 3.

Mean Effective Gain - The performance of MIMO systems is also characterized by the mean effective gain (MEG) of the antennas. The MEG is a statistical measure of the antenna gain that can be defined as the ratio of the mean received power of the antenna and the total mean incident power. It can be expressed by the following equation as in [17]

$$MEG = \int_0^{2\pi}\int_0^{\pi}\left(\frac{XPR}{1+XPR}G_\theta(\theta,\varphi)P_\theta(\theta,\varphi) + \frac{1}{1+XPR}G_\varphi(\theta,\varphi)P_\varphi(\theta,\varphi)\right) sin\theta d\theta d\varphi \tag{3}$$

where P_θ and P_φ are the angular diversity functions of the incident power with respect to θ and φ directions respectively, G_θ and G_φ are the gains with respect to θ and φ directions respectively, and XPR represents the cross-polarization power gain which is defined as

$$XPR = \frac{\int_0^{2\pi}\int_0^{\pi} P_\theta(\theta,\varphi) sin\theta d\theta d\varphi}{\int_0^{2\pi}\int_0^{\pi} P_\varphi(\theta,\varphi) sin\theta d\theta d\varphi} \tag{4}$$

The MEG is a normalized measure of the received power, where the powers (either P_θ or P_φ) are normalized as [18]

$$\int_0^{2\pi}\int_0^{\pi} P_\theta(\theta,\varphi)\, sin\theta d\theta d\varphi = 1 \tag{5}$$

and the gains are normalized in such a way that

$$\int_0^{2\pi}\int_0^{\pi}\left(G_\theta(\theta,\varphi)+G_\varphi(\theta,\varphi)\right) sin\theta d\theta d\varphi = 4\pi \tag{6}$$

In the case where the antenna is located in a totally random channel environment, the value of XPR is 1 and $P_\theta = P_\varphi = 1/4\pi$. Then, MEG can be calculated using (3) as follows:

$$MEG = \int_0^{2\pi}\int_0^{\pi}\left(\frac{1}{1+1}\frac{1}{4\pi}G_\theta(\theta,\varphi) + \frac{1}{1+1}\frac{1}{4\pi}G_\varphi(\theta,\varphi)\right) sin\theta d\theta d\varphi = \frac{1}{2} \tag{7}$$

The MEG is then equal to the total antenna efficiency divided by two or -3 dB [19] and it is independent of the radiation patterns. In order to achieve good diversity gain, the ratio of the MEG between the two antennas should close to unity in order to ensure that average received power by each antenna is nearly equal [20].

Correlation Coefficient - The correlation coefficient is a parameter of great importance for the systems providing diversity. The signals received in the diversity systems can be correlated to some extent. The correlation coefficient is a mathematical and statistical tool that measures the degree of similarity among the received signals. Its modulus varies from 0 to 1. Ideally, diversity systems require a correlation coefficient of zero or low by default.

This parameter can be viewed by three ways: complex, envelope and power correlation coefficients. Complex correlation coefficient gives the complex measure of correlation between received signals at the antennas. It can be given as [18]

$$\rho_c = \frac{\int_0^{2\pi}\int_0^{\pi}\left(XPRE_{\theta k}(\theta,\varphi)E_{\theta l}^*(\theta,\varphi)P_\theta(\theta,\varphi)+E_{\varphi k}(\theta,\varphi)E_{\varphi l}^*(\theta,\varphi)P_\varphi(\theta,\varphi)\right)sin\theta d\theta d\varphi}{\sqrt{\sigma_k^2\sigma_l^2}} \tag{8}$$

where σ_k^2 and σ_l^2 represent the variances of kth and l^{th} branches and can be written mathematically as

$$\rho_c = \int_0^{2\pi}\int_0^{\pi}\left(XPRG_{\theta k}(\theta,\varphi)P_\theta(\theta,\varphi) + G_{\varphi k}(\theta,\varphi)P_\varphi(\theta,\varphi)\right) \tag{9}$$

also

$$G_{\theta k}(\theta,\varphi) = E_{\theta k}(\theta,\varphi)E_{\theta l}^*(\theta,\varphi) \tag{10}$$

$$G_{\varphi k}(\theta,\varphi) = E_{\varphi k}(\theta,\varphi)E_{\varphi l}^*(\theta,\varphi) \tag{11}$$

where $E_{\theta k}$ and $E_{\varphi k}$ are complex electric fields in the directions θ and φ respectively for the kth antenna. Similar expressions are valid for l^{th} antenna.

Usually, the envelope correlation is presented to evaluate the diversity capabilities of MIMO systems [21]. This parameter is always real and by definition gives the correlation among the amplitudes of the signals at antennas. For Rayleigh fading channel, the envelope correlation can be given as follows:

$$\rho_e = |\rho_c|^2 \tag{12}$$

It is clear that correlation should be preferably computed from 3D radiation patterns but it becomes tedious. However, assuming that the diversity system will operate in a uniform multipath environment, the correlation coefficient can be calculated from S-parameters using the following equation in [22]

$$\rho_e = \left|\frac{S_{11}^*S_{12}+S_{21}^*S_{22}}{\sqrt{1-|S_{11}|^2-|S_{21}|^2}\cdot\sqrt{1-|S_{22}|^2-|S_{12}|^2}}\right|^2 \tag{13}$$

It offers a simple procedure compared to the radiation pattern approach, but it should be emphasized that this equation is strictly valid when the following assumptions are fulfilled:

- Antennas should have high efficiency and no mutual losses.
- Antenna system is positioned in a uniform multipath environment which is not strictly the case in real environments; however, the evaluation of some prototypes in different real environments has already shown that there are no major differences in these cases.
- Load termination of the non-measured antenna is 50 Ω. In reality, the radio front-end module does not always achieve this situation, but the 50 Ω evaluation procedure is commonly accepted.

All these limitations are clearly showing that in real systems the envelope correlation calculated based on of the help of the S_{ij} parameters is not the exact value, but nevertheless is a good approximation. In addition, it should be noted that antennas with an envelope correlation coefficient less than 0.5 are recognized to provide significant diversity performance.

Diversity Gain - The diversity gain (DG) is a figure of merit used to quantify the performance level of diversity techniques. The DG is the slope of the error probability curve in terms of the received SNR in a log-log scale. However, the DG can also be defined as the increment of the SNR at a given probability, normally 1% or 10% [23]. Such DG can easily be calculated by looking at the cumulative distribution function (CDF) curves of the SNR, and comparing the combined SNR using some specific diversity technique with the SNR of an un-coded SISO communication system. Mathematically, it can be expressed as

$$DG = \frac{(SNR)_c}{(SNR)_r} \tag{14}$$

where indices "c" and "r" are used for the combined and the reference. In this context, DG can be defined as the difference between a combined CDF as compared to a reference CDF at a certain level of CDF [24]. Depending on the reference CDF, it is possible to write three definitions for the diversity gain:

- Apparent diversity gain - Difference between power levels in dB (at certain CDF level), between CDF of combined signal, and CDF of signal at the port with the strongest average signal levels.
- Effective diversity gain - Difference between power levels in dB (at certain CDF level), between CDF of combined signal, and CDF of signal at the port of an ideal single antenna (corresponding to radiation efficiency of 100%), measured in the same environment.
- Actual diversity gain - Difference between power levels in dB (at certain CDF level), between CDF combined signal, and CDF of signal at the port of an existing practical single antenna that is to be replaced by the diversity antenna under test, measured at the same location (for example, relative to a head phantom).

The DG is also related to the correlation coefficient. The relation between DG and correlation coefficient can be given approximately by

$$DG = 10\sqrt{1 - |\rho|^2} \tag{15}$$

This relationship clearly shows that the lower the correlation coefficient the higher will be the diversity gain. Therefore, high isolation is required between the antennas otherwise the DG will be low. Further, whatever the combining method is being used, the maximum diversity gain is obtained when the correlation coefficient is zero. The measurements in different types of environments (urban, suburban, rural and motorways) have made possible to write empirical relationships [25] for each type of combination scheme. These formulae of diversity gain have been given as follows:

For Selecting Combining scheme,

$$DG(dB) = 5{,}71\, exp\left(-0.87\sqrt{\rho_e} - 0.16\Delta\right) \tag{16}$$

For Equal Gain Combining method,

$$DG(dB) = -8.98 + 15.22\, exp\left(-0.20\sqrt{\rho_e} - 0.04\Delta\right) \tag{17}$$

And for Maximal Ratio Combining method,

$$DG(dB) = 7.14\, exp\left(-0.59\sqrt{\rho_e} - 0.11\Delta\right) \tag{18}$$

Thus, in ideal conditions ($\rho_e = 0$; $\Delta = 0$), MRC (Maximum Ratio Combining) scheme gives the best diversity gain, i.e., 7.41 dB.

Total Active Reflection Coefficient - The reflection coefficient does not accurately characterize the radiation efficiency and bandwidth of a MIMO antenna. Instead of simple reflection coefficient, the array's total active reflection coefficient (TARC) can be used so that it accounts for both coupling and random signal combination. Thus, TARC provides a more meaningful measure of MIMO efficiency. For a desired port excitation, summation of the available power at all excitation ports is assumed as incident power, radiated power as transferred power, and the difference between these two as reflected power. The square root of the ratio of reflected power and incident power is defined as the TARC [26], mathematically given by

$$\Gamma_a^t = \frac{available\ power - radiated\ power}{available\ power} \tag{19}$$

For instance, TARC for a lossless N-port antenna can be described as

$$\Gamma_a^t = \sqrt{\sum_{i=1}^{N} |b_i|^2} \Big/ \sqrt{\sum_{i=1}^{N} |a_i|^2} \tag{20}$$

where $[b] = [S] \cdot [a]$, a_i is the incident signal vector with randomly phased elements and b_i is the reflected signal vector.

Furthermore, for 2×2 network, the scattering matrix can be written as

$$\begin{pmatrix} b_1 \\ b_2 \end{pmatrix} = \begin{pmatrix} s_{11} & s_{12} \\ s_{21} & s_{22} \end{pmatrix} \begin{pmatrix} a_1 \\ a_2 \end{pmatrix} \tag{21}$$

It can be assumed that the reflected signal will be randomly phased with independent and identical distributed Gaussian random variables because MIMO channels are assumed as Gaussian and multipath spread in the propagation channel. Since sum or difference of independent Gaussian random variables is Gaussian, reflected signals are characterized as

$$b_1 = s_{11}a_1 + s_{12}a_2 = s_{11}a_0e^{i\theta_1} + s_{12}a_0e^{i\theta_2} = a_1\left(s_{11} + +s_{12}e^{i\theta}\right) \tag{22}$$

$$b_2 = s_{21}a_1 + s_{22}a_2 = s_{21}a_0e^{i\theta_1} + s_{22}a_0e^{i\theta_2} = a_1\left(s_{21} + +s_{22}e^{i\theta}\right) \tag{23}$$

Therefore, TARC for two-port antenna can be described as follows [27]:

$$\Gamma_a^t = \sqrt{\frac{\left|S_{11}+S_{12}e^{i\theta}\right|^2+\left|S_{21}+S_{22}e^{i\theta}\right|^2}{2}} \tag{24}$$

The TARC of MIMO antenna is calculated by applying different combinations of excitation signals to each port. There is no need to define the TARC as a complex number since the phase reference plane does not have any physical meaning for a multiport antenna. The TARC is a real number between zero and one. When the value of the TARC is equal to zero, all the delivered power is radiated and when it is equal to one, all the power is either reflected back or goes to the other ports.

2.3. Summary on UWB MIMO antenna characteristics

In context of UWB where the whole band approved by FCC is required to be covered in one shot, the design of antenna becomes challenging enough. The characteristics of the antennas are required to be stable for the wide frequency band. Moreover, time domain measurements like dispersion and group delay become significant in addition to conventional frequency domain characteristics. Furthermore, the development of future UWB-MIMO communication systems brings more challenges for the antenna design. MIMO antennas are required to be characterized for mutual coupling, correlation and diversity gain. However, a detailed study on characterization of MIMO antennas for UWB is among the current hot topics of research. Also, the design of UWB-MIMO antenna system is always confronted with the same constraints like cost, size, ease of fabrication and integration with other circuits as in the case of single antenna design. Having the specific parameters used essentially for the analysis of UWB and MIMO antennas, the current research orientations with a state of the art are now detailed.

3. Techniques to reduce mutual coupling and to enhance isolation

Theoretical work has proved that mutual coupling has a significant effect on MIMO channel capacity in rich scattering environments. The degree to which coupling induced correlation degrades MIMO channel capacity depends on the multipath's angular power spectral density. Another problem resulting from an increase in mutual coupling is the subsequent decrease of the array's radiation efficiency due to impedance mismatch. Hence, the reduction of mutual coupling becomes very important. Similarly, poor isolation also degrades the array's radiation efficiency due to the leakage of transmitted power from the excited antenna to the port of non-excited antenna. Therefore, the need of good isolation is imperative. The different techniques for this purpose are presented in the literature. Most of these techniques are vital only for narrowband MIMO systems however some of them are presented for UWB-MIMO systems. These techniques are discussed in the sub-sections as follows, and some examples illustrate them. A summarized state of art corresponding to the presented MIMO-UWB antennas is presented in Table 2 detailing the used method, the reference, the antenna layout and the analyzed parameters.

3.1. Using Decoupling and Matching Networks (DMN)

Narrowband MIMO systems - The achievement of low mutual coupling and good isolation using decoupling and matching networks is well explained by S. Dossche et al. in [28]. As earlier described in previous section, the envelope correlation can be calculated from the far-field radiation patterns as well as from the scattering parameters of the antenna system, assuming uniform propagation channel. The envelope correlation can be written as in (13) for a reciprocal and symmetrical antenna system:

$$\rho_e = \frac{|2Re\{S_{11}^*S_{12}\}|^2}{(1-|S_{11}|^2-|S_{12}|^2)^2} \quad (25)$$

From above equation, it is clear that by changing the magnitude and phase of either S_{11} or S_{12}, the correlation between the two antennas can be decreased. In practice, this can be achieved by using a matching network for connecting the antennas. From the system point of view, it is also important to consider the value of $1 - |S_{11}|^2 - |S_{12}|^2$ that takes into account the effective radiated power by the antenna system, and it is maximized by minimizing $|S_{11}|$ and $|S_{12}|$. Thus, two matching networks can be used at both sides to minimize S_{11} and S_{22} while a decoupling network can be used to make S_{11} in quadrature with S_{12} i.e., S_{12} is pure imaginary and thus the real part of mutual impedance Z_{12} is equal to zero. This can be achieved by using a lossless decoupling network.

Also, Weber et al. have used passive DMNs and studied them in detail [29]. In this context, the method for derivation of the admittance matrices of the DMN is explained. An important feature of this method is the formation of predefined orthogonal system port patterns. The admittance matrix can be converted to an actual circuit layout of the DMN in terms of capacitors and inductors. Recently, in [30], hybrid circuit is used as decorrelation circuit. This circuit provides a straightforward, frequency-independent, and feasible solution if the elements are symmetrically placed. This hybrid circuit introduces a 180° phase shift between the signals from the two antenna branches. A little variant of this technique can be observed in [31] where several parasitic elements have been employed between the radiating elements to reduce the isolation. The reduction is dependent on the length and number of the parasitic elements. At least 10 dB improvement, in isolation is noticed.

UWB-MIMO Systems - It can be noticed that lot of work has been presented to get better isolation using DMNs. However, this technique is not tractable for UWB-MIMO systems. The matching networks to design and to realize for multiband, wideband and ultra wideband MIMO systems are enough difficult. Thus, this technique is not employed yet for UWB-MIMO systems in the literature to the best of our knowledge.

3.2. Using Electromagnetic Band Gap (EBG) structures

As noted by Sievenpiper [32], an electromagnetic band gap (EBG) structure behaves as a high impedance surface. This structure consists of an array of metal protrusions on a flat metal sheet. They can be visualized as mushrooms or thumbtacks protruding from the surface. If the protrusions are small compared to the wavelength, their electromagnetic properties can be described using lumped circuit elements – capacitors and inductors. The proximity of the neighboring metal elements provides the capacitance, and the long conducting path linking them together provides the inductance. They behave as parallel resonant LC circuits, which act as electric filters to block the flow of currents along the sheet. This is the origin of the high electromagnetic surface impedance. Because of its unusual impedance, the surface wave modes on this structure are very different from those on a smooth metal sheet. In this way, EBG structures have the ability of suppressing surface waves propagation in a frequency band which makes them very useful to improve the ports isolation in printed antennas.

Narrowband MIMO Systems - In [33], such structures are used to increase isolation in patch antennas by using very simple EBG structures with the help of a multilayer substrate containing a high and a low permittivity layers. A planar EBG consisting of a double squared ring is printed on high permittivity layer while antenna is printed on low permittivity layer. The isolation is enhanced by approximately 10 dB. Further, a simple line fed microstrip patch array designed on a relatively thick substrate gives very good port isolation by using three periods of mushroom EBG elements in addition to variable offset superstrates in [34]. The isolation was improved by 10 dB. However, bandwidth was increased by 50 MHz by using additional superstrates. Recently, in [35], a mushroom-type EBG structure is designed to minimize the loading effects between the two slot antennas without significantly modifying the radiation pattern and input impedance profile. When the EBG reflector is utilized, the insertion loss between the antennas is increased due to the suppression of the parallel-plate modes in the band-gap. The reduction in antenna coupling at some specific frequency is observed by more than 30 dB in comparison to the prototype without the EBG reflector.

UWB-MIMO Systems - Though this technique is widely used for narrowband MIMO systems, yet it has some constraints. The method is not viable for wideband systems because a large number of mushroom-like EBG structures will be required to cover the wide range of frequency. As a result, antennas will require large area to embed these structures for UWB-MIMO systems. Further, an intricate process is required to fabricate such structures. They involve an intricate fabrication process with cells shorted to the ground through vias.

3.3. Using neutralizing line

Narrowband MIMO Systems - The technique of using neutralizing line is based on the concept to neutralize two antennas operating in the same frequency band to enhance the isolation. Originally, this technique is proposed by C. Luxey et al. in [16]. They have used a suspended neutralization strip line physically connected to the antenna elements. This line samples a certain amount of the signal on one antenna element and delivers to the other antenna element in order to cancel out the existing mutual coupling, thus increasing total efficiency. In other words, an additional coupling path is created to compensate for the electrical currents owing on the PCB from one antenna to another. Initially for UMTS PIFAs, this technique worked in both cases, i.e., the line is inserted either between shorting pins of PIFAs or between feeding pins of PIFAs. Later, the same technique has been tested for two square patch antennas in [35]. Recently, the same research group is presented a novel implementation of neutralizing line in [36] based on the same concept. In this new form, space between antennas is not occupied rather folded lines are used between the ground plane and the side of each PIFA without disturbing their initial resonant frequency. This idea is also employed to enhancing isolation by many other researchers of mobile companies due to its simplicity, e.g., Nokia [37], LG Electronics [38], and Samsung Electronics [39].

UWB-MIMO Systems - Although this technique is very attractive and has provided good results, yet it is not employed to UWB MIMO antennas so far. It could be very difficult to couple the elements operating over the wide range of bandwidth in such a way that they cancel out mutual coupling.

3.4. Using Defected Ground Structure (DGS)

The researchers have found that the defected ground structure (DGS) is also able to provide a bandstop effect due to the combination of inductance and capacitance [40]. The defects on the ground plane store a fraction of propagating energy and that can be modeled in terms of a simple equivalent reactive circuit as was explained in detail in [41]. DGS has been applied to antenna designs to suppress harmonics, cross polarization of a patch antenna, and to increase the isolation between antennas.

Narrowband MIMO Systems - In [42], a defected ground structure (DGS) consisting of concentric circular rings in different configurations is presented and its stop band characteristics are examined. Later, this DGS is being employed to reduce mutual coupling between two cylindrical dielectric resonator antennas. About 5 dB suppression has been obtained near the operating frequency around 3.3 GHz. Other variants of this technique could be embedding of slits [43] or meander lines [44] in the ground plane. In [43], the ground plane structure consisting of five pairs of slits etched into the middle of a ground plane of two closely packed planar inverted-F antennas is proposed. These slits are interleaved with metal strips and these strips could be thought of as capacitors. At the same time, some inductance is introduced along the central small connecting strip. Therefore, the structure behaves as a bandstop filter based on a parallel resonator. As a result, such a pattern etched onto the ground plane effectively suppresses mutual coupling. A significant improvement up to 20 dB in isolation is observed in the case of monopole antennas. In [44], it has been demonstrated that meander line embedded ground plane provides better isolation as compared with slitted ground plane. Recently, a combination of two techniques, i.e., DGS and EBG, is presented in [45]. A slitted pattern is etched on the ground plane and three mushroom photonic band gap (PBG) are etched on each wall. Using two techniques together, isolation between the ports of closely-packed antenna elements is increased by 30 dB.

UWB-MIMO Systems - In [46], a diversity antenna operating at a frequency range of 3.1-5.8 GHz is designed consisting of two orthogonal half circles with the radiators placed symmetrically with respect to a protruded T-shaped ground plane, which has a slot at the upper center portion of the ground plane. This slot helps in enhancing the isolation and matching the impedance. In [47], this technique is used in real sense where circular slot antenna with a stepped ground plane is proposed. A stepped ground plane generates non-planar connections and discontinuous interfaces between the elements and the system ground planes. This strategy has effectively decreased the mutual coupling and provided 10 dB enhancements in isolation characteristics. The antenna consists of four radiating elements and operates over the range of 2-6 GHz.

3.5. Using spatial and angular variations

Narrowband MIMO Systems - The technique of using spatial and angular variations relative to the antenna elements of array is very commonly used to reduce mutual coupling. It is well demonstrated that by increasing the space between the radiating elements decorrelates them and even the spacing greater than or equal to $\lambda/2$ gives mutual coupling less than -20 dB, where λ is free space wavelength at the center frequency [48]. However, the spacing becomes less than $\lambda/2$ in the case of compact MIMO antennas for portable devices, thus it requires considering the mutual coupling effects to be compensated [49]. Therefore, in addition to separating the radiating elements by some distance, positioning of the radiating elements at different angles with respect to each other helps to reduce mutual coupling by exploiting the diversity in polarization. Chae et al. [27] has presented the detailed study using this technique. Further, the same technique is described and employed in [50].

UWB-MIMO Systems - Being very simple technique, it has not some specific constraints relating to the bandwidth but with size of the antenna. First of all, this technique is used for UWB diversity antenna by Wong et al. in [51] where the antenna consists of two truncated square monopoles orthogonally and symmetrically printed on the two sides of a T-shaped protruded ground plane as shown in Table 2. This antenna operates over 2.3-7.7 GHz giving isolation more than 20 dB. Using T-shaped ground plane also indicates that the modification of ground plane is an additional technique used together with polarization diversity to enhance the isolation. Recently, Chen et al. have used the similar technique [52]. It presents very compact UWB diversity antenna exploiting polarization diversity. The antenna elements are fed orthogonally and are designed for the lower band of UWB, i.e., 3.1-4.8 GHz. The isolation between two antennas is greater than 20 dB across the bandwidth. Also, the same research group has presented a detailed analysis of two suspended UWB plate antennas operating over 3.0-6.0 GHz in [53] for UWB-MIMO systems. They tested two configurations; (i) when shorting walls are vertically positioned (ii) when shorting walls are horizontally positioned. The effects of the variation of distance between antennas on mutual coupling, isolation and impedance matching are presented.

3.6. Inserting stubs

Narrowband MIMO Systems - The technique of using stubs to get better isolation also deals with the ground plane instead of the radiating elements. One or more stubs are inserted to enhance the isolation. To the best of our knowledge, there is no scientific publication presenting the use of this technique for narrowband MIMO systems.

UWB-MIMO Systems - The method of inserting stubs is mainly found in the literature for UWB-MIMO antennas. For instance, in [54], two elements diversity planar antenna, with three stubs on the ground plane to improve the isolation, has been proposed particularly for PDA phone. The 10 dB return loss bandwidth is achieved from 2.27 GHz to 10.2 GHz and isolation is always more than 15 dB. Similarly, another printed UWB diversity antenna consisting of two square radiators and a cross stub placed between them on the ground is presented in [55]. The 10 dB return loss bandwidth of the antenna ranges from 3.1 GHz to 10.6 GHz and the isolation between the two ports is higher than 18 dB within 3.3 GHz to 10.5 GHz.

3.7. Using heterogeneous elements

Narrowband MIMO Systems - The method of using heterogeneous elements is sometimes used for multi-band antennas in narrowband systems. The objective is then relatively different than to realize a MIMO channel.

UWB-MIMO Systems - In [56], a vector antenna system has been presented. This vector antenna system comprises a center-fed loop antenna and two orthogonal bow-tie antennas in the plane of the loop. This antenna system has large form factor and operates in the frequency range of 3.6-8.5 GHz. The isolation between the antennas is more than 15 dB and reduced mutual coupling is obtained exploiting the advantage of orthogonal components of electric field. The capability of antenna system for UWB operations is authenticated by time domain measurements. It is shown that the vector antenna can provide nearly the same capacity as a traditional spatial array.

4. Some contributions towards UWB-MIMO antennas

4.1. Introduction

A lot of UWB antennas and MIMO antennas have already been presented in the literature. But few publications have been presented on the design and characterization of MIMO antennas for UWB applications as presented in the previous section. This section deals solely with our contributions towards UWB-MIMO antennas. It presents the defined objectives and consequently the followed approaches to achieve these goals. The designs and structures of the proposed different types of MIMO antennas for UWB applications exploiting spatial, polarization and pattern diversities are described. The analysis and evaluation of performance of these proposed designs are provides taking the special parameters into account which are necessary to characterize UWB-MIMO antennas. Finally, a solution to enhance isolation with reduced size antenna is presented.

Technique	Reference	Antenna layout	Analyzed parameters
Using Defected Ground Structure (DGS)	Two cone-shaped radiating elements [46]	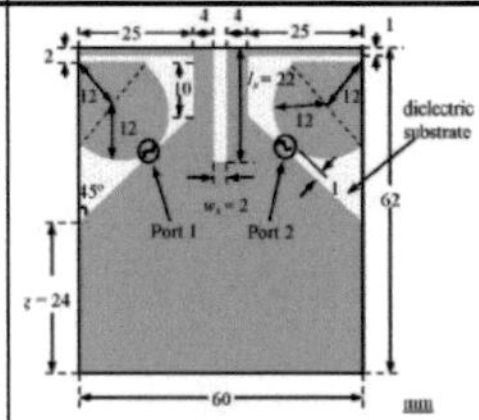	-substrate RO4003, 60×62 mm^2 -BW: 3.1-5.8 GHz -radiation patterns -gain variation: 2 dBi -efficiency: 75-93% -S_{21} < - 20 dB
Using Defected Ground Structure (DGS)	Four elements arranged as 2 × 2 array [47]	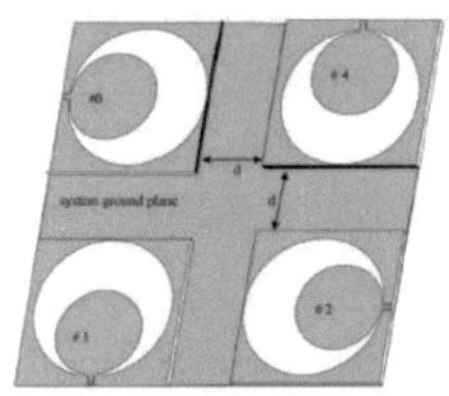	-substrate FR4, G×G mm^2 -BW: 2-6 GHz -radiation patterns -gain variation: 2.7 dBi -current distributions -S_{21} < - 25 dB
Using Spatial and Angular Variations	Two truncated square monopoles [51]	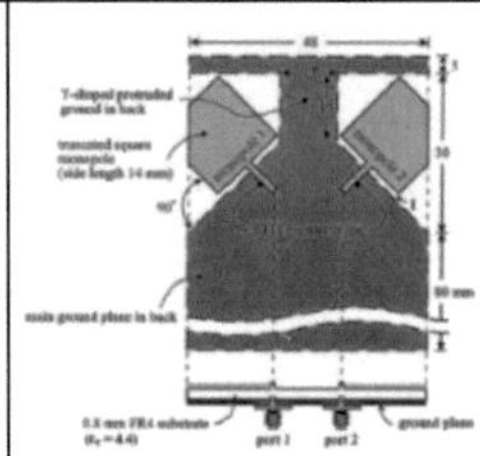	-substrate FR4, 80×48 mm^2 -BW: 2.3-7.7 GHz -radiation patterns -gain variation: 1.1 dBi -S_{21} < - 20 dB
Using Spatial and Angular Variations	Two notched triangular radiator [52]	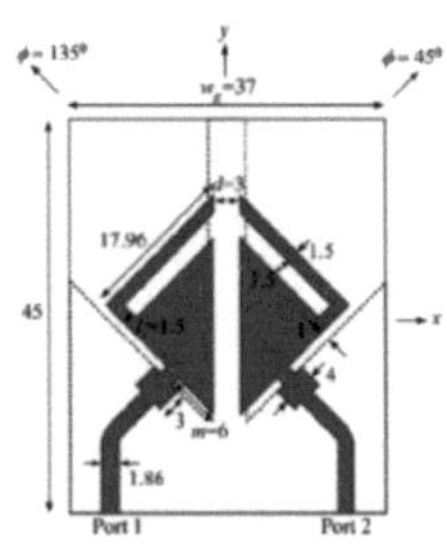	-substrate: R04003, 45×37 mm^2 -BW: 3.1-5.0 GHz -radiation patterns -gain variation: 3.0 dBi -efficiency: 70-93% -current distributions -S_{21} < - 20 dB -correlation < - 25dB -time-domain performance
Using Spatial and Angular Variations	Two suspended UWB plate antennas with horizontal and vertical configurations [53]	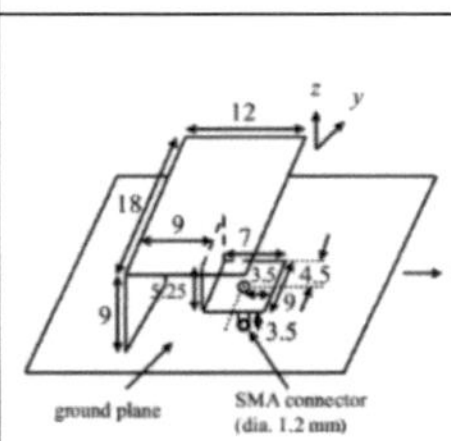	-substrate 45×100 mm^2 -BW: 3-6 GHz -correlation < - 8dB
Inserting stubs	Two Y-shaped radiators with three stubs [54]	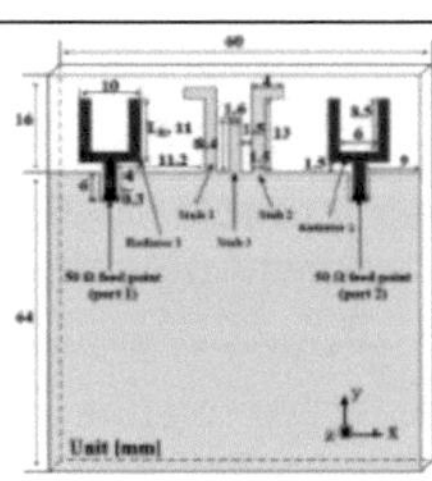	-substrate FR4, 64×60 mm^2 -BW: 2.27-10.2 GHz -radiation patterns -gain variation: 1.5 dBi -current distributions -S_{21} < - 20 dB

Technique	Reference	Antenna layout	Analyzed parameters
Inserting stubs	Two square patches with cross-shaped stub [55]	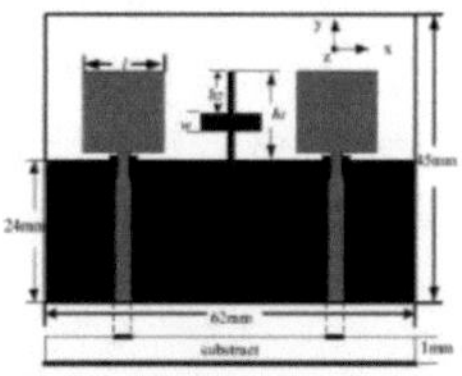	-substrate: $\varepsilon_r = 2.65$, 45×62 mm^2 -BW: 3.3-10.5 GHz -radiation patterns -gain variation: 3.2 dBi -current distributions -$S_{21} < -18$ dB
Using heterogeneous elements	Vector antenna with one loop and two bow-ties [56]	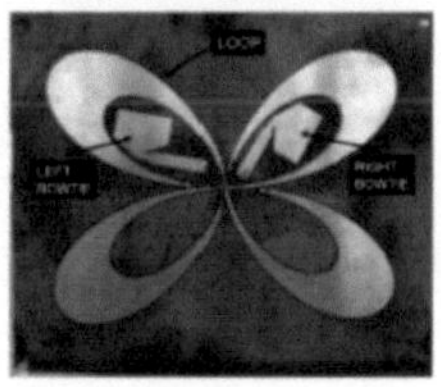	-substrate: $\varepsilon_r = 2.6$ -BW: 3.6-8.5 GHz -radiation patterns -$S_{21} < -15$ dB -impulse response / -phase response

Table 2. Summarized state of the art in UWB-MIMO antennas.

The objective is to design a UWB-MIMO antenna that covers the entire range of frequency approved by the FCC for UWB systems (i.e., 3.1-10.6 GHz) with minimum mutual coupling between the constituent antenna elements, thus attaining good diversity performance. The detailed specifications for the desired MIMO antenna for UWB applications are given in Table 3. To achieve the goal, two types of MIMO antennas are envisaged, i.e., homogeneous and heterogeneous MIMO antennas and are being designed. In wireless communications, the antennas are expected to be embeddable or easy to be integrated into wireless devices in system design; therefore the antennas directly printed onto a PCB/substrate are the most promising designs. Such antennas are usually constructed by etching the radiators onto the dielectric substrate of PCB slabs and a ground plane around the radiators. Taking this major argument into account, printed and planar monopoles have been selected as the constituent radiating elements for UWB-MIMO antennas.

Parameter	Value
Operating BW	3.1-10.6 GHz (± 100 MHz acceptable)
Gain variation	Not more than 4 dBi
Radiation efficiency	High (> 70%) and variation not more than 25%
Group delay	Not more than 2 ns
Isolation	Not less than 14 dB
Correlation coefficient	Not more than – 15 dB
TARC	Not more than – 10 dB
Design profile	Compact, printed and easy to fabricate

Table 3. Design specifications for UWB-MIMO antenna.

4.2. Presentation of proposed designs

Different types of MIMO antenna systems have been proposed for UWB applications. These antennas are categorized in two groups: homogeneous UWB-MIMO antennas and heterogeneous UWB-MIMO antennas. The term "homogeneous" refers to the fact that the identical radiating elements constitute MIMO antenna and the term "heterogeneous" is used to indicate that the constituent elements are not identical. A lot of UWB antennas have been presented in the literature. Among these antennas, printed and planar monopoles are very attractive for their efficient UWB attributes. Therefore, we have also selected printed monopoles to develop UWB-MIMO antennas.

Among the proposed and studied antennas, three homogeneous 2-elements UWB-MIMO antennas are designed using identical circular disc monopoles (system-1 and system-2) and identical stepped rectangular monopoles (system-3) and one heterogeneous 2-element UWB-MIMO antenna (system-4) is designed using stepped rectangular and circular ring monopole. The geometries of these antennas are shown in Figure 1.

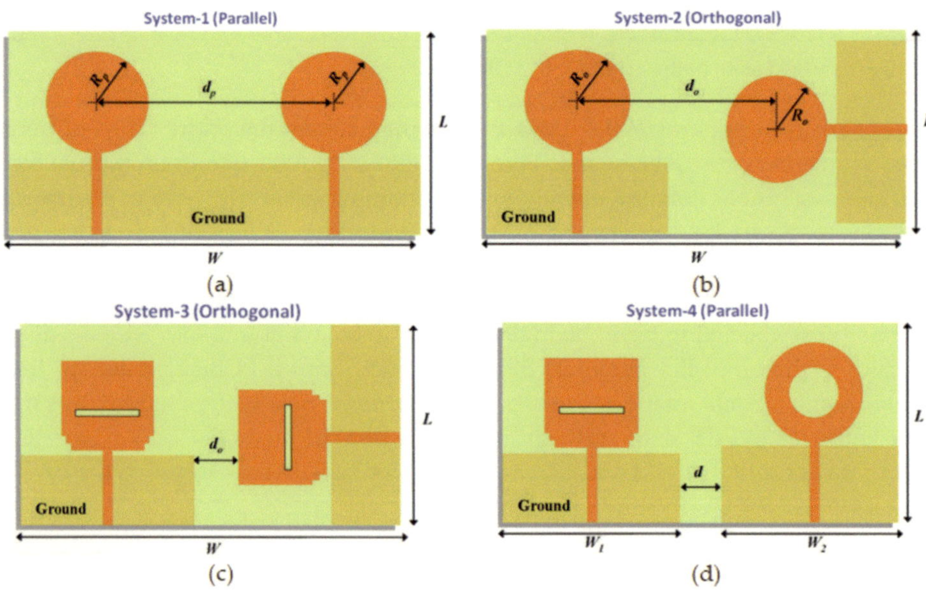

Figure 1. Geometries of designed UWB-MIMO antenna.

All the antennas are fabricated on the FR4 substrate of dielectric permittivity of 4.4, dielectric loss tangent of 0.02 and thickness of $0.8\,mm$. The constituent printed UWB antennas i.e. stepped rectangular, circular disc and circular ring monopoles have already been presented in [57], [58] and [59] respectively. The selection of these antennas to design MIMO antennas can be justified by their good performance, size and ease of integration. However, these antennas have been redesigned to adapt the changes in substrate and thereafter optimized to reduce their dimensions as compared with those presented. It is found that partial ground plane and feed gap play vital role in matching the impedance thus increase the BW if optimally sized. The radiating elements are fed by 50 Ω microstrip lines.

First of all, system-1 comprising two identical circular disc monopoles is designed as shown in Fig. 1a. The radiating elements have common ground plane of length of 12 mm and width of 80 mm. The dimensions of the system-1 are: $= 80\,mm\ (0.83\,\lambda_0),\ L = 43\ mm\,(0.44\,\lambda_0),$

$R = 13.7\ mm\ (0.14\,\lambda_0)$, and $d = 43.5\ mm\ (0.45\,\lambda_0)$; where λ_0 is the free space wavelength corresponding to the lower edge frequency. System-1 is designed on the basis of using spatial diversity. The radiating elements are separated by such a value of distance that mutual coupling becomes less than $-10\ dB$. On the other hand, system-2 and system-3 exploit spatial and polarization diversities by placing the radiating elements orthogonally separated by some distance as shown in Fig. 1b and Fig. 1c respectively. The optimized dimensions of system-2 are: $L = 40\ mm$, $W = 80\ mm$, $R = 10.7\ mm$ and $d = 37.8\ mm$. Similarly, the optimized dimensions of system-3 are: $L = 30\ mm$, $W = 68\ mm$ and $d = 11\ mm$. System-4 is developed to exploit the spatial and pattern diversity therefore it consists of different radiating elements. The dimensions of system-4 after optimization are: $L = 35\ mm$, $W = 85\ mm$, $d = 20\ mm$. Both radiating elements have their own ground planes unlike system-1. It is worth mentioning that CST Microwave Studio is being used for designing and simulating the antennas. Figure 3 illustrates the UWB-MIMO performance of system-1. Figure 3a shows the simulated reflection coefficients of the left (S_{11}) and the right (S_{22}) monopoles of system-1. As illustrated in figure, the - 10 dB bandwidth ranges less than 3.1 GHz to more than 10.6 GHz that confirms the UWB characteristics. Through simulations as well as from geometrical point of view, it is clear that symmetry exists for the antenna elements, i.e. S_{22} is the same as S_{11}. The radiation patterns (Figure 3c) are nearly omnidirectional at lower frequencies in H-plane. The pattern also follows donut shape at lower frequencies in E-plane, but it becomes more directional at higher frequencies. The transition of the radiation patterns from a simple donut pattern at the first resonance to the complicated patterns at higher resonances indicates that this antenna must have gone through some major changes in its behavior. The maximum absolute gain values and the total efficiency of the left and right monopoles are presented in Figure 3d and Figure 3e respectively. It can be recalled that the total efficiency of an antenna takes into account all the losses in the antenna such as reflections due to mismatch between transmission lines and the antenna, conduction and dielectric losses. From the plots, it can be noticed that gains as well as efficiencies of both elements are the same and it again confirms that the symmetry holds in system-1. The variation of the gain values along the wide range of frequencies is found to be less than 3.5 dBi for both radiating elements. The total efficiency is always more than 75% and the variation is less than 15% throughout the bandwidth of interest. The radiating elements are also characterized for their time domain performance to confirm their capability for UWB operations. The fifth derivative of Gaussian pulse is used to excite the antenna elements as this covers the FCC's defined UWB spectrum efficiently. The width of the pulse to excite the radiating elements is 0.13 ns where the pulse width is measured at 50% of the maximum amplitude.

The time domain impulse response is determined by placing the probes in the far-field zone. Figure 3f shows the impulse response for both elements. It can be seen that the pulse of 0.32 ns wide is received. Figure 3g shows the group delay for both monopoles. A face-to-face arrangement with 500 mm distance between them is made to determine the group delay. The group delay is nearly within 2.2 ns throughout the whole of the required pass band. To attain better diversity performance it is important to keep mutual coupling between radiating elements as minimum as possible. The distance of 43.5 mm for system-1 is being selected on the basis of an optimization. Figure 3b shows the curves of S_{21} and S_{12}. There exists the reciprocity, so S_{21} and S_{12} are equal. Isolation is less than – 10 dB. Figure 3h presents the surface current distributions when the left monopole of system-1 is excited while the right monopole is terminated with the matched impedance. It can be observed that very low amount of current is coupled to the right monopole at first resonance and second resonance, and it is justified by the value S_{21} of at these frequencies. However, other two resonances have not also good isolations and current is coupled to some extent. Because of the symmetry of system-1, it is not required to show the current distributions when the right monopole is excited. To evaluate diversity performance into detail, the correlation coefficient is calculated from S-parameters as well as from 3D pattern (figure 3i). Finally TARC is calculated and shown in Figure 3j. It meets the requirement giving the values less than – 10 dB.

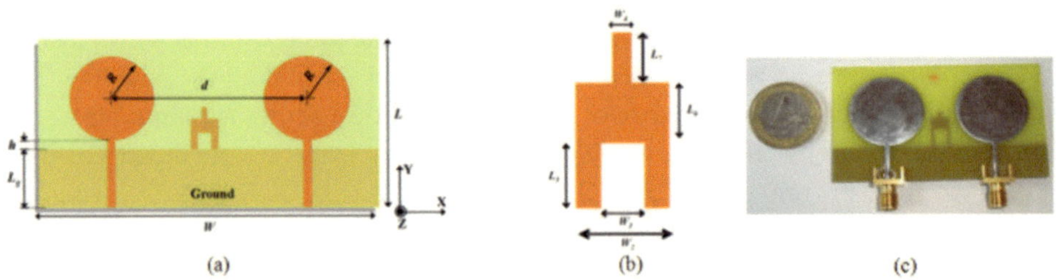

Figure 2. (a) Layout of UWB MIMO antenna (b) Detailed layout of inverted-Y shaped stub (c) photography of the prototype.

It can be noticed that the system-1 is not capable of meeting the specifications defined in Table 3 for group delay as well as isolation. The other solutions, presented previously, have been envisaged both this reason but also in order to improve the compactness. System-2 exploits spatial diversity as well as polarization diversity. The orthogonal configuration results in decorrelating the radiating elements: radiation pattern and isolation are improved while dispersion is mitigated. System-2 using orthogonal topology shows better results as it exploits the polarization diversity, however physical constraint lies regarding the antenna feeding. With comparable characteristics, system-3 constituted by two identical stepped patch monopoles presents a more compact size. System-4 is designed by integrating two non-similar radiating elements, i.e., stepped patch and circular ring monopole. The exploitation of pattern diversity eliminates the need to print the radiating elements orthogonally. Circular ring monopole nearly behaves the same as circular disc because the current distribution in the center of disc is negligible. The performance of these three solutions is presented with more details in [60], [61] and [62]. So far, the combinations of diversities to reduce the mutual coupling or to improve the isolation have been exploited. The antennas can be reduced in size if some special technique is used to overcome the problems of coupling and isolation between the radiating elements. From system-1, a novel UWB-MIMO antenna (called system-5) has been designed with enough compact dimensions taking advantage of a stub which is inserted on the ground plane. The design of stub, a sort of inverted-Y shape, is initiated from the idea of microstrip LC filters. The introduced stub behaves like a stopband LC filter; therefore it suppresses efficiently the currents from the excited port to the inactive port. The best position of the stub is found to be the middle of the ground plane. The performance has been evaluated numerically and experimentally [63]. Figure 4 displays the measured and simulated reflection coeffcients for the antennas with stub and without stub. It is noticed that the measured results

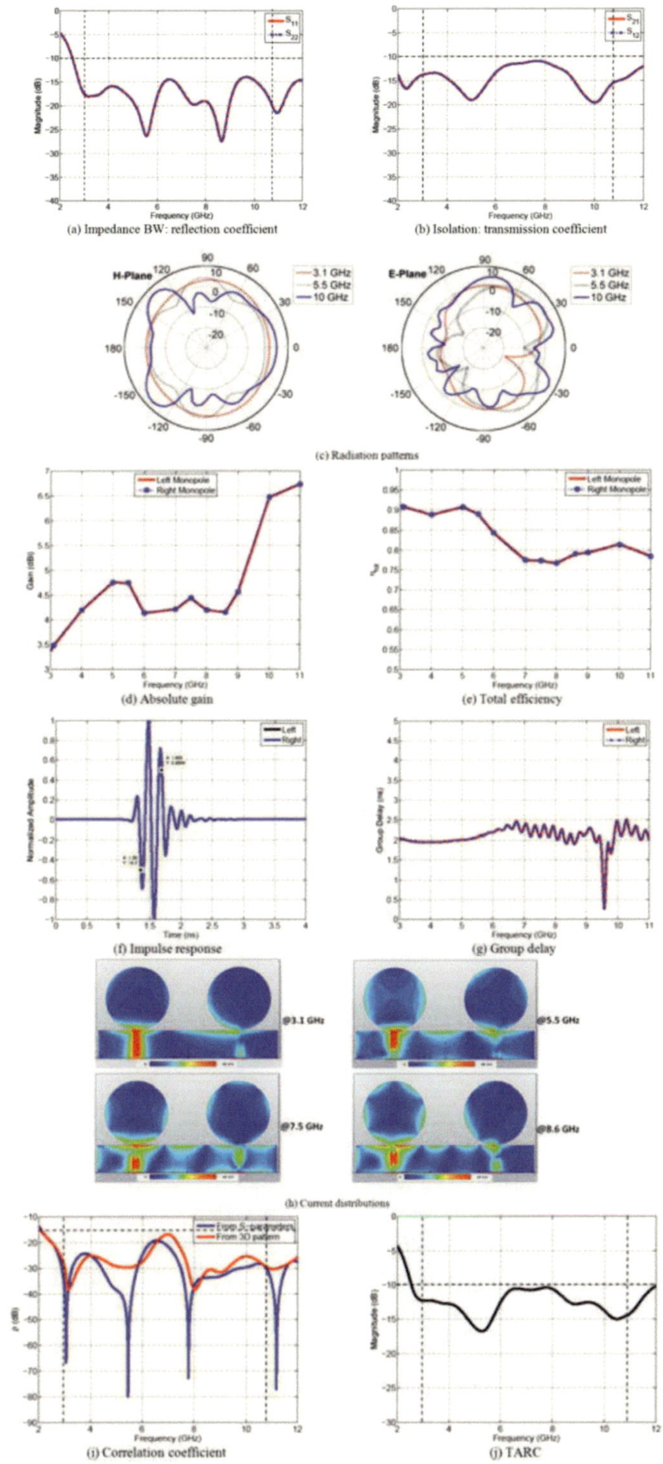

Figure 3. UWB-MIMO antenna performance of system-1.

agree with the simulated ones. The important point to be noticed is that the impedance BW remains the same, i.e., 3.2-10.6 GHz. Further, figure 5a gives the measured and the simulated port isolations for the case when there is stub and figure 5b depicts the results when there is no stub. It is clear from the measurements that the insertion of stub has played a vital role in enhancing the port isolation. System-5 is more compact and efficient as compared to system-1, and verifies the constraints given by table 3.

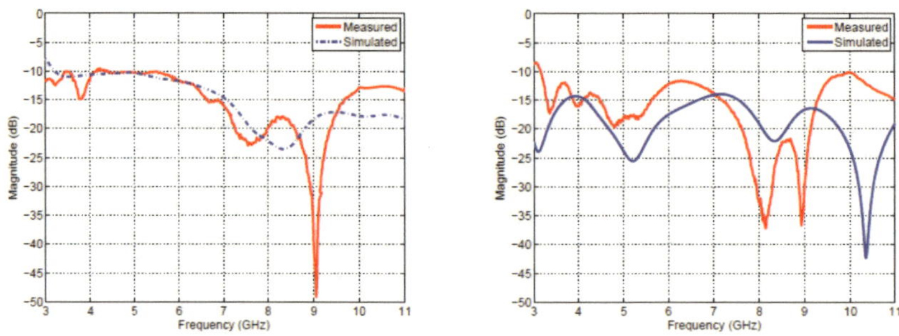

(a) S_{11} (with stub) (b) S_{11} (without stub)

Figure 4. Measured impedance characteristics.

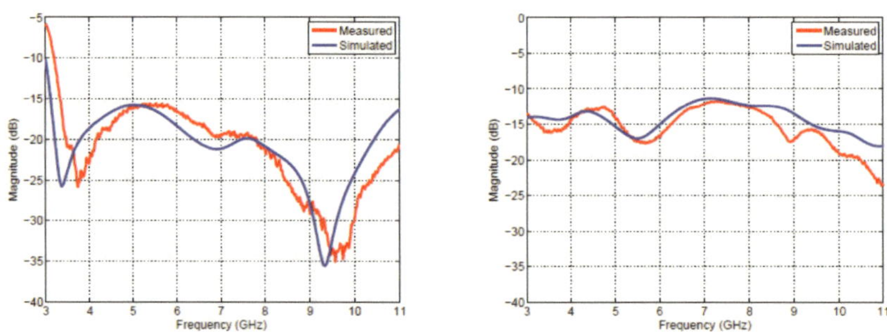

(a) S_{21} (with stub) (b) S_{21} (without stub)

Figure 5. Measured diversity performance of UWB-MIMO antenna.

5. Conclusion

Taking a little overview of UWB and MIMO, it makes easier to understand the idea of implementing MIMO technique in UWB communications systems. As per FCC rules, extremely low power is being allowed to be transmitted, i.e. – 41.3 dBm/MHz, and it impedes the development of UWB communication systems with higher data rates or covering longer distances. To overcome this bottleneck, MIMO technique has been considered to be one of the solutions that will improve the reliability and the capacity of UWB systems. However, a number of challenges arise to shape this solution physically. In this chapter, we took the challenges into account related to antennas as their properties play a key role in determining MIMO system performance. Table 4 summarizes the presented UWB-MIMO antennas and compares the performance.

Topology	Size (mm²)	Impedance BW (GHz)	Radiation pattern	Variations in gain (dBi)	Efficiency (%)	Group delay (ns)	Isolation (dB)	Correlation (dB)	TARC (dB)
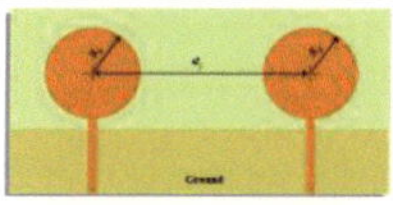 system-1 [60]	43×80	3.1-10.6	Distorted	3.5	75	2.2	11	- 17	- 10
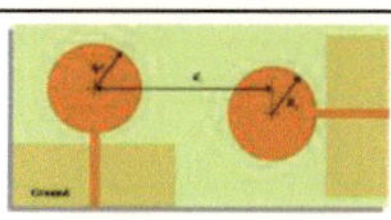 system-2 [60]	40×80	3.1-10.6	Nearly original	3.5	80	1.7	15	- 28	- 10
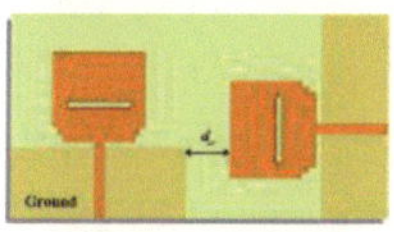 system-3 [61]	30×68	3.2-10.6	Nearly original	2.5	74	0.8	15	- 20	- 10
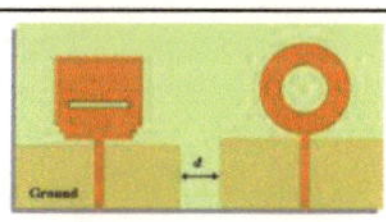 system-4 [62]	35×85	3.1-10.6	Nearly original	2.5	73	1.2	14	- 20	- 10
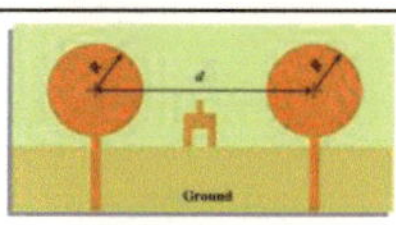 system-5 [63]	40×68	3.2-10.6	Nearly original	2.2	78	1	15	- 20	- 9.5
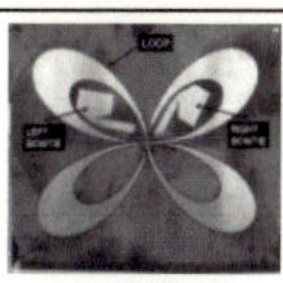 [56]	125×125	3.6-8.5	Nearly original	--	--	--	15	--	--
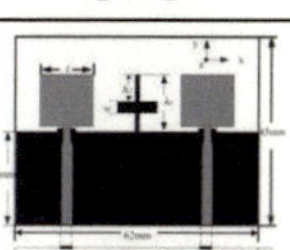 [55]	45×62	3.3-10.5	Nearly original	3.2	--	--	18	--	--
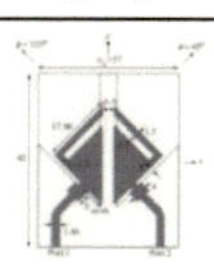 [52]	45×37	3.1-5.0	Nearly original	3.0	--	--	20	--	--

Table 4. Summary: performance comparisons of presented UWB-MIMO antennas.

Author details

Ali Imran Najam, Yvan Duroc and Smail Tedjini
Grenoble Institute of Technology, France

6. References

[1] Federal Communication Commission (FCC). Revision of part 15 of the commission's rules regarding ultra-wideband transmission systems. *First report and order, ET Docket 98-153, FCC 02-48, adopted: Feb. 2002, released in April 2002.*

[2] Xu H., Yang L. Ultra-wideband technology: Yesterday, today and tomorrow. *Proceedings of IEEE Radio and Wireless Symp.*, (715-718), Jan. 2008.

[3] Balakrishnan J., Batra A., Dabak A. A multi-band OFDM system for UWB application. *Proceedings of IEEE Conf. Ultra Wide-Band Systems and Technologies*, (354-358), Nov. 2003.

[4] Win M.Z., Scholtz R.A.. Impulse radio: How it works. *IEEE Communication Let.*, 1998; 2 (2) 36–38.

[5] Foschini G. J., Gans M. J. On Limits of Wireless Communications in a Fading Environment when using Multiple Antennas. *Wireless Personal Communications* 1998; 6 311-335.

[6] Telatar I.E. Capacity of multi-antenna Gaussian channels. *European Trans. Telecommunications* 1999; 10 (6) 585-595.

[7] Kupferschmidt C., Dimitrov E., Kaiser T. Multiple antenna UWB systems – WP3 of the EUWB project. *Proceedings of IEEE International Conf. Ultra Wideband*, Vol. 3, (143-144), Sep. 2008.

[8] Abou-Rjeily C., Daniele N., Belfiore J-C. Space-time coding for multiuser Ultra-wideband communications. *IEEE Trans. Communications* 2006; 54 (11) 1960-1972.

[9] Yazdandoost K.Y. & Kohno, R. Ultra Wideband antenna. *IEEE Communication Magazine* 2004 ; 42 (6) 30-38.

[10] Balanis C.A. Antenna theory: Analysis and design. *Wiley-Interscience, third Edition, John Wiley & Sons*, 2005.

[11] Vuong T.P., Ghiotto A., Duroc Y., Tedjini S. Design and characteristics of a small U-slotted planar antenna for IR-UWB. *Microwave and Optical Technology Let.* 2007; 49 (7) 1727-1731.

[12] Licul S., Noronha J.A.N., DavisW.A., Sweeney D.G., Anderson C.R., Bielawa, T.M. A parametric study of time-domain characteristics of possible UWB antenna architectures. *Proceedings of IEEE 58th Vehicular Technology Conf.*, Vol. 5, (3110-3114), Oct. 2003.

[13] Wallace J.W., Jensen M.A. Mutual coupling in MIMO wireless systems: a rigorous network theory and analysis. *IEEE Trans. Wireless Communications* 2004 3 (4) 1317-1325.

[14] Svantesson T., Ranheim A. Mutual coupling effects on the capacity of multielement antenna systems. *Proceedings of IEEE international Conf. Acoustics, Speech, and Signal Processing*, 2001, (2485-2488).

[15] Kildal P.R., Rosengren K. Correlation and capacity of MIMO systems and mutual coupling, radiation efficiency, and diversity gain of their antennas: simulations and

measurements in a reverberation chamber. *IEEE Communications Mag.* 2004; 42 (12) 104-112.

[16] Diallo A., Luxey C., Thuc P.L., Staraj R., Kossavias G. Enhanced diversity antennas for UMTS handsets. *Proceedings of 1st European Conf. Antennas and Propagation*, Nov. 2006.

[17] Hui H.T. Practical dual-helical antenna array for diversity/MIMO receiving antennas on mobile handsets. *IEEE Proceedings of Microwave, Antennas and Propagation* 2005; 152 (5) 367-372.

[18] Pedersen G.F., Andersen J.B. Handset antennas for mobile communications: integration, diversity, and performance. *Review of Science Radio 1996-1999, Wiley – IEEE Press* 1999; 119-138.

[19] Kalliola K., Sulonen K., Laitinen H., Kivekas O., Krogerus J., Vainikainen P. Angular power distribution and mean effective gain of mobile antenna in different propagation environments. *IEEE Trans. Vehicular Technology* 2002; 52 (5) 823-838.

[20] Chiau C.C., Chen X., Parini C.Q. A compact four-element diversity antenna array for pda terminals in a MIMO system. *Microwave and Optical Technology Let.,* 2005; 44 (5) 408-412.

[21] Ko S.C.K., Murch R.D. Compact integrated diversity antenna for wireless communications. *IEEE Trans. Antennas and Propagation* 2001; 49 (6) 954-960.

[22] Salonen I., Vainikainen P. Estimation of signal correlation in antenna arrays. *Proceedings of 12th International Symp. on Antennas,* Vol. 2, (383-386), Nov. 2002.

[23] De Flaviis F., Jofre L., Romeu J., Grau A. Multiantenna systems for MIMO communications. *Morgan & Claypool Publishers,* 2008.

[24] Volakis J.L. Antenna engineering handbook. 4th Edition, Mc Graw Hill Companies, 2007.

[25] Turkmani A.M.D., Arowojolu A.A., Jefford P.A., Kellett C.J. An experimentation evaluation of the performance of two-branch space and polarization diversity schemes at 1800 MHz. *IEEE Trans. Vehicular Technologies* 1995; 44 (2) 318-326.

[26] Mallahzadeh A.R., Eshaghi S., Alipour A. Multiport characteristics of a wide-band cavity backed annular patch antenna for multipolarization operations. *Progress in Electromagnetics Research* 2009; 187-203.

[27] Chae S.H., Oh S.K., Park S.O. Analysis of mutual coupling, correlations, and TARC in Wibro MIMO array antenna. *IEEE Antennas and Wireless Propagation Let.* 2007; 6 122-125.

[28] Dossche S., Blanch S., Romeu J. Optimum antenna matching to minimize signal correlation on a two-port antenna diversity system. *IEEE Electronics Let.* 2004; 40 (19) 1164-1165.

[29] Weber J., Volmer C., Blau K., Stephan R., Hein M.A. Miniaturized antenna arrays using decoupling networks with realistic elements. *IEEE Trans. Microwave Theory and Techniques* 2006; 54 (6) 2733-2740.

[30] Shanawani M., Paul D.L., Dumanli S., Railton C. Design of a novel antenna array for MIMO applications. *Proceedings of 3rd International Conf. Information and Communication Technologies: From Theory to Applications,* Apr. 2008.

[31] Min K.S., Kim D.J., Moon Y.M. Improved MIMO antenna by mutual coupling suppression element. *Proceedings of European Conf. Wireless Technology*, (125-128), Oct. 2005.

[32] Sievenpiper D., Zhang L., Broas R.F.J., Alexopolous N.G., Yablonovitch E. High impedance electromagnetic surfaces with a forbidden frequency band. *IEEE Trans. Microwave Theory and Techniques* 1999; 47 (11) 2059-2074.

[33] Rajo-Iglesias E., Quevedo-Teruel O., Incla-Sanchez L. Study of mutual coupling reduction in single and stacked multilayer patch antennas by using planar EBG structures. *Proceedings of IEEE International Symp. Antennas and Propagation Society*, (393-396), Jun. 2007.

[34] Michailidis E., Tsimenidis C., Chester G. Mutual coupling reduction in a linear two element patch array and its effect on theoretical MIMO capacity. *Proceedings of Conf. Antennas and Propagation*, (457-460), Mar. 2008.

[35] Ranvier S., Luxey C., Suvikunnas P., Staraj R., Vainikainen P. Mutual coupling reduction for patch antennas array. *Proceedings of IEEE International Symp. Antennas and Propagation Society*, 3632-3635, Jun. 2007.

[36] Chebihi A., Luxey C., Diallo A., Le Thuc P., Staraj R. A novel isolation technique for closely spaced PIFAs for UMTS mobile phones. *IEEE Antennas and Wireless Propagation Let.* 2008; 7 665-668.

[37] Rahola J., Ollikainen J. Analysis of isolation of two-port antenna systems using simultaneous matching. *Proceedings of 2nd European Conf. Antennas and Propagation*, Nov. 2007.

[38] Chung K., Yoon J.H. Integrated MIMO antenna with high isolation characteristics. *IET Electronics Letters* 2007; 43 (4) 199-201.

[39] Byun J., Jo J.H., Lee B. dual-band diversity antenna for mobile handset applications. *Microwave and Optical Technologies Let.* 2008; 50 (10) 2600-2604.

[40] Ahn D., Park J.S., Kim C.S, Kim J., Qian Y., Itoh T. A design if the low-pass filter using the novel microstrip defected ground structure. *IEEE Trans. Microwave Theory and Techniques* 2001; 49 (1) 86-93.

[41] Caloz C., Okabe H., Iwai T., Itoh T. A simple and accurate model for microstrip structures with slotted ground plane. *IEEE Microwave Wireless Components Let.* 2004; 14 (4) 133-135.

[42] Guha D., Biswas S., Biswas M., Siddiqui J.Y., Antar Y.M.M. Concentric ring-shaped defected ground structures for microstrip applications. *IEEE Antennas and Wireless Propagation Let.* 2006; 5 402-405.

[43] Chiu C.Y., Cheng C.H., Murch R.D., Rowell C.R. Reduction of mutual coupling between closely-spaced antenna elements. *IEEE Trans. Antennas and Propagation* 2007; 55 (6) 1731-1738.

[44] Coles K., Kerby K.C., Bernhard J.T. Comparisons of ground plane techniques for the isolation of co-sited microstrip antennas. *Proceedings of 29th General Assembly of URSI*, (129-130), Aug. 2008.

[45] Jolani F. ,Daggarpour A.M., Dadashzadeh G. Reduction of mutual coupling between dual-element antennas with new PBG techniques. *Proceedings of 13th International Symp. Antennas Technology and Applied Electromagnetics*, Feb. 2009.

[46] Liu L., Zhao H., See T.S.P., Chen Z.N. A printed ultra-wideband diversity antenna. *Proceedings of IEEE International Conf. Ultra Wideband*, Vol. 2, (351-356), Sept. 2006.

[47] Lin S.Y., Huang H.R. Ultra-wideband MIMO antenna with enhanced isolation. *Microwave and Optical Technologies Let.* 2009; 51 (2) 570-573.

[48] Carrasco H, Hristov H.D., Feick R., Cofre D. Mutual coupling between planar inverted-F antennas. *Microwave and Optical Technologies Let.* 2004; 42 (3) 224-227.

[49] Ciattaglia M., Marrocco G. Investigation on antenna coupling in pulsed arrays. *IEEE Trans. Antennas and Propagation* 2006; 54 (3) 835-843.

[50] Thaysen J., Jakobsen K.B. Design considerations for low antenna correlation and mutual coupling reduction in multi antenna terminals. *European Trans. Telecommunications* 2007; 18 (3) 319-326.

[51] Wong K.L., Su S.W., Kuo Y.L. A printed ultra-wideband diversity monopole antenna, *Microwave and Optical Technologies Let.*, Vol. 38, No. 4, 2003, (257-259).

[52] See T.S.P., Swee A.M.L., Chen Z.N. Correlation analysis of UWB MIMO antenna system configurations. *Proceedings of IEEE International Conf. Ultra Wideband*, (105-108), Sep. 2008.

[53] See T.S.P., Chen Z.N. An ultra-wideband diversity antenna, *IEEE Trans. Antennas and Propagation* 2009; 57 (6) 1597-1605.

[54] Hong S., Lee J., Choi, J. Design of UWB diversity antenna for PDA applications. *Proceedings of 11th International Conf. Advanced Communication Technology*, (583-585), Feb. 2008.

[55] Chen Y., Lu W.J., Cheng C.H., Cao W., Li Y. Printed diversity antenna with cross shape stub for ultra-wideband applications. *Proceedings of 11th IEEE International Conf. Communication Systems*, (813-816), Nov. 2008.

[56] Rajagopalan A., Gupta G., Konanur A.S., Hughes B., Lazzi G. Increasing channel capacity of an ultrawideband MIMO system using vector antennas. *IEEE Trans. Antennas and Propagation* 2007; 55 (10) 2880-2887.

[57] Liang J., Chiau C.C., Chen X., Parini C.G. Printed circular disks monopole antenna for ultra-wideband applications. *IEEE Electronics Let.* 2004; 40 (20) 1246-1247.

[58] Choi S.H., Park J.K., Kim S.K., Park J.Y. A new ultra-wideband antenna for UWB applications. *Microwave and Optical Technologies Let.* 2004; 40 (5) 399-401.

[59] Liang J., Chiau C.C., Chen X., Parini C.G. Printed circular ring monopole antennas. *Microwave and Optical Technologies Let.* 2005; 45 (5) 372-375.

[60] Najam A.I., Duroc Y., Tedjini, S. Design and analysis of MIMO for UWB communications. *Proceedings of European Conf. Antennas and Propagation*, Apr. 2010.

[61] Najam A.I., Duroc Y., Tedjini, S. A novel collocated antennas system for UWB-MIMO applications. *Proceedings of IEEE Radio and Wireless Symp.*, (368-371), Jan. 2009.

[62] Najam A.I., Duroc Y., Tedjini, S. Design and characterization of an antenna system for UWB-MIMO communications Systems. *Proceedings of European Conf. Antennas and Propagation*, Apr. 2010.

[63] Najam A.I., Duroc Y., Tedjini, S. UWB-MIMO antenna with novel stub structure. *Progress In Electromagnetics Research J. C* 2011; 19 245-257.

Chapter 11

Creating Band-Notched Characteristics for Compact UWB Monopole Antennas

Y.F. Weng, S.W. Cheung, T.I. Yuk and L. Liu

Additional information is available at the end of the chapter

http://dx.doi.org/10.5772/50190

1. Introduction

In 2002, the FCC in the US authorized the unlicensed use of the ultrawideband (UWB) frequency spectrum for commercial applications in the range from 3.1 to 10.6 GHz with an emission limit of -41.3 dBm/MHz which is near to the thermal noise floor [1]. UWB communication systems operating in such a wide frequency band and low power emission level could easily be interfered by the existing nearby communication systems such as the Wireless Local Area Networks (WLANs) operating in the frequency bands of 2.45-GHz (2.4–2.484 GHz), 5.25-GHz (5.15–5.35 GHz) and 5.75-GHz (5.725–5.825 GHz) [2] and the Worldwide Interoperability for Microwave Access (WiMAX) systems operating in the 2.35-GHz (2.3-2.4 GHz), 2.6-GHz (2.5–2.69 GHz), 3.35-GHz (3.3-3.4 GHz), 3.5-GHz (3.4–3.6 GHz), 3.7-GHz (3.6-3.8 GHz) and 5.8-GHz (5.725–5.85 GHz) bands[3]. Many countries such as the UK, Canada, France, Germany, Argentina and India etc. will allow at least four of these bands in operations [3]. In such cases, the UWB systems could be affected by several interference signals. These interference signal could be suppressed by using RF filtering.

Traditional filtering is implemented using lumped elements, which however increases the cost and system complexity and occupies more space in the wireless devices. Another feasible solution is to design the UWB antennas with band-notched characteristics to suppress the interference signals [4, 5]. Figure 1 shows a general design concept for a band-notched UWB antenna. An UWB antenna, as shown in Figure 1(a), has an impedance bandwidth from f_L to f_H, which are the lowest and highest -10-dB cut-off frequencies, respectively, of the S_{11}. A bandstop resonant structure, as shown in Figure 1(b), also has a bandwidth from f_L to f_H, but with a resonant frequency at f_N to stop the undesired signal passing through. Combining the UWB antenna with the resonant structure as shown in Figure 1(c), a band-notched antenna is formed. The band-notched antenna will not interfere with other communication systems nearby using the same frequency band at f_N.

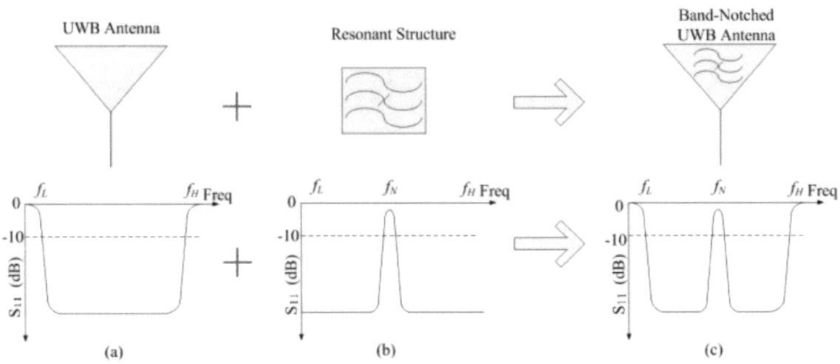

Figure 1. Basic design concept for band-notched UWB antenna

In general, the design procedure for a band-notched antenna can be described as follows. An UWB antenna without band-notched function is designed to have good impedance matching over the UWB, which is used as a reference antenna. Proposed resonant structures are added to the reference antenna to create notches at some specific frequencies. The dimensions of the resonant structures can be used to control the center frequencies and bandwidths of the notches. Different designs have been proposed to realize the band-notched characteristic for UWB planar monopole antennas [5-18]. These include using parasitic elements [6], folded strips [7], split-ring resonators (SRRs) [8], quarter-wavelength tuning stubs [9], meander-ground structures [10, 11], resonated cells on the coplanar-waveguide (CPW) [12], fractal tuning stub [13], slots on the radiator [14-16] or ground [17], and slots or folded-striplines along the antenna feed line [18]. However, most of these designs targeted at creating a single-notched band and only one design targeted at a triple-notched band using meander lines [11].

In this chapter, we study the applications of CPW resonators, λ/4-resonators and MLs to design single, dual, triple and quadruple band-notched characteristics for compact UWB monopole antennas. The studies are carried out using computer simulations and the simulated results are verified using the antenna measurement system, Satimo Starlab. The simulated and measured results on the return loss, radiation pattern, peak gain and efficiency agree well. The pulse responses and fidelities of the single, dual, triple and quadruple band-notched antennas are also measured and compared with those of the reference antenna.

2. Dual band-notched antenna design using CPW resonators

2.1. Design of dual band-notched UWB antenna

The compact CPW antenna used for studies is shown in Figure 2. It has a semi-circular radiator fed by a 50-Ω CPW. The antenna is designed on a Rogers substrate, RO4350B, with an area of 32 mm×35 mm, a relative dielectric constant of ε_r=3.48, a thickness of 0.762 mm and a loss tangent of 0.0037. The width, *S*, of the central-strip conductor and the distance, *W*, between the feed line and ground plane are 3 and 0.3 mm, respectively, in order to have a characteristic impedance of 50 Ω. The ground plane is rectangular in shape plus a half ellipse with a minor-radius-to-major-radius ratio of 0.5 to reduce the beam tilt of the antenna [12]. Detail dimensions of the dual band-notched antenna are listed in Table 1.

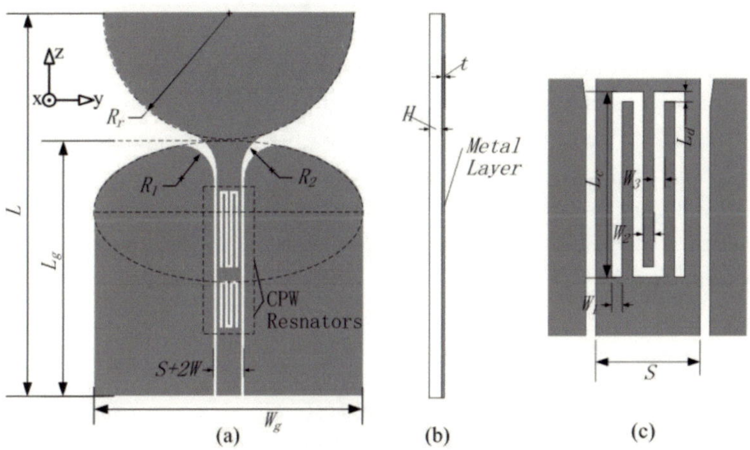

Figure 2. (a) Layout and (b) side view of CPW antenna with two CPW resonators, and (c) layout of CPW resonator

Parameter	Value (mm)	Parameter	Value (mm)
L	35	L_d	0.3
W_g	32	W_1	0.3
L_g	20	W_2	0.3
R_r	15	W_3	0.3
R_1	4	S	3
R_2	4.5	W	0.3
t	0.0035	L_c[a]	5.3
H	0.762	L_c[b]	9

(a) Notched band at 5.5GHz
(b)Notched band at 3.5GHz

Table 1. Antenna dimensions for dual-band notch

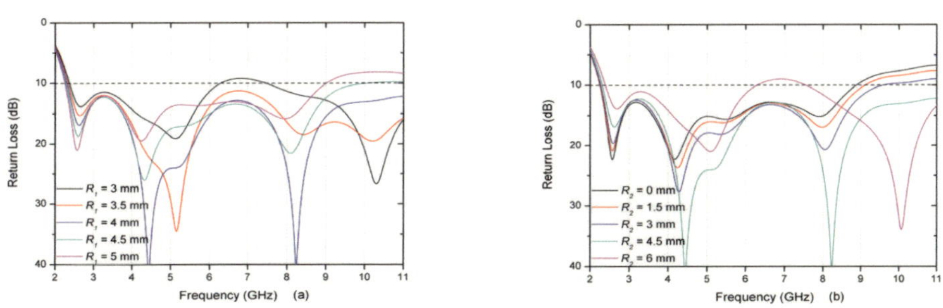

Figure 3. Simulated return losses of reference UWB antenna for different values of (a) R_1 and (b) R_2

The small area connecting the CPW and the semi-circular radiator is quite critical for impedance matching and so is smoothed by using two arc shapes with radii R_1 and R_2. Figures 3(a) and 3(b) show the effects of different values of R_1 and R_2 on the return loss. It can be seen from Figure 3(a) that, when R_1 is small, say $R_1 = 3$ mm, the bandwidth (for return loss > 10 dB) of the reference antenna covers only few GHz, from around 2.5 to 6.3 GHz. Increasing R_1 improves the impendence matching in the lower frequency band, but degrades it in the higher frequency band. For R_2, Figure 3(b) shows an opposite effect. That is, decreasing R_2 improves the impendence matching in the lower frequency band, but degrades it in the higher frequency band. Increasing R_2 improves the bandwidth towards the high frequency, but if R_2 is too large, say $R_2 = 6$ mm, the return loss at around 7 GHz reduces to less than 10 dB.

2.2. Design of simple CPW resonator

$\lambda_g/2$-open-ended CPW resonators and $\lambda_g/4$-CPW resonators can be used to design bandstop filters [19, 20]. However, the sizes of these resonators are too large to be integrated onto the compact UWB antennas. Here we propose a new simple structure having a much smaller size, as shown in Figure 2(c), for the CPW resonator. The resonator is a simple rectangular slot with three open stubs from the opposite sides. With such a small structure, two CPW resonators with a separation of 2 mm can be etched on the feed line, as shown in Figure 2(a), to generate 2 band notches for the antenna. The lower CPW resonator is at a distance 8 mm from the lower edge of the ground plane. The resonance of the CPW resonator is determined by the length L_c and the small gap L_d as indicated in Figure 2(c). The narrowest microstrip line that we can make using the prototype-machine in our laboratory is 0.1 mm. So for convenience in our design process, we fix the stub width W_2 and stub spacing W_3 of the CPW resonator to 0.3 mm. The simulated return loss and insertion loss of a CPW resonator, with $W_1 = W_2 = W_3 = L_d = 0.3$ mm and $L_c = 9$ mm, are shown in Figure 4. It can be observed that the CPW resonator has a bandstop characteristic of about 27 dB at 3.5 GHz.

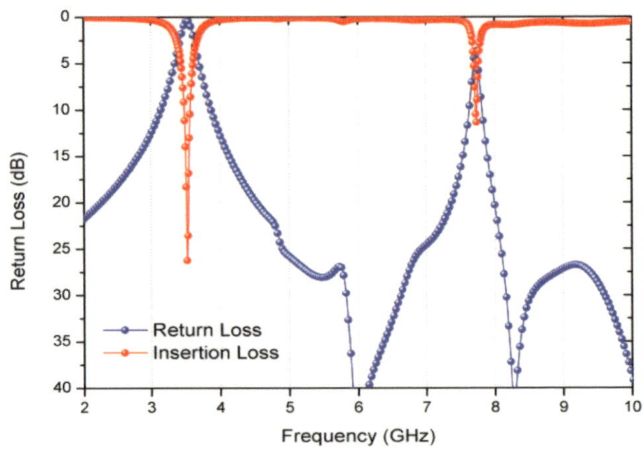

Figure 4. Simulated return loss and insertion loss of CPW resonator

2.3. Parametric study on dual band-notched CPW antenna

In the notched antenna shown in Figure 2(a), two CPW resonators are placed in series, known as a dual-band CPW resonator, and used to create two different notched bands. Simulation results on the return loss with different values of L_c^a and L_d^a in the CPW resonator using computer simulation are shown in Figures 5(a) and 5(b), respectively. It can be seen that the resonant frequency varies with L_c^a and L_d^a, but the return loss in the rest of the UWB remains almost unchanged. This property provides the antenna designers a great freedom to set the frequency of the notched bands. It should be noted that a spurious response at about 7.9 GHz is observed in Figure 5, which is due to the 1st odd-harmonic resonance.

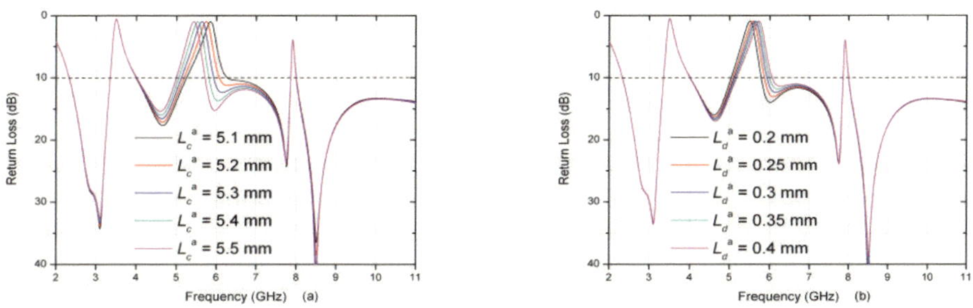

Figure 5. Simulated return loss of dual band-notched antenna with different values of (a) L_c^a and (b) L_d^a

To better understand the antenna operation, the surface-current distributions on the antenna in the passband and notched bands are studied using computer simulation and results are shown in Figure 6. In the passbands of 3 GHz and 11 GHz, Figures 6(a) and 6(d) show that majority of the current flows through the CPW into the radiator and then radiates to free space. However, at the notched frequencies of 3.5 GHz and 5.5 GHz, Figures 6(b) and 6(c) show that the energies are well confined in the upper and lower CPW resonators, respectively, and do not get radiated, while the radiator remains cool.

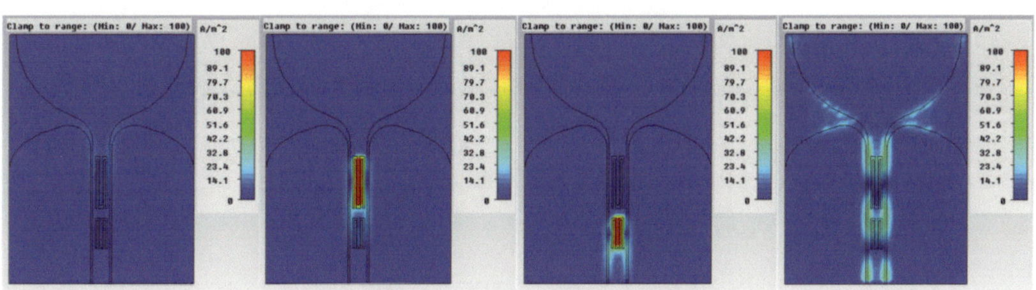

(a) (b) (c) (d)

Figure 6. Distribution of surface current at (a) 3 GHz, (b) 3.5 GHz, (c) 5.5 GHz and (d) 11 GHz

2.4. Results and discussions

To validate the simulation results, the antenna is fabricated on a Roger substrate, RO4350B, as shown in Figure 7, and measured using the antenna measurement system, Satimo Starlab.

Figure 7. Photograph of CPW antennas without and with dual-band CPW resonator

2.4.1. Frequency-domain performance

The return loss, efficiency and peak gain across the UWB, and the radiation patterns in the passbands and notch frequencies of the reference antenna and dual band-notched antenna are all studied using computer simulation and measurement.

The return losses of the two antennas are shown in Figure 8. Across the UWB, excluding the notched bands, both the simulated and measured return losses of the reference antenna are larger than 10 dB which satisfies the UWB requirement. In the notched bands, the return loss of the notched antenna is substantially smaller than 10 dB. The measured results in Figure 8 show that the two notches at the frequencies of 3.5 and 5.5 GHz have the bandwidths of 585 and 758 MHz, respectively. A spurious response at about 7.9 GHz is observed in Figure 8, which is due to the 1st odd-harmonic of the resonant frequency at 3.5 GHz. The discrepancies between the simulated and measured results are due to the tolerances in prototype fabrication and measurements and also partly due to the SMA connector which is not included in our simulation.

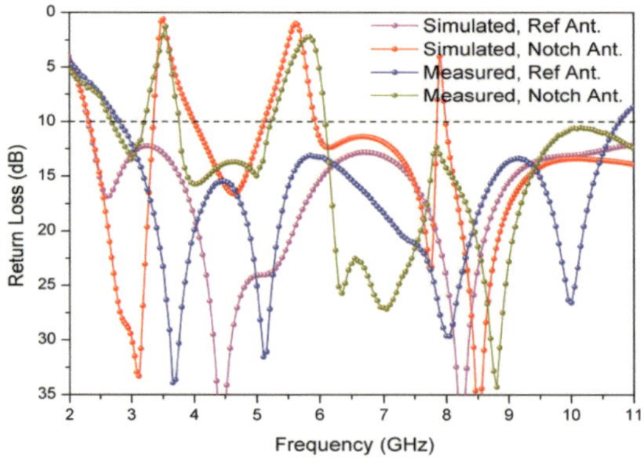

Figure 8. Simulated and measured return losses of reference and dual band-notched antennas

The simulated and measured results on the peak gain and radiation efficiency of the two antennas are shown in Figures 9(a) and 9(b), respectively. At the notch frequencies of 3.5 and 5.5 GHz, the measured gain is suppressed to -7.6 and -4.3 dBi, respectively, with the corresponding efficiency substantially reduced to 8.3% and 9.7%. Thus the dual-band resonator can work effectively to generate a dual band-notched characteristic for the UWB antenna. The spurious stopband at around 7.9 GHz causes a 10% drop in efficiency, as can be seen in Figure 9(b). Note that there are substantial discrepancies between the simulated and measured results in peak gain and efficiency, especially at low frequencies. This is mainly due to the small ground plane of the antenna, which results in leakage current flowing from the ground plane to the outer conductor of the feeding coaxial cable [21-23].

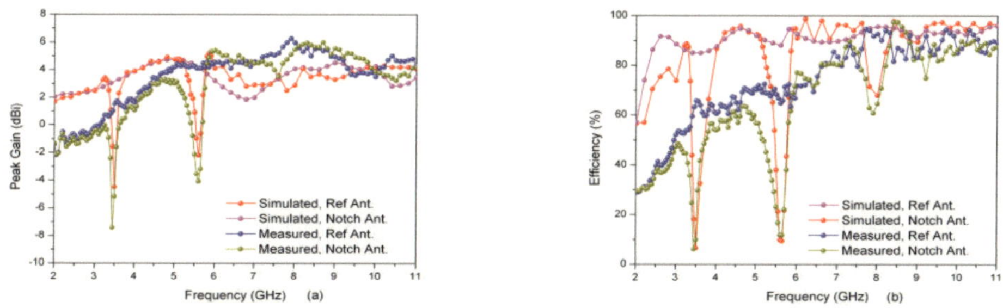

Figure 9. Simulated and measured (a) peak gains and (b) efficiencies of antennas

The simulated and measured radiation patterns of the dual band-notched antenna at the frequencies of 3, 3.5, 5.5 and 11 GHz and in the vertical and horizontal cuts, i.e., in the x-z and x-y planes, respectively, are shown in Figure 10. It can be seen that the measured radiation patterns agree well with the simulated results. For UWB applications, an omnidirectional radiation pattern is normally preferred (i.e., in the x-y plane). The results of Figures 10(a) and 10(g) show that the radiation patterns in the passbands satisfy this requirement well. In the x-z plane, the radiation patterns in Figures 10(b) and 10(h) show two nulls occurring at the positive and negative z-directions, which is typical for monopole antennas. For the radiation patterns at the notch frequencies of 3.5 and 5.5 GHz, the gain is almost evenly suppressed in all directions by the resonators and the average gain is about -10 dBi.

2.4.2. Time-domain performance

UWB radio systems typically employ pulse modulation where extremely narrow (short) bursts of RF energy are used to convey information [1]. Antennas with notches will introduce distortion to these bursts. To investigate this, the pulse response of the proposed antenna in the time domain is studied as follows [24]. Two antennas of the same type are placed side-by-side or face-to-face at a distance of 50 cm (to ensure in the far field region) inside the quiet zone of an anechoic chamber. The antennas are connected using coaxial cables to the two ports of the Agilent PNA N5230C. The transfer function (or S_{21}) of the two-antenna setup is measured in the frequency domain. The time response is then obtained by using the time-domain function (inverse FFT) of the PNA.

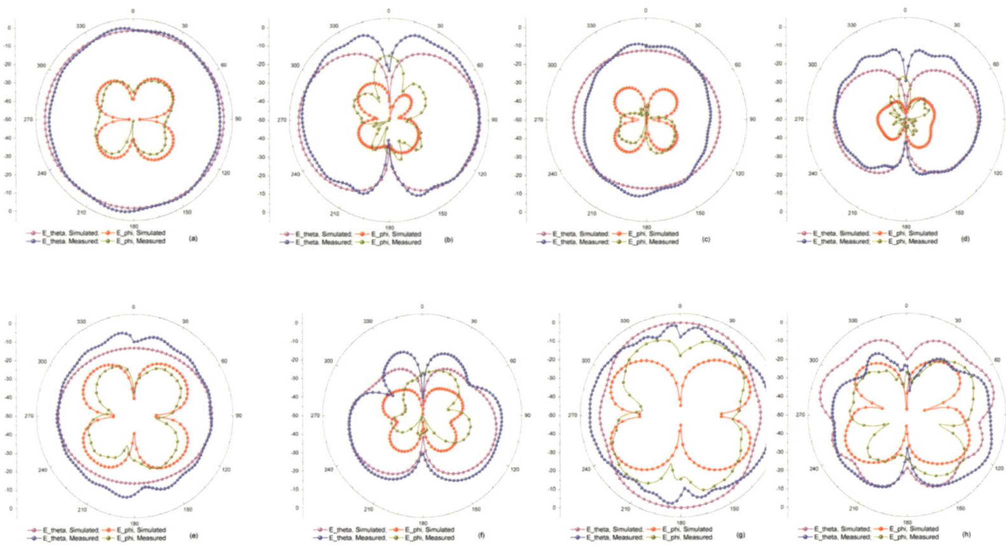

Figure 10. Simulated and measured radiation patterns with dual-band CPW resonator. (a) 3 GHz in x-z plane; (b) 3 GHz in x-y plane; (c) 3.45 GHz in x-y plane; (d) 3.45 GHz in x-y plane; (e) 5.5 GHz in x-z plane; (f) 5.5 GHz in x-y plane; (g) 11 GHz in x-y plane; and (h) 11 GHz in x-y plane

To fully utilize the FCC's UWB, it would be better to select the transmitted pulse with spectrum as close as possible to the FCC's emission limit mask [25]. However, due to the limitation of the equipment, i.e. Agilent PNA N5230C, used in our laboratory, we only manage to generate pulses with a rectangular spectrum from 3.1 to 10.6 GHz as the transmitted pulses.

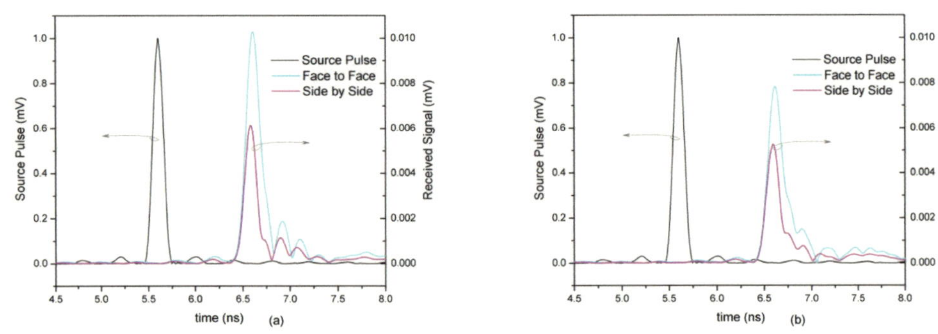

Figure 11. Measured pulse responses for (a) reference UWB antenna and (b) proposed dual band-notched antennas

For comparison, the measured results on pulse responses of the reference antenna and dual band-notched antenna are plotted in Figure 11. It can be seen that, in both cases, the magnitudes of the received pulses are larger in the face-to-face arrangement than in the side-by-side arrangement. This is because, at higher frequencies, the planar structure of the monopole antennas causes the radiation patterns to become slightly directional at the face direction. Figure 11(b) shows that the received pulse for the dual band-notched antenna has late time ringing (or distortion) and lower power.

To evaluate the quality of the received pulses, we define the fidelity as [22, 26]:

$$F = \max_{\tau} \int_{-\infty}^{\infty} f(t)\, r\left(t+\tau\right) dt \tag{1}$$

where $f(t)$ and $r(t)$ are the transmitted and received pulses, respectively, normalized to have unity energy, and τ is the time delay chosen to maximize the integral term. The calculated fidelities F given by (1), for using the reference and notched antennas, are shown in Table 2. It can be seen that the fidelities in the face-to-face and side-by-side arrangements for both antennas used are about the same, which should be the case for antennas with an omnidirectional radiation pattern. As expected, the reference antenna achieves the fidelity of more than 97%, very close to the transmit signal. Table 2 also shows that the antenna with a dual band-notch still can achieve a fidelity of more than 95%.

	Reference Antenna	*Dual Band-Notched Antenna*
Face-to-Face	*0.9825*	*0.9570*
Side-by-Side	*0.9744*	*0.9531*

Table 2. Calculated Fidelity for Reference and Band-Notched Antennas

3. Design of band-notched microstrip monopole antenna using $\lambda/4$-resonator

3.1. Design of UWB monopole antenna

In our design of a single band-notched antenna for UWB applications, we propose to use a planar monopole antenna with microstrip feed to achieve a compact size for applications in wireless devices. The geometry of the design is shown in Figure 12 which consists of an elliptical radiator fed by a 50-Ω microstrip line, and a rectangular ground plane on the other side of the substrate. The antenna is designed on a polytetrafluoroethylene (PTFE) substrate PCB with a dimension of 30 × 39.3 mm², a relative dielectric constant of $\varepsilon_r = 3.5$, a thickness of 0.8 mm and a loss tangent of 0.003.

The parameters r_2 (*gap*) and w_2 of the antenna shown in Figure 12 are optimized for wideband operation using computer simulation and results are shown in Figure 13. The impedance bandwidth of a monopole can be increased by widening the radiator shape. In our case, Figure 13(a) shows that with r_2 = 3 mm (i.e., a thin vertical elliptical radiator), the antenna has two distinct narrow bands, resonating at about 2.8 and 7.4 GHz. The overall bandwidth is less than the UWB. When the width r_2 of the radiator increases, the bandwidth improves. However, with r_2 = 15 mm, the radiator is too wide which reduces the return loss at high frequencies. Figure 13(b) shows that the effects of the distance, *gap*, between the elliptical radiator and the upper edge of ground on the return loss. It can be seen that *gap* is quite sensitive to impedance matching (same as in the previous design of the CPW antenna). With smaller values of *gap*, the antenna has low return loss at high frequencies. With larger values of *gap*, it has low return loss at low frequencies. In our design, the optimized value for *gap* is 0.5 mm. Figure 13(c) shows the return loss for different values of w_2, the width of the microstrip feed-line at the radiator end. Using the characteristic parameters of the substrate, the width w_1 needs to be 1.73 mm in order to achieve a 50-Ω characteristic impedance for the microstrip line [27]. It can be seen that, if $w_1 = w_2 = 1.73$ mm, the impedance bandwidth (for return loss > 10 dB) of the antenna cannot cover the whole UWB. Thus, the width of the upper 6 mm of the microstrip feed-line is tapered linearly to improve matching. With $w_2 = 0.6$ mm, the antenna has an impedance bandwidth (10-dB return loss) from around 2.5 GHz to over 12 GHz, which fully satisfies the FCC requirements for the UWB. This final design is used as a reference UWB antenna for comparison with our band-notched antenna design. The dimensions are listed in Table 3.

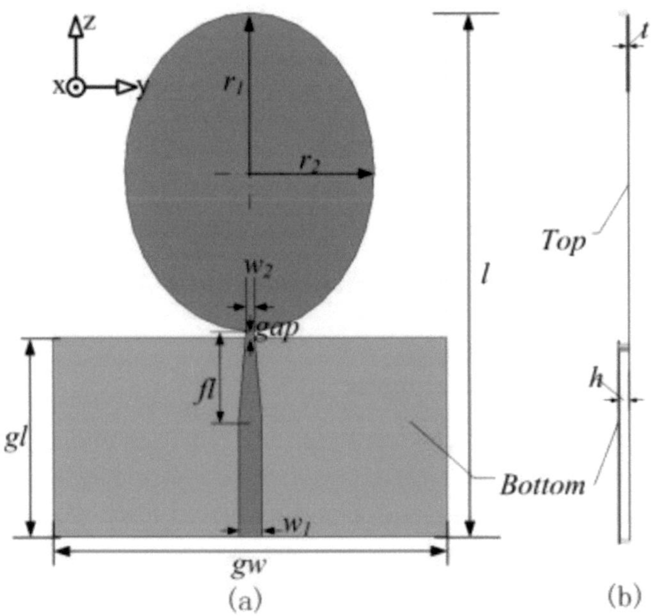

Figure 12. Layout of UWB antenna: (a) top view and (b) side view

Parameter	*Value*	*Parameter*	*Value*	*Parameter*	*Value*
gl	15	*h*	0.762	w_1	1.73
gw	30	*gap*	0.5	w_2	0.6
l	39.3	s_1	0	R_{via}	0.3
r_1	12	s_2	0.5	*fl*	6
r_2	9	s_3	9		
t	0.035	s_4	1.0		

Table 3. Antenna dimensions (in mm)

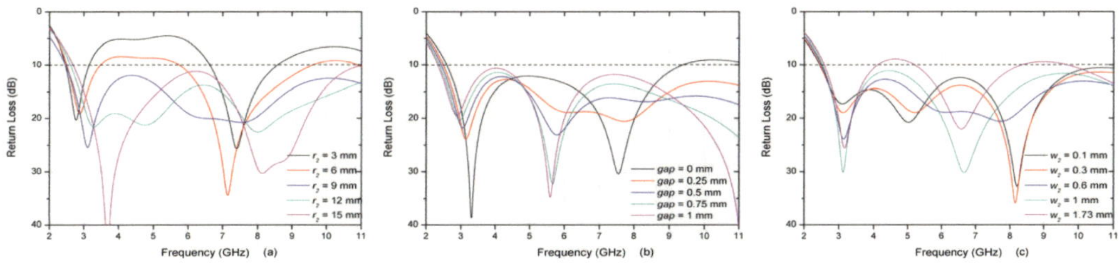

Figure 13. Simulated return loss of the planar monopole antenna with different values of (a) r_2, (b) *gap*, and (c) w_2

3.2. λ/4-resonator

To generate a band-notched characteristic for the antenna, we propose to use a pair of coupled-fed λ/4-resonators (a λ/4-microstrip line with a shorting end) which has the advantages of simple structure, easy design and fabrication, and low cost. To create an effective notch for the antenna, the λ/4-resonators should be placed in the positions where majority of the current passes through before radiating into space. Figure 14 shows the simulated current distribution of the UWB antenna at 5.5 GHz. The current is mainly distributed at the edges of the microstrip line, the upper edges of the ground plane and the edges of the radiator. Thus, we should place the λ/4-resonators along these edges to create a notch for the antenna. Figure 15 shows our proposed design, where the two λ/4-resonators are symmetrically placed at a distance s_4 from the center of the microstrip feed line and are connected to the ground through a via.

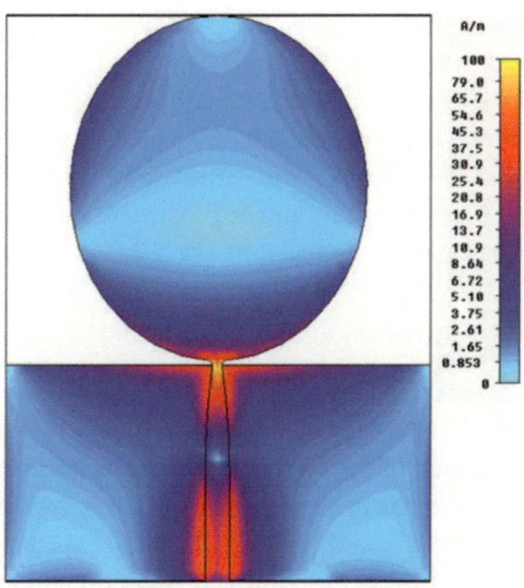

Figure 14. Current distribution of proposed UWB antenna at 5.5 GHz

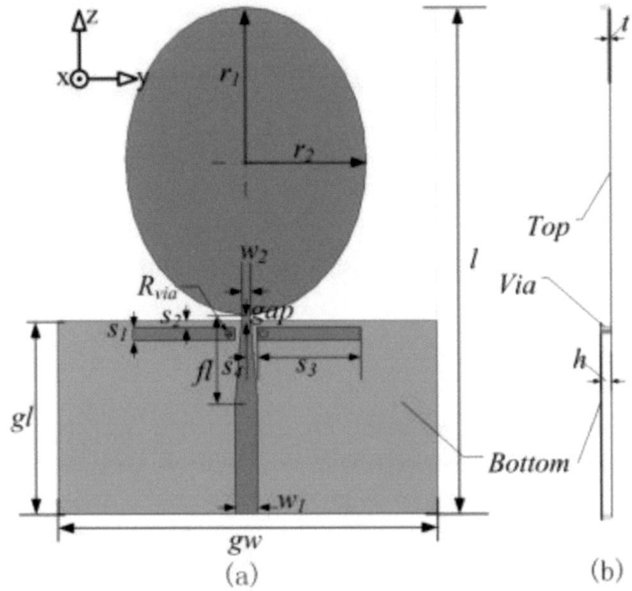

Figure 15. Layout of single band-notched UWB antenna: (a) top view and (b) side view

The guided wavelength of the waves propagating along a substrate can be approximated by [27]:

$$\lambda_g \approx \lambda_0 / \sqrt{(\varepsilon_r + 1)/2} \tag{2}$$

with λ_0 being the free space wavelength and ε_r the relative permittivity of the substrate. In our design, the PTFE substrate used has a relative dielectric constant of $\varepsilon_r = 3.5$. To design a notch at 5.5 GHz which is the center frequency of the IEEE 802.11a WLAN band [28], applying (2) yields $\lambda_g \approx 36.34$ mm or $\lambda_g/4 \approx 9.085$ mm. While in our simulation studies, the required length of the $\lambda/4$-resonator for a notch at 5.5 GHz is $s_3 = 9$ mm. Thus the difference between numerical calculation using (2) and simulation (or practical implementation) is only about 4.5%, which may be caused by the approximation of the expression for λ_g in (2). The dimensions of the $\lambda/4$-resonators are listed in Table 3.

3.3. Current flow on $\lambda/4$-resonators

To better understand the working principle of the $\lambda/4$-resonators on the antenna, the currents flowing around the areas of the resonators and feed line at the notch frequency of 5.5 GHz and passband frequency of 3.5 GHz are simulated and shown in Figure 16. At 5.5 GHz, Figure 16(a) shows that the current is coupled from the feed line and the upper edge of the ground plane to the resonators and then flows to ground through the vias. This stops the energy on the feed line flowing into the elliptical radiator and radiating into free space. However, at the passband of 3.5 GHz, as shown in Figure 16(b), majority of the current flows to the radiator and radiates to free space.

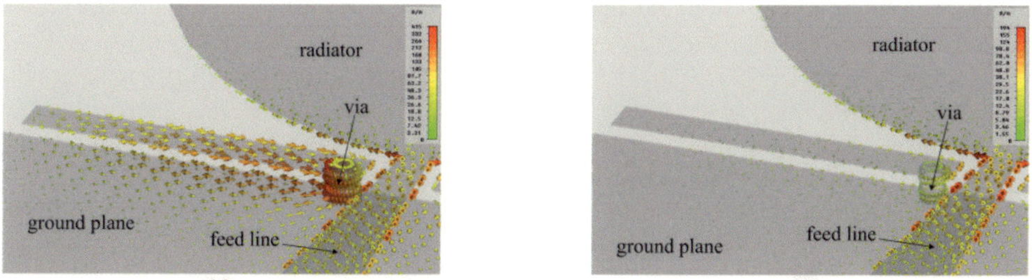

(a) (b)

Figure 16. Current flow around λ/4-resonators and feed line of single band-notched antenna at (a) notch frequency 5.5 GHz and (b) passband frequency 3.5 GHz

3.4. Parametric study on $\lambda/4$-resonators

The width s_1 of the resonators, and the distances s_2 between the resonators and the upper edge of ground plane and s_3 between the feed line and resonators, determine the capacitance of the resonators. A parametric study of the single band-notched UWB antenna is carried out using computer simulation to explore how the dimensions of the $\lambda/4$-resonators affect the characteristic of the band notch. Simulation results on the effects of the dimensions s_1, s_2 and s_3 on the return loss of the antenna are shown in Figures 17. Figures 17(a) and 17(c) show that s_1 and s_3 determine the notch frequency. s_3 is more sensitive and so can be used for coarse adjustment of the notch frequency, while s_1 can be used for fine adjustment. Figure 17(b) shows that s_2 mainly affects the notch bandwidth. These plots also reveal that when the values of s_1, s_2 and s_3 are changed, the return loss in the rest of the UWB band remains nearly unchanged. This property provides the designers with a great freedom to select the notched-band frequency and bandwidth for the antenna.

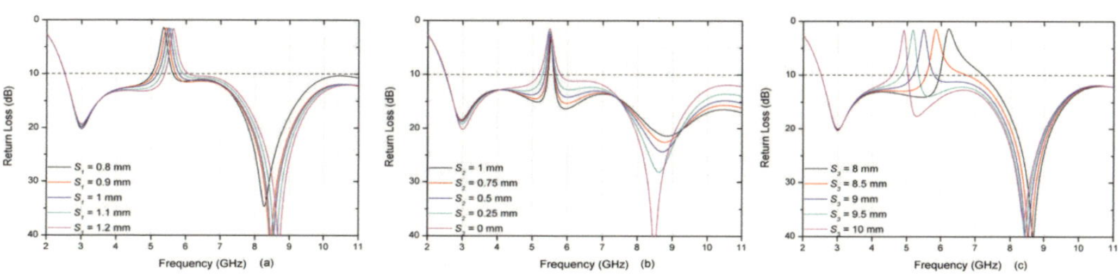

Figure 17. Simulated return loss of single band-notched antenna with different values of (a) s_1 , (b) s_2, and (c) s_3

3.5. Current distribution

Figures 18(a) – 18(d) show the surface current distributions of the antenna at 3.5, 5.5, 7 and 12 GHz, respectively. At the passband frequencies, i.e., 3.5, 7 and 12 GHz, Figures 18(a), 18(c) and 18(d) show that majority of the current flows from the microstrip line to the radiator and so finally radiates into free space. However, at the notch frequency 5.5 GHz, Figure 18(b) shows that the current is confined much more around the areas near the $\lambda/4$-

resonators than those in the main radiator of the antenna. As a result, the energy does not get radiated.

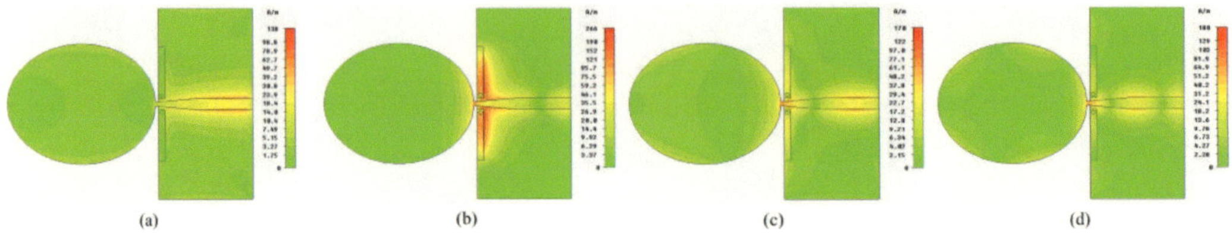

Figure 18. Surface current distribution of single band-notched antenna at (a) 3.5, (b) 5.5, (c) 7, and (d) 12 GHz

3.6. Results and discussions

The design of the band-notched antenna is fabricated using a Rogers substrate, RO4350B, as shown in Figure 19. The return loss, peak gain and efficiency across the UWB band, and the radiation patterns at 3, 5.5, 7 and 12 GHz are simulated and measured using the Satimo Starlab measurement system. The pulse responses between a pair of the antennas placed face-to-face and side-by-side are investigated.

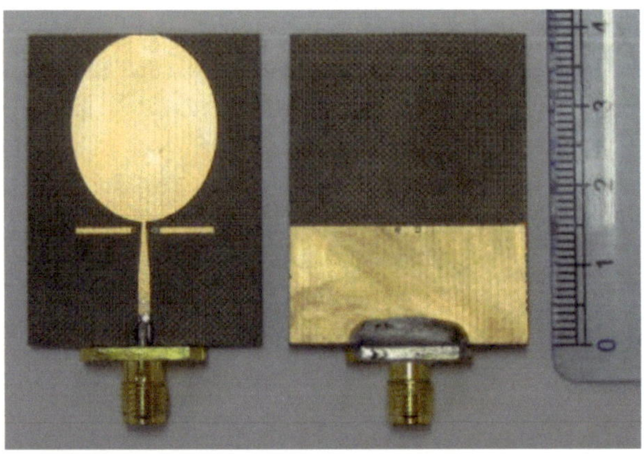

Figure 19. Photograph of single band-notched antenna

3.6.1. Frequency-domain performance

The simulated and measured return losses of the single band-notched antenna are shown in Figure 20. It can be seen that, the antenna can operate from 2.57 GHz to over 12 GHz with return loss > 10 dB, except in the WLAN band from 5.18 to 6.23 GHz, where the measured return loss is substantially less than 10 dB.

The simulated and measured radiation patterns of the antenna at the frequencies of 3, 5.5, 7 and 12 GHz in the two principle planes, the x-y and x-z planes, are shown in Figure 21. At 3, 7 and 12 GHz, Figures 21(a), 21(e) and 21(g) show that the antenna has approximately omnidirectional radiation patterns in the x-y plane. In the x-z plane, Figures 21(b), 21(f) and

21(h) show that there are two nulls in the z-direction, which is typical for monopole antennas. The radiation patterns in Figures 21(c) and 21(d) at the notch frequency of 5.5 GHz indicate that the gain is almost evenly suppressed in all directions by the $\lambda/4$-resonators with an average peak gain of about -10 dBi.

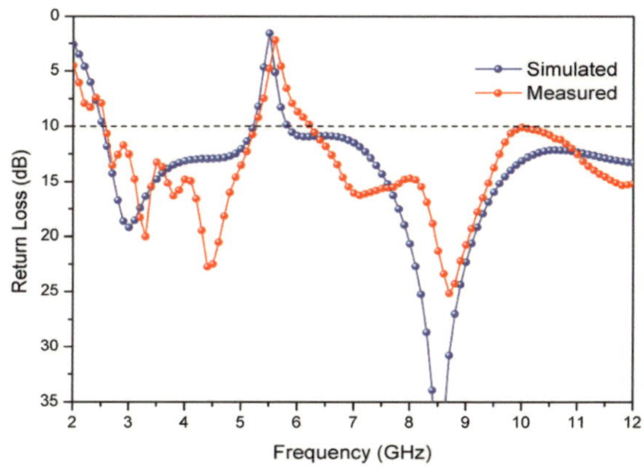

Figure 20. Simulated and measured return losses of single band-notched antenna

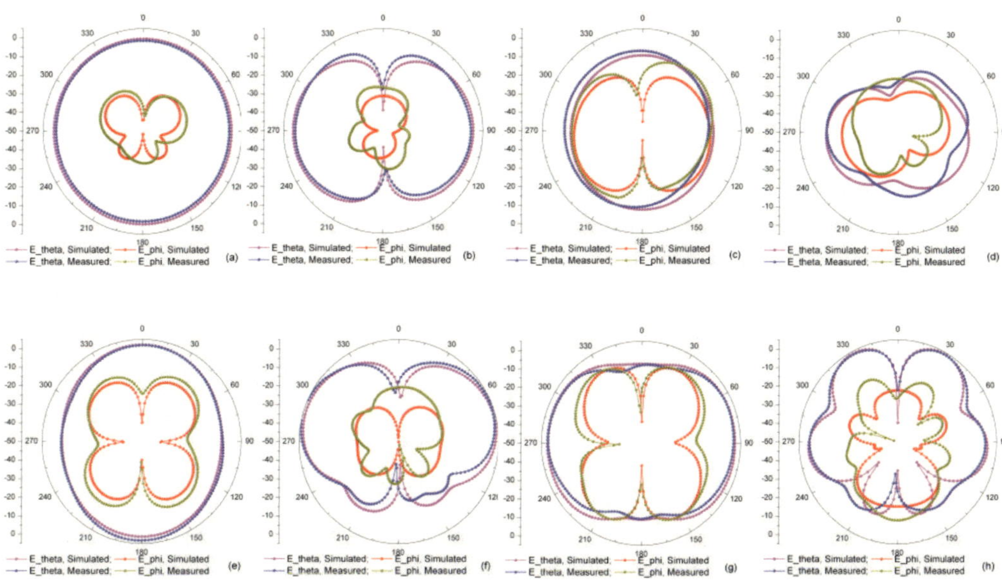

Figure 21. Simulated and measured radiation patterns of single band-notched antenna. (a) 3 GHz in x-y plane, (b) 3 GHz in x-z plane, (c) 5.5 GHz in x-y plane, (d) 5.5 GHz in x-z plane, (e) 7 GHz in x-y plane, (f) 7 GHz in x-z plane, (g) 12 GHz in x-y plane, and (h) 12 GHz in x-z plane

The peak gain and efficiency of the antenna are shown in Figures 22(a) and 22(b), respectively. The measured average peak gain over the UWB, computed by excluding the notched band, is around 3.5 dBi. However, at the notched band, significant reductions in gain and radiation efficiency can be observed. The measured peak gain is suppressed from about 2.5 dBi to -5.4 dBi and the radiation efficiency is reduced from about 80% to 15%. These results indicate that the $\lambda/4$-resonators effectively generate a single band-notched characteristic for the antenna.

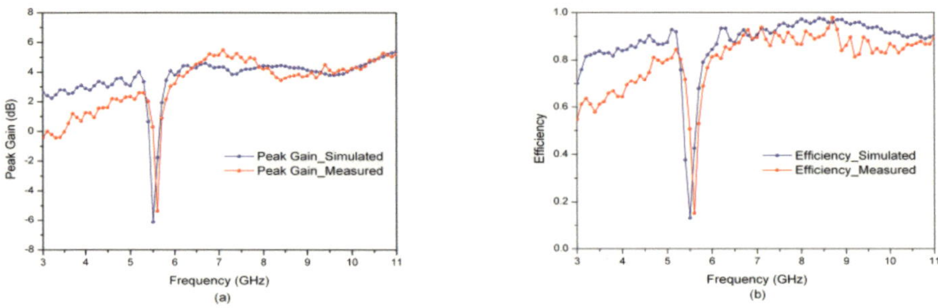

Figure 22. Simulated and measured (a) peak gains and (b) efficiencies of proposed antenna

3.6.2. Time-domain performance

The measurement procedure for the time-domain performance is described in the previous section. For comparison, the pulse responses of the reference UWB antenna (without having the notched characteristic) are also measured and shown in Figure 23. It can be seen that, the pulse responses in the face-to-face arrangements have larger amplitudes than those in the side-by-side arrangements. This agrees with the radiation patterns shown in Figure 21 where radiations in the x-direction are slightly larger than those in the y-direction in most of the frequencies tested. The pulse responses for the reference antenna are only slightly larger than those for the notched antenna in the same arrangements. Late time ringing (or distortion) and lower power are observed in the received pulses for both antennas.

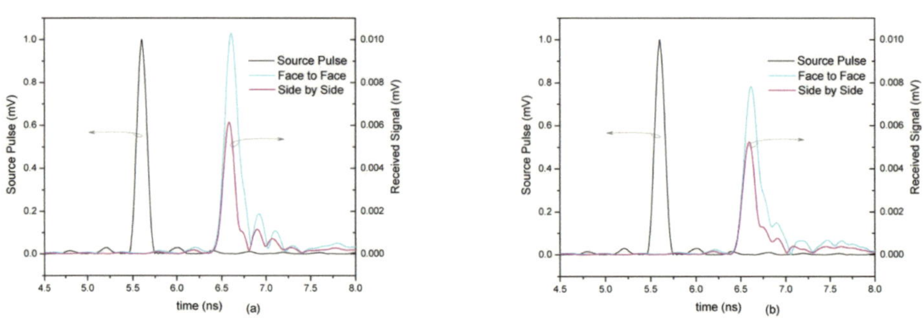

Figure 23. Measured pulse responses for (a) reference and (b) single-band notched antennas

The fidelities F of the time responses using (1) are computed and shown in Table 4. In both the face-to-face and side-by-side arrangements, the fidelities are about the same. As expected, the reference UWB antenna has the fidelity of more than 97%, very close to the source signal. The results in Table 4 show that the antenna with a single-band notch can achieve the fidelity of more than 94%.

	Reference Antenna	*Single Band-Notched Antenna*
Face-to-Face	*0.9825*	*0.9481*
Side-by-Side	*0.9744*	*0.9413*

Table 4. Calculated Fidelity for Reference and Band-Notched Antennas

4. Multiple band-notched characteristics using meander lines (MLs)

4.1. Design multiple band-notched antennas using MLs

Meander line (ML), also known as serpentine line, consisting of a number of transmission lines closely packed and jointed to each others, as shown in Figure 24, is an effective way for size reduction of a transmission line [29-36]. The idea behind meandering is to increase the electrical length per unit area of circuit board space when the signal is propagating along the ML.

Figure 24. Typical ML with 8 segments

To design multiple band-notched characteristics for compact UWB antennas, the $\lambda/4$-resonator used in section 3 is too large to be used. By folding the $\lambda/4$-resonator into a ML, we can obtain a compact structure. Due to the mutual coupling between the adjacent segments of the ML, the total physical resonator length is no longer $\lambda/4$ long. Studies have shown that we can make the electrical length of the MLs to be $\lambda/4$ with a smaller size. With the compact structures of MLs, we can easily place several pairs of MLs in different positions of the antenna to obtain a multiple band-notched characteristic. In our proposed design, two different types of feeding techniques, known here as parallel-coupled feeding (PCF) and direct-connected feeding (DCF), as shown in Figures 25(a) and 25(b), respectively, are employed. In the PCF ML, the signal is coupled from the transmission line to the ML, while in the DCF ML, the signal is fed directly to the ML. Details of the different band-notched designs for a compact UWB antenna using MLs are described in the following sections.

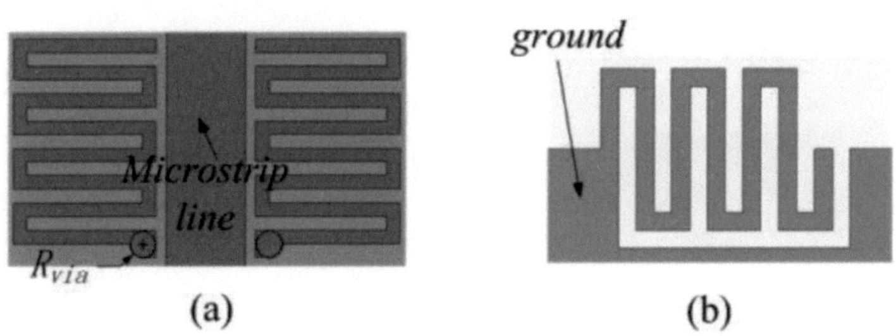

Figure 25. MLs with (a) parallel-coupled fed and (b) direct-connected fed

The design procedure for our band-notched antennas is as follows. An UWB antenna without any ML is designed and used as a reference antenna for comparison. Several pairs of MLs are then added to the reference antenna to make it a multiple band-notched antenna. (A single notched-band antenna requires only one pair of MLs.) The dimensions of the individual MLs are adjusted to achieve the desirable center frequencies and bandwidths for the notches. For convenience, we use the same antenna in section 2 as the reference antenna for our band-notched antennas design here.

As previously described and shown in Figure 14, the surface current concentrates most at the edges of the feed line, the ground plane and the radiator. So we can place the resonators implemented using MLs along these edges as shown in Figures 26(a), 26(b), 26(c) and 26(d) to create a single-, dual-, triple- and quadruple-band notches, respectively, for the antenna. Two different types of feeding techniques, i.e. PCF and DCF, are used. The MLs with PCF are etched on the same side as the radiator on the substrate and connected to ground through vias, while the MLs with DCF are etched on the ground plane on the other side of the substrate.

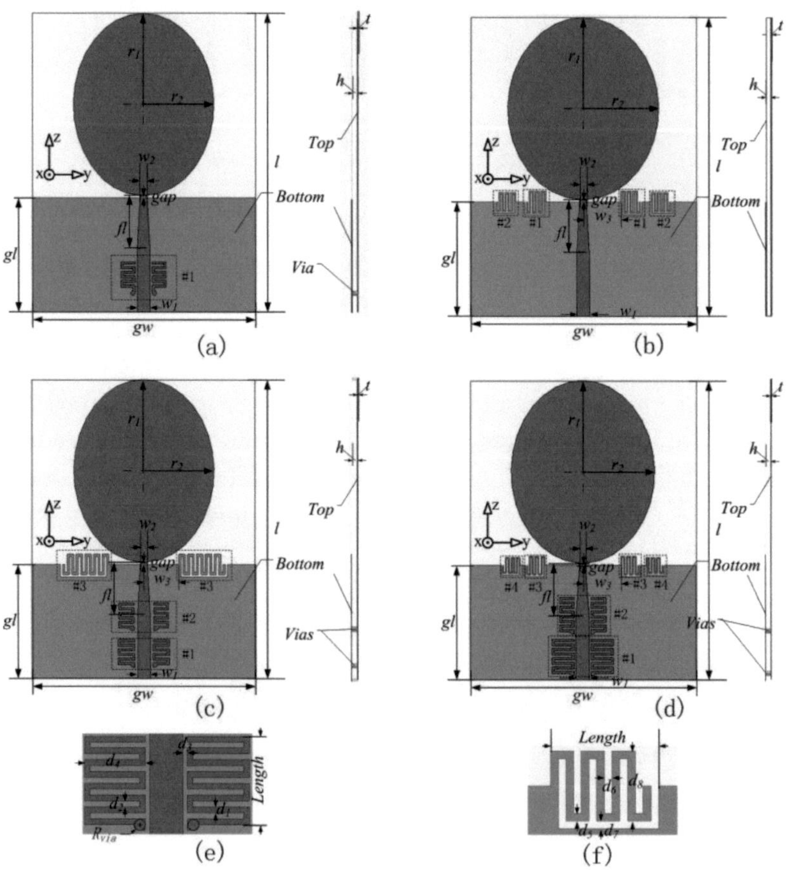

Figure 26. Top views and side views of (a) single, (b) dual, (c) triple, (d) quadruple band-notched antennas, (e) PCF and (f) DCF MLs

In the single and dual band-notched designs shown in Figures 26(a) and 26(b), respectively [10, 11, 37], the notches are designed at the frequencies of existing communication systems such as the WiMax system, upper and lower bands of the WLAN systems. In the triple band-notched design of Figure 26(c), the notches are designed at 3.5, 5.25 and 5.775 GHz. In the quadruple band-notched antenna design of Figure 26(d), we select four notch frequencies with around 1 GHz separation to show the flexibility in designing the notch frequencies. To avoid duplicated description, the quadruple band-notched antenna is used here to explain creating notches using MLs. In Figure 26(d), the four pairs of MLs, labeled as MLs #1, #2, #3 and #4, are designed to create the notches at 3.14, 4.34, 5.4 and 6.4 GHz, respectively. Detailed dimensions of the MLs for single, dual, triple and quadruple band-notched antennas are listed in Tables 5 to 8, respectively.

ML	d_1	d_2	d_3	d_4
#1	0.3	0.3	0.4	1.9

Table 5. Dimensions of PCF MLs and DCF MLs for Single Band-Notched Antenna

ML	d_5	d_6	d_7	d_8	w_3
#1	0.25	0.25	0.25	2.9	5
#2	0.25	0.25	0.25	2.6	8.5

Table 6. Dimensions of PCF MLs and DCF MLs for Dual Band-Notched Antenna

ML	d_1	d_2	d_3	d_4	*ML*	d_5	d_6	d_7	d_8	w_3
#1	0.3	0.3	0.25	2.2	#3	0.3	0.3	0.3	2.7	5.2
#2	0.3	0.3	0.3	2						

Table 7. Dimensions of PCF MLs and DCF MLs for Triple Band-Notched Antenna

ML	d_1	d_2	d_3	d_4	*ML*	d_5	d_6	d_7	d_8	w_3
#1	0.3	0.3	0.25	3	#3	0.2	0.2	0.2	2.5	5.2
#2	0.3	0.3	0.25	2.1	#4	0.2	0.2	0.2	2.1	8.5

Table 8. Dimensions of PCF MLs and DCF MLs for Quadruple Band-Notched Antenna

4.2. Parallel-coupled fed (PCF) and direct-connected fed (DCF) MLs

The PCF MLs, i.e., MLs #1 and #2 in Figure 26(d), on both sides of the feed line have open circuit at one end and short circuit to the ground through a via at the other end. With an electrical length of $\lambda_g/4$, where λ_g is the guided wavelength approximately given by (2), the MLs are $\lambda_g/4$-resonators with parallel-coupled fed. At resonant frequencies, the signal on the feed line is coupled to the MLs and then flowing through the vias to the ground plane. This creates high impedance for the signal and prevents the signal from flowing into the radiator. A $\lambda_g/4$-resonator operating in the fundamental mode will have a $\lambda_g/4$-length standing wave formed along it. Computer simulation has been carried out to study the surface current on ML #2 at the resonant frequency of 4.34 GHz and results show that, at any instance, the currents on the whole ML always have the same phase (in the same direction), i.e. a standing wave is formed on the ML which is expected for a $\lambda_g/4$-resonator operating in the fundamental mode. This confirmed that MLs #1 & #2 are $\lambda_g/4$-resonators. Figure 27(a) shows a snap-shot from the simulation result on the surface current of ML #2. It can be seen that, the current is quite small at the open end of the ML but substantially larger near the via, which is typical for a $\lambda_g/4$-resonator.

The DCF MLs, i.e., MLs #3 and #4 in Figure 26(d), at the upper edges of the ground plane are stubs having open circuit at one end and directly connect to ground at the other end. With an electrical length of $\lambda_g/4$, the MLs are $\lambda_g/4$-resonators with direct-connected fed. At resonance, the MLs prevent the signal from passing through, creating high input impedance. This causes severely mismatching to the antenna and reduces the return loss. As mentioned before, a $\lambda_g/4$-resonator operating in the fundamental mode has a $\lambda_g/4$-length standing wave formed along it. Computer simulation has also been carried out on the surface current of ML #3 at the resonant frequency of 5.4 GHz. Results show that, at any instance, the currents along the ML all have the same phase (in the same direction), as expected for a $\lambda_g/4$-resonator operating in the fundamental mode. This again confirms that MLs #3 & #4 are $\lambda_g/4$-resonators. Figure 27(b) shows a snap shot from the simulation result on the surface current of ML #3. The current is substantially smaller at the open end of the ML than that at the connected end, which is expected for a $\lambda_g/4$-resonator.

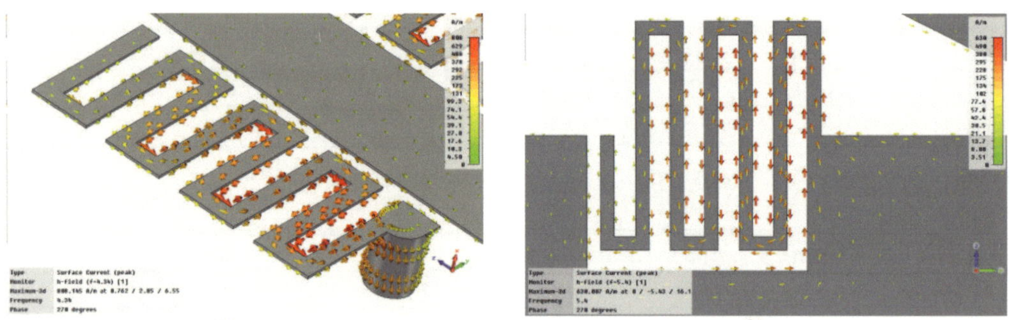

(a) (b)

Figure 27. Current distribution on (a) ML #2 at 4.34 GHz and (b) ML #3 at 5.4 GHz

4.3. Discussions on electrical lengths of PCF and DCF MLs

For a $\lambda_g/4$-resonator implemented using a straight microstrip line, if ε_r in (2) is known, the physical length of the microstrip line, for the $\lambda_g/4$-resonator to work at a particular frequency, can easily be calculated. Unfortunately, this does not work with the MLs in our design. For example, the total lengths measured along the internal perimeters of MLs #1 and #2 in Figure 26(d) for the resonant frequencies of 3.14 and 4.34 GHz are 26.1 and 17.7 mm, respectively, which are quite different from the computed $\lambda_g/4$=15.95 and 11.54 mm using (2). Then for MLs #3 and #4, the internal perimeters for the resonant frequencies of 5.4 and

6.4 GHz computed using Table 8 are 16.2 and 15 mm, respectively, which again are very different from the corresponding lengths $\lambda_g/4$=9.27 and 7.825 mm when computed using (2). These results show that (2) cannot be used in our MLs to determine the notch frequencies for the antennas. The reason can be explained as follows.

(2) is derived using the distributed-elements model and can be applied quite accurately to a straight microstrip line. For complicated structures which are many times, say 10, less than the free space wavelength, the lumped-elements model should be used [38]. In our design, the sizes of MLs #1, #2, #3 and #4 are 3×4.5 mm^2, 2.1×4.5 mm^2, 2.8×2.7 mm^2, 2.8×2.3 mm^2, respectively. The maximum lengths of these structures are less than 1/10th of the corresponding guided wavelengths of 51.18, 38.82, 29.76 and 25.11 mm at the resonant frequencies of 3.14, 4.34, 5.4 and 6.4 GHz, respectively, thus both the distributed-elements model and the lumped-elements model together should be used [38-40]. In the lumped-element model, there will be capacitance formed by the shunt capacitance between the ML and the substrate, series capacitance formed between segments of the ML, inductance formed by the series inductance along the ML plus shunt inductance formed between the segments of the ML, and inductance formed by the via [41], so the models are extremely complicated [35, 42-44]. These models will have different resonant modes of operation. In the fundamental mode, only the largest capacitances and inductances are in effect and form an *LC* resonator which will have the fundamental resonant frequency f_r proportional to $1/\sqrt{L_{eq}C_{eq}}$, i.e.,

$$f_r \propto \frac{1}{\sqrt{L_{eq}C_{eq}}} \tag{3}$$

where L_{eq} and C_{eq} are the equivalent inductance and capacitance, respectively, resulted from the self inductance and self capacitance of the ML.

For the PCF MLs, at resonance, the signals will be electrically coupled from the feed line to MLs #1 and #2 via a mutual coupling capacitor C_m between them [38]. This coupling will detune the resonant frequency as will be shown later. For the DCF MLs, the signals will be directly fed to MLs #3 and #4. Since it is not easy, if not impossible at all, to relate L_{eq} and C_{eq} to the dimensions of the MLs, full-wave simulation needs to be used for the design.

4.4. Setting the notch center frequencies and bandwidths

Parametric study of the quadruple band-notched antenna is carried out using computer simulation to explore how the dimensions of the MLs affect the characteristics of the band notches.

For the PCF resonators, ML #1 is used for studies. The smallest dimension we can make for our antenna using the prototype-machine in our laboratory is 0.1 mm. So for convenience in our design process, we fix the segment width d_1 and segment spacing d_2 of the ML to be the same value. Simulation results show that, with $d_1 = d_2 < 0.2$ mm, the notch bandwidth is too large (more than 1 GHz). Thus for having a low percentage of tolerance and compact size, we fix d_1 and d_2 to a slightly larger value of 0.3 mm. The segment length d_4 and distance d_3 between the ML and the feed line are then used for parametric studies using computer simulation. Results show that d_4 and d_3 affect the notch frequency and bandwidth, respectively. The effect of d_4 on the notch frequency is shown in Figure 28(a), indicating that d_4 is inversely proportional to the notch frequency. This is because increasing the segment length d_4 increases the series inductance and capacitance along the ML and hence lowering down the resonant frequency through (3). With d_4 increased from 2.8 to 3.2 mm, the resonant frequency shifts from 2.85 to 3.23 GHz, which is quite significant. ML #1 is parallel-coupled fed via the mutual capacitance between the feed line and the ML. At the resonant frequency, the signal travelling along the feed line is coupled to the ML. When d_3 is increased, the coupling effect (or the mutual capacitance) between them is reduced, leading to a lower reactance and hence a lower quality factor Q. For a given resonant frequency f_r, the bandwidth is given by f_r/Q. Thus d_3 is inversely proportional to the notch bandwidth and so can be used to adjust the notch bandwidth. The effect of d_3 on the notch bandwidth is shown in Figure 28(b). With d_3 increased from 0.1 to 0.5 mm, the 10-dB bandwidth changes from 0.31 to 0.11 GHz. It should be noted that the notch frequency is dominated by the ML's equivalent self-capacitance C_{eq} and inductance L_{eq} given by (3) and so d_3 will only slightly affect the notch frequency, as can be seen in Figure 28(b).

For the DCF resonators, ML #3 in Figure 28(d) is used for parametric studies. Again, for convenience, we fix the segment width d_6 and segment spacing d_5 of the ML to have the same value. Simulation results show that a smaller value of 0.2 mm for the segment width d_6 and segment spacing d_5 can be used without having a too large notch bandwidth, so we fix $d_6 = d_5 = 0.2$ mm for having a smaller size. The segment length d_8, the distance d_7 between the ML and the ground, and the distance w_3 between the center of the feed line and the ML are used for parametric studies. Simulation results show that d_8 affects the notch frequency, for the same reason as described for the PCF resonators. The effect of d_8 on the notch frequency is shown in Figure 28(c). With d_8 increased from 2.3 to 2.7 mm, the notch frequency shifts from 5.71 to 5.04 GHz.

Simulation results show that both d_7 and w_3 in ML #3 affects the notch bandwidth, with w_3 having significantly higher effects, so w_3 is used for studies and results are shown in Figure 26(d). MLs #3 and #4 are $\lambda_g/4$-resonators with DCF. There is series inductance on the ground plane between the feed line and the ML. This inductance is connected in series with the ML and increases with the current flowing through it [45]. Since the current density is higher in the region closer to the feed line than far away from it, the inductance is larger with a shorter distance w_3, leading to a larger reactance and a higher quality factor Q. For a given resonant frequency f_r, the bandwidth given by f_r/Q is therefore smaller. This is confirmed by the simulation results of Figure 28(d) which shows that, with w_3 increased from 4.6 to 5.8 mm, the 10-dB bandwidth changes from 0.74 to 0.28 GHz. Thus w_3 can be used to adjust the notch bandwidth. Similar to the case for the PCF resonators, the notch frequency is dominated by the self-capacitance and inductance of the ML, so w_3 will only slightly affect the notch frequency as can be seen in Figure 28(d). It should be noted that, if ML #3 is too close to the feed line, e.g. w_3 = 2 mm, the notch bandwidth will be too wide and, together with the adjacent notches, form a very wide stopband. Moreover, the small distance *gap* is very critical for impedance matching of the antenna. Placing the ML too close to this gap will affect the matching performance and reduce the impedance bandwidth, as evident in Figure 28(d).

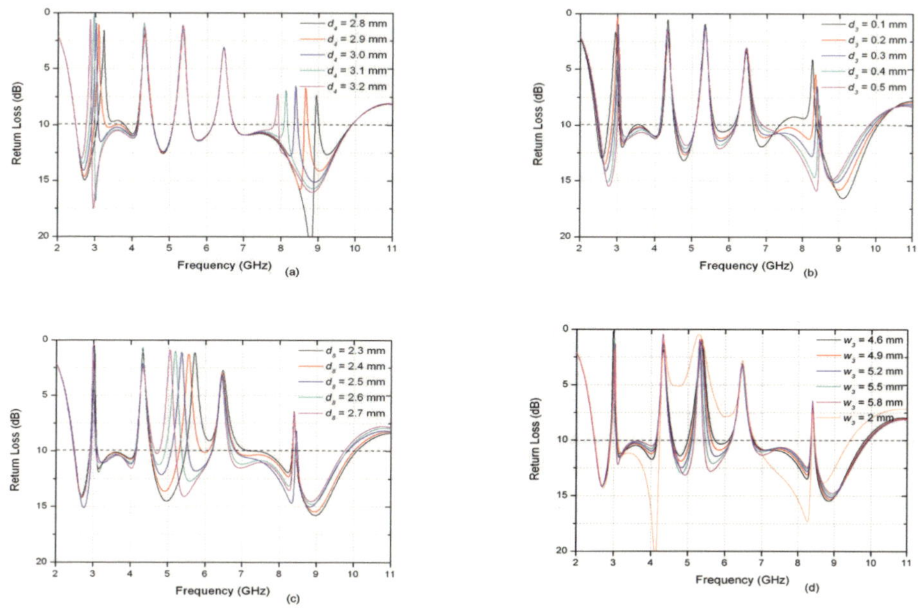

Figure 28. Return loss with different values of (a) d_4 and (b) d_3 for ML #1, and (c) d_8 and (d) w_3 for ML #3 of the quadruple band-notched antenna

For a parallel-coupled fed $\lambda_g/4$-resonator implemented using a straight microstrip line, the first spurious response always occurs at three times the resonant frequency [46]. ML #1 is a parallel-coupled fed resonator at around 3 GHz. Figures 28(a) and 28(b) show that it has a spurious response at about 8.5 GHz, slightly less than three times the resonant frequency. In this respect, ML #1 again behaves like a parallel-coupled fed resonator. Note that ML #2 also has the first spurious response at about three times the resonant frequency, which is not shown in Figure 28(a) due to the small scale.

These results also reveal that when the values of d_4, d_3, d_8 and w_3 are changed, the return loss in the rest of the UWB band remains about the same. This provides the designers with a great freedom to select the notched-band frequency and bandwidth for the antenna.

4.5. Results and discussions

4.5.1. Frequency-domain performance

The return loss, efficiency and peak gain across the UWB, and the radiation patterns at the passband and also the notch frequencies of all the four band-notched antennas are studied using computer simulation. To validate the simulation results, the antennas are also fabricated on PCBs with PTFE substrate with parameters described previously, as shown in Figure 29, and measured using the antenna measurement system, Satimo Starlab.

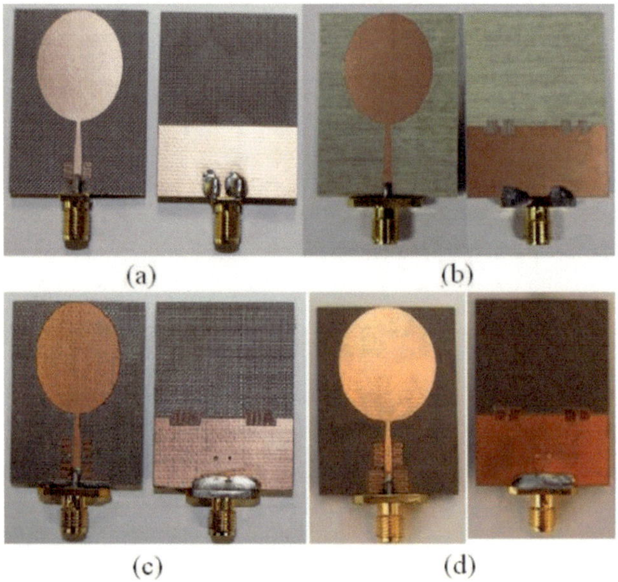

Figure 29. Top and bottom views of prototyped (a) single, (b) dual, (c) triple and (d) quadruple band-notched antennas

4.5.2. Single band-notched antenna

The simulated and measured return losses of the single band-notched antenna of Figure 29(a) are shown in Figure 30. It can be seen that, the antenna can operate from 2.76 GHz to over 12 GHz with return loss ≥ 10 dB which fully satisfies the UWB requirement. In the WLAN band from 4.98 to 5.93 GHz, the measured return loss is substantially lower than 10 dB.

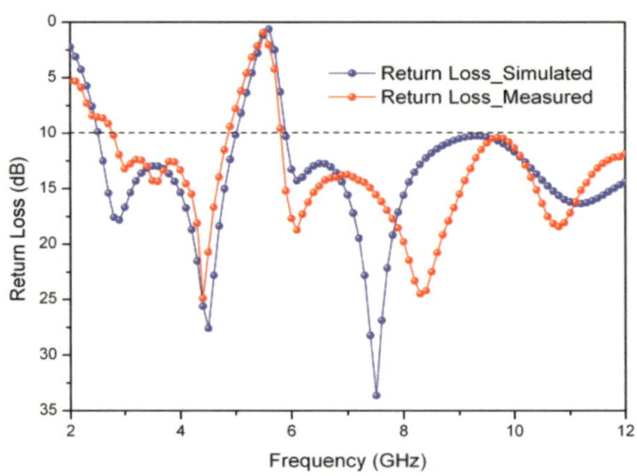

Figure 30. Simulated and measured return losses of single band-notched antenna

The simulated and measured radiation patterns of the single band-notched antenna at the frequencies of 3, 5.5, 7 and 10 GHz in the two principle planes, the x-y and x-z planes, are shown in Figure 31. At 4, 7 and 10 GHz, Figures 31(a), 31(e) and 31(g) show that the antenna has approximately omnidirectional radiation patterns in the x-y plane. In the x-z plane, Figures. 31(b), 31(f) and 31(h) show that there are two nulls at the z-direction, typical for monopole antennas. At the notch frequency of 5.5 GHz, the radiation patterns in Figures 31(c) and 31(d) indicate that the gain is almost evenly suppressed in all directions by the pair of MLs and the average gain drops to about -10 dBi.

The simulated and measured peak gains of the single band-notched antenna are shown in Figure 32(a) and the efficiencies are shown in Figure 32(b). The average gain over the UWB, computed by excluding the notched band, is about 3.5 dBi. However, at the notched band of about 5.5 GHz, the antenna gain is suppressed from about 3 dBi to -7.1 dBi and the radiation efficiency is reduced from about 85% to 8.7%. These phenomena indicate that the MLs work effectively to introduce a single band-notched characteristic for the antenna.

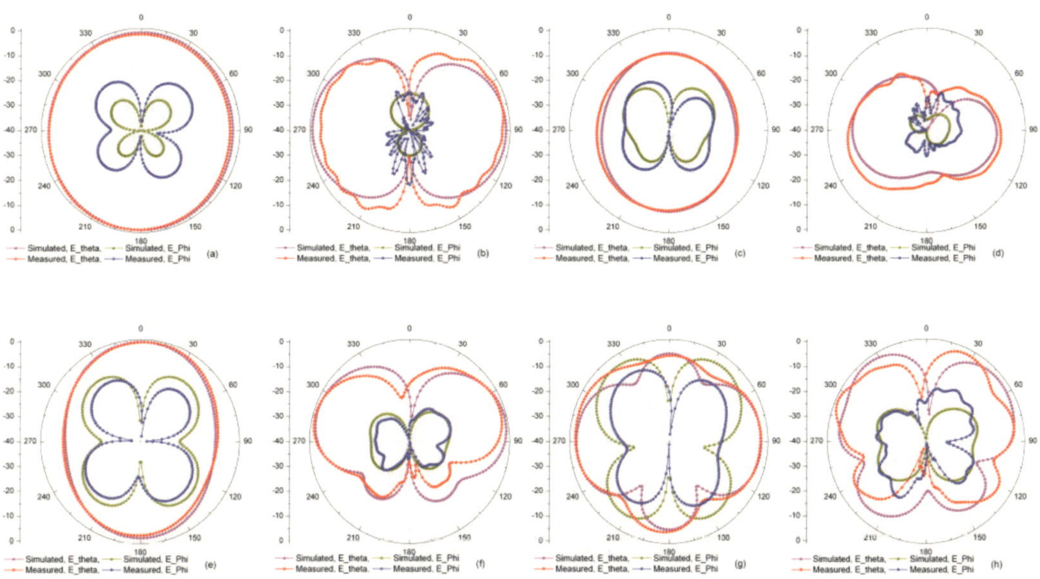

Figure 31. Simulated and measured radiation patterns of single band-notched antenna. (a) 3 GHz in x-y plane; (b) 3 GHz in x-z plane; (c) 5.5 GHz in x-y plane; (d) 5.5 GHz in x-z plane; (e) 7 GHz in x-y plane; (f) 7 GHz in x-z plane; (g) 10 GHz in x-y plane; and (h) 10 GHz in x-z plane

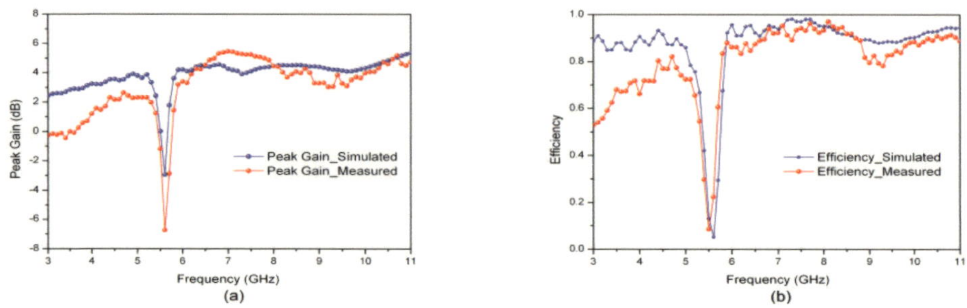

Figure 32. Simulated and measured (a) peak gains and (b) efficiencies of single band-notched antenna

4.5.3. Dual band-notched antenna

The simulated and measured return losses of the dual band-notched antenna of Figure 29(b) are shown in Figure 33. It can be seen that, the antenna can operate from 2.95 GHz to about 11.8 GHz with return loss ≥ 10 dB. In the lower and upper WLAN bands from 5.01 to 5.48 GHz and 5.63 to 5.92 GHz, respectively, the measured return loss for the lower WLAN band is much lower than 10 dB. In the upper WLAN band, the return loss is larger. This is because the MLs are further away from the center feed line and so couple less energy compared to the MLs for the lower WLAN band.

The simulated and measured radiation patterns of the antenna at 3.5 and 9 GHz in the two principle planes, the x-z and x-y planes, are shown in Figure 34. The antenna has approximately omnidirectional radiation patterns in the x-y plane. The x-z plane patterns show two nulls at the z-direction, which are similar to that of a typical monopole antenna.

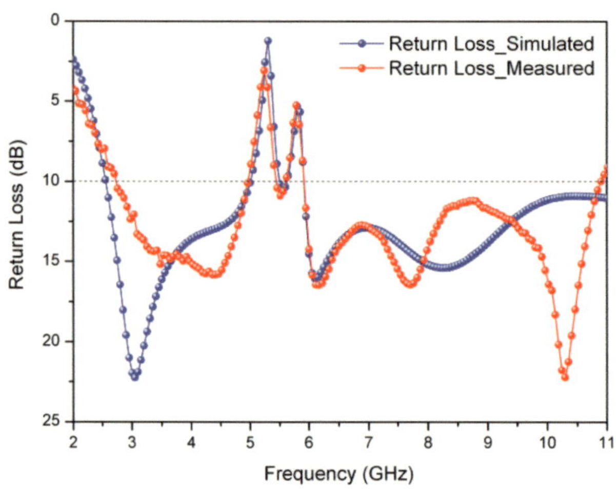

Figure 33. Simulated and measured return losses of dual band-notched antenna

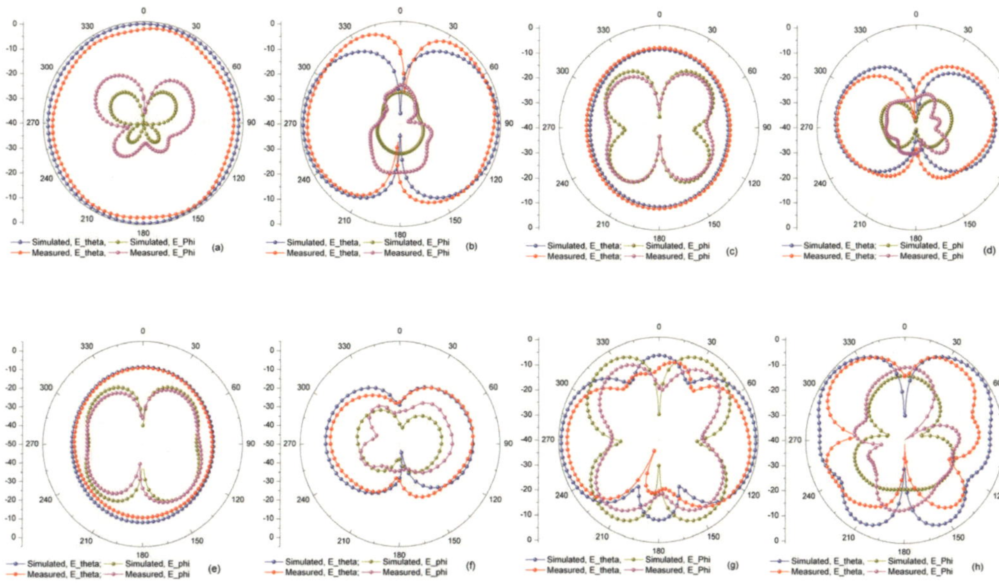

Figure 34. Simulated and measured radiation patterns of dual band-notched antenna. (a) 3.5 GHz in x-y plane; (b) 3.5 GHz in x-z plane; (c) 5.25 GHz in x-y plane; (d) 5.25 GHz in x-z plane; (e) 5.775 GHz in x-y plane; (f) 5.775 GHz in x-z plane; (g) 9 GHz in x-y plane; and (h) 9 GHz in x-z plane

The measured peak gain and efficiency of the dual band-notched antenna are shown in Figure 35. The average antenna gain over the UWB, computed by excluding the notched bands, is about 3 dBi. While at the notch frequencies, a significant gain and radiation efficiency reductions can be seen. At the notch frequencies of 5.25 and 5.83 GHz, the peak gain drops to -6 dBi and -2 dBi and the radiation efficiency reduces to below 20% and 45%, respectively. These phenomena indicate that by introducing two pairs of MLs, a dual band-notched characteristic can be achieved.

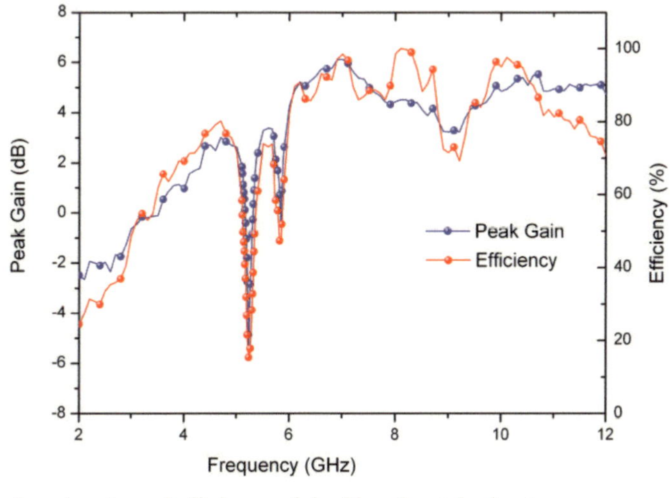

Figure 35. Measured peak gain and efficiency of dual band-notched antenna

4.5.4. Triple band-notched antenna

The simulated and measured return losses of the triple band-notched antenna of Figure 28(c) are shown in Figure 36. The antenna can operate from 2.68 GHz to 11.15 GHz with return loss ≥ 10 dB. In the three desired notched bands, the measured return loss is substantially less than 10 dB.

The simulated and measured radiation patterns of the antenna at the frequencies of 2, 3.45, 5.25, 5.775 and 11 GHz in the x-y and x-z planes are shown in Figure 37. At 2 and 11 GHz, Figures 37(a) and 37(i) show that the antenna has approximately omnidirectional radiation patterns in the x-y plane. In the x-z plane patterns, Figures 37(b) and 37(j) show that there are two nulls in the positive and negative z directions. The radiation patterns in Figures 37(c) and 37(d) for 3.45 GHz, in Figures. 37(e) and 37(f) for 5.25 GHz, and in Figures 37(g) and 37(h) for 5.775 GHz indicate that the gain is almost evenly suppressed in all directions in the three notched bands by the MLs and the average gain is about -10 dBi.

The simulated and measured peak gains and efficiencies of the antenna are shown in Figures 38(a) and 38(b), respectively. The measured peak gain is between 2 to 4.75 dBi over the UWB, except in the notched bands. In the three notched bands, i.e., the WiMax, lower WLAN and upper WLAN frequency bands, the peak gain is suppressed to -3.4 dBi, -2.3 dBi and -2.1 dBi, respectively. The radiation efficiency is between 55% - 99%, but substantially reduced to 14.6%, 15.2% and 22.8%, respectively, in these frequency bands. These results indicate that the MLs work effectively to introduce a triple band-notched characteristic for the UWB antenna.

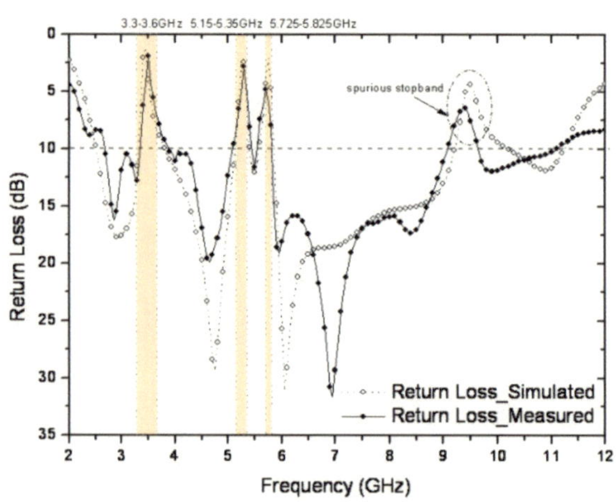

Figure 36. Simulated and measured return losses of the triple band-notched antenna

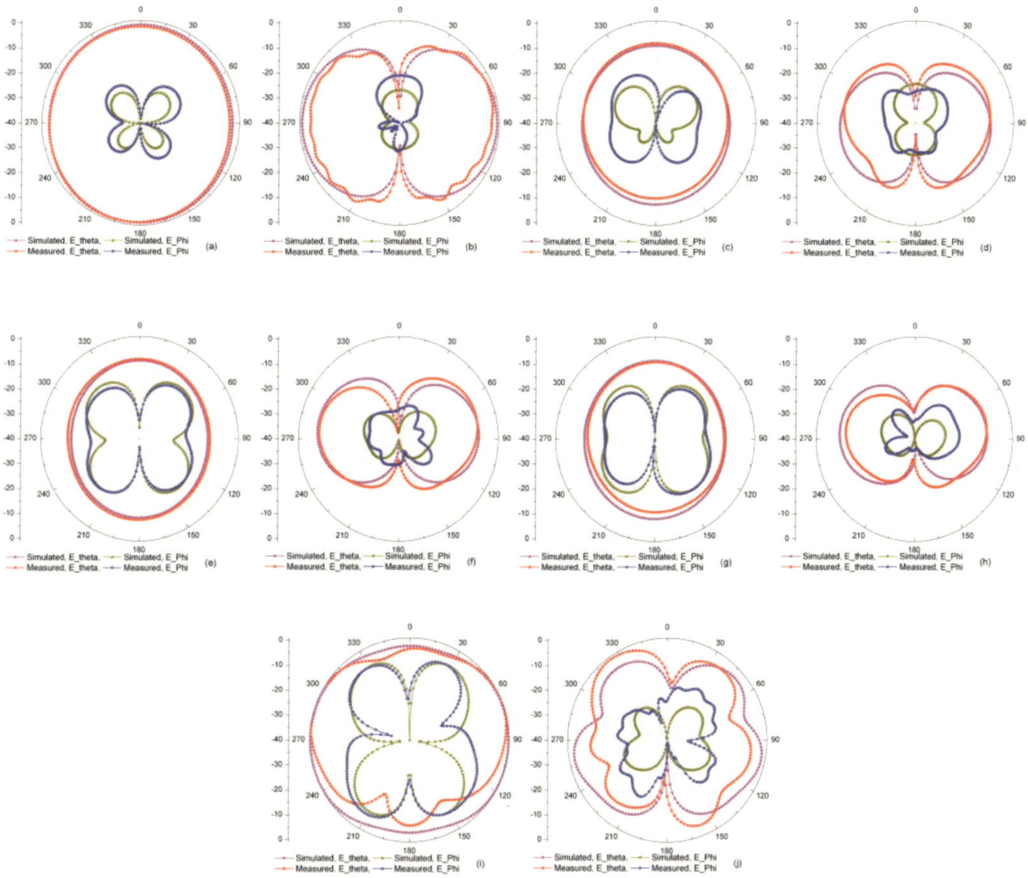

Figure 37. Simulated and measured radiation patterns of triple band-notched antenna. (a) 2 GHz in x-y plane; (b) 2 GHz in x-z plane; (c) 3.45 GHz in x-y plane; (d) 3.45 GHz in x-z plane; (e) 5.25 GHz in x-y plane; (f) 5.25 GHz in x-z plane; (g) 5.775 GHz in x-y plane; (h) 5.775 GHz in x-z plane; (i) 11 GHz in x-y plane; and (j) 11 GHz in x-z plane

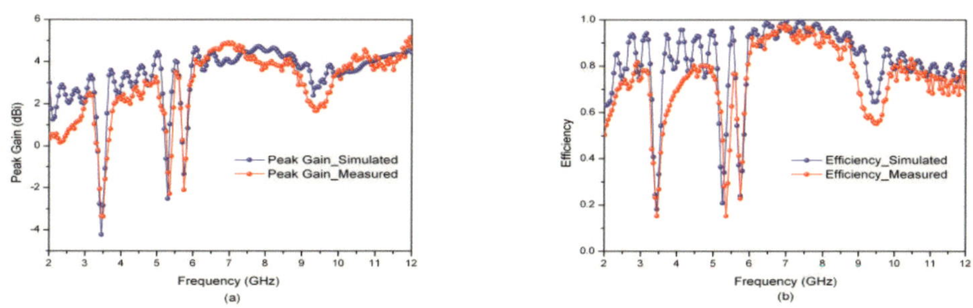

Figure 38. Simulated and measured (a) peak gains and (b) efficiencies of triple band-notched antenna

4.5.5. Quadruple band-notched antenna

The return losses of the reference and quadruple band-notched antennas are shown in Figure 39. The simulation and measurement return losses agree well and are larger than 10 dB across the UWB. In the notched bands, the return loss is substantially smaller than 10 dB. In the rest of the UWB, the return loss is larger than 10 dB, indicating good radiation performance. Figure 39 also shows that ML #1, being a parallel-coupled resonator at 3.14 GHz, has a spurious response at about 8.65 GHz, which is caused by the harmonic responses of the resonator. The measured results in Figure 39 show that the four notches at the frequencies of 3.14, 4.34, 5.4 and 6.4 have the bandwidth of 178, 374, 495 and 862 MHz, respectively.

The simulated and measured results on the peak gain and radiation efficiency of the antenna are shown in Figures 40(a) and 40(b), respectively. A relatively flat peak gain and constant efficiency are observed over the UWB, except in the notch bands. At the notch frequencies of 3.14, 4.34, 5.4 and 6.4 GHz, the measured gain is suppressed to -5.4, -4.1 and -3.7 and -2.5 dBi, respectively, with the corresponding efficiency substantially reduced to 13.2%, 26.5%, 26.8% and 35.4%. Thus the MLs work effectively to generate a quadruple band-notched characteristic for the UWB antenna. Comparing the PCF MLs to the DCF MLs in terms of peak gain and efficiency reduction, the PCF MLs perform better than the DCF MLs. This is because the PCF MLs are closer to the feed line and thus couple more energy. There are discrepancies between the simulated and measured peak gains and efficiencies, especially in the lower frequency bands. This is mainly due to the small ground plane of the antenna, which results in current flowing back from the ground plane to the outer conductor of the feeding coaxial cable [21, 22].

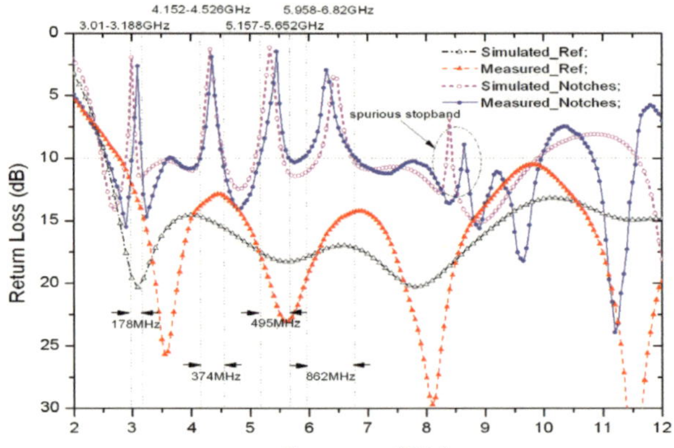

Figure 39. Simulated and measured return losses of reference antenna and quadruple band-notched antenna

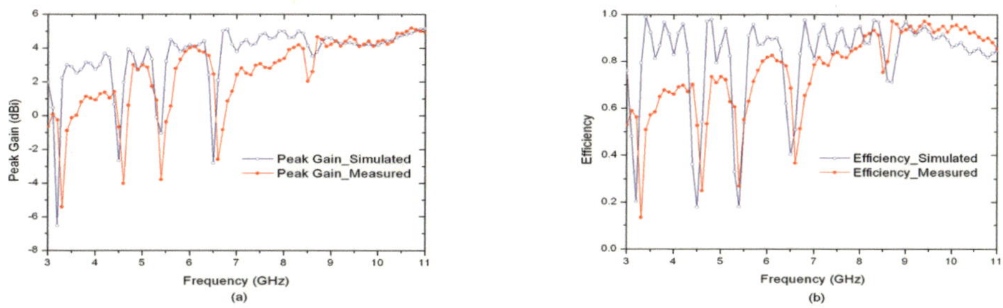

Figure 40. Simulated and measured (a) peak gains and (b) efficiencies of quadruple band-notched antenna

The simulated and measured radiation patterns of the quadruple band-notched antenna at the notch frequencies of 3.14 and 6.4 GHz and passband frequencies of 2.5, 7 and 11 GHz in the x-y and x-z planes are shown in Figure 41. At all the frequencies tested, Figures 41(a), 41(c), 41(e), 41 (g) and 41(i) show that the radiation pattern for co-polarization is approximately omnidirectional and stable in the x-y plane. While Figures 41(b), 41(d), 41(f), 41(h) and 41(j) show two nulls occurring in the positive and negative z-directions, which is typical for monopole antennas. At the notch frequencies of 3.14 and 6.4 GHz, Figures 41(c), 41(d), 41(e) and 41(f) show that the gain is almost evenly suppressed in all directions. The cross-polarization level is quite low, except at the highest frequency of 11 GHz, as can be seen in Figure 41.

4.5.6. *Time-domain performance*

For comparison, the studies of time-domain performances are carried out as described in previous sections on the single, dual, triple and quadruple band-notched antennas and results are all plotted in Figure 42. The single, dual and triple band-notched antennas have notches at 5.5 GHz, 5.25 & 5.775 GHz, 3.5 & 5.25 & 5.775 GHz, respectively, while the quadruple band-notched antenna has notches at 3.14 & 4.34 & 5.4 & 6.4 GHz. It can be seen that, as the number of notches increases, more late time ringing (distortion) and lower powers are observed in the received pulses. To evaluate the quality of the received pulses, the fidelity F is used. The calculated fidelities F for the four band-notched antennas are shown in Table 9. It can be seen that the fidelities in the face-to-face and side-by-side arrangements are about the same, which should be the case for monopole antennas. As expected, the reference UWB antenna has the fidelity of more than 97%, the best among the antennas tested. With quadruple notches, the proposed antenna has the fidelity of more than 85%. It was expected that the fidelities F would drop when the number of notches increased. However, Table 9 shows that the fidelity of the single band-notched antenna is less than that of the dual band-notched antenna. This is because the dual band-notched antenna has the two notches in the lower and upper WLAN bands of 5.15-5.35 GHz and 5.725-5.825 GHz, respectively. While the single band-notched antenna has the single notched-band from 5.15 GHz to 5.825 GHz, which is larger than the total of the two notches in the dual band-notched antenna. Thus the fidelity depends not only on the number of notches, but also the bandwidths and locations of the notches.

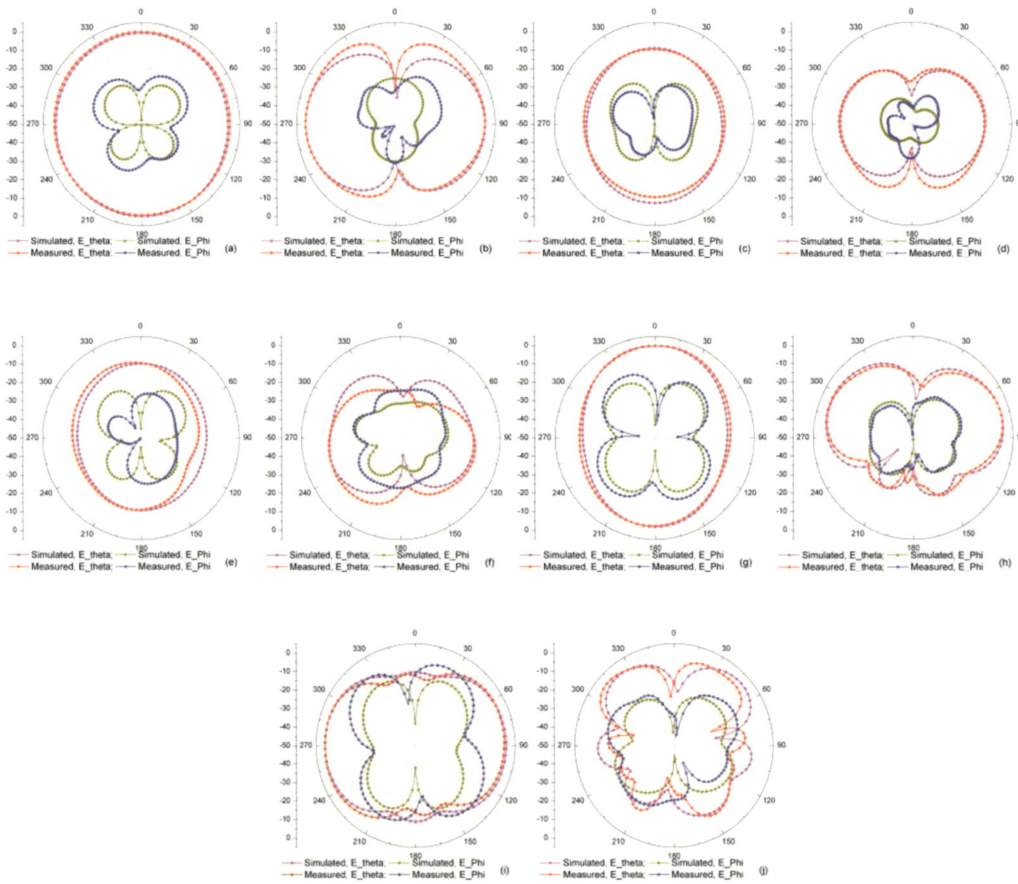

Figure 41. Simulated and measured radiation patterns of quadruple band-notched antenna. x-y plane: (a) 2.5 GHz, (c) 3.14 GHz, (e) 6.4 GHz, (g) 7 GHz and (i) 11 GHz. x-z plane: (b) 2.5 GHz, (d) 3.14 GHz, (f) 6.4 GHz, (h) 7 GHz and (j) 11 GHz

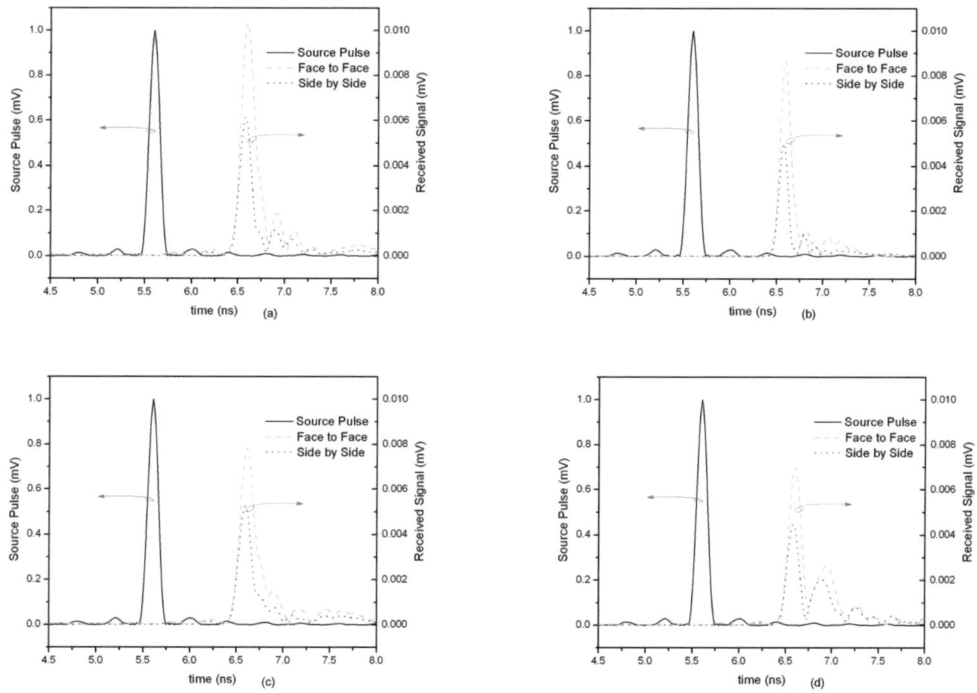

Figure 42. Measured transmit pulses with spectrum from 3.1 to 10.6 GHz and received pulses of (a) single, (b) dual, (c) triple and (d) quadruple band-notched antennas

	Reference Antenna	*Single Band-Notched Antenna*	*Dual Band-Notched Antenna*	*Triple Band-Notched Antenna*	*Quadruple Band-Notched Antenna*
Face-to-Face	*0.9825*	*0.9444*	*0.9590*	*0.9184*	*0.8918*
Side-by-Side	*0.9744*	*0.9463*	*0.9557*	*0.9281*	*0.8538*

Table 9. Calculated Fidelity For Different Band-Notched Antennas

5. Conclusions

In this chapter, the designs of band-notched characteristics for compact UWB antennas have been presented. The single, dual, triple and quadruple band-notched UWB antennas using different resonator structures including the CPW resonators, the $\lambda/4$-resonators and the MLs have been designed and studied using computer simulation. For verification of the simulation results, these antennas have been fabricated and measured using the antenna measurement system, Starlab. The bandwidths and center frequencies of all these notches can be controlled independently by adjusting the dimensions of the resonators. In the frequency domain, the simulated and measured return losses of the antennas agree well. The UWB monopole antennas have approximately omnidirectional radiation patterns with good band-notched characteristics. The simulated and measured results have shown that substantial reductions in efficiency and peak gain can be achieved at the notch frequencies. In the time domain, the pulse responses of these notched antennas have been measured inside the quiet zone of the anechoic chamber in The University of Hong Kong. Fidelity has been used to evaluate the time-domain performance of these antennas. Results have shown that all these band-notched antennas designed have the fidelities of more than 85%, compared to 97% for the reference UWB antenna.

Author details

Y.F. Weng, S.W. Cheung, T.I. Yuk and L. Liu
Department of Electrical and Electronic Engineering, The University of Hong Kong, Hong Kong

6. References

[1] [1] FCC, "First Report and Order: Revision of Part 15 of the Commission's Rules Regarding Ultra-Wideband Transmission Systems," Apr. 2002.
[2] Available: http://www.fcc.gov/pshs/techtopics/techtopics10.html
[3] Available: http://www.wimaxforum.org/
[4] A. Kerkhoff and H. Ling, "Design of a planar monopole antenna for use with ultra-wideband (UWB) having a band-notched characteristic," in *Antennas and Propagation Society International Symposium, 2003. IEEE*, 2003, pp. 830-833 vol.1.
[5] H. G. Schantz, *et al.*, "Frequency notched UWB antennas," in *Ultra Wideband Systems and Technologies, 2003 IEEE Conference on*, 2003, pp. 214-218.
[6] K. H. Kim, *et al.*, "Band-notched UWB planar monopole antenna with two parasitic patches," *Electronics Letters,* vol. 41, pp. 783-785, 2005.
[7] T. G. Ma and S. J. Wu, "Ultrawideband Band-Notched Folded Strip Monopole Antenna," *Antennas and Propagation, IEEE Transactions on,* vol. 55, pp. 2473-2479, 2007.
[8] J. Kim, *et al.*, "5.2 GHz notched ultra-wideband antenna using slot-type SRR," *Electronics Letters,* vol. 42, pp. 315-316, 2006.
[9] A. J. Kerkhoff and L. Hao, "Design of a Band-Notched Planar Monopole Antenna Using Genetic Algorithm Optimization," *Antennas and Propagation, IEEE Transactions on,* vol. 55, pp. 604-610, 2007.
[10] Y. F. Weng, *et al.*, "UWB antenna with single or dual band-notched characteristic for WLAN band using meandered ground stubs," in *Antennas & Propagation Conference, 2009. LAPC 2009. Loughborough*, 2009, pp. 757-760.
[11] Y. F. Weng, *et al.*, "Triple band-notched UWB antenna using meandered ground stubs," in *Antennas and Propagation Conference (LAPC), 2010 Loughborough*, 2010, pp. 341-344.
[12] S. W. Qu, *et al.*, "A Band-Notched Ultrawideband Printed Monopole Antenna," *Antennas and Wireless Propagation Letters, IEEE,* vol. 5, pp. 495-498, 2006.
[13] W. J. Lui, *et al.*, "Compact frequency notched ultra-wideband fractal printed slot antenna," *Microwave and Wireless Components Letters, IEEE,* vol. 16, pp. 224-226, 2006.

[14] Y. J. Cho, *et al.*, "A miniature UWB planar monopole antenna with 5 GHz band-rejection filter," in *Microwave Conference, 2005 European*, 2005, p. 4 pp.

[15] J. Zhang, *et al*, "CPW-coupled-fed Elliptical Monopole UWB Antenna with Dual-band Notched Characteristic," presented at the Progress In Electromagnetics Research Symposium 2012 (PIERS2012), Kuala Lumpur, Malaysia, 2012.

[16] X. L. Sun, *et al*, "CPW-Coupled-Fed Elliptical Monopole UWB Antenna with Simple Dual-Band Notch Design for WLAN Band," presented at the 2012 IEEE International Symposium on Antennas and Propagation (2012 IEEE AP-S), Chicago, Illinois USA, 2012.

[17] Y. D. Dong, *et al.*, "Analysis of Planar Ultrawideband Antennas With On-Ground Slot Band-Notched Structures," *Antennas and Propagation, IEEE Transactions on,* vol. 57, pp. 1886-1893, 2009.

[18] E. Pancera, *et al.*, "Novel Design of UWB Antenna with Band - Notch Capability," in *Wireless Technologies, 2007 European Conference on*, 2007, pp. 48-50.

[19] M. Houdart, "Coplanar Lines : Application to Broadband Microwave Integrated Circuits," in *Microwave Conference, 1976. 6th European*, 1976, pp. 49-53.

[20] X. Y. Wu, *et al.*, "Quality factors of coplanar waveguide resonators," in *Microwave Conference, 1999 Asia Pacific*, 1999, pp. 670-673 vol.3.

[21] D. H. Kwon and Y. Kim, "Suppression of Cable Leakage Current for Edge-Fed Printed Dipole UWB Antennas Using Leakage-Blocking Slots," *Antennas and Wireless Propagation Letters, IEEE,* vol. 5, pp. 183-186, 2006.

[22] Z. N. Chen, *et al.*, "Small Printed Ultrawideband Antenna With Reduced Ground Plane Effect," *Antennas and Propagation, IEEE Transactions on,* vol. 55, pp. 383-388, 2007.

[23] L. Liu, *et al*, "Cable Effects on Measuring Small Planar UWB Monopole Antennas," in *Ultra Wideband*, M. Matin, Ed., ed: InTech, 2012.

[24] J. H. Reed, *An Introduction to Ultra Wideband Communicaiton Systems*: Prentice Hall, 2005.

[25] Z. N. Chen, *et al.*, "Considerations for source pulses and antennas in UWB radio systems," *Antennas and Propagation, IEEE Transactions on,* vol. 52, pp. 1739-1748, 2004.

[26] D. Lamensdorf and L. Susman, "Baseband-pulse-antenna techniques," *Antennas and Propagation Magazine, IEEE,* vol. 36, pp. 20-30, 1994.

[27] D. M. Pozar, *Microwave engineering, 3rd edition*: John Wiley & Sons, Inc., 2005.

[28] Available: http://transition.fcc.gov/pshs/techtopics/techtopics10.html

[29] B. J. Rubin and B. Singh, "Study of meander line delay in circuit boards," *Microwave Theory and Techniques, IEEE Transactions on,* vol. 48, pp. 1452-1460, 2000.

[30] M. Yamaguchi, *et al.*, "Fabrication and basic characteristics of dry-etched micro inductors," *Magnetics, IEEE Transactions on,* vol. 26, pp. 2014-2016, 1990.

[31] A. C. Reyes, *et al.*, "Coplanar waveguides and microwave inductors on silicon substrates," *Microwave Theory and Techniques, IEEE Transactions on,* vol. 43, pp. 2016-2022, 1995.

[32] C. H. Ahn and M. G. Allen, "A comparison of two micromachined inductors (bar- and meander-type) for fully integrated boost DC/DC power converters," *Power Electronics, IEEE Transactions on,* vol. 11, pp. 239-245, 1996.

[33] G. Marrocco, "Gain-optimized self-resonant meander line antennas for RFID applications," *Antennas and Wireless Propagation Letters, IEEE,* vol. 2, pp. 302-305, 2003.

[34] T. J. Warnagiris and T. J. Minardo, "Performance of a meandered line as an electrically small transmitting antenna," *Antennas and Propagation, IEEE Transactions on,* vol. 46, pp. 1797-1801, 1998.

[35] I. Kelander, *et al.*, "Modeling of spiral and meander lines in multilayer passive integration," in *Electrical Performance of Electronic Packaging, 2002,* 2002, pp. 83-86.

[36] H. T. Su, *et al.*, "Highly miniature HTS microwave filters," *Applied Superconductivity, IEEE Transactions on,* vol. 11, pp. 349-352, 2001.

[37] Y. F. Weng, *et al.*, "Band-notched characteristic using meandered ground stubs for compact UWB antennas," in *Ultra-Wideband (ICUWB), 2010 IEEE International Conference on,* 2010, pp. 1-4.

[38] J. S. Hong and M. J. Lancaster, *Microstrip filters for RF microwave applications.* NY: John Wiley & Sons, Ltd., 2001.

[39] R. A. Pucel, "Design Considerations for Monolithic Microwave Circuits," *Microwave Theory and Techniques, IEEE Transactions on,* vol. 29, pp. 513-534, 1981.

[40] E. Pettenpaul, *et al.*, "CAD models of lumped elements on GaAs up to 18 GHz," *Microwave Theory and Techniques, IEEE Transactions on,* vol. 36, pp. 294-304, 1988.

[41] M. E. Goldfarb and R. A. Pucel, "Modeling via hole grounds in microstrip," *Microwave and Guided Wave Letters, IEEE,* vol. 1, pp. 135-137, 1991.

[42] K. M.-J. Ho, Ellis, G. A., Ooi, B.-L. and Leong, M.-S., "Modeling of coplanar waveguide meander-line inductors," *International Journal of RF and Microwave Computer-Aided Engineering,* vol. 12, pp. 520-529, 2002.

[43] R. Murji and M. J. Deen, "Accurate modeling and parameter extraction for meander-line N-well resistors," *Electron Devices, IEEE Transactions on,* vol. 52, pp. 1364-1369, 2005.

[44] M. H. Kerinami and O. M. Ramahi, "Effects of segment length and number of turns on designing a precise meander delay line," in *Electromagnetic Compatibility, 2003. EMC '03. 2003 IEEE International Symposium on,* 2003, pp. 1121-1124.

[45] C. L. Holloway and E. F. Kuester, "Net and partial inductance of a microstrip ground plane," *Electromagnetic Compatibility, IEEE Transactions on,* vol. 40, pp. 33-46, 1998.

[46] F. E.H. and Z. R.A, *Microwave Engineering Using Microstrip Circuits*: Prentice Hall, 1990.

Chapter 12

Cable Effects on Measuring Small Planar UWB Monopole Antennas

L. Liu, S.W. Cheung, Y.F. Weng and T.I. Yuk

Additional information is available at the end of the chapter

http://dx.doi.org/10.5772/46080

1. Introduction

Since the US-FCC assigned the ultrawide band (UWB) for unlicensed use in 2002, UWB technology has attracted much attention in both the commercial and academic domains. The characteristics of ultrawide bandwidth from 3.1 to 10.6 GHz and low power emission of -41.3 dBm/MHz make it a promising candidate for different applications such as high speed communications and radar imaging systems. However, the requirements for impedance matching, constant gain, constant radiation patterns and high radiation efficiency over such a wide bandwidth for the UWB antenna become great challenges to antenna designers.

With the increasing demand for smaller wireless devices, planar antenna, with the advantages of compact size, low profile, low cost, ease of fabrication and ease of integration with RF circuits, appears to be more preferable for UWB applications. Among different planar antennas, monopole antenna has the simplest structure, compact size and omnidirectional radiation pattern and so is one of the best candidates. For a planar monopole antenna to work properly, the ground plane is required to be electrically large enough to approximate an infinite-ground plane and so occupies a large portion of the overall antenna size. Thus to design a compact UWB monopole antenna, the ground plane is usually the one to be minimized. Designing a planar monopole antenna with a small ground plane to cover the UWB is not a difficult task and can be achieved through different techniques [1-7]. The design is usually done using computer simulation. In carrying out the design in simulation, the antenna is fed directly with a signal source without using a feeding cable. However, when the final design is completed and prototyped for measurements, a feeding cable is normally used to connect the antenna to the measurement system. The small ground plane cannot approximate an infinite ground plane well and causes the currents to flow back to the outer surface of the feeding cable, resulting in secondary radiation. This leads to discrepancies between the simulated and measured performances of the antenna and creates uncertainties to the design of the antenna.

To resolve the problem, a sleeve balun can be placed at the end of the cable to prevent currents from flowing back to the feeding cable [8,9]. A sleeve balun is a metal tube with a length of quarter-wavelength to provide an open circuit for the signal. Although sleeve baluns can be designed to possess good choking characteristics, they are narrowband devices and so are not suitable for UWB antennas. For wideband and high-frequency operation, the feeding cable can be covered with an EMI suppressant material to absorb unwanted EM radiation [10]. By using this method, the shape of the measured and simulated radiation patterns of the antenna will be similar, but the measured efficiency and gain will be lower due to the energy absorbed by the EMI suppressant material. The discrepancies again produce uncertainties to the design of the antenna.

In this chapter, the effects of ground-plane size and feeding cable on the measurements of small UWB monopole antennas are investigated. A group of nine planar UWB monopoles with an identical elliptical radiators but different ground-plane sizes are designed using computer simulation where no feeding cable is used. These antennas are also prototyped and measured using the antenna measurement system, Satimo Starlab, where a feeding cable is used [11]. The simulated and measured performances show large discrepancies at low frequencies. To investigate the discrepancies, two different types of feeding cables, a high-frequency coaxial cable and a high-frequency coaxial cable with EMI suppressant tubing, are studied. The simulation models for the two cables are developed and used in computer simulation. With the application of the two cable models, the simulated and measured performances show good agreements. The results show that the feeding cable without EMI suppressant tubing causes many ripples on the 3D-radiation patterns of the antenna.

2. UWB monopole antennas

Dipole antenna, often called dipole, is one of the simplest but most widely used types of antennas. The typical structure of a dipole consists of two thin-wire conductors normally having equal length L as shown in Figure 1(a) where it is assumed $L=\lambda/2$ with λ being the wavelength at the resonant frequency. At resonance, the currents form the standing waves on both conductors, as shown in Figure 1(a), giving rise to electromagnetic (EM) radiations. Monopole antenna, often called monopole, has half the size of a dipole. An ideal monopole normally consists of a single thin-wire conductor perpendicularly mounted on an infinite ground plane as shown in Figure 1(b). At resonance, the current forms a standing wave on the conductor which radiates EM fields. The EM fields incident on the infinite ground plane are reflected as if they were radiated from the monopole image having the same current distribution as that of the lower conductor of the dipole in Figure 1(a). Thus a monopole can be viewed as the corresponding double-length center-fed linear dipole [12]. A monopole radiates energy into only the upper half space. So for a given input power, a monopole has the radiated power and hence the gain twice as much as the corresponding dipole.

A thin-wire monopole has a simple structure, but a very narrow bandwidth, making it unsuitable for UWB applications. To broaden the impedance bandwidth, the thin-wire conductor can be made flat to become a planar element and then laid parallel to the ground plane to form a low-profile planar monopole. The planar element can take on different shapes as shown in Figure 2 [13].

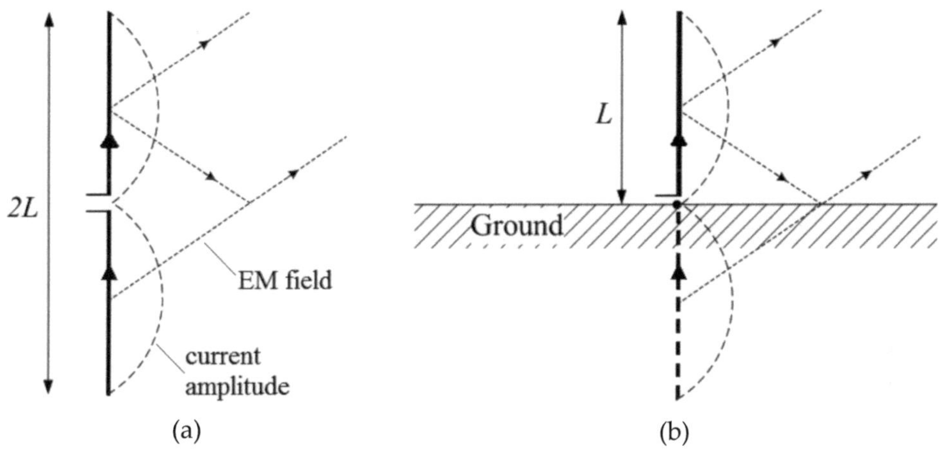

Figure 1. (a) Center-fed dipole and (b) vertical monopole above infinite ground

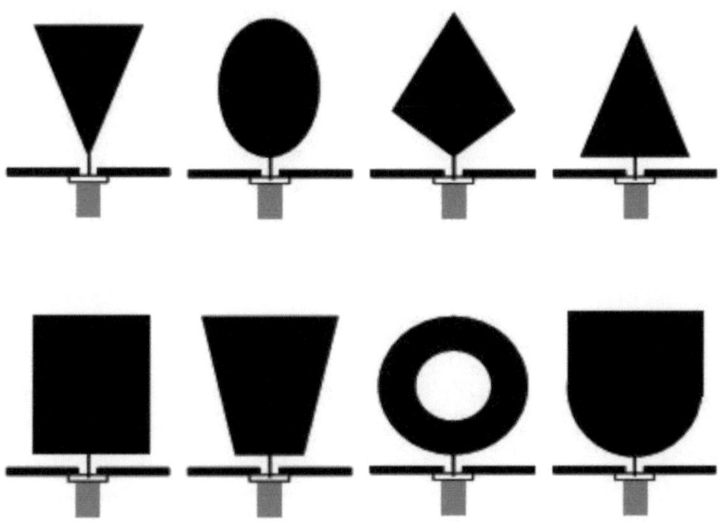

Figure 2. Planar monopoles using different radiator shapes [13]

3. Effects of ground plane on small UWB monopoles measurements

With the increasing demand for smaller wireless devices, planar monopole antennas with small ground planes have attract much attention. However, in the design of such an antenna, very often, after the antenna performance in terms of gain, efficiency and return loss has been optimized using computer simulation, the measured performance of the prototyped antenna does not agree with the simulated performance. Large discrepancies usually occur at lower frequencies. This creates uncertainties and doubts in the design of the antenna. As will be shown in the following sections, the discrepancies at low frequencies are mainly caused by the feeding cable used to connect the antenna to the measurement system.

3.1. Cable effects on measuring monopoles with small ground planes

Nowadays, the design of antennas is usually done by using computer simulation. In simulation, normally the antenna is directly fed from a signal source and no feeding cable is used. However, when the antenna is fabricated and measured in a practical situation, a feeding cable is always used to connect the antenna to the measurement system and the signal is fed through the feeding cable to the antenna. In such arrangement, the cable could affect the measured results in two possible ways [14] as illustrated in Figure 3.

Since the feeding cable is quite near to the radiator and so is in the near field region of the antenna, the radiated EM fields incident on the cable will be scattered and reflected as shown in Figure 3(a). The feeding cable becomes a parasitic element [15]. Due to the small size of the cable, this cable effect on the measurement results is relatively small.

If the antenna is a planar monopole with a small ground plane, some EM fields will not be reflected as in the case of having an infinite ground plane shown in Figure 1(b). Instead, the EM fields arriving at the edges of the small ground plane will be diffracted. This induces surface currents to flow back on the outer surface of the feeding cable, resulting in secondary radiation as shown in Figure 3(b). This effect on measurements could be quite significant, depending on the electrical size of the ground plane. Computer simulation is carried out to study the effects of using large and small ground planes of a thin-wire monopole antenna fed by a coaxial cable on the Electric fields (E-fields). The large and small ground planes have a circular shape with the radii of 2.06λ and 0.41λ, respectively, and a thickness of 0.0008λ, where λ is the wavelength at the resonant frequency. The length of feeding cable is 1.64λ. Figure 4(a) shows a snap-shot of the E-fields radiated from the monopole antenna using the small ground plane. It can be seen that the E-fields arriving at the edges of the ground plane are quite strong. The ground-plane edges diffract the strong incident E-fields in all directions [16, 17]. After diffraction, some of the E-fields go to the upper free space and others go to the lower free space with respect to the ground plane. A significant portion of the E-fields diffracts onto the bottom surface of the ground plane, which will induce surface currents. The surface currents will flow towards the feeding cable at the center of the ground plane and onto the outer conductor-surface of the cable. Figure 4(a) shows that a standing wave is formed on the feeding cable and this will result in "secondary radiation" and affect the measured results. When the large ground plane with a radius of 2.06λ is used, Figure 4(b) shows a snap-shot of the E-fields radiating from the monopole antenna. It can be seen that the E-fields arriving at the edge of the ground plane are quite weak. As a result, the induced currents on the bottom surface of the ground plane and hence the currents flowing back onto the outer conductor-surface of the feeding cable are very weak. In this case, secondary radiation is much less.

The simulated 3D-radiation patterns for the two cases are shown in Figures 4(c) and (d). It can be seen that both radiation patterns have peak-radiation at elevation from the horizontal ground plane, typical for monopoles with finite ground planes. For the antenna with the large ground, peak-radiation is stronger and at a smaller elevation angle than those for the antenna with the small ground plane. Peak-radiation in the lower hemisphere of the radiation pattern is much weaker for the antenna with the large ground than for the antenna with the small ground plane. For the antenna with the small ground plane, Figure 4(c) shows that ripples occur in both the upper and lower hemispheres of the pattern. However, for the antenna with the large ground plane, Figures 4(d) shows no ripple in the upper hemisphere of the pattern, but many ripples with much smaller magnitudes in the lower hemisphere.

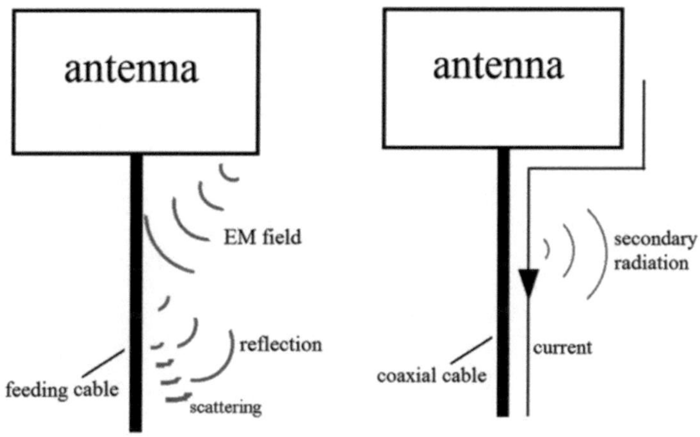

(a) (b)

Figure 3. Illustration of two possible ways affecting measured results of antenna: (a) reflections of EM fields from antenna and (b) currents flowing back to feeding cable

Ripples on a 3D-radiation pattern are the results of EM fields with different phases adding together constructively and destructively in different spatial directions. To study the causes of these ripples on the 3D-radiation patterns of Figures 4(c) and (d), computer simulation is carried out on the same antenna with the same large and small ground planes but without using the feeding cable. Results at resonant frequency are shown in Figure 5. For the antenna with the small ground plane, Figure 5(a) shows that the ripples disappear. Thus the ripples in Figure 4(c) are mainly caused by the feeding cable. In fact, this agrees with Figure 4(a) which shows that, at the resonant frequency, a standing wave is developed on the feeding cable which gives out EM radiation. The EM fields are added together constructively and destructively in different spatial directions, producing ripples in the 3D-radiation pattern in Figure 4(c). For the antenna with the large ground plane and without using the feeding cable, Figure 5(b) shows that the 3D-radiation pattern is about the same as that in Figure 4(d) using the feeding cable. This agrees with Figure 4(b) which shows no standing wave developed on the feeding cable and so no radiation from the cable. Thus the feeding cable has no effect on the radiation pattern. However, Figure 5(b) shows that the ripples still occur in the lower hemisphere of the pattern. This indicates that the ripples are mainly caused by diffraction of EM fields at the edges of the ground plane. The reason can be explained as follows. The EM fields radiated from the monopole are diffracted at the edge of the ground plane into free space below the ground plane. Since the ground plane has a diameter of 4.12λ, the diffracted EM fields are added together constructively and destructively in different spatial directions in the lower hemisphere, forming many ripples in the radiation pattern.

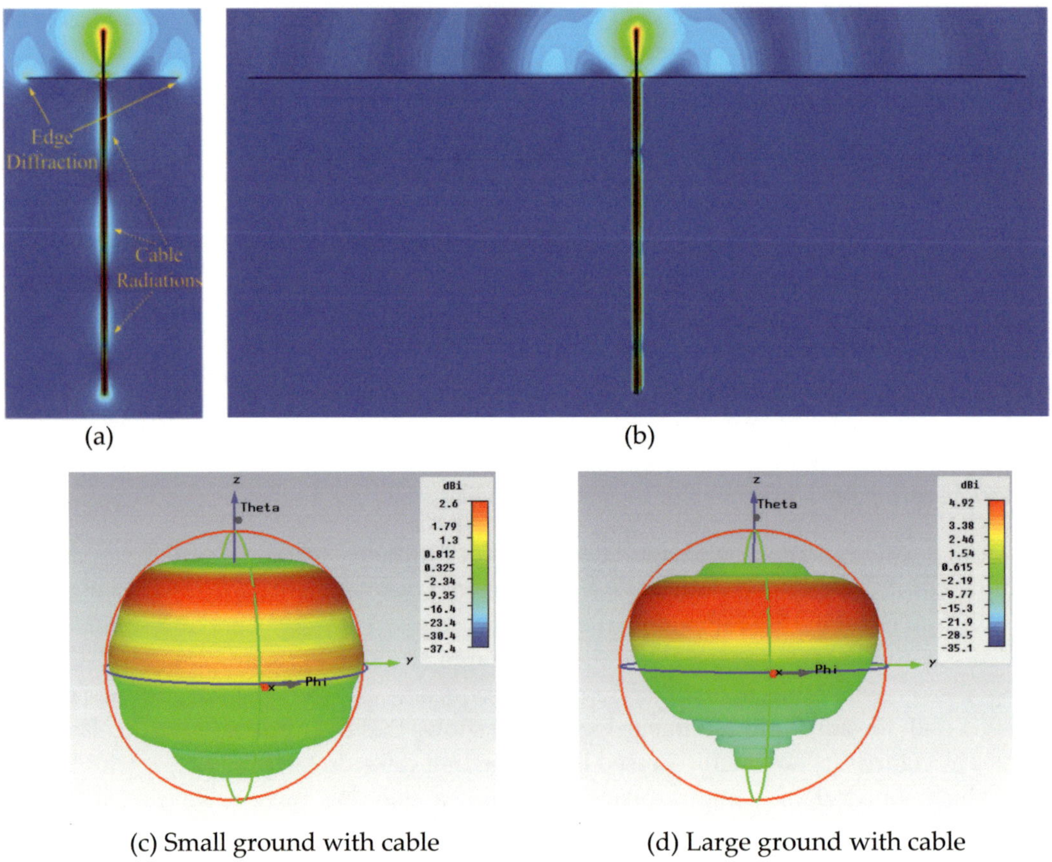

Figure 4. E-field radiation of thin-wire monopole antenna fed using coaxial cable for (a) small ground plane and (b) large ground plane, and corresponding radiation patterns (c) and (d).

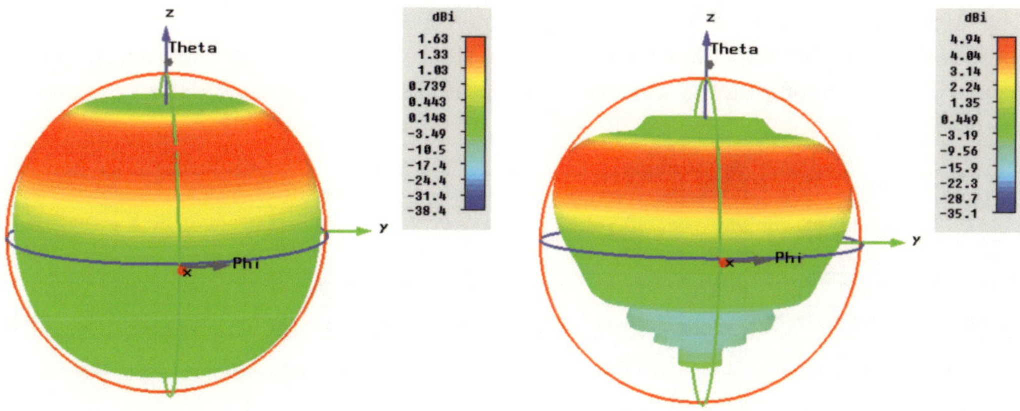

(a) (b)

Figure 5. Radiation patterns of thin-wire monopole antenna without using feeding cable (a) small ground plane, and (b) large ground plane

3.2. Studies of UWB monopole antennas with different ground-plane sizes

3.2.1. Structure of the antennas

To investigate the effects of ground-plane size on measurements of small UWB monopole antennas, a group of nine antennas, Ants 1, 2, …, 9, are used. These antennas have an identical elliptical-shaped radiator printed on one side of the substrate but a ground plane with different sizes on the other side of the substrate [18]. They are designed on the Rogers substrate, RO4350, with a relative dielectric constant of 3.48, a thickness of 0.762 mm and a loss tangent of 0.0037, as shown in Figure 6.

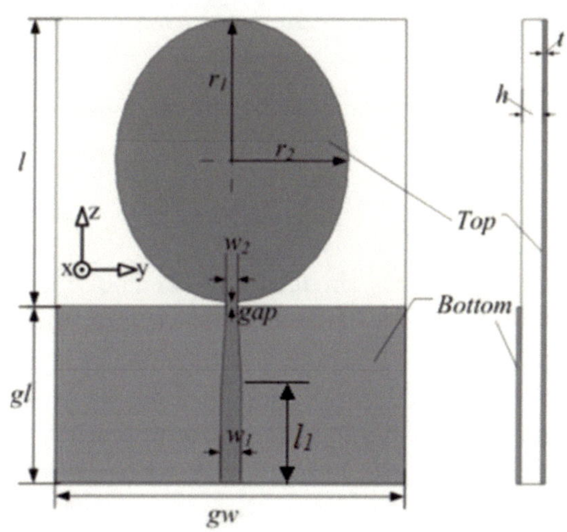

(a) (b)

Figure 6. Structure of UWB antennas: (a) top view and (b) side view

The tapered microstrip feed line and the gap between the radiator and the ground plane are important factors for impedance matching and so are optimized for maximum impedance bandwidth using computer simulation. The optimized dimensions of these nine antennas are listed in Table 1.

Antenna	*gl×gw*	w_1	w_2	l_1	*gap*	r_1	r_2	*Antenna*	*gl×gw*	w_1	w_2	l_1	*gap*	r_1	r_2
Ant 1	15×30	1.73	0.9	8	0.3	12	11	*Ant 6*	30×80	1.73	0.83	23	0.3	12	11
Ant 2	15×50	1.73	0.76	5	0.2	12	11	*Ant 7*	50×30	1.73	0.94	44	0.3	12	11
Ant 3	15×80	1.73	0.72	3	0.3	12	11	*Ant 8*	50×50	1.73	0.69	43	0.3	12	11
Ant 4	30×30	1.73	0.9	23	0.3	12	11	*Ant 9*	50×80	1.73	0.72	38	0.3	12	11
Ant 5	30×50	1.73	0.83	23	0.3	12	11								

Table 1. Ground-plane dimensions of antennas (unit: mm)

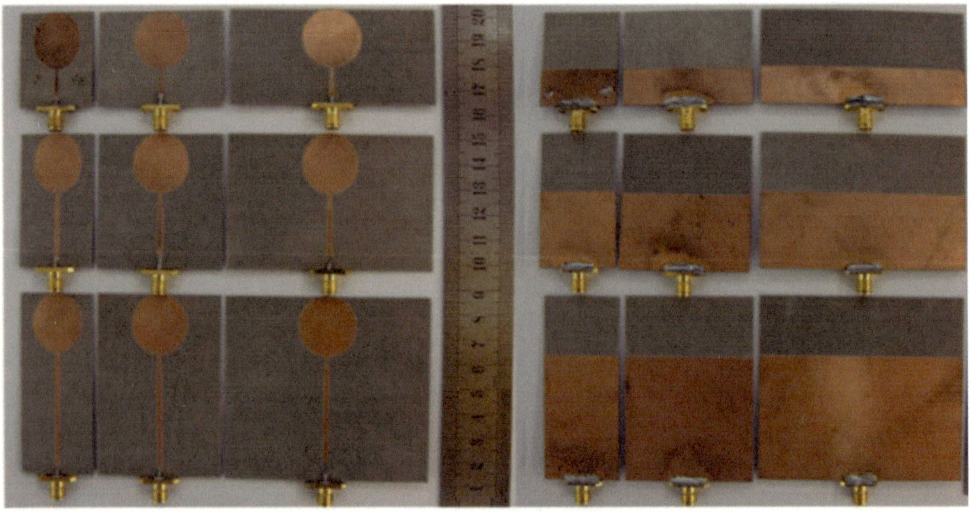

(a) (b)

Figure 7. Prototypes of nine planar monopole antennas with different ground-plane sizes: (a) top view and (b) bottom view.

3.2.2. Results and discussions

The performances of the nine antennas, in terms of S_{11} and efficiency, are studied by computer simulation. In simulation, no feeding cable is used and the antennas are fed directly by the signal source. Using the optimized dimensions in Table 1, the nine antennas are also prototyped using the Rogers substrate, RO4350, as shown in Figure 7, and measured using the antenna measurement system, Satimo Starlab, shown in Figure 8. In measurements, of course, a feeding cable with an SMA connector (provided by Satimo) is used to connect the antennas to the Starlab system. The cable is enclosed by an EMI suppressant tube to absorb EM radiation. The simulated and measured S_{11} and efficiencies of the antennas are shown in Figure 9. It can be seen that the simulated and measured impedance bandwidths (S_{11}<-10 dB) for all antennas show good agreements. However, for efficiency, the measured results are always lower than the simulated results. The discrepancies are more obvious for antennas with smaller ground planes and at lower frequencies.

To examine the effects of ground-plane size on the discrepancy of efficiency, we divide the whole frequency band from 2 to 12 GHz into three sub-bands, i.e. 2-4 GHz, 4-6 GHz, and 6-12 GHz, and compute the average discrepancy in the whole band and in each of the sub-bands. Results are listed in Table 2, where each row has the same ground-plane width and increasing length and each column has the same ground-plane length and increasing width. From Figure 9 and Table 2, we can observe the following phenomena:

1. The lower cut-off frequency reduces with increasing ground-plane length (gl).

In Figures 9(a), (b) and (c), the ground planes have the same length of 15 mm but different widths. The lower cut-off frequencies (S_{11}=-10 dB) of the antennas are all at about 2.8 GHz. This phenomenon is also observed in Figures 9(d), (e) and (f), and in Figures 9(g), (h) and (i).

In Figures 9(a), (d) and (g), the ground planes have the same width. As the ground-plane length increases from 15 to 30 and 50 mm, the lower cut-off frequency decreases from 2.76 to 2.38 to 2.21, respectively. This phenomenon is also observed in Figures 9(b), (e) and (h), and Figures 9(c), (f) and (i). Thus the lower cut-off frequency reduces with increasing ground-plane length (*gl*).

For dipole antenna, the lower cut-off frequency is inversely proportional to the length of the radiator. The results in Figure 9 show that the monopole antennas with small ground planes behave like asymmetric dipole antennas [19] and the lower cut-off frequency is inversely proportional to the length of the ground plane.

2. Discrepancy reduces with increasing frequency

Figure 9 shows that the discrepancy is larger at lower frequencies and smaller at higher frequencies. This phenomenon can also be observed in Table 2 which shows the discrepancy is always smallest in the higher sub-band and largest in the lower sub-band. This is because at higher frequencies, the ground plane becomes electrically larger.

3. Discrepancy reduces with increasing ground-plane width and ground-plane length.

Table 2 shows the discrepancy reduces with increasing ground-plane width and ground-plane length.

4. The width of ground plane has more effect on the efficiency than the length.

Each row in Table 2 represents the ground planes of the same widths but different lengths. Table 2 shows the average discrepancy through 2-12 GHz decreases significantly with increasing ground-plane length. Each column in Table 2 represents the ground planes of the same length but different widths and the results show the discrepancy does not change much. This can also be seen in Figure 9 by comparing the simulated and measured efficiencies of the corresponding antennas. These results show that increasing the ground-plane width has more effect on reducing the discrepancy than increasing the ground-plane length. This is because a small ground plane serves as a radiator and the width of the radiator improves the impedance bandwidth.

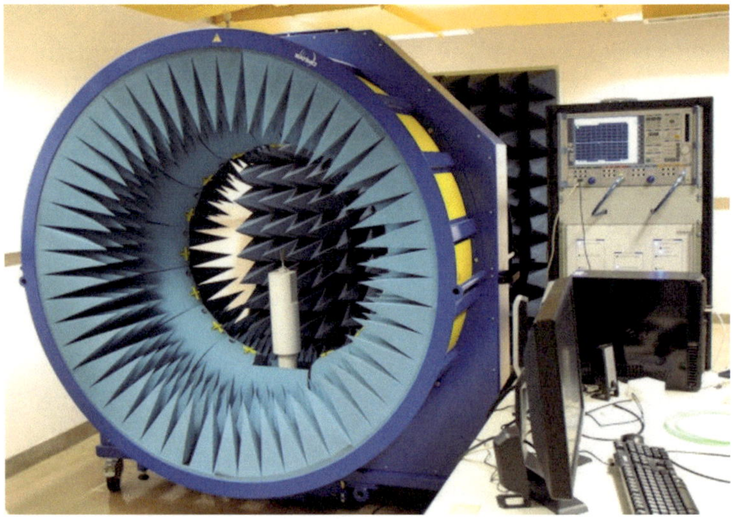

Figure 8. Antenna in Starlab system for measurement

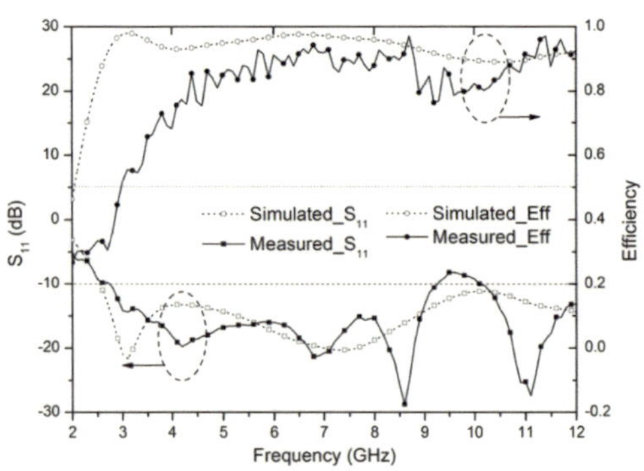

(a) Ant 1: 15×30 mm²

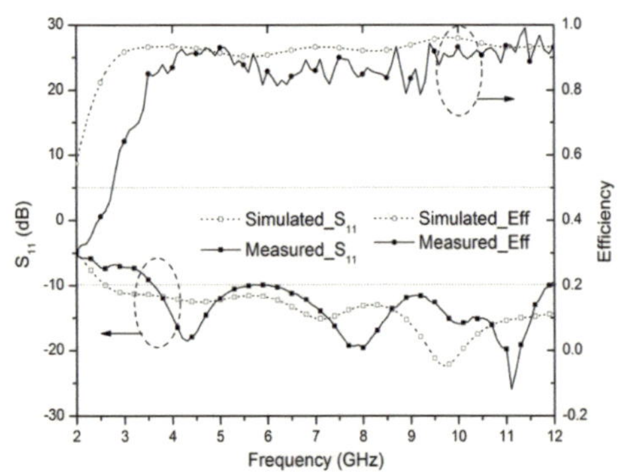

(b) Ant 2: 15×50 mm²

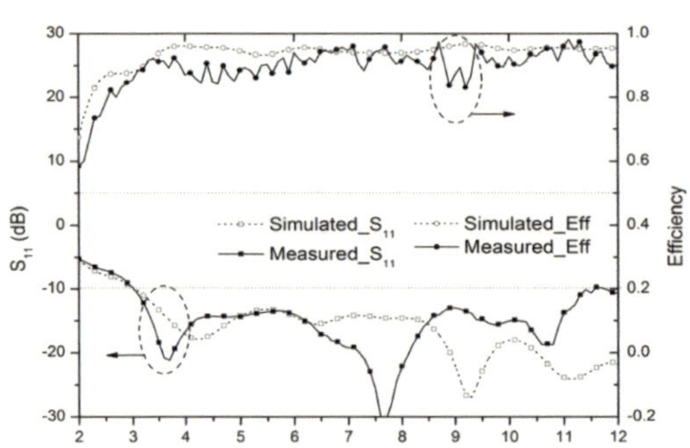

(c) Ant 3: 15×80 mmc²

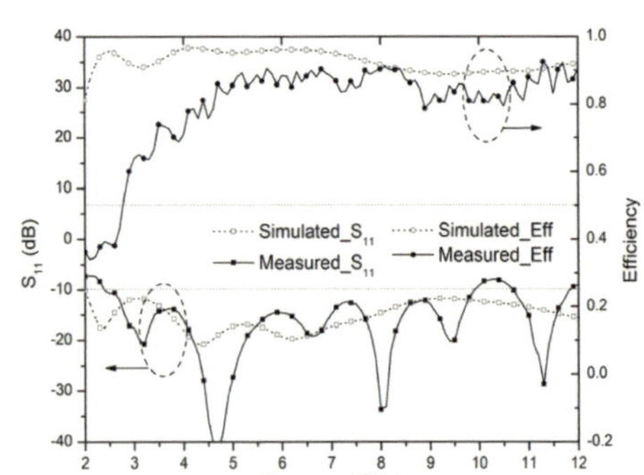

(d) Ant 4: 30×30 mm²

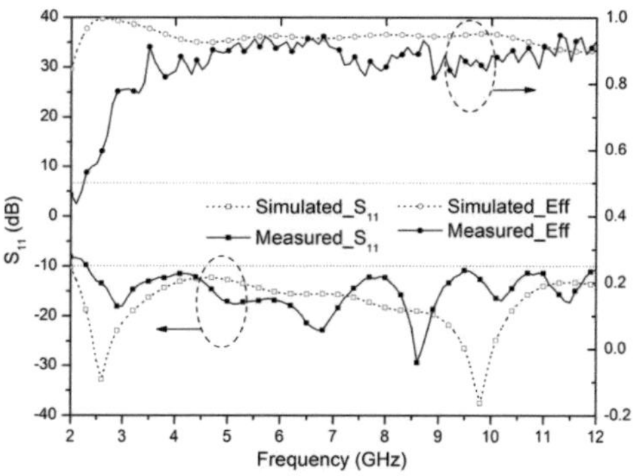

(e) Ant 5: 30×50 mm^2

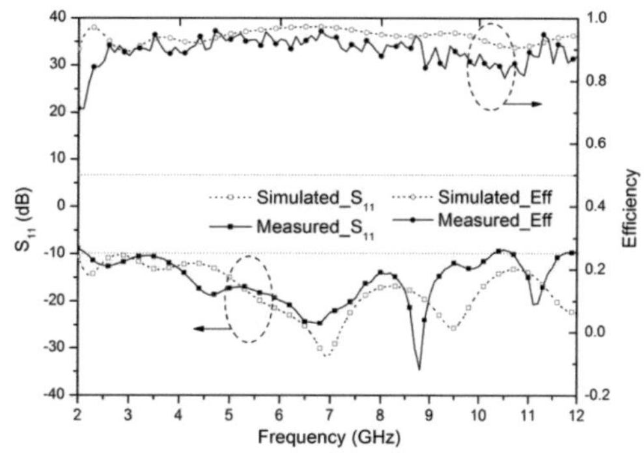

(f) Ant 6: 30×80 mm^2

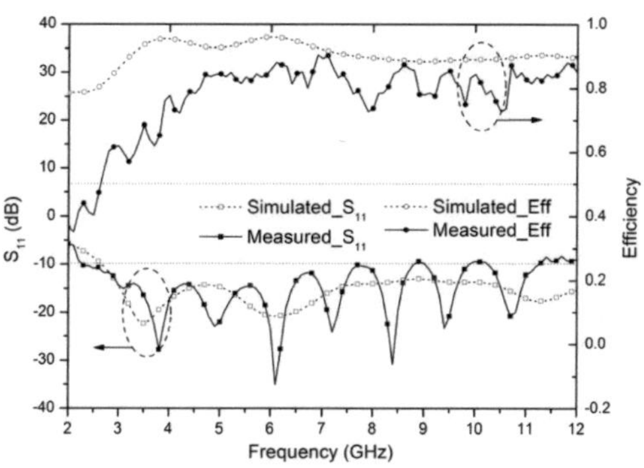

(g) Ant 7: 50×30 mm^2

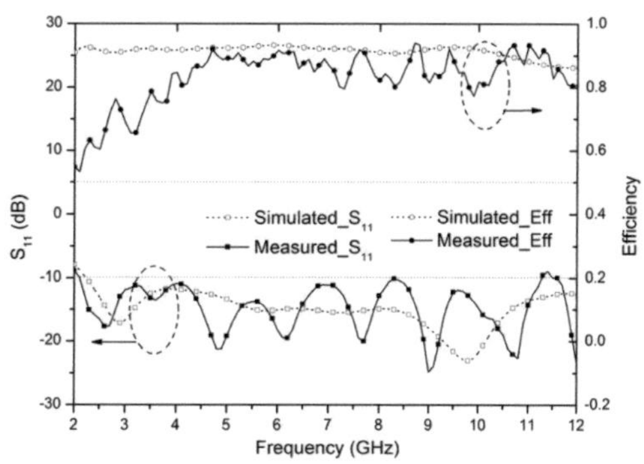

(h) Ant 8: 50×50 mm^2

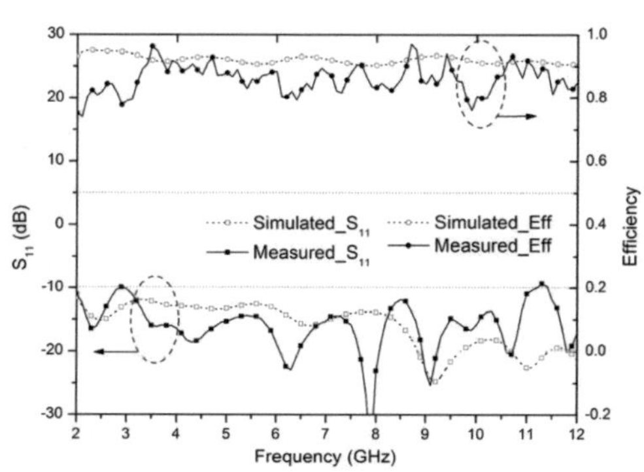

(i) Ant 9: 50×80 mm^2

Figure 9. Simulated and measured S_{11} and efficiencies of nine antennas with different ground-plane sizes *gl*×*gw*

(Ant) GP dimension	(1) 15 (*gl*) × 30 (*gw*)=450			(2) 15 (*gl*) × 50 (*gw*)=750			(3) 15 (*gl*) × 80 (*gw*)=1200		
Avg. 2-12 GHz	0.110			0.074			0.040		
2-4 GHz, 4-6GHz, 6-12 GHz	0.244	0.121	0.060	0.172	0.026	0.055	0.050	0.051	0.034
(Ant) GP dimension	(4) 30 (*gl*) × 30 (*gw*)=900			(5) 30 (*gl*) × 50 (*gw*)=1500			(6) 30 (*gl*) × 80 (*gw*)=2400		
Avg. 2-12 GHz	0.102			0.081			0.049		
2-4 GHz, 4-6GHz, 6-12 GHz	0.225	0.114	0.055	0.180	0.056	0.054	0.050	0.052	0.038
(Ant) GP dimension	(7) 50 (*gl*) × 30 (*gw*)=1500			(8) 50 (*gl*) × 50 (*gw*)=2500			(9) 50 (*gl*) × 80 (*gw*)=4000		
Avg. 2-12 GHz	0.151			0.072			0.061		
2-4 GHz, 4-6GHz, 6-12 GHz	0.186	0.116	0.058	0.130	0.057	0.056	0.069	0.058	0.061

Table 2. Average discrepancy of efficiency between simulated and measured results (GP: Ground plane)

4. Effects of feeding cable on small UWB monopoles measurements

As explained before, the feeding cable used in measurements will scatter, reflect and radiate EM fields, causing interference to the measured results of antennas. Here the effects of the feeding cable are investigated by using simulation and measurement. In the antenna measurement system, Starlab, the feeding cable is enclosed by an EMI suppressant material which is highly lossy. The EM fields incident on it and radiated from it will be absorbed. This significantly reduces unwanted interference to the measured radiation patterns. However, absorbing the EM radiation leads to reduced efficiency. That is why the measured radiation efficiencies are always lower than the simulated results.

4.1. Modeling of feeding cables

Here, we describe the simulation models for two types of feeding cables, denoted here as cables A and B as shown in Figures 10(a) and (b), respectively [20], and use the models in our simulation to study their effects on the measurements of the antenna performances. Cable A is just an ordinary coaxial cable, having a center conductor with a radius of 0.45 mm, and an outer conductor with inner and outer radii of 1.5 and 1.8 mm, respectively. Both cables have a length of about 250 mm. Figure 11(a) shows the cross section of cable A. The space between the center and outer conductors is filled with a dielectric Teflon having a permittivity of 2.08. For cable B shown in Figure 10(b), it is a coaxial cable provided by Satimo for use with the antenna measurement system, Starlab. The cross section of the cable is shown in Figure 11(b) which is identical to cable A, except that the cable has an EMI suppressant tube with a thickness of 1.25 mm on the surface. The property of the tubing material is quite complicated and hard to express precisely. Simulation results show that, by setting both the permittivity and the permeability to 5, and the electric and magnetic loss tangents to 0.004 and 0.5, respectively, the discrepancies between the simulated and measured S_{11} and efficiency are much reduced, thus these parameters are used in our simulation model for cable B.

(a) Cable A

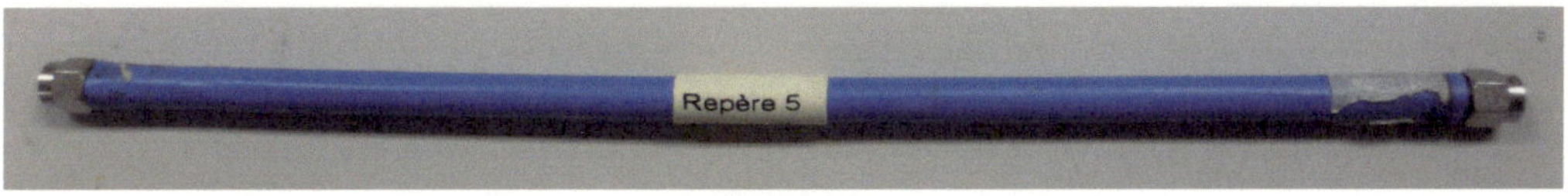

(b) Cable B

Figure 10. Two types of feeding cables used for studies.

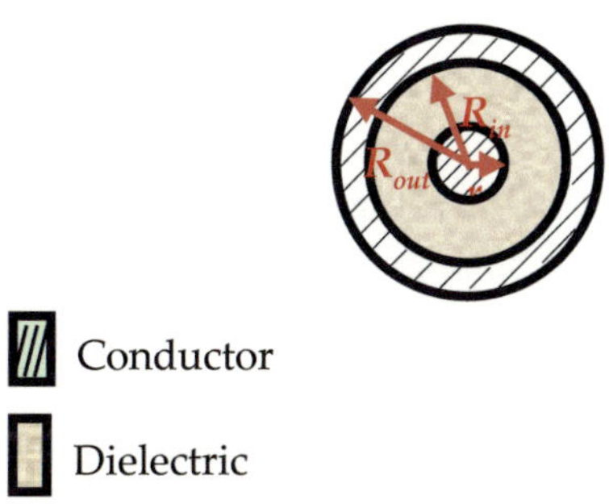

(a) Cable A

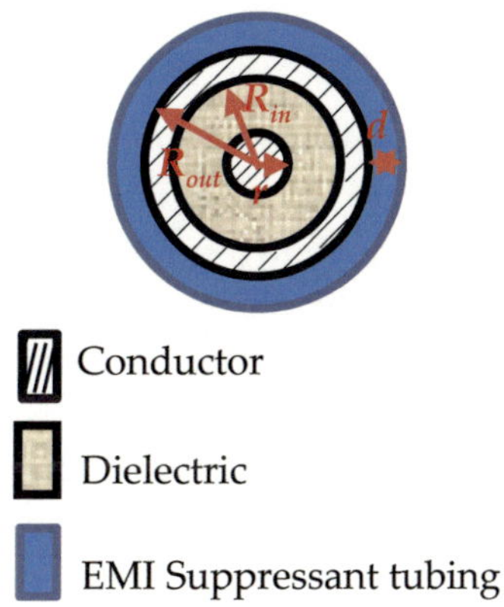

(b) Cable B

Figure 11. Cross sections of cables used for studies

4.2. Antennas used for studies

Three of the nine UWB antennas, Ants 4, 5 and 9 with ground-plane sizes ($gl \times gw$) of 30 × 30 mm^2, 30 × 50 mm^2, and 50 × 80 mm^2, respectively, shown in Figure 7 are selected for investigation of the cable effects using computer simulation. The simulated models developed for cables A and B are used in simulation to feed the signal to the antenna as shown in Figures 12(a) and (b), respectively. In the Starlab system, there is a system cable with EMI suppressant tubing (similar to that used in cable B) used to connect the feeding cable (cable A or cable B) to the network analyzer of the system. The system cable has a length of about 3-4 m long. However, to reduce the simulation time, we only use a total length of 400 mm in our simulation models. Thus when cable A with a length of 250 mm is used, we append a 150-mm cable B to it, making it a total length of 400 mm, as shown in Fig. 12(a). When cable B with a length of 250 mm is used, we actually use a total length of 400 mm, instead of 250 mm. In Figure 12, a metal brick with a size of 6.5 × 6.5 × 13.5 mm^3 is used to model the SMA connector.

(a)

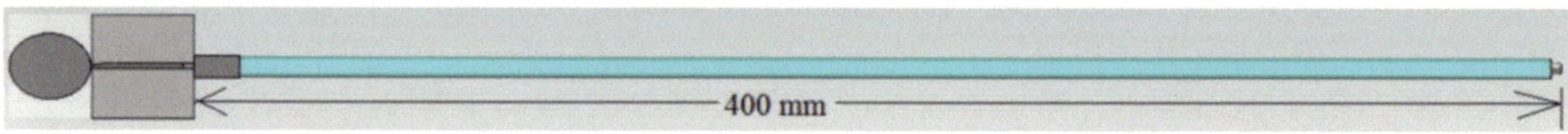

(b)

Figure 12. Simulation models of antenna connected to (a) cable A, and (b) cable B

4.3. Results and discussions

4.3.1. Effects of ordinary coaxial cable (cable A)

With the use of cable A as the feeding cable, the simulated and measured S_{11} and efficiencies of the three antennas are shown in Figure 13. For comparison, the simulation results without a feeding cable are also shown in the same figure. It can be seen in Figure 13 that at high frequencies the simulated efficiencies of the antenna with and without using the cable model are about the same. This is because at high frequencies the ground planes are electrically large, leading to little cable effects on measurements. As the frequency reduces, the ground planes become smaller and discrepancies occur. For impedance bandwidth (S_{11} = -10 dB), all results agree well. This seems to indicate that the cable does not have much effect on the measurements, which as shown later is not true. In fact, at low frequencies, the current flows back from the small ground plane to the surface of the feeding cable, as described previously, giving rise to EM radiation, and then get measured by the system. The measured and simulated results on 3D-radiation patterns reveal that the feeding cable has serious effects on measurements.

The simulated and measured 3D-radiation patterns of the antenna with the ground-plane size of 30 x 30 mm^2 at the frequencies of 3, 7 and 11 GHz, are shown in Figure 14. Without using the feeding cable, the simulated result in Figure 14(a) shows that the antenna has an "apple-shape" radiation pattern at the frequency of 3 GHz which is typical for monopole antennas. At higher frequencies of 7 and 11 GHz, the radiation patterns become slightly directional due to operating in the higher modes. However, when cable A is used, the simulated radiation patterns in Figure 14(b) show many ripples, particularly serious at the lower frequency of 3 GHz. This is because, at 3 GHz, the ground-plane size of 30 x 30 mm^2 (only about half wavelength) is too small to serve as an infinite ground for the monopole. As showed previously, with a small ground plane, the EM fields radiated from antenna are diffracted at the edges and induce currents to flow back to the feeding cable. This can be seen in Figure 15(a) which shows a snap-shot of the simulated surface-current on the feeding cable at 3 GHz. A standing wave is developed along the feeding cable. This produces secondary EM radiation and causes the ripples on the 3D-radiation patterns of Figure 14(b). At 11 GHz, the ground plane is electrically larger and so the 3D-radiation

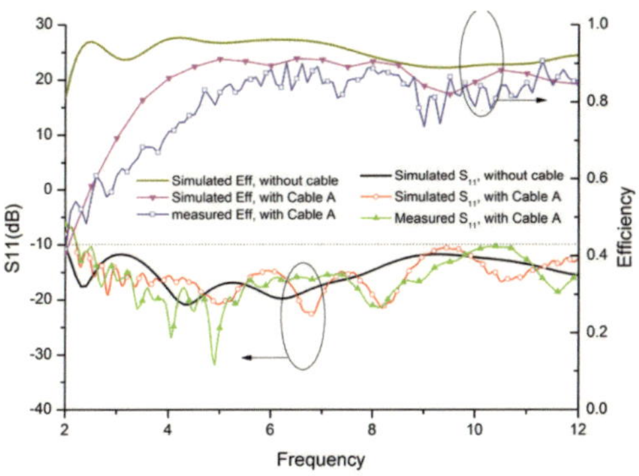

(a) Ant 4: 30 × 30 mm²

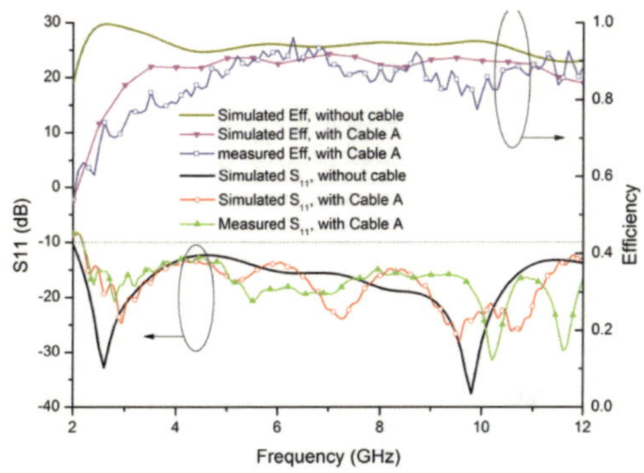

(b) Ant 5: 30 × 50 mm²

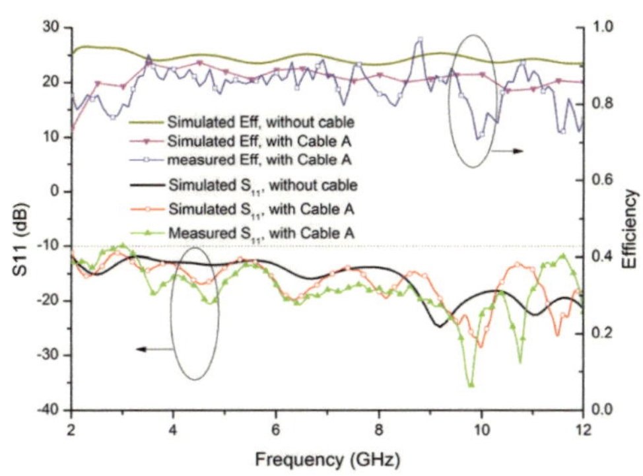

(c) Ant 9: 50 × 80 mm²

Figure 13. Simulated and measured S11 and efficiencies using cable A with different ground-plane sizes.

pattern becomes very similar to the corresponding radiation pattern in Figure 14(a) without using the cable. This can also be seen in Figure 15(b) which shows the simulated surface-current on the feeding cable at 11 GHz. The standing wave on the feeding cable becomes insignificant. The measured 3D-radiation patterns using cable A at 3, 7 and 11 GHz are shown in Figure 14(c), indicating very good agreements with the corresponding simulated radiation patterns in Figure 14(a). These results verify the validity of our simulation mode for the cable.

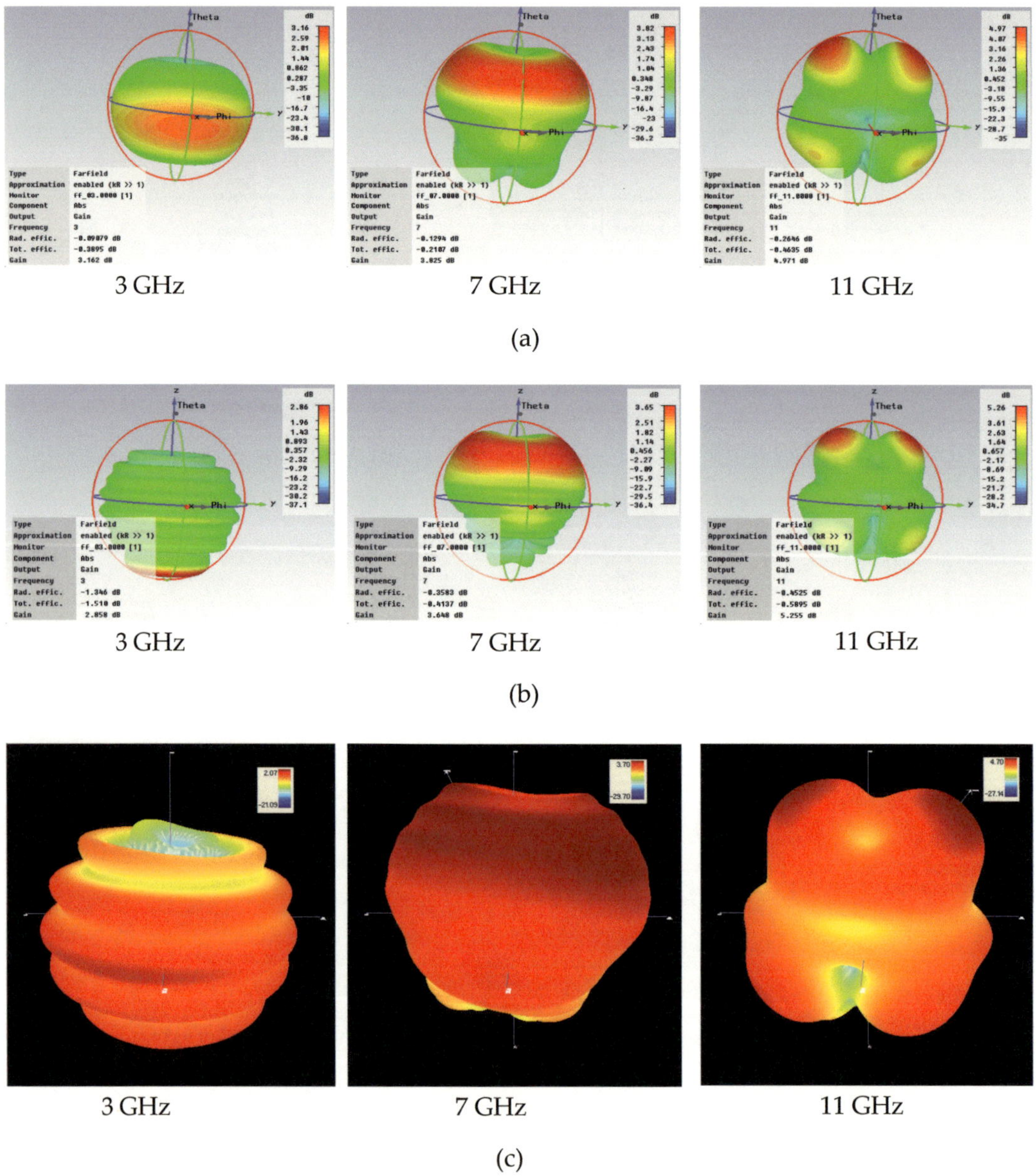

Figure 14. 3D-radiation patterns at 3, 7 and 11 GHz. (a) Simulation without cable, (b) simulation using cable A, and (c) measurement using cable A. Ground-plane size: 30 × 30 mm^2.

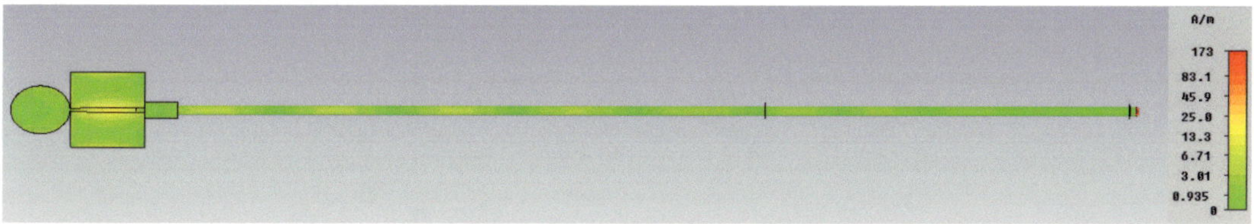

(a)

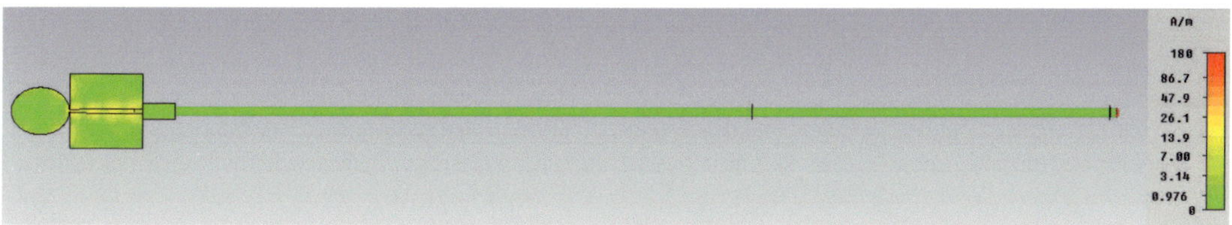

(b)

Figure 15. Surface current distributions of antenna using cable A at (a) 3 GHz, and (b) 11 GHz

4.3.2. Effects of feeding cable with EMI suppressant tubing (cable B)

Figure 16 shows the simulated and measured results for using cable B. It can be seen that, with the use of the simulation model for cable B, the measured and simulated S_{11} and efficiencies have very good agreements for the three antennas even at lower frequencies. These results confirm the accuracy of our simulation model for the feeding cable used in the antenna measurement system. The simulated results without using the feeding cable are also shown in the same figure for comparison. It can be seen that the simulated efficiency without using the feeding cable at low frequencies is much higher than the simulated or measured efficiencies using the feeding cable. This is because at low frequencies, the current flows back to the feeding cable and causes secondary radiation which is mostly absorbed by the EMI suppressant tube enclosing the cable.

The 3D-radiation patterns of the antenna with a ground-plane size of 30 x 30 mm^2 at the frequencies of 3, 7 and 11 GHz are shown in Figure 17. With the use of cable B, Figures 17(b) and (c) show no ripple on the 3D-radiation patterns. The simulated 3D-radiation patterns with and without using cable B agree quite well, indicating the effectiveness of using EMI suppressant tubing for the feeding cable. The measured 3D-radiation patterns in Figure 17(c) are similar to the corresponding simulated radiation patterns shown in Figure 17(b). Figure 18 shows the simulated current distributions on the outer surface of the feeding cable at 3 and 7 GHz. Compared with those in Figure 15, it can be seen that the surface current is very small even at 3 GHz because of the EMI suppressant material. It should be noted that, at low frequencies, since the EM fields radiated from the feeding cable are mostly absorbed by EMI suppressant tubing, the efficiency and hence the gain are much reduced.

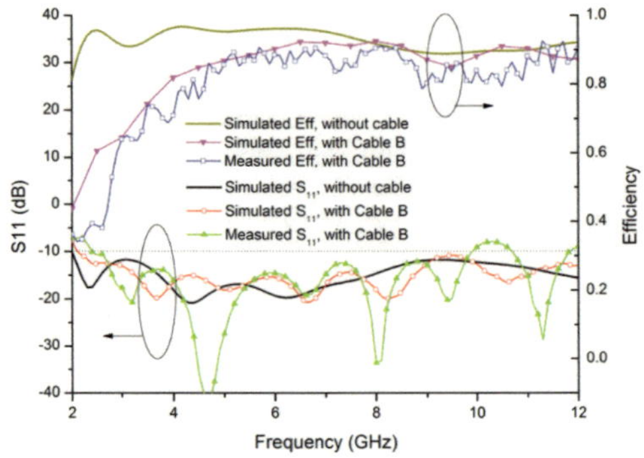

(a) Ant 4: 30 × 30 mm²

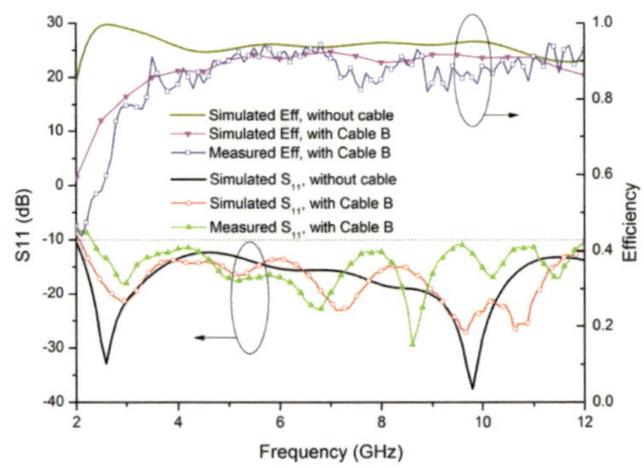

(b) Ant 5: 30 × 50 mm²

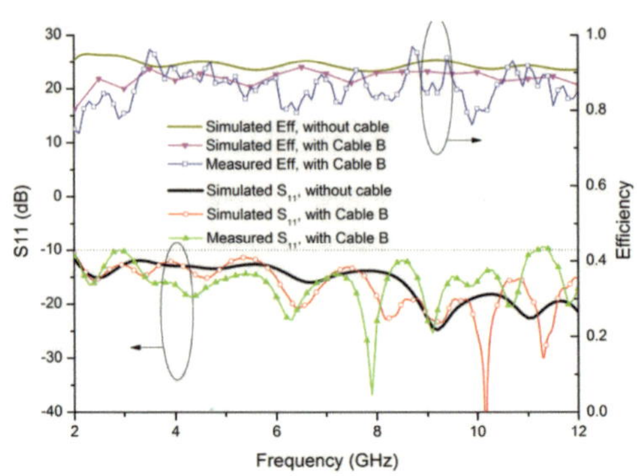

(c) Ant 9: 50 × 80 mm²

Figure 16. Simulated and measured S_{11} and efficiencies using cable B with different ground-plane sizes

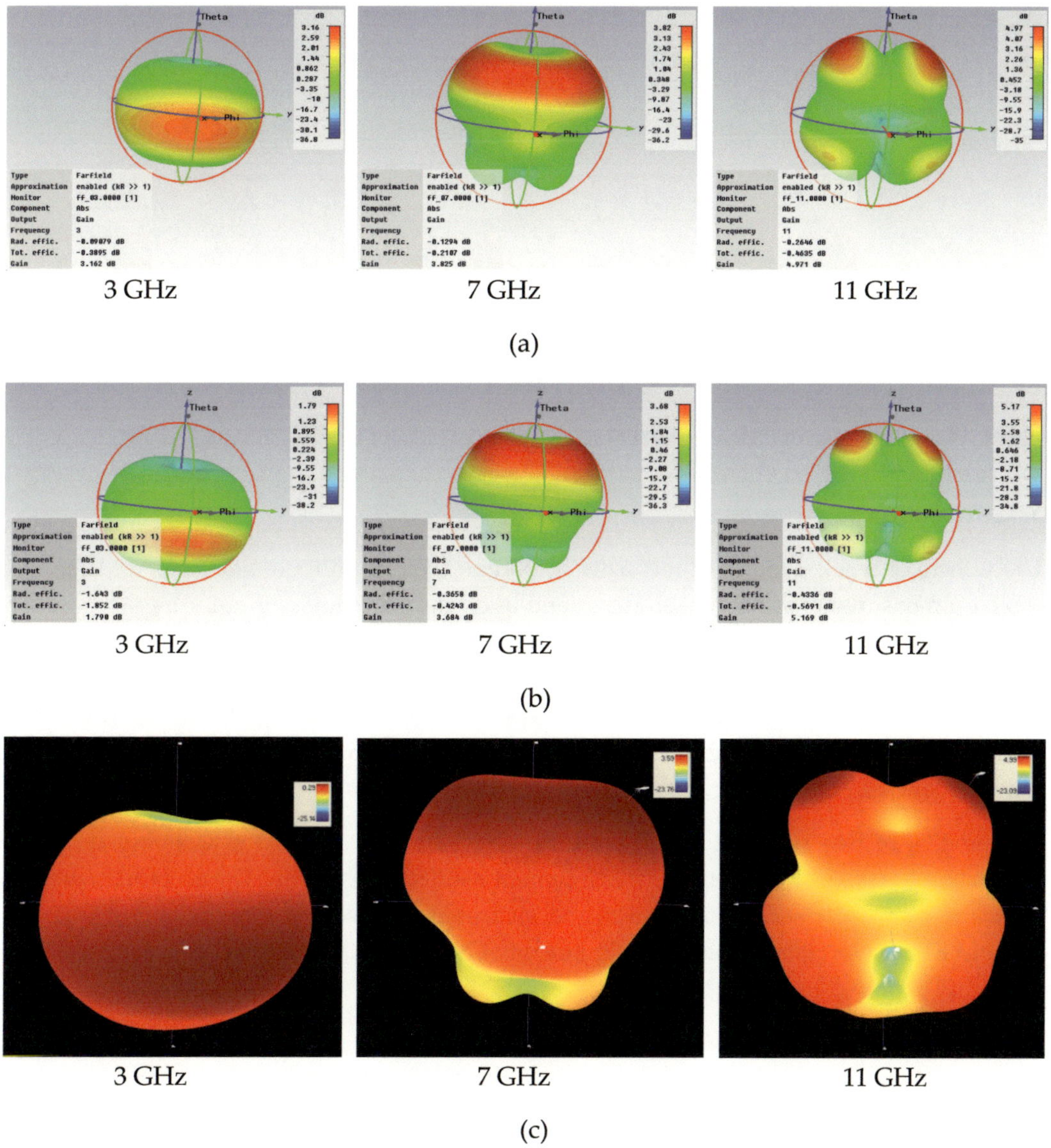

Figure 17. 3D-radiation patterns at 3, 7 and 11 GHz. (a) Simulation without cable, (b) simulation using cable B, and (c) measurement using cable B. Ground-plane size: 30×30 mm^2.

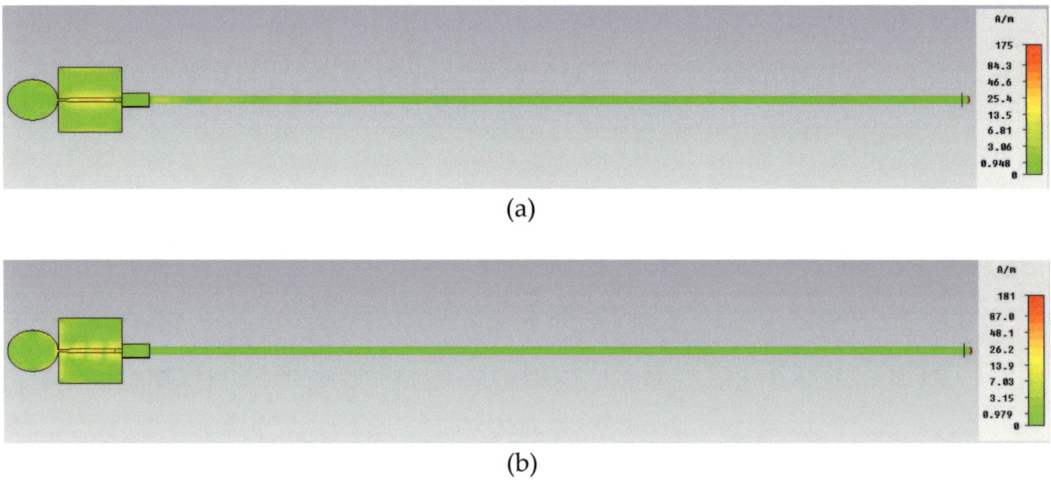

Figure 18. Surface current distribution of antenna using cable B at (a) 3 GHz and (b) 11 GHz

5. Conclusions

The effects of small ground-plane sizes of planar UWB monopole antennas and feeding cable on measurements have been described and studied using computer simulation and measurement. A group of nine UWB antennas with the same radiator but different ground-plane sizes have been used for studies. Results have shown that the widths of the ground planes have more effects on the measured efficiencies. There are large discrepancies between the simulation and measured performances of these antennas at low frequencies.

The models of two practical feeding cables have been developed for studying the cable effects using computer simulation. Measurement results have verified the accuracies the two simulation models. Both the simulation and measured results have shown the feeding cable without EMI suppressant tubing has significant effects on measurements.

Author details

L. Liu, S.W. Cheung, Y.F. Weng and T.I. Yuk
Department of Electrical and Electronic Engineering, The University of Hong Kong, Hong Kong

6. References

[1] Liu L, Cheung SW, TI Yuk (2011) Bandwidth Improvements Using Ground Slots for Compact UWB Microstrip-fed Antennas. Progress In Electromagnetics Research Symposium (PIERS) 2011. Suzhou, China.

[2] Cheung SW, Liu L, Azim R, Islam MT (2011) A Compact Circular-Ring Antenna for Ultra-Wideband Applications. Microwave and Optical Technology Letters. 53: 2283–2288.

[3] Liu L, Cheung SW, Yuk TI (2011) Bandwidth Improvements Using Ground Slots for Compact UWB Microstrip-fed Antennas. Progress In Electromagnetics Research Symposium (PIERS) 2011. Suzhou, China.

[4] Sun YY, Cheung SW, Yuk TI (2011) Studies of Planar Antennas with Different Radiator Shapes for Ultra-wideband Body-centric Wireless Communications. Progress In Electromagnetics Research Symposium (PIERS) 2011. Suzhou, China.

[5] Zhang J, Sun XL, Cheung SW, Yuk TI (2012) CPW-Coupled-Fed Elliptical Monopole Antenna for UWB Applications. IEEE Radio Wireless Week 2012 (RWW2012), Santa Clara, CA, USA

[6] Sun YY, Islam MT, Cheung SW, Yuk TI, Azim T, Misran N (2011) Offset-fed UWB Antenna with Multi-slotted Ground Plane. IEEE International Workshop on Antenna Technology (iWAT2011), 2011

[7] Sun YY, Cheung SW, Yuk TI (2012) Planar Monopole Ultra-wideband Antennas with Different Radiator Shapes for Body-centric Wireless Networks. Progress In Electromagnetics Research Symposium 2012 (PIERS2012), Kuala Lumpur, Malaysia

[8] Saario SA, Lu JW, Thiel DV (2002) Full-wave analysis of choking characteristics of sleeve balun on coaxial cables. Electronics Letters. 38: 304-305.

[9] Chow YL, Tsang KF, Wong CN (1999) An accurate method to measure the antenna impedance of a portable radio. Microwave and Optical Technology Letters. 23: 349-352.

[10] The Capcon website. Available: http://www.capconemi.com/st4page1.html

[11] The Satimo website. Available: http://www.satimo.com/content/products/starlab

[12] Gandhi OP, Lazzi G, Furse CM. Monopole Antennas. Available: http://www.ece.utah.edu/~ece3300/Labs/lab3/MONOPOLE%20ANTENNAS.pdf

[13] Chen ZN, Chia MYW (2006) Broadband Planar Antennas Design and Applications. West Sussex, England: John Wiley & Sons, Ltd.

[14] Icheln C (2001) Methods for measuring RF radiation properties of small antennas. PhD thesis, Helsinki University of Technology, Espoo, Finland, Nov.

[15] DeMarinis J (1988) The antenna cable as a source of error in EMI measurements. IEEE 1988 International Symposium on Electromagnetic Compatibility Symposium Record. pp 9-14.

[16] Weiner MM, (2003) Monopole Antennas. NY: Marcel Dekker, Inc.

[17] Awadalla K, Maclean T (1979) Monopole antenna at center of circular ground plane: Input impedance and radiation pattern. IEEE Transactions on Antennas and Propagation. 27: 151-153.

[18] Weng YF, Cheung SW, Yuk TI (2010) Effects of ground-plane size on planar UWB monopole antenna. TENCON 2010. pp. 422-425.

[19] Icheln C, Ollikainen J, Vainikainen P (1999) Reducing the influence of feed cables on small antenna measurements. Electronics Letters. 35: 1212-1214.

[20] Liu L, Weng YF, Cheung SW, Yuk TI Foged LJ (2011) Modeling of cable for measurements of small monopole antennas. Loughborough Antennas & Propagation Conference (LAPC) 2011, Loughborough, UK.

Chapter 13

UWB-Bandpass Filters with Improved Stopband Performance

Mongkol Meeloon, Sarawuth Chaimool and Prayoot Akkaraekthalin

Additional information is available at the end of the chapter

http://dx.doi.org/10.5772/48736

1. Introduction

In wireless communication systems, microwave planar bandpass filters are employed in most applications. The broadband and multiband applications are renewing the interest in the design of planar broadband filters with low loss, compact size, high suppression of spurious responses, and improved stopband performances. However, since the ultra-wideband (UWB) system covers very wide frequency range of 3.1 to 10.6 GHz and then may be interfered with the existing undesired narrow band from the 5.2 GHz or 5.8 GHz wireless local area network (WLAN) radio signals. Moreover, the WiMAX (3.5 GHz) and RFID (6.8 GHz) communications may interfere with the UWB system within the range defined by the FCC. Therefore, UWB bandpass filters with single- and multi-narrow notched bands are needed to avoid being interfered by the exiting RF signals. In order to obtain operations, several techniques have been reported in literatures based on UWB-bandpass filters with slotted resonators (Meeloon et al., 2007, 2008, 2009), UWB-bandpass filters with slotted resonators and embedded slotted feed (Meeloon et al., 2009, 2011), and UWB-bandpass filters with slotted resonators and embedded fold-slot feed (Meeloon et al., 2010, 2011).

In this chapter, many advanced UWB-bandpass filters are presented based on slotted linear tapered-line resonator (SLTR) and slotted step-impedance resonator (SSIR) structures for size reduction and improved stopband performances. A comprehensive treatment of slotted resonators and both ends of the resonator with interdigital coupled lines is described. The design concept is demonstrated using two filter examples including one with an SLTR and another one with an SSIR. These filters have not only compact size but also a wider upper stopband resulting from resonator bandstop characteristics. Single-SLTR and single-SSIR filters are designed and constructed and their performances are extensively investigated in simulation and measurement. The proposed filters demonstrate their capabilities in suppression of spurious responses. Also, two-SLTR and two-SSIR filters are designed and fabricated to prove that they improve the passband and upper stopband performances with sharpened rejection skirts outside the passband and widened upper stopband.

Then, UWB-bandpass filters based on SLTR and SSIR with embedded slot feed structure for notched band are presented. The embedded slot feed at the end of resonators will be comprehensively described. The proposed filters have narrow notches in the passband, resulting from the embedded slot feed. The center frequencies and bandwidths of the notched band can be easily adjusted by tuning the length and width of the embedded slot parameters. The wider upper stopbands caused by resonator characteristics have been also obatained.

After that, UWB-bandpass filters with single-notched and dual-notched bands and improved stopband performance are proposed using SLTR and SSIR as multi-mode resonator (MMR) and embedded slotted feed. To avoid the existing interferences in the UWB passband, two different embedded slotted feed are employed to obtain two narrow-notched bands. The center frequency and bandwidth of the notched bands can be controlled by adjusting the dimensions of the embedded slotted feed. To further suppress the upper stopband, the defected slot in $\lambda/2$ stepped impedance resonator fed by interdigital coupled line is introduced. Very good agreements between the measured and simulated filter characteristics have been obtained validating the proposed filter prototypes.

Finally, UWB-bandpass filters based on SLTR and SSIR with embedded fold-slot are presented. The proposed filters have narrow notches in the passband and size reduction, resulting from the embedded fold-slot. The length and width of the embedded fold-slot parameters resulting in their performances have been also studied.

2. UWB-bandpass filters with slot resonator

2.1. Interdigital coupled line characteristics

Fig. 1(a) shows a conventional interdigital coupled line which has been widely used as a capacitive coupling element in multi-stage bandpass filters. The optimized interdigital coupled lines must be performed to achieve design-specified coupling factor between two adjacent line resonators. The usual procedure is to reduce both strip and slot widths in order to achieve a tight coupling and lower insertion. However, it may introduce some difficulties into the design procedure and fabrication process as the coupling response is sensitive to the strip/slot widths configuration. In this work, we redesign and optimize the interdigital coupled line, as a result shown in Fig. 1(b). An RT/Duroid3003 substrate, which has a given dielectric constant of 3.0, a thickness of 1.524 mm, and a loss tangent of 0.0013 is used for designing the new interdigital coupled lines at a central frequency of 6.85 GHz and a fractional bandwidth of 100%. The conventional and new coupled lines are evaluated by using an electromagnetic simulation program, IE3D, which is based on the method of moments and proven to be quite accurate in its prediction. The response curves of both coupled lines are demonstrated in Fig. 2. It can be noticed that both interdigital coupled lines have almost the same resonant frequency of 6.85 GHz. Nevertheless, the new coupled line has superior performances with better S_{11} and S_{21} in the passband. This means that it is more suited for use as a bandpass filter element than the conventional one.

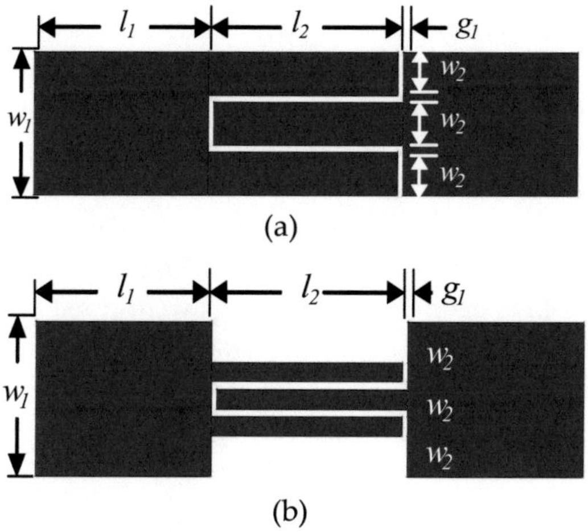

Figure 1. Interdigital coupled lines: (a) conventional one with w_1 = 4.0mm, l_1 = 6.0mm, l_2 = 6.45mm, g_1 = 0.2mm, and w_2=1.2mm and (b) optimized one with w_1=4.0mm, l_1=6.0mm, l_2 = 6.45mm, g_1 = 0.2mm, and w_2 = 0.5mm.

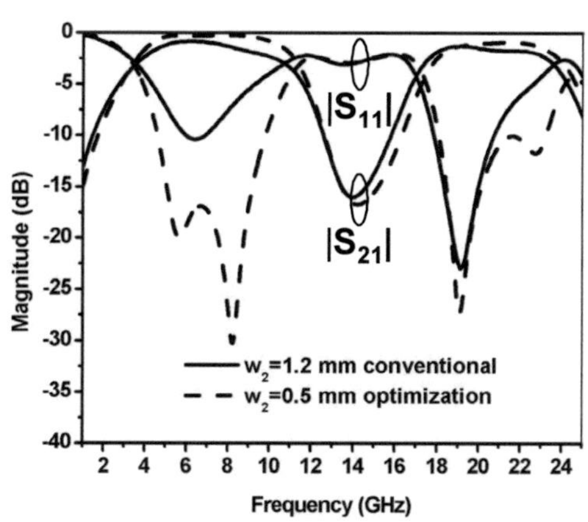

Figure 2. Compared responses of interdigital coupled lines with w_2 = 1.2mm for the conventional one and w_2 = 0.5mm for the optimized one.

2.2. SLTR and SSIR characteristics

Fig. 3(a) shows a conventional λ/2 microstrip linear tapered-line resonator (LTLR). This resonator has inherently spurious resonant frequencies at $2f_o$ and $3f_o$, where f_o is the fundamental frequency, which may be too close to the desired wide passband. A microstrip SIR has been proposed as shown in Fig. 3(b) for higher stopband performances. In this chapter, microstrip SLTRs as shown in Fig. 3(c) and (d), composed of a microstrip tapered-line with slots are proposed. A microstrip SSIR consisting of a microstrip stepped impedance line with a slot has been also investigated, as shown in Fig. 3(e). Fig. 4 shows the current densities of resonators at $3f_0$ about 21 GHz (stopband frequency). We can notice that in Fig. 4(a) and (b), the current densities pass through the resonators. For the proposed resonators in Fig. 4(c) and (d), the current densities cannot pass through the resonators but stop at the slot. It means that the stopband frequency occurs at $3f_0$, which is about 21 GHz. We then perform the parameter study for the proposed SLTR and SSIR. The same substrate with the interdigital coupled lines in previous section has been employed. The IE3D program had been used to determine the frequency response of S_{21}. The input and output ports have been defined at both ends of the proposed resonator. Fig. 5(a) shows the bandstop characteristics of S_{21} when varying w_3 of the SLTR from 4.5 to 7.5 mm. It can be found that a stopband center can move from the frequency of 24 GHz down to 15.5 GHz. As we can see that the slot is longer, lower stopband center can be obtained due to the slot length affects the distance of current distribution in resonator. When varying g_2 of the SLTR, a stopband center is slightly shifted as shown in Fig. 5(b). Fig. 6(a) and (b) shows bandstop characteristics of the SSIR when varying w_3 and g_2. The results are same with responses of the SLTR but better S_{21} magnitudes.

In order to obtain good stopband characteristics without passband perturbations of the desired UWB-bandpass filters, slot length w_3 =5.5mm and slot width g_2 =0.2mm have been chosen as optimized parameters. It can be clearly noticed that the conventional microstrip resonators and the SIR have not obtained stopband characteristics while the proposed slotted resonators have stopband responses with various resonant frequencies. With these stopband characteristics, superior suppression of the spurious responses in the upper band could be obtained when they have been applied to the proposed bandpass filters.

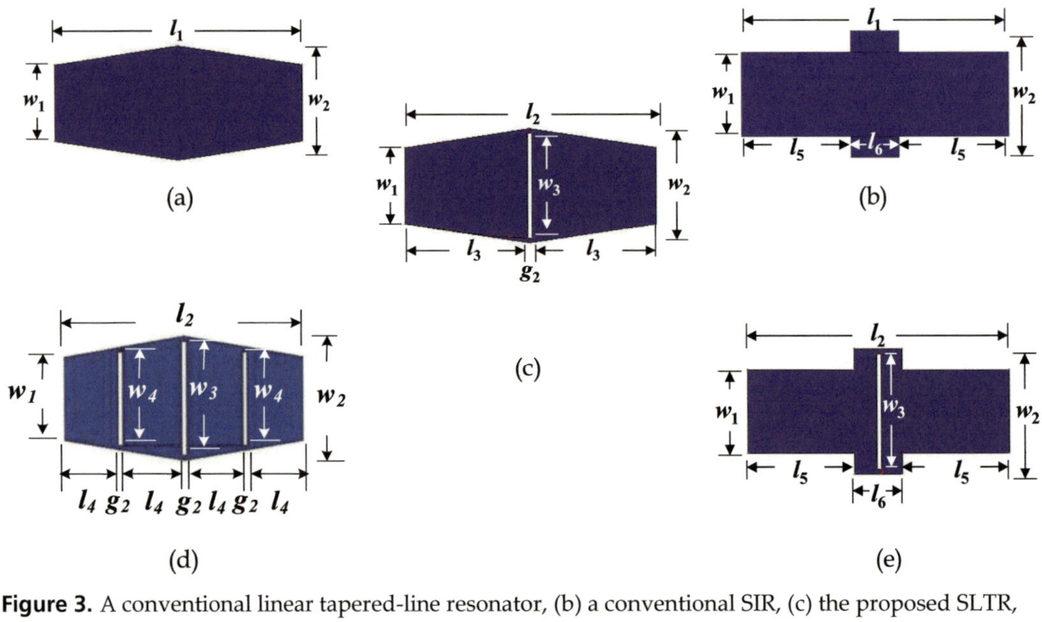

Figure 3. A conventional linear tapered-line resonator, (b) a conventional SIR, (c) the proposed SLTR, (d) the proposed SLTR with three slots, and (e) the proposed SSIR. The dimensions are as follows: l_1 =14.0mm, l_2 =11.0mm, l_3 =5.4mm, l_4 =3.47mm, l_5 = 4.5mm, l_6 = 2mm, w_1 = 4mm, w_2 = 6mm, w_3 = 5.5mm, w_4 = 4.5mm, and g_2 = 0.2mm

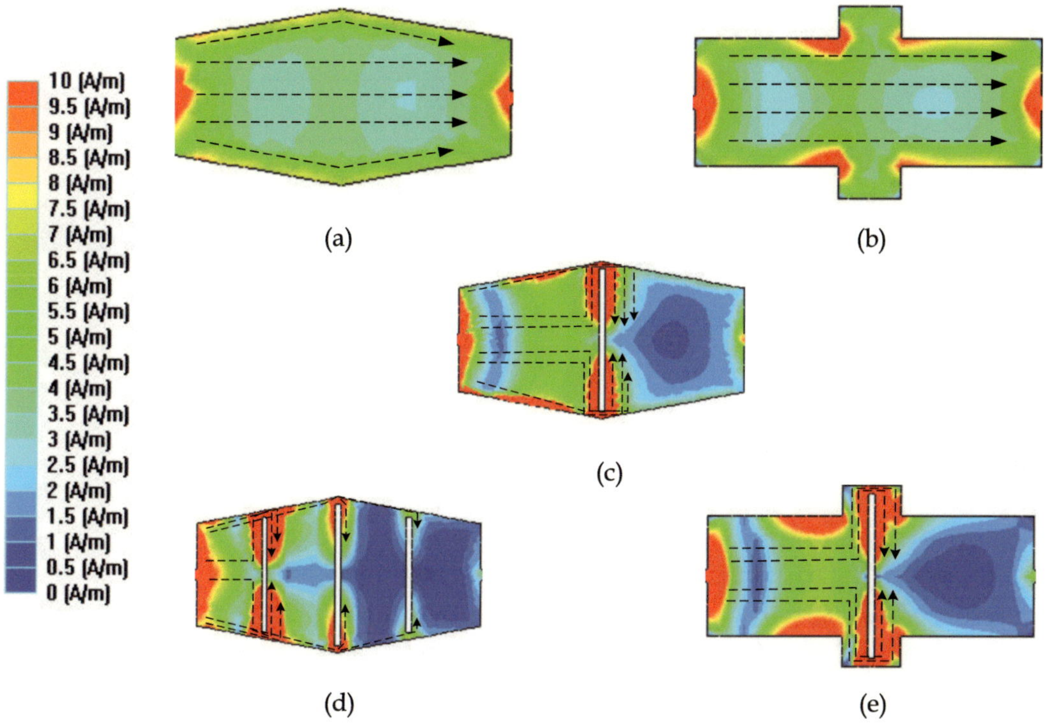

Figure 4. Current densities of resonators: (a) a conventional linear tapered-line resonator, (b) a conventional SIR, (c) the proposed SLTR, (d) the proposed SLTR with three slots and (e) the proposed SSIR

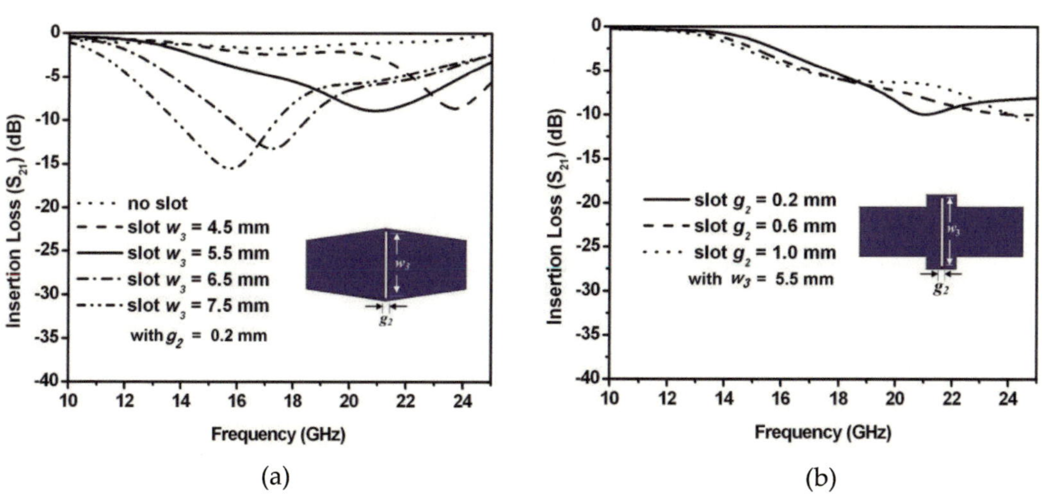

Figure 5. Bandstop characteristics at $3f_o$: (a) SLTR with varied w_3 and (b) SLTR with varied g_2

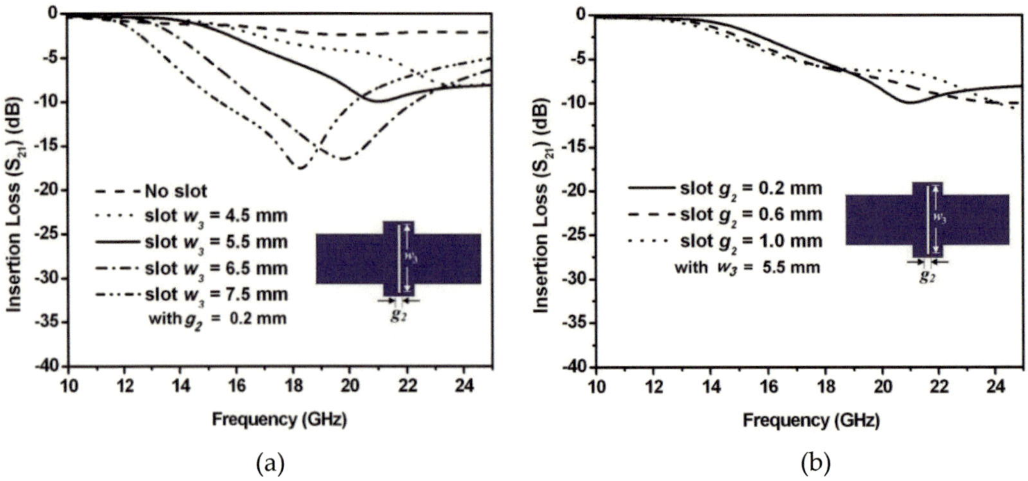

Figure 6. S_{21} responses: (a) SSIR with varied w_3 and (b) SSIR with varied g_2

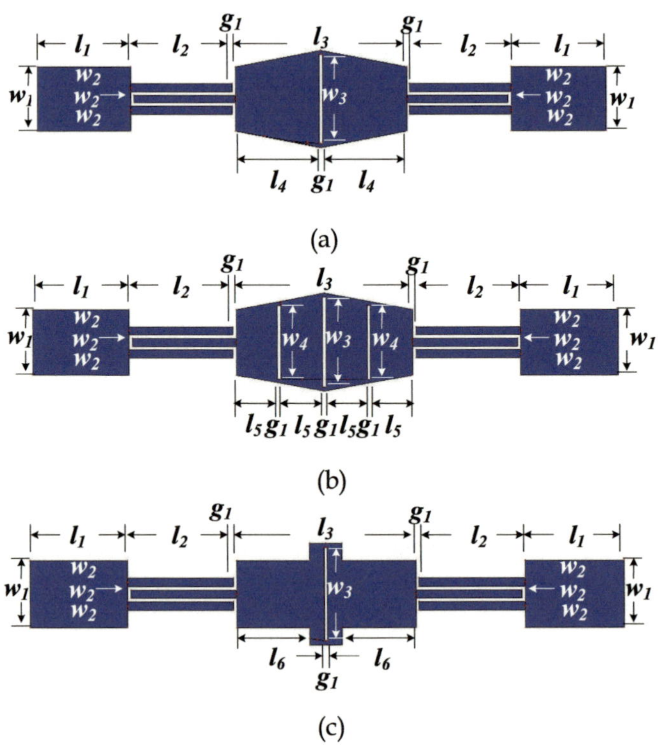

Figure 7. Bandpass filters with a single resonator using: (a) the SLTR (b) the SLTR with three slots, and (c) the SSIR. The dimensions as follows: l_1=6.0mm, l_2=6.45mm, l_3=11.0mm, l_4=5.4mm, l_5=3.47mm, l_6 = 4.5mm, w_1 = 4.0mm, w_2 = 0.2mm, w_3 = 5.5mm, w_4 = 4.5mm, and g_1 = 0.2mm

2.3. Filter designs and measured results

2.3.1. *Single-SLTR filter*

In the following, the two UWB-bandpass filters have been built using the MMR conventional λ/2 resonators and the proposed SLTR and SLTR with three slots fed by λ/4 interdigital coupled lines at both ends with a central frequency of 6.85 GHz and a fractional bandwidth of 100% as shown in Fig. 7(a) and (b). The RT/Duroid 3003 substrate with a dielectric constant of 3.0, a thickness of 1.524mm and a loss tangent of 0.0013 has been used. The optimized dimensions of the resonators and the interdigital coupled lines have been obtained in the previous section. Their electrical performances are then simulated by using IE3D program.

2.3.2. *Single-SSIR filter*

Fig. 7(c) shows schematics of the the proposed single-microstrip SSIR bandpass filters. The optimized dimensions of the resonators and the interdigital coupled lines are the same as resulting from the previous section.

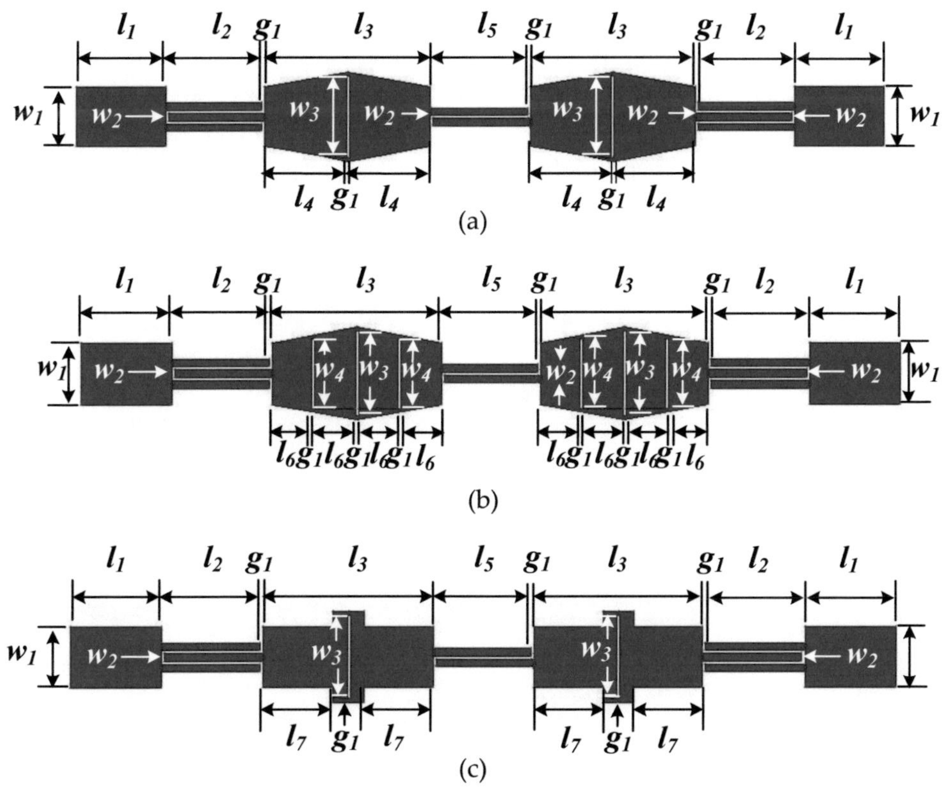

Figure 8. Two-resonator bandpass filters: (a) the SLTR with one slot, (b) the SLTR with three slots, and (c) the SSIR. The dimensions are as follows: l_1=6.0mm, l_2=6.45mm, l_3=11.0mm, l_4=5.4mm, l_5=6.5mm, l_6=3.47mm, l_7=4.5mm, w_1 = 4.0mm, w_2 = 0.2mm, w_3 = 5.5mm, w_4 = 4.5mm, and g_1 = 0.2mm

2.3.3. Two-SLTR filter

Fig. 8 depicts schematics of the bandpass filters with two linear tapered-line resonators connected in cascade. Fig. 8 (a) and (b) was the proposed SLTR filters with single- and three-slotted structures, respectively. All dimensions of the resonators and interdigital coupled lines have been shown.

2.3.4. Two-SSIR filter

Fig. 8(c) depicts schematics of the proposed two-SSIR bandpass filter. The dimensions of the resonators and the interdigital coupled lines are the same.

2.4. Experimental verification

In this section, six UWB-bandpass filters are implemented on the RT/Duroid 3003 with a substrate thickness of 1.524 mm, and a dielectric constant of 3.0 at a central frequency of 6.85 GHz and a fractional bandwidth of 100%. Fig. 9 and Fig. 10 shows photographs of the fabricated SLTR and SSIR filters. Fig. 11 shows comparisons of measured and simulated responses of the SLTR and SSIR filters. In Fig. 11(a), (b) and (c), it can be found that the measured results agree very well with the simulation expectations, confirming that the proposed UWB-SLTR and SSIR bandpass filter is capable of reducing the insertion losses within the passband and also widening the upper stopband. The measured return and insertion losses are found to be higher than 14 dB and less than 2 dB over the UWB-passband, respectively. In Fig. 11(d), (e) and (f), two-SLTR and SSIR filter shows the improved upper stopband performance, with the return losses higher than 15 dB inside the passband. The lower and upper band skirts get sharpened to a great extent, while the upper stopband with the insertion losses above 15 dB occupies an enlarged range of 11.3–25 GHz. Also, the proposed two-SLTR filter with three slots has an improved upper stopband performance, as shown in Fig. 11 (e). A two- SSIR filter has improved the upper stopband performances, with the return losses higher than 15 dB outside the UWB-passband, and

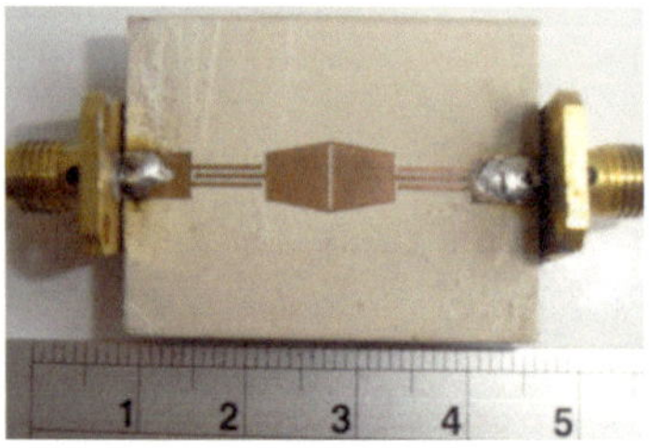

(a)

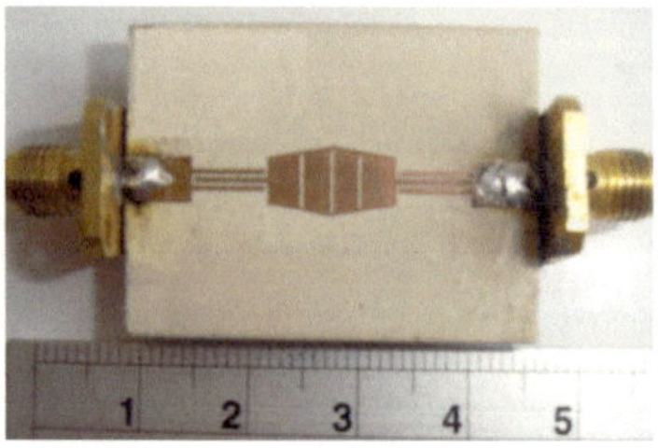

(b)

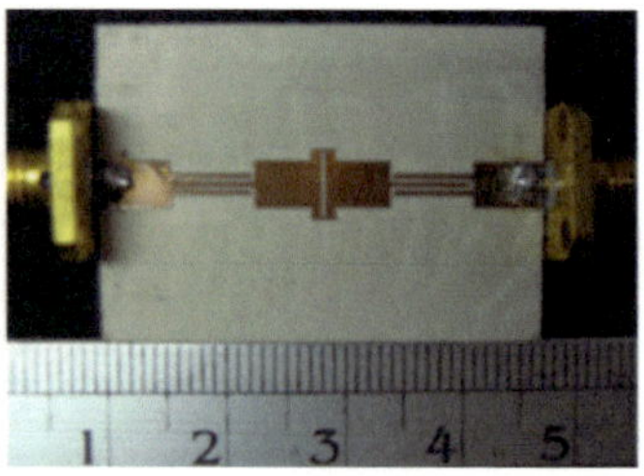

(c)

Figure 9. Photographs of the fabricated single-filters: (a) single-SLTR, (b) single-SLTR with three slots, and (c) single-SSIR

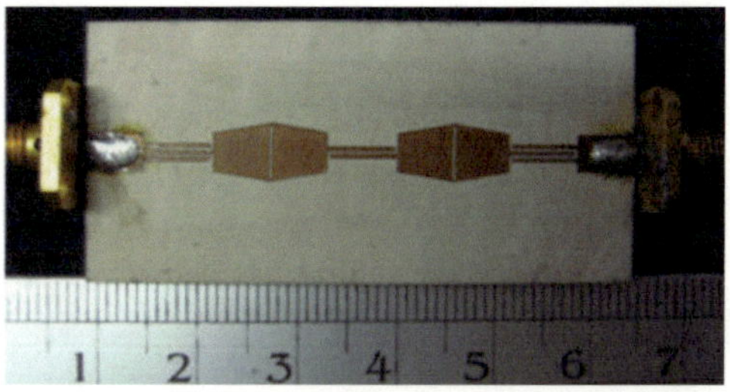

(a)

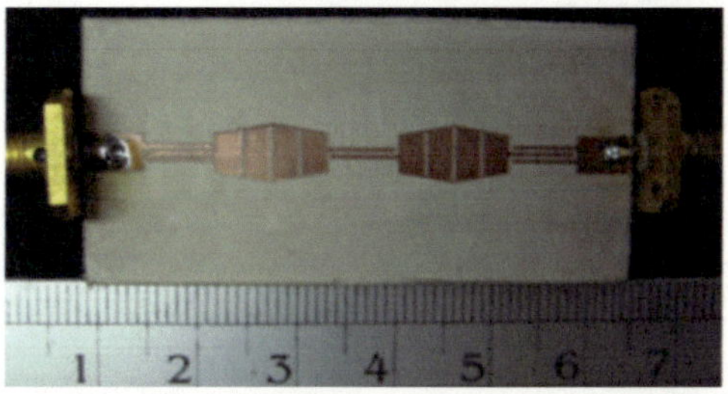

(b)

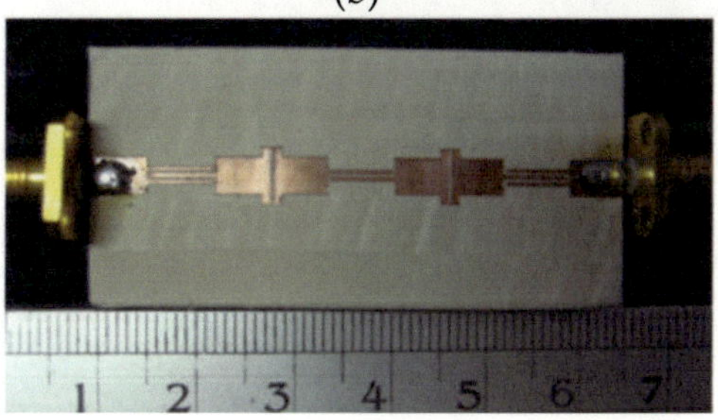

(c)

Figure 10. Photographs of the fabricated two-filters: (a) two-SLTR, (b) two-SLTR with three slots, and (c) two-SSIR

insertion losses above 25 dB in a range of 19–25 GHz and above 47 dB at 24 GHz as shown in Fig. 11 (f). However, the measured passband insertion loss is higher than the simulation result due to the dimension of the conductor and further the conductivity slightly deviated from the design. The group delay of both filters slightly varies between 0.2 and 0.3 ns in the passband.

3. UWB-bandpass filters with embedded slot

This section proposes new UWB-bandpass filters using slotted linear tapered-line resonators (SLTR) and slotted step-impedance resonator (SSIR) structures driven by interdigital coupled lines at both ends of the resonators for improving the stopband performances. Also, using embedded slot structure in the input and output feed line can create a notched band.

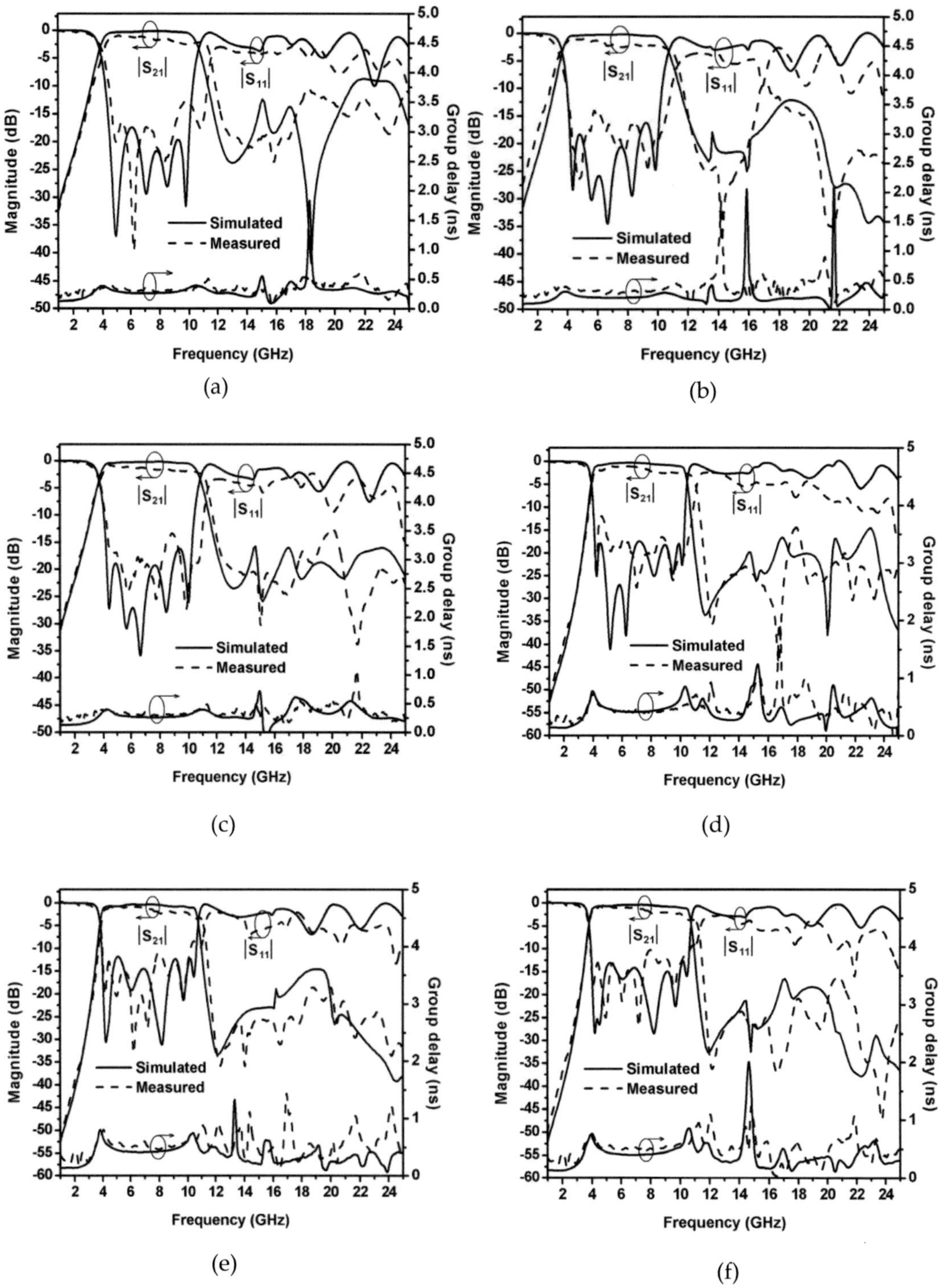

Figure 11. Comparisons of measured and simulated responses of the filters: (a) single-SLTR, (b) single-SLTR with three slots, (c) single-SSIR, (d) two-SLTR, (e) two-SLTR with three slots and (f) two-SSIR

3.1. Embedded slot feed characteristics

The embedded slot at the feed line has been proposed in order to form a notch band. The RT/Duroid 3003 substrate has been used in this study. Fig.12 (a) shows the part of embedded slot feed and its frequency responses of S_{21} when varying the length L of the embedded slot from 5 to 10 mm. It can be found that a center frequency of notched frequency can be adjusted from 9 GHz down to 5 GHz. When increasing the width W of the embedded slot while keeping the center frequency to be the same, the bandwidth of notched band is increased as shown in Fig. 12 (b). The summary of geometry parameters for the embedded slot structure when varying the width W of the embedded slot from 0.6 to 1.6 mm and varying the length L of the embedded slot from 8.593 to 8.393 mm is shown in Table.1. It can be clearly seen that at 3 dB bandwidth of the notched band is increased from 220 to 810 MHz. Therefore, by tuning the length and width of the embedded slot, center frequencies and bandwidth of notched band can be easily adjusted. Embedded slot is thus suitable for use in the UWB-bandpass filter when a notched band is required. To creat the notched band at 5.6 GHz, the dimensions of the proposed embedded slot feed include l = 8.543 mm, w = 0.8 mm, and g = 0.2 mm.

To verify the notched mechanism, the current distributions of embedded structure at 5.6 GHz notch frequency are shown in Fig.13. We can notice that in Fig. 13 (a) the current distribution passes through the conventional feed line but it cannot pass through the proposed structure as shown in Fig.13 (b).

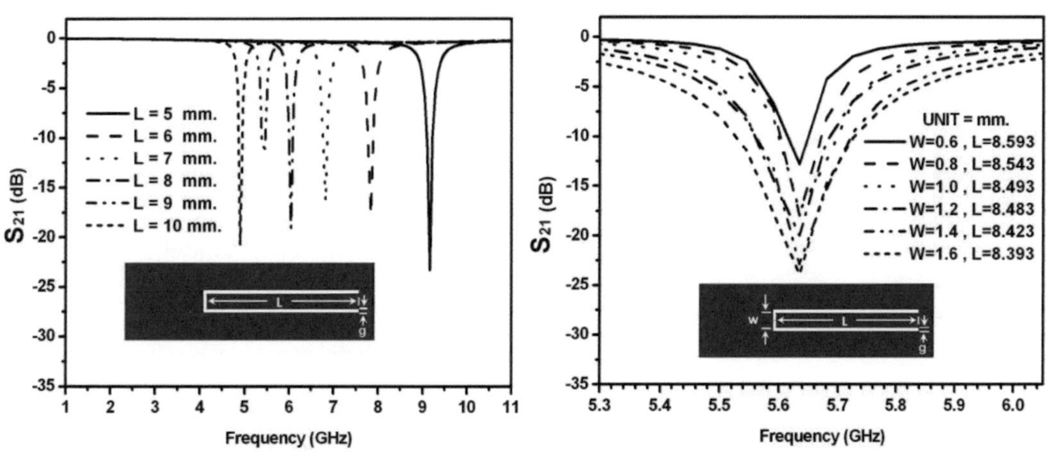

(a) (b)

Figure 12. S_{21} magnitude responses of the embedded slot structure with a slot of 0.2 mm when: (a) varying L (b) varying W

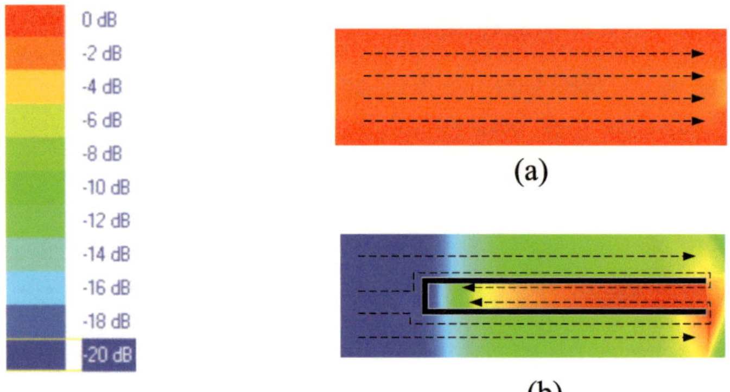

Figure 13. Current distribution at 5.6 GHz of embedded structure: (a) a conventional feed line and (b) the proposed embedded structure

W(mm)	L(mm)	BW(MHz)<-3 dB
0.6	8.593	220
0.8	8.543	250
1.0	8.493	360
1.2	8.483	470
1.4	8.423	720
1.6	8.393	810

Table 1. Geometry parameters for the embedded slot structure

3.2. Filter designs and measured results

3.2.1. SLTR filter with three slots and embedded slot

The UWB-bandpass filters using slotted linear tapered-line resonators (SLTR) with three slots are proposed. The filters consists of the interdigital coupled lines at both ends of the resonators for improving the stopband performances. Also, using embedded slot structure in the input feed line can create a notched band. Fig.14 (a) shows the SLTR with three slots and one embedded slot at the input feed for notched band. The two-cascaded SLTR with three slots and one embedded slot at input feed is also shown in Fig. 14 (b). All dimensions of the UWB-bandpass filters have been shown.

3.2.2. SSIR filter with embedded slot

Fig.14 (c) shows the SSIR with one embedded slot at the input feed for notched band. The SSIR with embedded slot at input and output feed for dual-notched is also shown in Fig.14(d). This section proposes UWB-bandpass filters using slotted step-impedance resonator (SSIR) driven by interdigital coupled lines at both ends of the resonators for improving the stopband performances. Also, using embedded slot structure in the feed line can create a notched band.

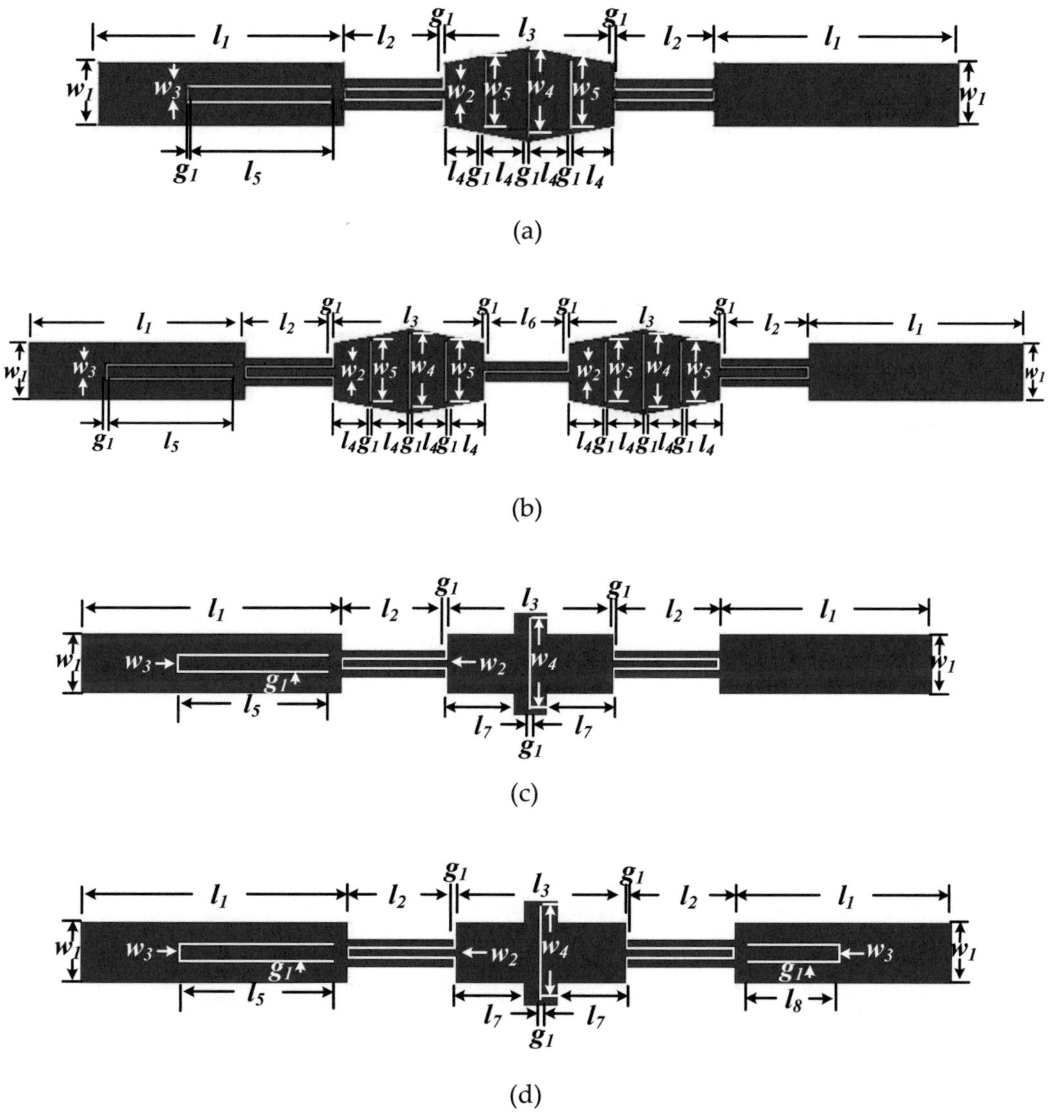

Figure 14. The proposed filters for notched band: (a) single-SLTR, (b) two-SLTR, (c) single-SSIR and (d) single-SSIR with dual-notched band. The dimensions are as follows: l_1=16.0mm, l_2=6.45mm, l_3=11.0mm, l_4=3.47mm, l_5=8.543mm, l_6=6.5mm, l_7=4.5mm, l_8=5.54mm, w_1 = 4.0mm, w_2 = 0.5mm, w_3 = 0.8mm, w_4 = 5.5mm, w_5 = 4.5mm and g_1 = 0.2mm

3.3. Experimental verification

Fig.15 (a) shows photograph of the fabricated single-SLTR for notched band. Fig. 16 (a) shows a comparison of measured and simulated responses of the single-SLTR filter with a notched band. The measured results are agreed very well with the simulation predictions, confirming that the proposed SLTR filter with a notch is capable of narrow notched band, good insertion losses within the passband and also widening the upper stopband. The measured return and insertion losses are found to be lower than 10 dB and higher than 2 dB over desired UWB passband, respectively. Fig. 15 (b) shows photograph of the fabricated two-SLTR filter. Fig. 16 (c) shows a comparison of measured and simulated responses of the two-SLTR filter, which a good agreement has been obtained for passband and sharp rejection skirt in upper stop band. The measured return and insertion losses of the two-SLTR filter are found to be higher than 2 dB and lower than 10 dB at the notched frequencies about 5.6 GHz which the bandwidth of notched band are about 276 MHz. Fig. 15 (c) and (d) show photograph of the fabricated SSIR filter with single and dual-notched band. Fig. 16 (c) and (d) show a comparison of measured and simulated responses of the SSIR filters with single and dual-notched band at the notched frequencies about 5.6 GHz and 8.3 GHz which the bandwidth of notched band are about 276 MHz and 300 MHz, respectively. The proposed filters show improved upper stopband performance with high insertion loss. The upper stopband with the insertion loss lower than 15 dB occupies an enlarged range of 12 to 18 GHz. As we can see that the proposed filter exhibits notched and dual- notched band, a wide upper stopband with values of S_{21} lower than 30 dB at 15 GHz, 10 dB at about 18 GHz. These superior upper stopband performances are caused by the stopband characteristics of the proposed slotted resonator structure and narrow notched band by embedded slot structure. The group delay of fourth filters slightly varies between 0.2 and 0.3 ns in the passband.

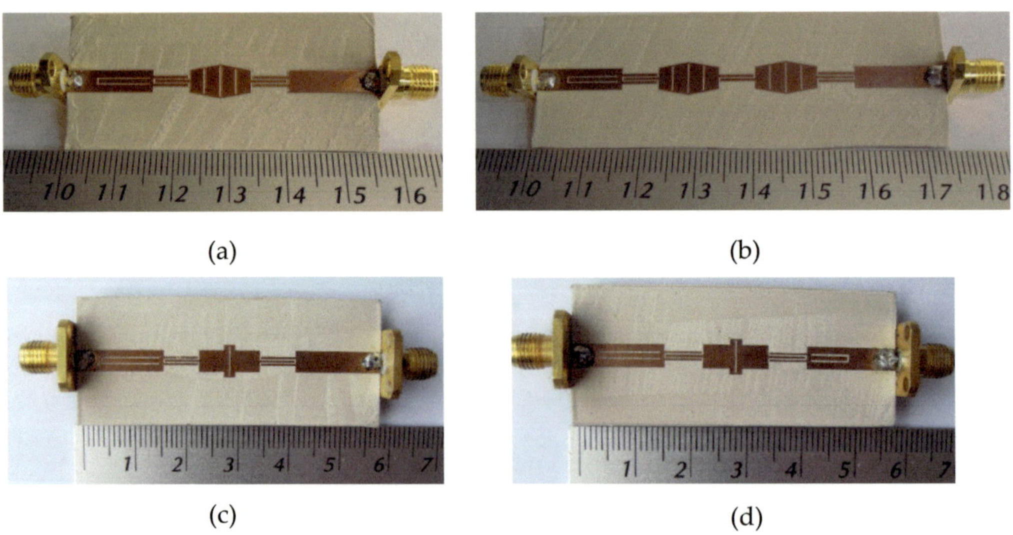

Figure 15. Photographs of fabricated UWB filters for notched band: (a) single-SLTR, (b) two-SLTR, (c) single-SSIR and (d) single-SSIR with dual-notched band

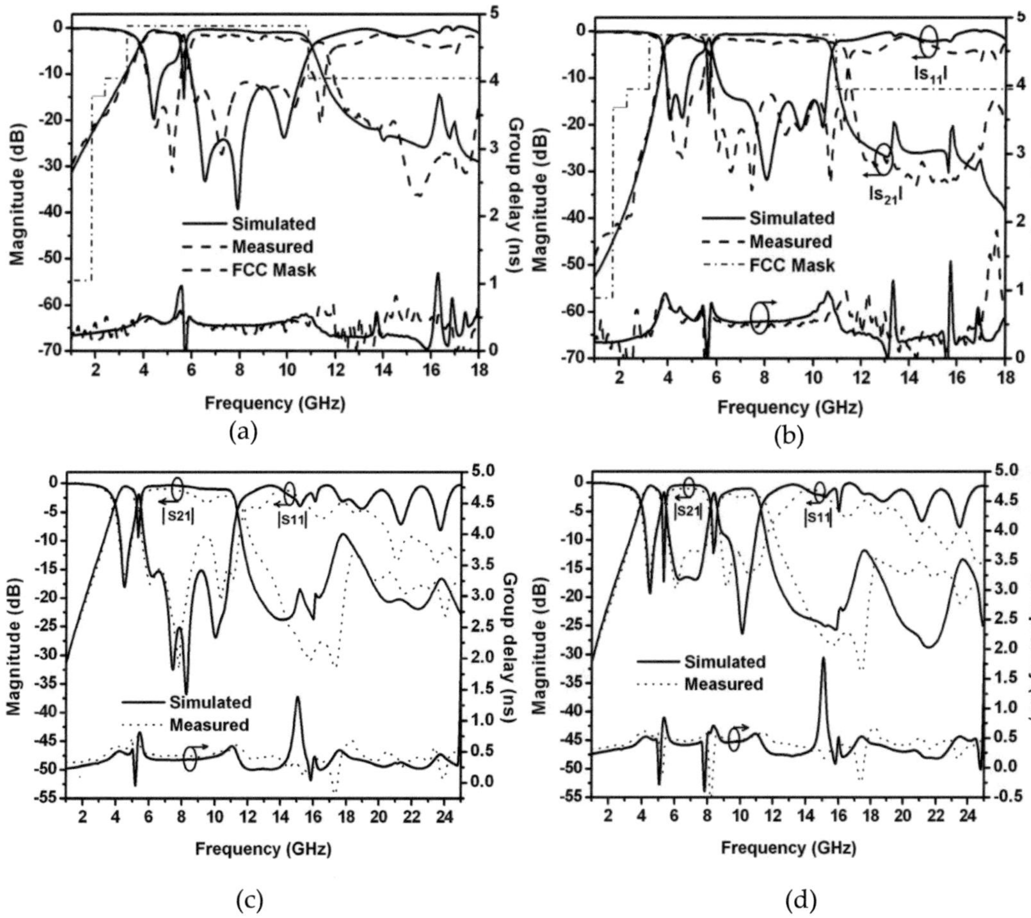

Figure 16. Comparisons of measured and simulated responses of the filters: (a) single-SLTR, (b) two-SLTR, (c) single-SSIR and (d) single-SSIR with dual-notched band

4. UWB-bandpass filters with embedded fold-slot

By modifying the UWB-bandpass filters with embedded slot, the embedded slot can be reduce size using embedded fold-slot. It usies slotted linear tapered-line resonator (SLTR) and slotted step-impedance resonator (SSIR) driven by interdigital coupled lines at both ends of the resonator to improve the stopband performances.

4.1. Embedded fold-slot feed characteristics

The embedded fold-slot at the input feed has been proposed in order to form a notch band. The RT/Duroid 3003 substrate has been used in this study. Therefore, by tuning the length and width of the embedded slot from previous section, center frequencies and bandwidth of notched band can be easily adjusted. Embedded fold-slot is thus suitable for use in the

UWB- bandpass filter when a notched band is required. To creat the notched band at 5.6 GHz, the dimensions of the proposed embedded fold-slot feed include l_{11} = 5.94 mm, w_5 = 0.8 mm, and g_1 = 0.2 mm. To verify the notched mechanism, the current distributions of embedded fold-slot structure at 5.6 GHz notch frequency are shown in Fig. 17. We can notice that in Fig. 17 (a) and (b) the current distribution cannot passes through the embedded slot feed line and the embedded fold-slot feed line.

4.2. Filter designs and measured results

4.2.1. SLTR filter with embedded fold-slot

The UWB-bandpass filters using slotted linear tapered-line resonators (SLTR) driven by interdigital coupled lines at both ends of the resonators for improving the stopband performances and using embedded fold-slot structure in the input feed line can create a notched band are proposed. Fig.18 (a) and 18 (b) show the SLTR and SLTR filters with three slots, using embedded fold-slot at the input feed for notched band. The two-cascaded SLTR with embedded fold-slot at input feed is also shown in Fig. 18 (c). All dimensions of the UWB-bandpass filters have been shown.

4.2.2. SSIR filter with embedded fold-slot

Fig. 18 (d) shows the SSIR with embedded fold-slot at the input feed for notched band.

This section proposes a new UWB-bandpass filter with simple structures using slotted step-impedance resonator (SSIR) driven by interdigital coupled lines at both ends of the resonators for improving the stopband performances. Also, using embedded fold-slot in the input feed line can create a notched band.

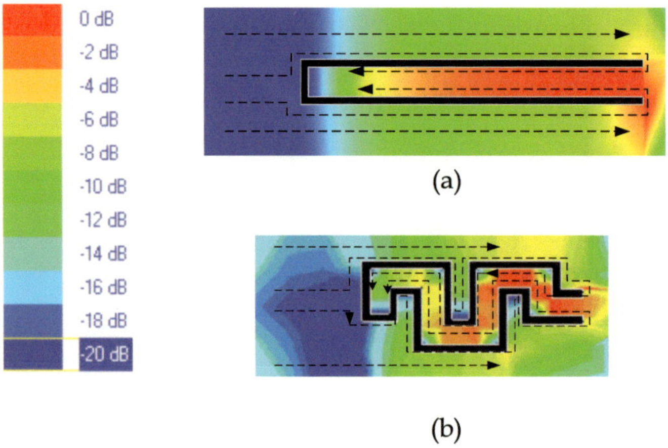

Figure 17. Current distribution at 5.6 GHz of embedded structure: (a) the embedded structure, (b) the proposed embedded fold-slot structure

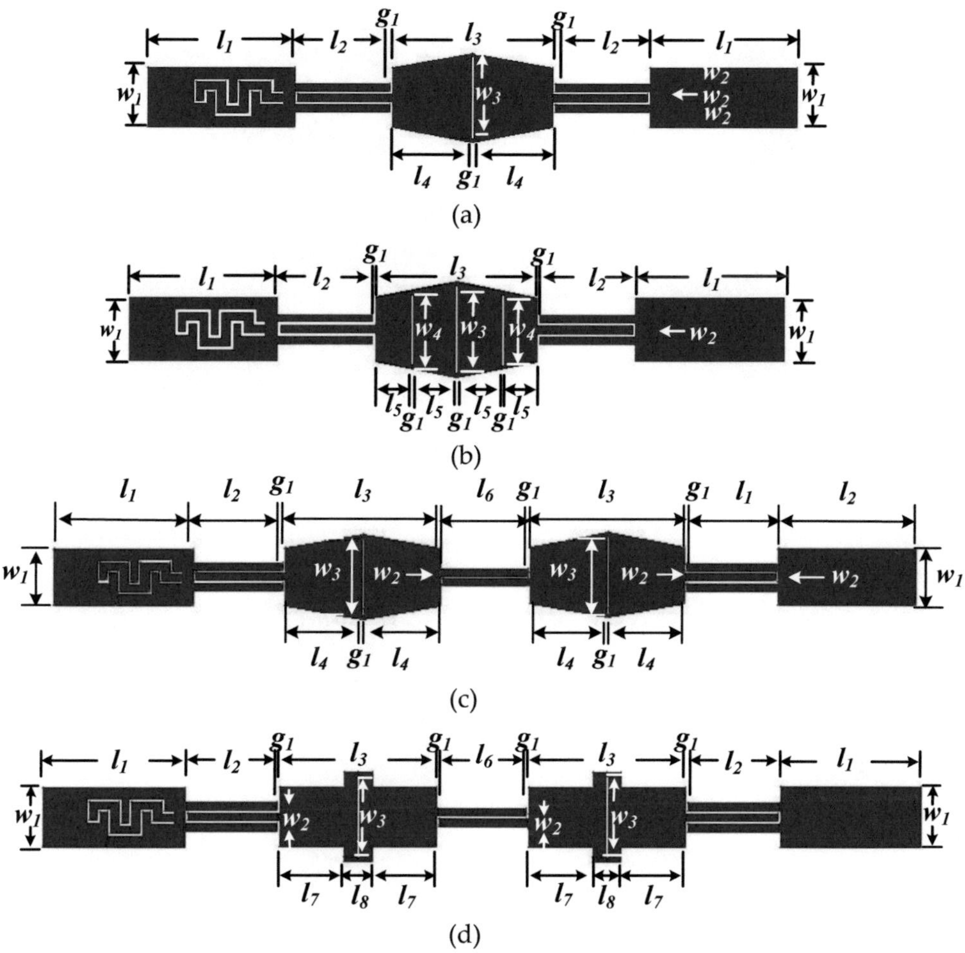

Figure 18. The proposed filters for notched band: (a) single-SLTR, (b) single-SLTR with three slot (c) two-SLTR, and (d) two-SSIR. The dimensions are as follows: l_1=10.0mm, l_2=6.45mm, l_3=11.0mm, l_4=5.4mm, l_5=3.47mm, l_6=6.5mm, l_7=4.5mm, l_8=2.0mm, l_9=2.12mm, l_{10}=1.47mm, l_{11}=5.94mm, w_1=4.0mm, w_2=0.5mm, w_3=5.5mm, w_4=4.5mm, w_5=0.5mm, w_6=2.1mm, w_7=0.52mm and g_1=0.2mm

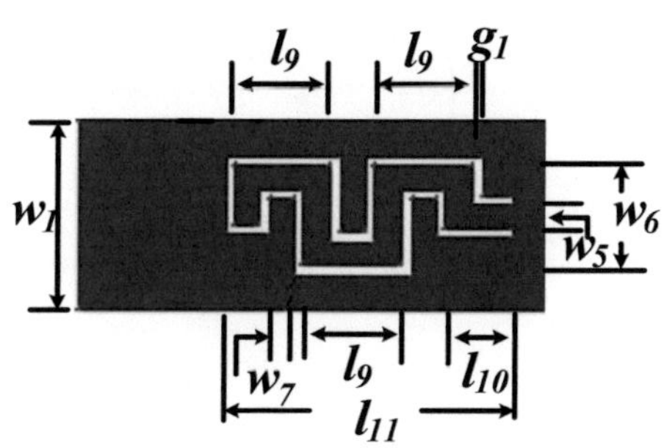

Figure 19. Embedded fold-slot feed

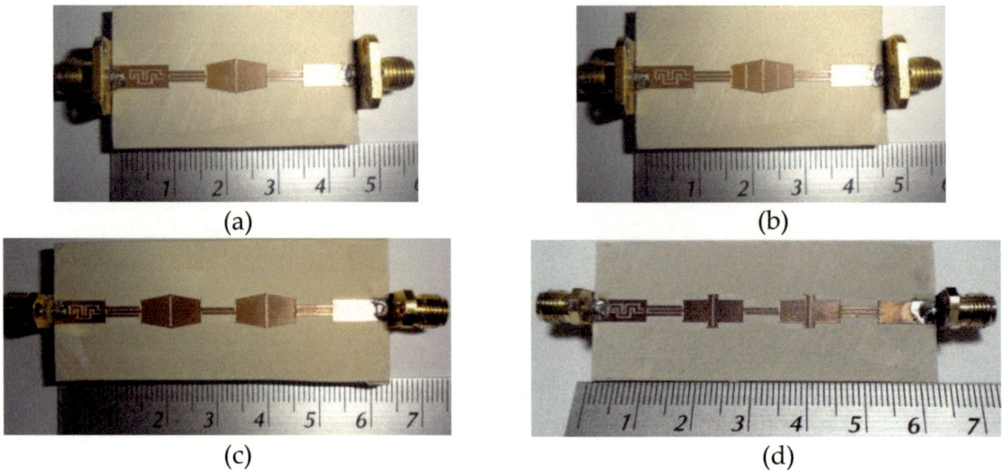

Figure 20. Photographs of fabricated UWB-filters for notched band: (a) single-SLTR, (b) two-SLTR, (c) two-SLTR and (d) two-SSIR

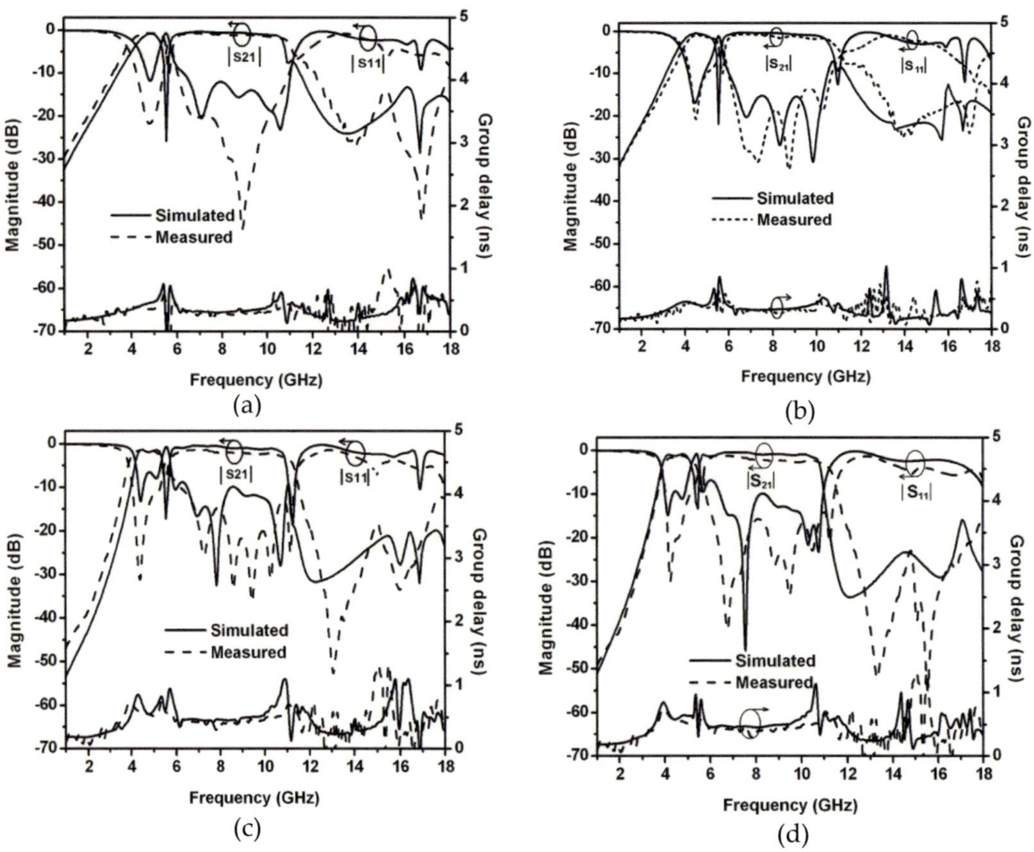

Figure 21. Comparisons of measured and simulated responses of the filters: (a) single-SLTR, (b) single-SLTR with three slots, (c) two-SLTR, (d) two –SSIR

4.3. Experimental verification

Fig. 20 shows the photograph of the fabricated SLTR and SSIR filters for notched band. Fig. 21 shows a comparison of measured and simulated responses of the SLTR and SSIR filters with a notched band. The measured and simulated results have shown good agreement with a notch is capable of narrowing notched band, having good insertion losses within the passband and also widening the upper stopband. The measured return and insertion losses are found to be lower than 10 dB and higher than 2 dB, respectively over desired UWB-passband. The notched frequency of about 5.6 GHz has a bandwidth of about 276 MHz. The proposed filters show narrow notched band and improved upper stopband performance with high insertion loss. The upper stopband with the insertion loss lower than 10 dB occupies an enlarged range of 14 to 18 GHz. The group delay of both filters slightly varies between 0.2 to 0.3 ns in the passband. These superior stopband performances are caused by the stopband characteristics of the proposed slotted resonator structure, and narrow notched band is caused by embedded fold-slot structure.

5. Conclusion

In this chapter, the novel SLTR and SSIR UWB-bandpass filters with improved upper stopband performances have been presented and implemented. By properly forming SLTR and SSIR together with two interdigital coupled lines at both ends, the proposed filters are designed and constructed. The single-SLTR and SSIR filters show their performances in suppression of spurious responses. Also, two-SLTR and two-SSIR filters are designed and fabricated to show that they improve the passband and upper stopband performances with sharpened rejection skirts outside the passband and widened upper stopband. When comparing with SLTRs, we find that the SSIR structures are easier to design and fabricate and they also have better stopband characteristics. In addition, the SLTR and SSIR filters using embedded slot and embedded fold-slot with notched band, reduce size and improved upper stopband performances have been presented and implemented. The proposed filters demonstrate their capability in narrow notched band with the embedded slot, embedded fold-slot feed and suppression of spurious responses with slotted resonators. Also, the fabricated filters prove that they can create notched band and improve upper stopband performances with sharpened rejection and widen the upper stopbands.

Author details

Mongkol Meeloon
Department of Special Investigation (DSI), Thailand

Sarawuth Chaimool and Prayoot Akkaraekthalin
King Mongkut's University of Technology, North Bangkok, Thailand

6. References

Meeloon, M.; Akkaraekthalin, P. (2007). Broadband bandpass filters using slotted linear tapered-line resonators for improving upper stopband performances, *Proceedings of EECON-30 30th Electrical Engineering conference*, pp.1109-1112, Kanchanaburi, Thailand, 2007

Meeloon, M.; Chaimool, S.; Akkaraekthalin, P. (2007). Broadband bandpass filters using slotted resonators fed by interdigital couple line for improving upper stopband performances, *Proceedings of APMC 2007 Asia-Pacific Microwave conference*, Bangkok, Thailand, 2007

Meeloon, M.; Chaimool, S.; Akkaraekthalin, P. (2008). Broadband bandpass filters using slotted stepped impedance resonators for improving upper stopband performances, *Proceedings of ECTI-CON2008 5th International Conference on Electrical Engineering/Electronics, Computer, Telecommunications and Information Technology*, pp.265-268, Krabi, Thailand, 2008

Meeloon, M.; Chaimool, S.; Akkaraekthalin, P. (2009). Broadband bandpass filters using slotted resonators fed by interdigital coupled lines for improved upper stopband performances, *AEU International Journal of Electronics and Communications*, vol. 63, pp.454-463, 2009

Meeloon, M.; Chaimool, S.; Akkaraekthalin, P. (2009). A Broadband bandpass filter with a notch-band using step-impedance resonator and embedded slots, *Proceedings of APMC 2009 Asia-Pacific Microwave conference*, pp.1446-1449, Singapore, 2009

Meeloon, M.; Chaimool, S.; Akkaraekthalin, P. (2009). Broadband bandpass filters with notched bands using embedded slots, *Proceedings of EECON-32 32th Electrical Engineering conference*, Nakhon Nayok, Thailand, 2009

Meeloon, M.; Chaimool, S.; Akkaraekthalin, P. (2011). Ultra-wideband bandpass filters with notched using slotted linear tapered-line resonators and embedded slot feed, *Proceedings of ECTI-CON2011 8th International Conference on Electrical Engineering/Electronics, Computer, Telecommunications and Information Technology*, pp.244-247, Khon Kaen, Thailand, 2011

Meeloon, M.; Chaimool, S.; Akkaraekthalin, P. (2010). Broadband bandpass filters with notched bands using embedded fold-slot, *Proceedings of EECON-33 33th Electrical Engineering conference*, Chiang Mai, Thailand, 2010

Meeloon, M.; Chaimool, S.; Akkaraekthalin, P.; Leenaphet, A.; Pattanakun, R. (2011). An Ultra-wideband bandpass filter with notched using step-impedance resonators and embedded fold-slot, *Proceedings of ISAP-2011 International Symposium on Antennas and Propagation*, Jeju, Korea, 2011

Meeloon, M.; Jeeracheeweekul, V.; Mahatthanathawee, S.; Ngamkhum, J.; Chomputawat, A.; Chomputawat, S.; Rattarasarn, M. (2011). An Ultra-wideband bandpass filter with notched band using embedded fold-slot feed and three-slotted resonator, *Proceedings of ISPACS-2011 International Symposium on Intelligent Signal Processing and Communicaion Systems*, Chiang Mai, Thailand, 2011

Chapter 14

UWB Coexistence with 3G and 4G Cellular Systems

Ahmed Bazil Taha and Miguel Calvo Ramon

Additional information is available at the end of the chapter

http://dx.doi.org/10.5772/47365

1. Introduction

The Federal Communications Commission (FCC) agreed in February 2002 to allocate 7.5 GHz of spectrum, in the 3.1 GHz to 10.6 GHz frequency band, for unlicensed use of ultra wide band (UWB) devices for communication applications. The move represented a victory in a long hard-fought battle that dated back decades. With its origins in the 1960s, when it was called time-domain electromagnetics, UWB came to be denoting the operation of sending and receiving extremely short bursts of RF energy. With its outstanding ability for applications that require precision distance or positioning measurements, as well as high-speed wireless connectivity, the largest spectrum allocation ever granted by the FCC is unique because it overlaps other services in the same frequency of operation. Previous spectrum allocations for unlicensed use have opened up bandwidth dedicated to unlicensed devices based on the assumption that operation is subject to the following two conditions:

1. The device will not cause harmful interference to other systems. Thus, the UWB interferences should not seriously degrade, obstruct, or repeatedly interrupt other radio communication systems.
2. The device must accept any interference received from any licensed system, including interference that may cause undesired operation. This means that devices using unlicensed spectrum must be designed to coexist in an uncontrolled environment.

Devices using UWB spectrum operate according to similar rules, but they are subject to more stringent requirements because UWB spectrum underlays other existing licensed and unlicensed spectrum allocations. In order to optimize spectrum use and reduce interference to existing services, the FCC's regulations are very conservative and require very low emitted power.

The UWB spectrum consists of three different parts as given below:

- The main spectrum extending from 3.1 GHz up to 10.6 GHz.
- The lower residual spectrum extending from 0 Hz up to 3.1 GHz.
- The upper residual spectrum extending from 10.6 GHz upwards.

The main objective of this chapter is to study the UWB coexistence with the 3G and 4G Cellular Systems. UMTS in the 2 GHz and in the 450 MHz are two examples of the 3G cellular systems while the WiMAX system is one of 4G cellular systems.

WiMAX (Worldwide Interoperability for Microwave Access) is a 4G wideband cellular communication system that can provide up to 70 Mbps in 20 MHz bandwidth. The spectrum of WiMAX at 3.5 GHz lies between 3300 to 3800 MHz. Thus, WiMAX receivers are affected by UWB interference from the UWB main part spectrum. For WiMAX at 2.5 GHz, the spectrum lies between 2300 to 2700 MHz. In this case, WiMAX receivers are affected by the interference from the lower residual part of the UWB spectrum. Table 1 shows the WiMAX modulation schemes and the necessary Signal to Interference and Noise Ratio (SINR) required to support them.

The UMTS (Universal Mobile Telecommunications System) is a 3G cellular system that can support voice, data and video services. The downlink frequency used by the UMTS systems lies between 2110 to 2170 MHz.

Deployment of UWB systems creates a "forbidden zone" around the UWB transmitter in which the receivers of WiMAX or UMTS systems can be drastically affected. In practice, the radius of the forbidden zone should be the minimum possible. In our work we will consider a forbidden zone within 1 to 2 m radius (other values such as 0.5 m can be considered) assuming that the maximum accepted downlink range reduction of the WiMAX systems at any moment is 1%. The maximum accepted reduction of the capacity of UMTS systems is assumed to be also 1%.

Order	Modulation	Required SINR (dB)
1	BPSK 1/2	6.4
2	QPSK 1/2	9.4
3	QPSK 3/4	11.2
4	16 QAM 1/2	16.4
5	16 QAM 3/4	18.2
6	64QAM 1/2	22.7
7	64QAM 3/4	24.4

Table 1. WiMAX Modulation Schemes.

2. Related work

In (Hamalainen et al., 2002) the coexistence of the UWB system with GSM900, UMTS/WCDMA, and GPS has been investigated. They have evaluated the level of the interference caused by different UWB signals to the three above mentioned systems. Also they have evaluated the performance degradation of UWB systems in the presence of narrow bandwidth interference and pulsed jamming. They gave the bit error rate (BER) of the above mentioned systems for different pulse length.

In (Hamalainen et al., 2004) the coexistence of the UWB system with IEEE802.11a and UMTS in Modified Saleh-Valenzuela Channel has been studied as well as the UWB system performance in the presence of multiband interference. The interference sources considered were WiFi and UMTS operating simultaneously with their maximum system bandwidths. The UWB system under consideration was single band and single user operating at a data rate of 100 Mbps without error correction coding. They gave the bit error rate (BER) of the UWB system for different types of modulation (Direct Sequence and Time Hopping).

The interference between the UMTS and the UWB systems has been studied in (Giuliano et al, 2003). The free space propagation model was used to calculate the UWB signal propagation loss. It has been concluded that the minimum allowable central frequency value for UWB device, transmitting at 100 Mbps, has to be 3.5 GHz in order to avoid harmful interference with UMTS. In (Hamalainen et al., 2001a), the effect of the in band interference caused by different types of UWB signal to the UMTS/WCDMA uplink and downlink was investigated. UWB frequency spectra have been produced by using several types of narrow pulse waveforms. They have concluded that one can reduce interfering UWB power by using different waveforms and pulse widths avoiding the UMTS frequencies without any additional filtering. In (Hamalainen et al., 2001 b) the effect of the in band interference power caused by three different types of UWB signals to GPS L1 and GSM-900 uplink band was studied. UWB frequency spectra were generated again using several types of narrow pulse waveforms based on Gaussian pulse. In band interference power has been calculated over the IF bandwidth of the two victim receivers as a function of the UWB pulse width. Also the signal attenuation with distance was presented.

In (Ahmed et al., 2004) the effect of the UWB on the DCS-1800 and GSM-900 macrocell downlink absolute range, using the Line of Sight propagation model between the UWB transmitter and the mobile receiver, was studied (without taking into account the shadowing factor within the propagation loss model).

The effect of the UWB emission on the UMTS and CDMA-450 macrocell downlink performance (range and capacity) has been given in (Ahmed et al., 2008). The effect of the UWB emission on the WiMAX macrocell downlink range has been studied by (Ahmed et al., 2010). In (Chiani et. al., 2009) an overview about the coexistence between UWB and narrow-band wireless communication systems has been presented. In (Chóliz et. al., 2011) the coexistence between UMTS and UWB has been evaluated and cooperative mitigation techniques have been proposed and implemented. In (Das et. al., 2010) an interference cancellation schemes in UWB systems used in wireless personal area network based on wavelet based pulse spectral shaping have been presented.

The effect of the UWB on fixed service system (point to point and Fixed Wireless Access (FWA) systems in bands from 1 to 6 GHz) has been investigated in (ITU, 2003). It was concluded that, when the UWB transmitter is in LOS with the two systems antennas, the effect is very high when the UWB power density is higher than -41.3 dBm/MHz.

3. Effect of UWB Interference on the portable WiMAX downlink range

For each WiMAX downlink channel, the UWB interfering signal is due to only a given part of the total UWB spectrum. To account for UWB interference, an extra source of interference is added to the WiMAX noise. Here we consider the UWB interference as a Gaussian signal. The WiMAX technology is based on Orthogonal Frequency Division Multiplex (OFDM) technique. Thus we will calculate the Signal to interference plus noise (SINR) on a single subcarrier, not in the overall bandwidth.

The interference power is calculated by assuming an UWB interfering source at different distances from the WiMAX receiver. Therefore, the interference power generated by a UWB device, I_{UWB}, is given (in dBm) by:

$$I_{UWB} = P_{UWB} - L_{UWB}(d) + G_{RX\_WiMAX} \tag{1}$$

where:

- P_{UWB} is the UWB Effective Isotropic Radiation Power (EIRP) in dBm in the WiMAX bandwidth.
- $L_{UWB}(d)$ is the path-loss between the UWB device and the WiMAX receiver which varies with the separation distance d in m.
- G_{RX_WiMAX} is the antenna gain of the WiMAX system in the receiving end.

Taking into account that UWB devices are short range, the quasi free space path-loss model with shadowing is often most appropriate, especially when the distance between the UWB transmitter and the mobile receiver is lower than 8 m. Thus, in the WiMAX downlink frequency band, the UWB signal propagation loss $L_{UWB}(d)$, measured in dB at a distance d in meters from the UWB transmitter, is calculated as:

$$L_{UWB}(d) \approx 20\log_{10}\left(\frac{4\pi}{\lambda}\right) + 10\,n\,\log_{10}(d) + N(0,\sigma) \tag{2}$$

Where λ is the operating wavelength at the WiMAX frequency, n is the indoor propagation exponent (1.8 to 2.0) and N(0, σ) is a Gaussian variable of zero mean and a standard deviation of σ, representing the deviation from the path loss mean value (shadowing). Practical values of σ are in the range 1.8 to 3 dB in the line of sight LOS environment. Here we assume that the Gaussian variable N(0, σ) is truncated at ±4σ. In our case σ is assumed to be 2 dB.

In the calculation of the propagation loss of the WiMAX signal we use the two-slope propagation loss model. Thus, for a distance higher than 100 m, the WiMAX signal propagation loss in dB is given as:

$$L_{WiMAX} = A + 10\,\gamma\log_{10}\left(\frac{d}{d_o}\right) + s + L_{glass} + 6\log_{10}\left(\frac{f}{1900}\right) - 10.8\log_{10}\left(\frac{h_{RX}}{2}\right) \tag{3}$$

Where:

- A is the free space propagation loss at a distance of 100 m.
- d is the distance between the WiMAX transmitter and the WiMAX receiver.
- γ is the propagation exponent with a typical value of 3.9 to 4.7.
- s is the shadowing margin assumed to be 10 dB.
- L_{glass} is the wall insertion loss assumed to be 5 dB.
- f is the operating frequency of the WiMAX system given in MHz.
- h_{RX} is the WiMAX antenna height in the receiving end.

The thermal noise of the WiMAX receiver Nrec_sc per subcarrier is given by:

$$N_{rec\_sc}(dBm) = -114 + 10\log_{10}\left(B_c\right)_{MHz} + NF \quad (4)$$

where:

- B_c is the WiMAX band width of a single carrier.
- NF is the WiMAX receiver noise figure in dB assumed to be constant within the WiMAX bandwidth of 20 MHz.

The WiMAX received power per subcarrier S_{WiMAX_sc} is given as:

$$S_{WiMAX\_sc} = P_{WMAX\_sc} + G_{Tx\_WiMAX} - L_{WiMAX} + G_{Rx\_WiMAX} \quad (5)$$

where:

- P_{WiMAX_sc} is the WiMAX transmitted power per subcarrier.
- G_{TX_WiMAX} is the antenna gain of the WiMAX in the transmitting end assumed to be 18 dB (antenna for a macrocell with 3 sectors).

The WiMAX cochannel interference due to the macrocells using the same frequency band that exists within the three nearest clusters of 4 macrocells is given by:

$$I_{cc-WiMAX} \approx S_{WiMAX\_sc} + 10\log_{10} 3\left(\frac{d}{\sqrt{12}\,R}\right)^{\gamma} \quad (6)$$

where R is the radius of the WiMAX macrocell.

For the UWB system the propagation loss with 99.995% confidence is given by:

$$L_{UWB}(d) \approx 20\log_{10}\left(\frac{4\pi}{\lambda}\right) + 10\,n\log_{10}(d) - 4\,\sigma \quad (7)$$

For the WiMAX receiver, the signal to interference plus noise ratio SINR per subcarrier is given by:

$$SINR = 10\log_{10}\left(\frac{S_{WiMAX\_sc}}{I_{CC-WiMAX} + N_{rec\_sc} + I_{UWB}}\right) \quad (8)$$

where $I_{cc-WiMAX}$ is the WiMAX cochannel interference, N_{rec} is the receiver thermal noise, and I_{UWB} is the UWB interference all given in real numbers.

4. Effect of UWB Interference on the UMTS and CDMA-450 macrocell downlink performance

To account for the UWB interference, an extra source of interference is added linearly to the UMTS and the CDMA-450 intra-system interference. The interference power is calculated by assuming the UWB source to be at different distances from the UMTS receiver (the mobile station). Therefore, the interference power generated by a UWB device, I_{UWB}, is given by (in dBm):

$$I_{UWB} = P_{UWB} - L_{UWB}(d) + G_{UMTS} \tag{9}$$

Where:

- P_{UWB} is the mean UWB EIRP in dBm in the UMTS band.
- $L_{UWB}(d)$ is the path-loss between the UWB device and the UMTS receiver which varies with the separation distance, d in m, and
- G_{UMTS} is the UMTS antenna gain.

As the UWB devices are typically low power and short range devices the line-of-sight path-loss model is often most appropriate. Then the UWB signal propagation loss in dB is calculated as:

$$L_{UWB}(d) \approx 39.03 + 20\log_{10}(d) - 4\sigma \tag{10}$$

The effect of the UWB interference is to reduce the UMTS macrocell range or/and the macrocell capacity.

The normalized range is given as (Ahmed et. al., 2008):

$$\frac{R_{UMTS}}{R_{UMTS,o}} = \left(\frac{I_{UMTS}}{I_{UMTS} + I_{UWB}}\right)^{1/\gamma} \tag{11}$$

The normalized capacity C_n is given as (Ahmed et. al., 2008):

$$C_n = \left(\frac{I_{UMTS}}{I_{UMTS} + I_{UWB}}\right) \tag{12}$$

The interference power generated by a UWB device that affects the CDMA-450 receiver, IUWB, is given by (in dBm):

$$I_{UWB} = P_{UWB} - L_{UWB}(d) + G_{CDMA} \tag{13}$$

where:

- P_{UWB} is the UWB EIRP in dBm in the CDMA-450 band.
- $L_{UWB}(d)$ is the path-loss between the UWB device and the CDMA-450 receiver which varies with the separation distance, d in m, and
- G_{CDMA} is the CDMA-450 antenna gain .

In the frequency band used by CDMA-450, the UWB signal propagation loss in dB is calculated as:

$$L_{UWB}(d) \approx 25.7 + 20\log_{10}(d) - 4\sigma \tag{14}$$

The normalized range is now given by (Ahmed et. al., 2008):

$$\frac{R_{CDMA}}{R_{CDMA,o}} = \left(\frac{I_{CDMA}}{I_{CDMA} + I_{UWB}} \right)^{1/\gamma} \tag{15}$$

where:

- $R_{CDMA,o}$ is the CDMA-450 macrocell initial range without the UWB interference.
- R_{CDMA} is the CDMA-450 macrocell range with the existence of the UWB interference.

The normalized capacity of the CDMA-450 system Cn is given by (Ahmed et. al., 2008):

$$C_n = \left(\frac{I_{CDMA}}{I_{CDMA} + I_{UWB}} \right) \tag{16}$$

5. Results for a WiMAX system and a single UWB interferer

Fig. 1 represents the scenario of the studied WiMAX system. It shall be mentioned that the receiver is an indoor portable WiMAX. The UWB transmitter is also indoor within a distance of 0.5 to 5 m from the WiMAX receiver.

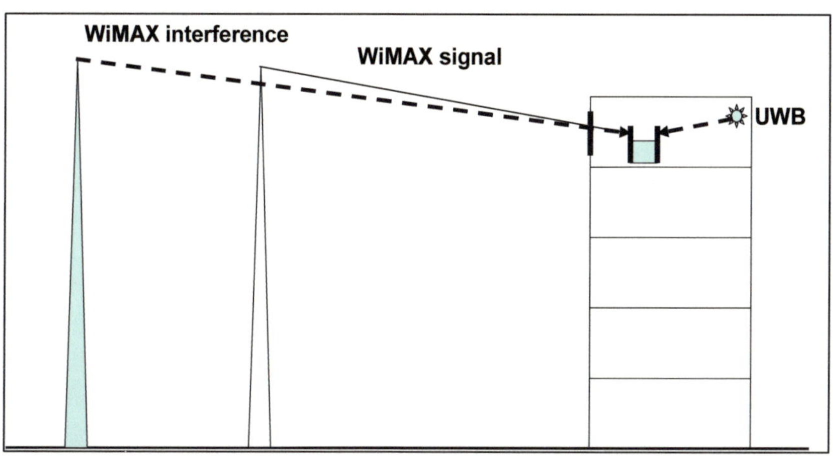

Figure 1. WiMAX studied scenario.

Let us study the case of 3.5 GHz WiMAX assuming that the WiMAX transmission power is 40 dBm/sector. Fig. 2 shows the WiMAX downlink modulation modes, as a function of distance between the WiMAX transmitter and receiver, for three different UWB power densities. It can be noticed that, without UWB interference, the WiMAX will have a range of 1481 m for the second modulation scheme. With a UWB power density of -88.5 dBm/MHz the range will be reduced by 2% and with a UWB power density of -41.3 dBm/MHz (recommended by FCC), the range will be only 213 m. Such a reduction drastically degrades the WiMAX performance.

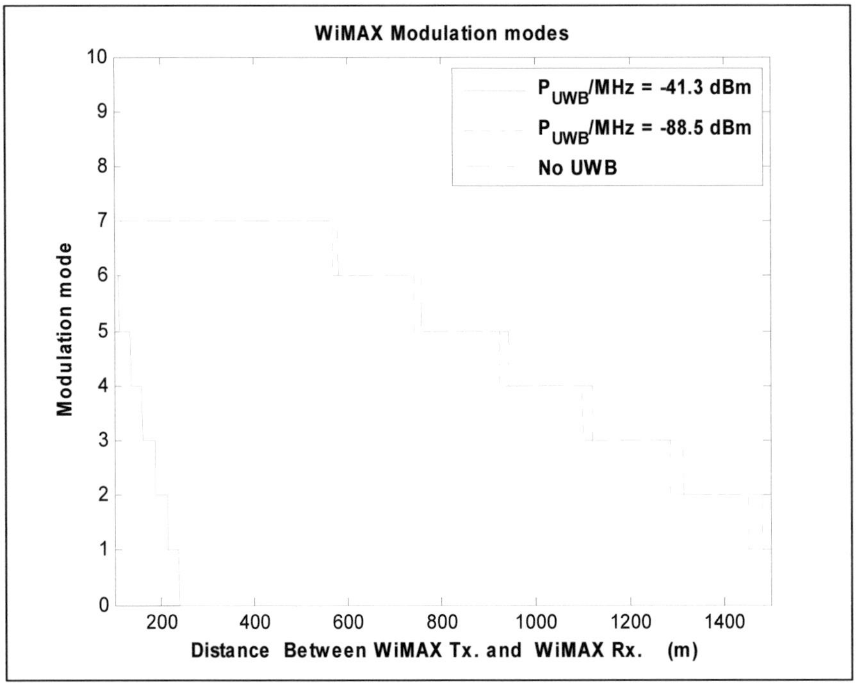

Figure 2. 3.5 GHz WiMAX modulation modes for different UWB power densities with a 1 m distance between the UWB transmitter and the WiMAX receiver assuming a WiMAX transmitted power of 40 dBm/sector.

Let us consider now the case when the WiMAX signal and also the interference are received through an open window. Fig. 3 shows the WiMAX downlink modulation modes, again as a function of distance between the WiMAX transmitter and receiver, for three different UWB power densities. As can be seen, without UWB interference, the WiMAX will have a range of 1930 m for the second modulation scheme. With a UWB power density of -88.5 dBm/MHz the range will be reduced by 2%. And for a UWB power density of -41.3 dBm/MHz (recommended by FCC), the range will be 310 m. Again the WiMAX range performance is drastically degraded.

Let us now study the case presented in Fig. 2 but assuming this time that the maximum allowed WiMAX reduction range is 1%. Fig. 4 shows the WiMAX downlink modulation modes as a function of distance between the WiMAX transmitter and receiver for three different UWB power densities. It is clearly seen that, without UWB interference, the WiMAX will have a range of 1481 m for the second modulation scheme. The range will be reduced by 1% when the interfering UWB power density is -91.5 dBm/MHz.

Let us consider now the case when the WiMAX system operates in the 2.5 GHz band. Fig. 5 shows the WiMAX downlink modulation modes as a function of distance between the

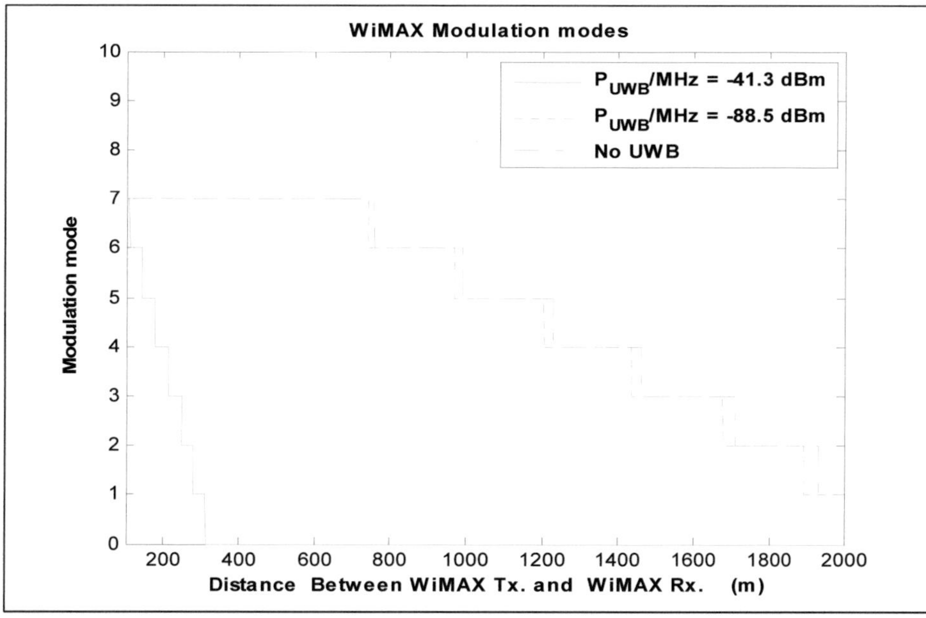

Figure 3. 3.5 GHz WiMAX modulation modes for different UWB power densities with a 1 m distance between the UWB transmitter and the WiMAX receiver assuming a WiMAX transmitted power of 40 dBm/sector and that the WiMAX signal and interference are received through an open window.

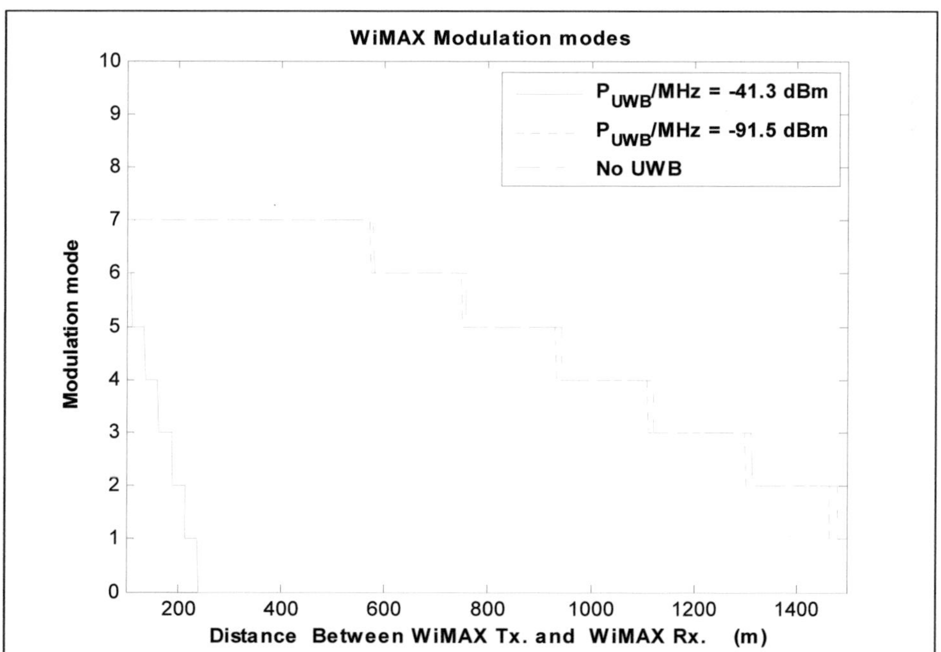

Figure 4. 3.5 GHz WiMAX modulation modes for different UWB power densities with a 1 m distance between the UWB transmitter and the WiMAX receiver assuming a WiMAX transmitted power of 40 dBm/sector.

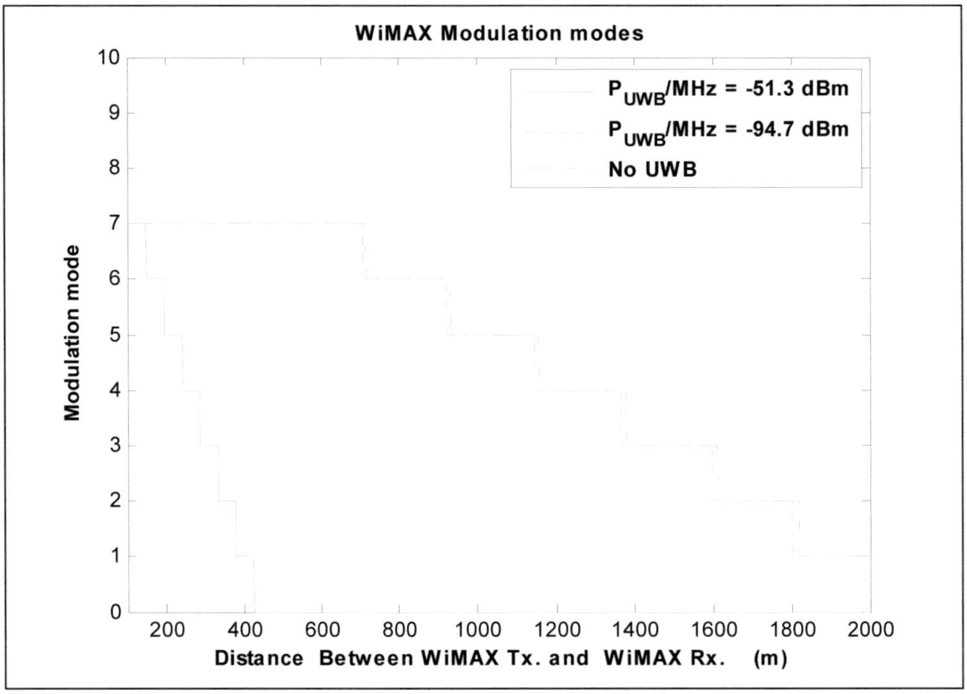

Figure 5. 2.5 GHz WiMAX modulation modes for different UWB power densities with a 1 m distance between the UWB transmitter and the WiMAX receiver assuming a WiMAX transmitted power of 40 dBm/sector.

WiMAX transmitter and receiver for three different UWB power densities. Notice that, without UWB interference, the WiMAX will have a range of 1817 m for the second modulation scheme. With a UWB power density of -94.7 dBm/MHz the range will be reduced by 1%. For a UWB power density of -51.3 dBm/MHz (recommended by FCC), the range will be 378 m and such a reduction represents a drastic degradation of the WiMAX performance. In this case an UWB with a power density of -91.5 dBm/MHz will reduce the WiMAX range by 2%.

6. Results for a WiMAX system and multi UWB interferers

We will consider now the case of multi-UWB transmitters, assuming the case that 4 UWB are located at a distance of 1m from the WiMAX receiver. Fig. 6 shows the WiMAX downlink modulation modes as a function of the distance between the WiMAX transmitter and receiver (WiMAX link length) for three different UWB power densities. It can be noticed that, without UWB interference, the WiMAX will have a range of 1481 m for the second modulation scheme. The range will be reduced by 1% when the UWB power density is higher than -97.5 dBm/MHz. In this case, an UWB power density of -94 dBm/MHz will reduce the WiMAX range by 2%.

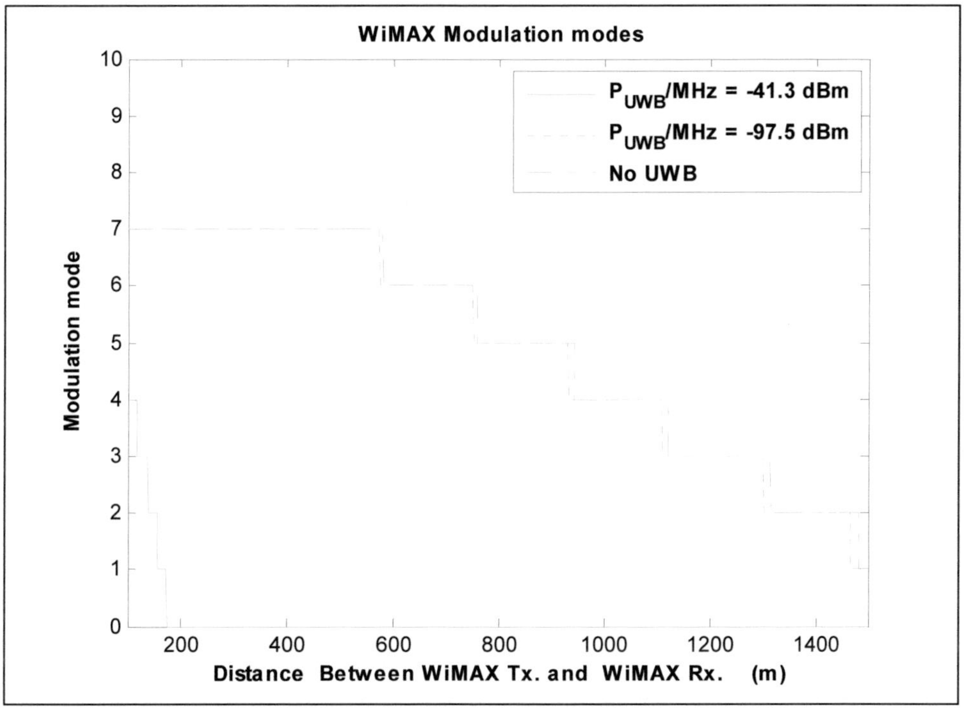

Figure 6. 3.5 GHz WiMAX downlink modulation modes versus distance between the WiMAX transmitter and receiver, for different UWB power densities from 4 UWB transmitters at 1 m distance to the WiMAX receiver and assuming a WiMAX transmitted power of 40 dBm/sector.

A band rejection up to 56 dB is needed for the DS-CDMA UWB system, while for the MB-OFDM UWB system a 51 dB band rejection is needed and can be obtained by nulling 16 subcarriers with a 40 dB notch filter.

We study now the same scenario but for the 2.5 GHz WiMAX. Fig. 7 shows the WiMAX downlink modulation modes as a function of the distance between the WiMAX transmitter and receiver for three different UWB power densities. It can be noticed that for the second modulation scheme without UWB interference, the WiMAX will have a range of 1817 m. At a UWB power density of -100.7 dBm/MHz, WiMAX range will be reduced by 1%.

In summary, from the results presented in Figures 2, 3, 4 and 5 it can be concluded that the power density of -41.3 dBm/MHz recommended by FCC, implies a very high range reduction, unless Detect and Avoid (DAA) techniques are implemented.

Fig. 8 represents the DAA requirement for Multiband OFDM UWB (MB-OFDM UWB) system and the Direct Sequence CDMA system (DS-CDMA UWB), with activity factors (fraction of the time they work at the 3.5 GHz band) of 32% and 100% respectively.

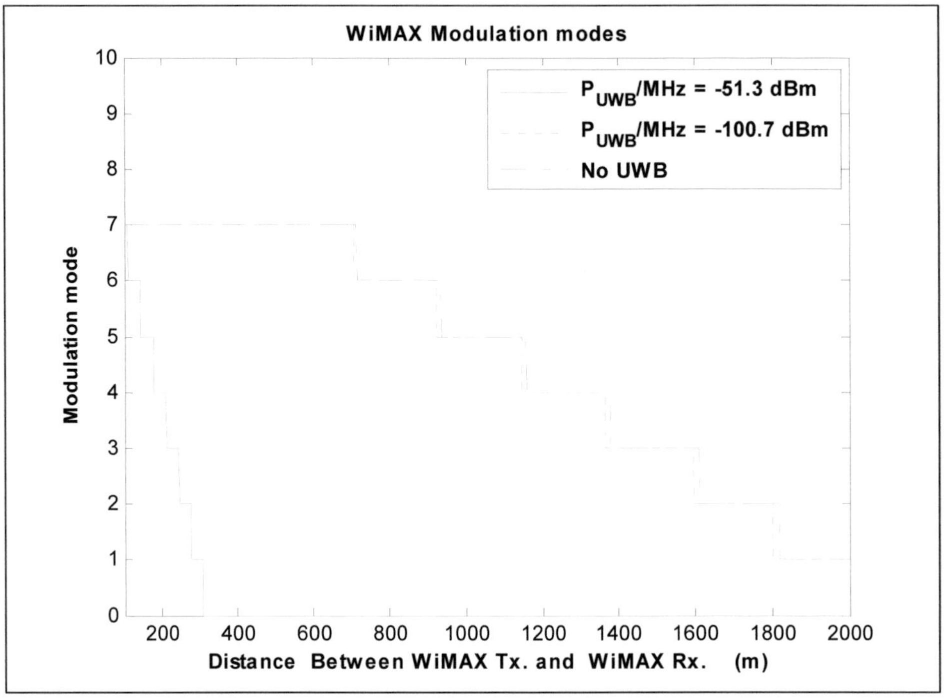

Figure 7. 2.5 GHz WiMAX modulation modes for different UWB power densities with a 1 m distance between the 4 UWB transmitters and the WiMAX receiver assuming a WiMAX transmitted power of 40 dBm/sector.

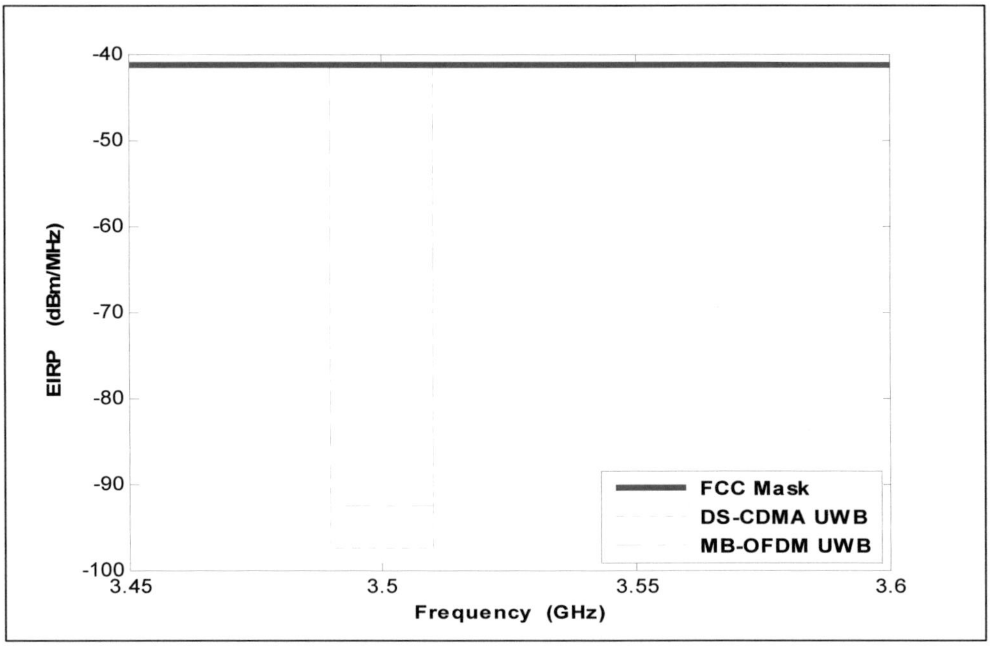

Figure 8. DAA requirements within the 3.5 GHz band.

7. Results for UMTS or CDMA-450 systems and single UWB interferer

Let us now study the coexistence of UWB systems with the UMTS (working at the 2 GHz band) and CDMA-450 systems. In the analysis we assume that the UWB data rate is higher than the UMTS or CDMA-450 chip rate. In this case, the UWB interference can be considered as a Gaussian noise. We address here the effect that the UWB system produces on the downlink of the UMTS and CDMA-450 systems. In Fig. 9, the UWB interference power on the UMTS downlink (i.e. interference as seen at the mobile) is plotted assuming a UWB power density (P_{UWB}) of -51.3 dBm/MHz within the UMTS bandwidth.

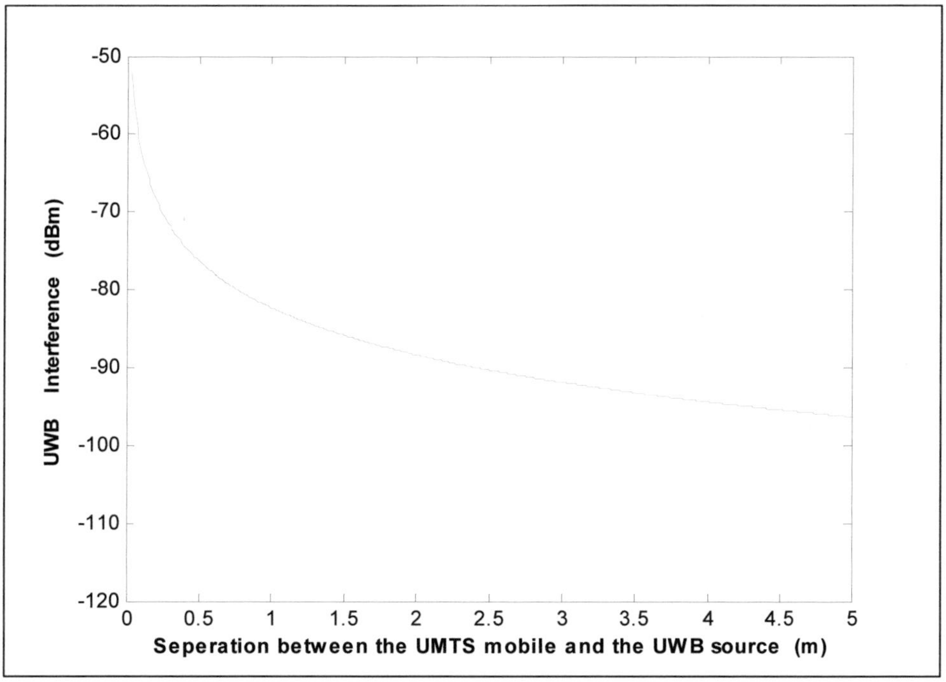

Figure 9. UWB interference as a function of the separation between the UWB transmitter and the UMTS mobile (P_{UWB} = -51.3 dBm/MHz).

Lets us study now the case of voice service [G_p = 256 and $(E_b/N_o)_{req}$ = 6 dB] assuming an UMTS interference of -88 dBm (14 dB Rise-Over-Thermal ROT). Fig. 10 shows the downlink macrocell normalized range as a function of the separation between the UMTS mobile and the UWB transmitter for three different values of the propagation exponent γ. It can be noticed that the UWB signal creates a high interference (which reflects a macrocell normalized range reduction of 35.6%) when the separation is 1 m. For larger separation, the interference is lower and thus the range reduction is also lower.

Fig. 11 shows the downlink macrocell normalized capacity as a function of the separation between the UMTS mobile and the UWB transmitter. It can be noticed that the UWB signal creates a high interference (which reflects a macrocell normalized capacity reduction of 78.6%) when the separation is 1 m. For larger separation, the interference is lower and thus the normalized capacity reduction is also lower.

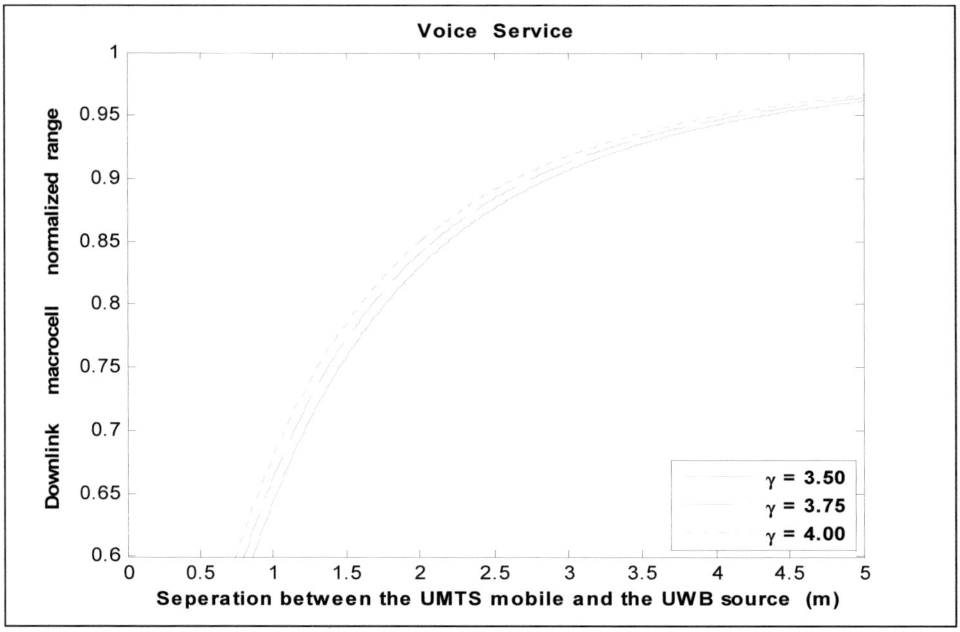

Figure 10. Effect of the UWB interference on the macrocell range as a function of the separation between the UWB transmitter and the UMTS mobile (P_{UWB} = -51.3dBm/MHz).

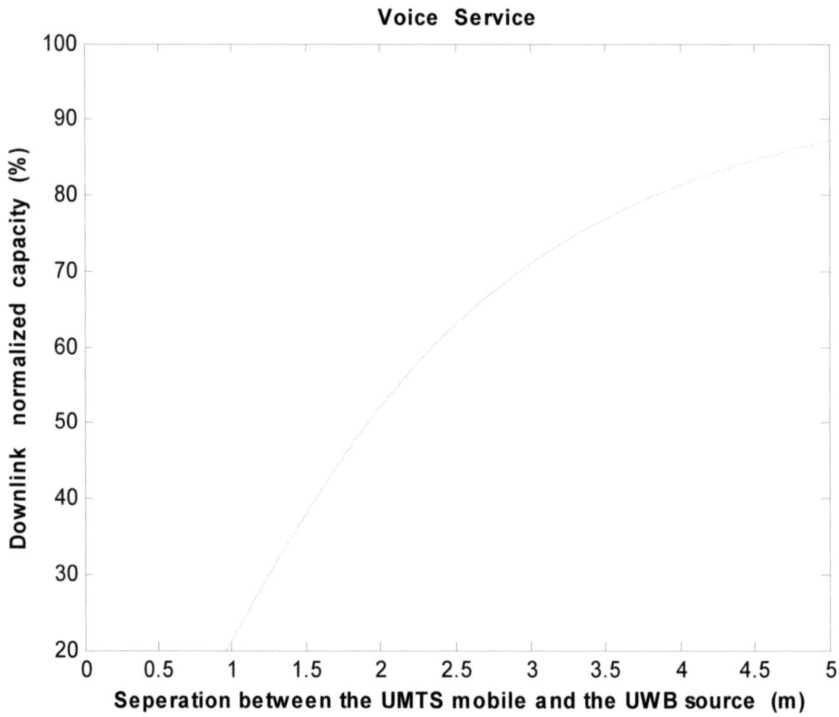

Figure 11. Effect of the UWB interference on the macrocell normalized capacity as a function of the separation between the UWB transmitter and the UMTS mobile (PUWB = -60 dBm/MHz).

Next let us study the data service case [G_p = 32 dB and $(E_b/N_o)_{req}$ = 5 dB] assuming an UMTS total interference of -92.5 dBm (9.5 dB Rise-Over-Thermal ROT), representing a highly loaded macrocell. Fig. 12 shows the downlink macrocell normalized range as a function of the separation between the UMTS mobile and the UWB transmitter for three different values of the propagation exponent s. It can be noticed that the UWB signal creates a high interference (which reflects a high macrocell normalized range reduction of 50.5%) when the separation is 1m. For larger separation, the interference is lower and thus the range reduction is also lower.

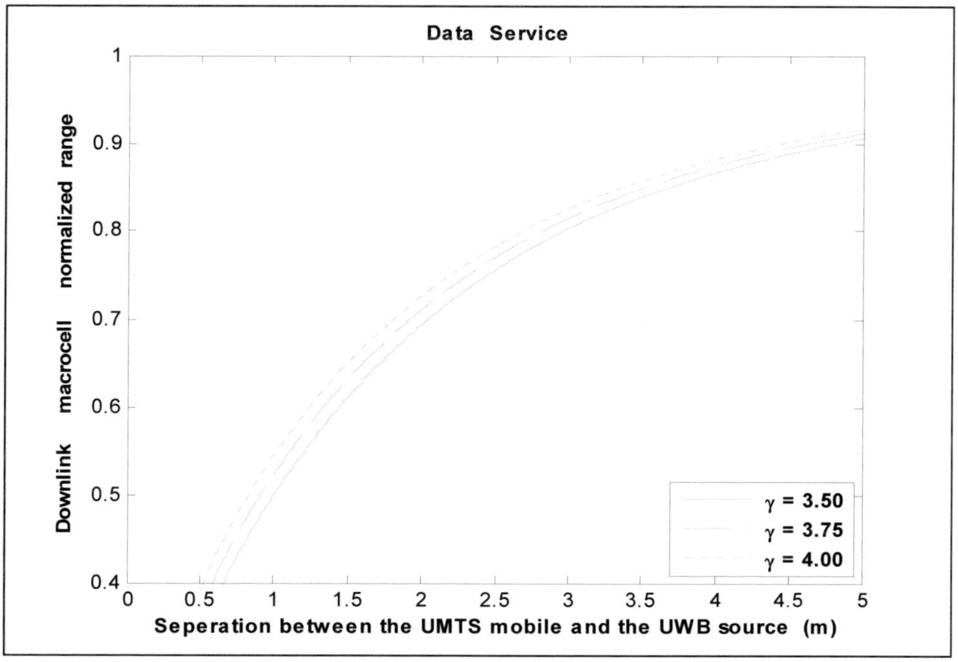

Figure 12. Effect of the UWB interference on the macrocell normalized range as a function of the separation between the UWB transmitter and the UMTS mobile (P_{UWB} = -51.3 dBm/MHz).

Fig. 13 shows the downlink macrocell normalized capacity as a function of the separation between the UMTS mobile and the UWB transmitter. It can be noticed that the UWB signal creates a high interference (which reflects a high macrocell normalized capacity reduction of 91%) when the separation is 1 m. For larger separation, the interference is lower and thus the normalized capacity reduction is also lower.

It is obvious that such reductions (in range and capacity) are unacceptable. Thus the EIRP power density should be reduced to get an acceptable range and capacity reduction.

Let us consider now the data service case assuming a P_{UWB} of -81.4 dBm/MHz. Fig. 14 shows the downlink macrocell normalized range as a function of the separation between the UMTS mobile and the UWB transmitter. It can be noticed that the UWB signal creates a high interference (which reflects a high macrocell normalized range reduction) when the separation is lower than 0.25 m. For larger separation, the interference is lower and at a distance higher than 1m, the effect of the interference is quasi null.

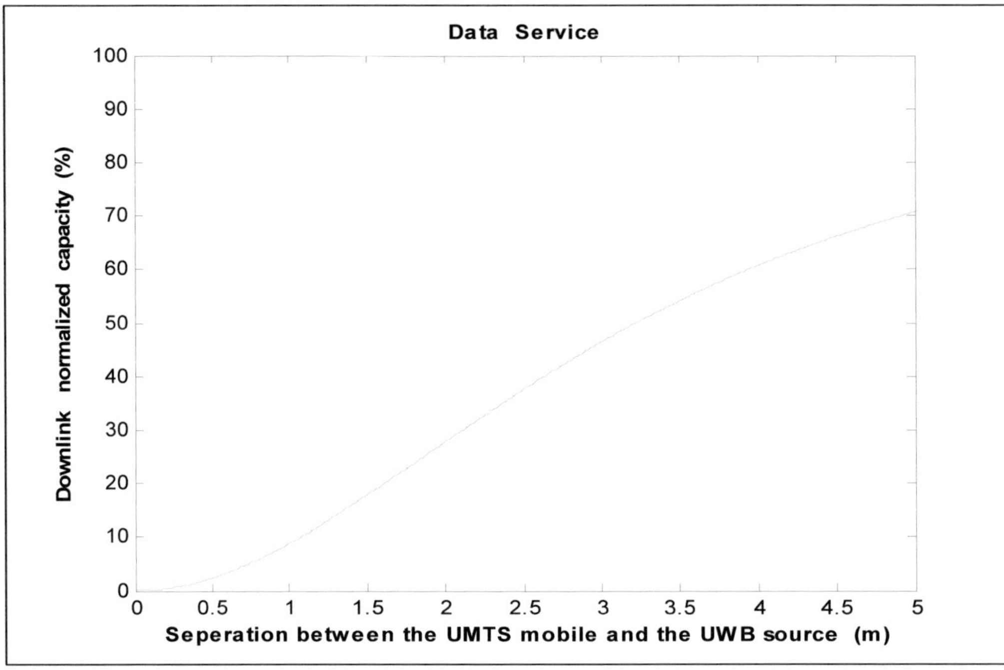

Figure 13. Effect of the UWB interference on the macrocell normalized capacity as a function of the separation between the UWB transmitter and the UMTS mobile (P_{UWB} = -51.3 dBm/MHz).

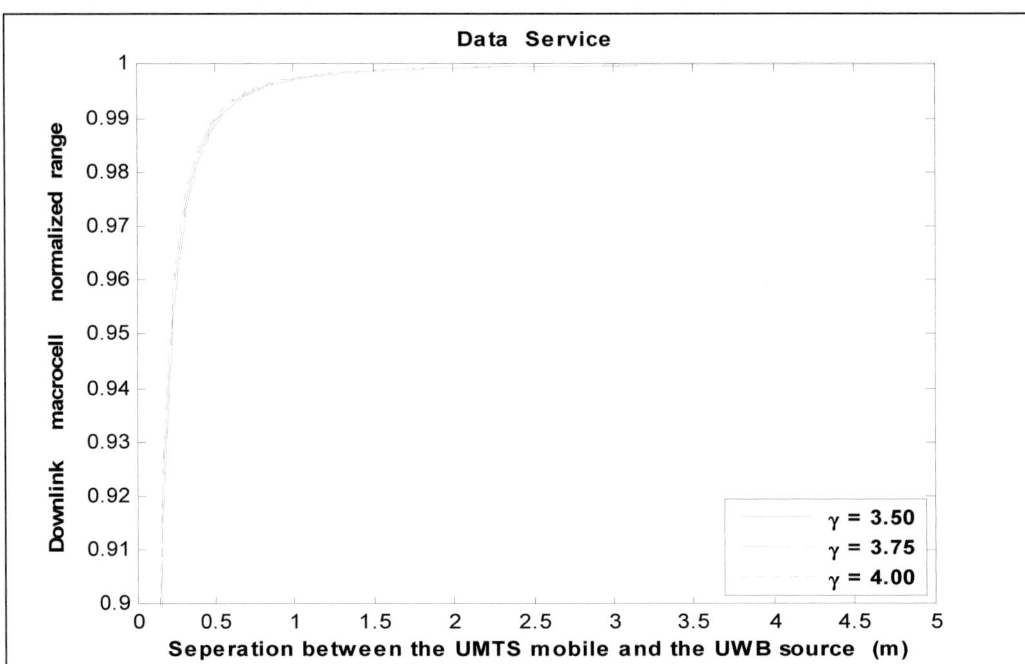

Figure 14. Effect of the UWB interference on the macrocell range as a function of the separation between the UWB transmitter and the UMTS mobile (P_{UWB} = -81.4 dBm/MHz).

Fig. 15 shows the downlink macrocell capacity as a function of the separation between the UMTS mobile and the UWB transmitter. It can be noticed that the UWB signal creates a high interference (which reflects a high macrocell capacity reduction) when the separation is less than 0.4 m. For larger separation, the interference is lower and at a distance higher than 1.0 m, the effect of the interference is to reduce the cell capacity by 1%.

Next we study the case of data service (G_p = 32 dB and $(E_b/N_o)_{req}$ = 5 dB) of the CDMA-450 3X assuming that the CDMA-450 total interference of -92.5 dBm (9.5 dB Rise-Over-Thermal ROT) and UWB power density of -95 dBm/MHz. The frequency of operation is assumed to be 450 MHz.

Fig. 16 shows the CDMA-450 downlink macrocell normalized range as a function of the separation between the CDMA mobile and the UWB transmitter. It can be noticed that the UWB signal creates a low interference when the separation is 1m which reflects a normalized range reduction of less than 0.3%.

Fig. 17 shows the CDMA-450 downlink macrocell normalized capacity as a function of the separation between the CDMA-450 mobile and the UWB transmitter. It can be noticed that the UWB signal creates a low interference when the separation is 1m which reflects a normalized capacity reduction of 1%.

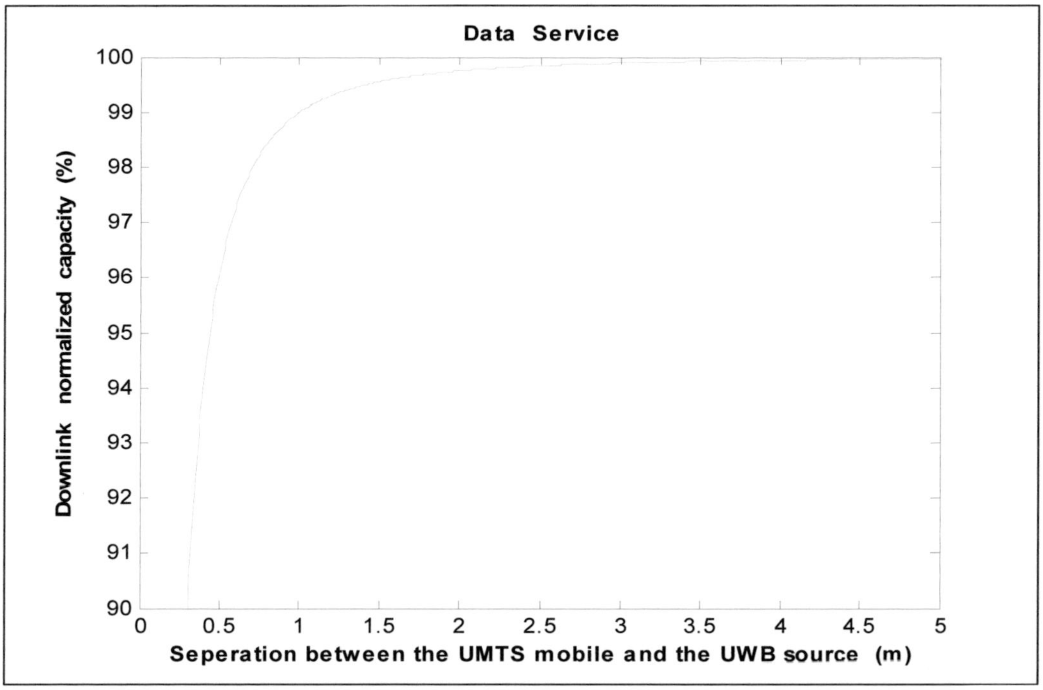

Figure 15. Effect of the UWB interference on the macrocell normalized capacity as a function of the separation between the UWB transmitter and the UMTS mobile (P_{UWB} = -81.4 dBm/MHz).

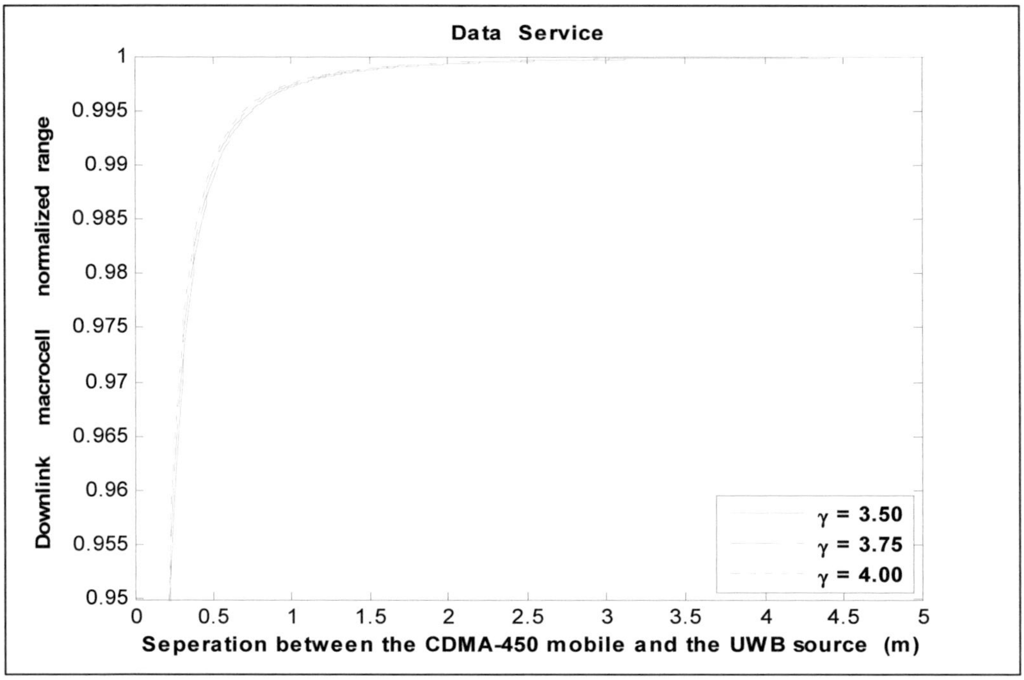

Figure 16. Effect of the UWB interference on the macrocell normalized range as a function of the separation between the UWB transmitter and the CDMA450 mobile (P_{UWB} = -95 dBm/MHz).

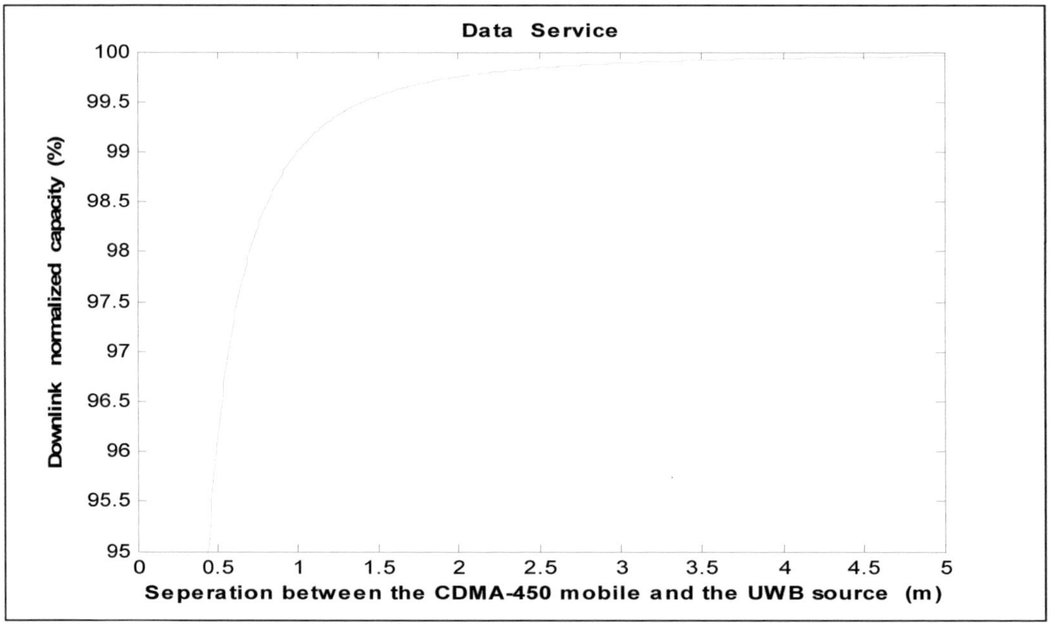

Figure 17. Effect of the UWB interference on the macrocell normalized capacity as a function of the separation between the UWB transmitter and the CDMA450 mobile (P_{UWB} = -95 dBm/MHz).

8. Results for a UMTS or CDMA-450 systems and multi UWB interferers

Then we study the case of multiple UWB transmitters with four UWB transmitters at a distance of 1m around the UMTS receiver. Fig. 18 shows the downlink macrocell normalized range as a function of the EIRP power density in dBm/MHz. It can be noticed that the cell range reduction is always lower than 1%.

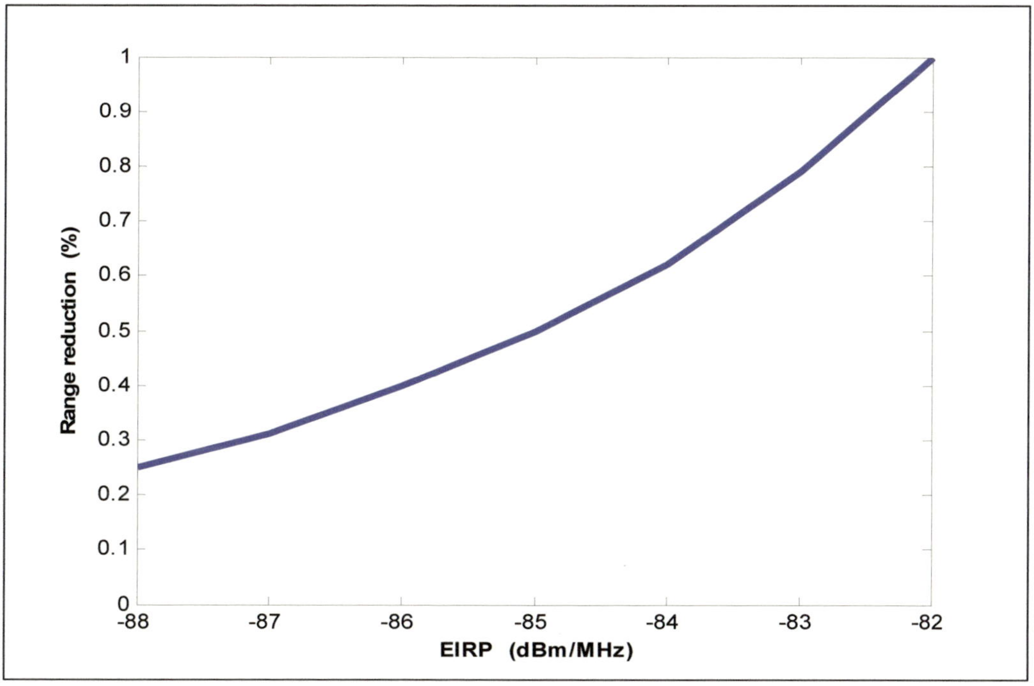

Figure 18. Range reduction as a function of the EIRP in (dBm/MHz) for multi UWB transmitters.

Fig. 19 shows the downlink macrocell normalized capacity as a function of the EIRP power density in dBm/MHz. It can be noticed that, for a capacity reduction of only 1%, EIRP should be -87.4 dBm/MHz. this represents a 6 dB reduction equal to [$10\log_{10}(4)$], where 4 is the number of the UWB sources. The conclusion is that, for the case of single UWB transmitter, the UMTS can easily tolerate the UWB interference when the UWB EIRP is lower than -81.4 dBm/MHz for 1m distance between the UWB transmitter and the UMTS mobile. For the multi UWB transmitter case, the UMTS can easily tolerate the UWB interference when the UWB EIRP is -87.4 dBm/MHz. When using a CDMA-450 system the maximum allowed EIRP reduces to -101 dBm/MHz.

Table 2 presents the maximum allowed EIRP for different frequency bands, for UWB activity factor of 100% and multi UWB transmitter scenario, for two different cases, (case A with 99.995% confidence and case B with 99% confidence respectively). Table 3 represents the maximum allowed EIRP for different frequency bands for UWB activity factor of 10% and multi UWB transmitter scenario for the two previous cases A and B.

It shall be mentioned that if the critical distance is reduced from the 1m already considered down to 0.5m, the maximum accepted UWB power densities should be decreased by 6 dB from the values given before.

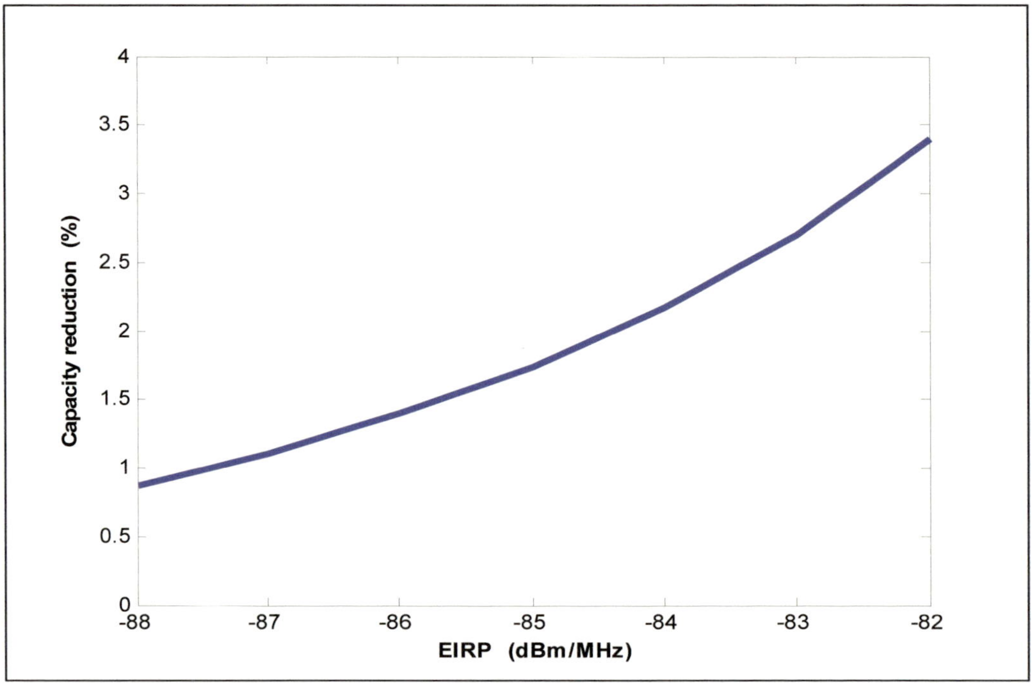

Figure 19. Capacity reduction as a function of the EIRP in (dBm/MHz) for multi UWB transmitters.

Frequency band in GHz	Maximum allowed UWB EIRP in (dBm/MHz) Case A		Case B
3.3-3.8	-98.0	(with DAA)	-94.7
2.5-2.7	-100.7		-97.4
2.1-2.2	-87.4		-84.1
0.43-0.47	-101.0		-97.7

Table 2. Maximum allowed EIRP for different frequency bands with UWB activity factor of 100%.

Frequency band in GHz	Maximum allowed UWB EIRP in (dBm/MHz) Case A		Case B
3.3-3.8	-88.0	(with DAA)	-84.7
2.5-2.7	-90.7		-87.4
2.1-2.2	-77.4		-74.1
0.43-0.47	-91.0		-87.7

Table 3. Maximum allowed EIRP for different frequency bands with UWB activity factor of 10%.

9. Conclusions

The coexistence of UWB with 3G and 4G Cellular Systems has been studied in this chapter. In particular UMTS in the 2 GHz and in the 450 MHz (CDMA-450) frequency bands have been selected as examples of 3G cellular systems and the WiMAX system as example of 4G.

The methodology used to account for the impact of UWB interference on the coverage range and capacity of the interfered systems has been explained in detail. Finally it has been applied in a set of study cases in scenarios involving the 3G and 4G selected systems.

From the above given results we can conclude that the spectrum mask proposed by the FCC for indoor application (-51 dBm/MHz in the UMTS band and -41 dBm/MHz for the CDMA-450 band) is very high and cannot be tolerated by the mobile systems. From the results obtained we conclude that another spectrum mask with lower UWB power density has to be used.

Author details

Ahmed Bazil Taha
Universidad Autonoma de Madrid, Spain

Miguel Calvo Ramon
Universidad Politecnica de Madrid, Spain

10. References

[1] Ahmed B. T., Ramon M. C. & Ariet L. H., 2004, "Impact of Ultra Band (UWB) on Macrocell Downlink of DCS-1800 and GSM-900 Systems", Radioenginnering, Vol. 14, No.1, pp. 51-55.

[2] Ahmed B. T., Ramón M. C., 2008, "On the Impact of Ultra-Wideband (UWB) on Macrocell Downlink of UMTS and CDMA-450 Systems", IEEE Electromagnetic Compatibility, Vol. 5, No. 2, pp. 406-412.

[3] Ahmed B. T., Campos JL. M., Cruz J. C, 2010, "Impact of Ultra Wide Band emission on WiMAX systems at 2.5 and 3.5 GHz", Computer Networks, Vol. 54, pp.1573-1583.

[4] Ciccoganini W., Durantini A., and Cassioli D., 2005, "Time domain propagation measurements of the UWB Indoor Channel Using PN-Sequence in the FCC-Compliant Band 3.6-6 GHz", IEEE trans. Antennas and Propagation, Vol. 53, No. 4, pp. 1542-1549.

[5] Giuliano R., Mazzenga F., Vatalaro F., "On the interference between UMTS and UWB systems", pp: 339 – 343, IEEE Conference on Ultra Wideband Systems and Technologies, 2003 , 16-19 Nov. 2003.

[6] Chiani M, and Giorgetti A, "Coexistence between UWB and Narrow-Band Wireless Communication Systems", Proceedings of the IEEE, Vol. 97, No. 2, February 2009, pp. 231-254.

[7] Chóliz J, Hernández A, Alastruey I and Valdovinos A, "Coexistence and interworking between UMTS and UWB: a performance evaluation of a UMTS/UWB interoperability platform", Telecomunication systems, Vol. 49, No. 4, pp. 409-420.

[8] Das B , Das Ch.S. and S. Das S, "Interference Cancellation Schemes in UWB Systems used in Wireless Personal Area Network based on Wavelet based Pulse Spectral shaping and Transmitted Reference UWB using AWGN Channel Model ", International Journal of Computer Applications Volume 2 – No.2, pp. 88.93, May 2010.

[9] Hamalinen M., Hovinen V., Iinatti J., Latva-aho M., 2001: "In band Interference Power Caused by Different Kinds of UWB Signals at UMTS/WCDMA Frequency Bands", , the 2001 IEEE Radio and Wireless Conference, RAWCON 2001, pp. 97-100, Waltham-Boston, Massachusetts, USA, Aug. , 2001.

[10] Hamalinen M., Iinatti J., Hovinen V., Latva-aho M., 2001," In band Interference of Three Kind of UWB Signals in GPS L1 Band and GSM900 Uplink Band", the 12th International Symposium on Personal, Indoor and Mobile Radio Communications, PIMRC2001, pp. D 76-80, USA, Sep - Oct , 2001.

[11] Hamalainen M. , Hovinrn V. , Tesi R., Iinatti J. & Latava-aho M., 2002, " On the UWB System Coexistance with GSM900, UMTS/WCDMA, and GPS", IEEE Journal on Selected Areas in Communications, Vol. 20, No. 9, pp. 1712-1721.

[12] Hamalinen M., Tesi R., Iinatti J.," UWB co-existence with IEEE802.11a and UMTS in modified Saleh-Valenzuela channel", Ultra Wideband Systems, 2004, pp. 45 – 49, May 2004.

[13] Holma H. , Toskala A., 2000, "WCDMA for UMTS", John Wiley & Sons.

[14] ITU Document 1-8/29-E, 2003, "Updating of Preleminary Study on Coexistance Betwwen UWB and the Fixed Service in Band from 1 to 6 GHz".

Chapter 15

Performance Assessment of UWB-Over-Fiber and Applications

João M.B. Oliveira, Luís M. Pessoa, Diogo Coelho,
Henrique M. Salgado and Jorge C.S. Castro

Additional information is available at the end of the chapter

http://dx.doi.org/10.5772/48748

1. Introduction

Since 2002, the Federal Communications Commission (FCC) authorized the use of ultra-wideband (UWB) signal transmissions for unlicensed use, in the range from 3.1 to 10.6 GHz, leading to a revived interest in research activities and to new opportunities for companies to explore and develop new broadband indoor and outdoor applications [1]. Moreover, UWB is seen as a promising technology for short range high speed wireless networks.

UWB signals are characterized by their huge bandwidth occupancy, high data rates, and very weak power density (−41.3 dBm/MHz), which gives them a noise-like signal characteristic, facilitating both interference mitigation and very low device power consumption. On the other hand, its very low intensity and high data rates limit the coverage to a few meters distance. Yet, by using radio-over-fiber (RoF) as a signal transportation technique, it is possible to deliver UWB signals over a fiber based network.

The radio-over-fiber (RoF) concept involves the transmission of RF signals by an optical fiber between a control station (CS) and a number of base stations (BSs). In the base stations, the RF signal is transmitted to end users by a wireless link. Integration of both optical and wireless broadband infrastructures into the same backhaul network leads to a significant simplification and cost reduction of BSs since all routing, switching and processing are shifted to the CS. This centralization of signal processing functions enables equipment sharing, dynamic allocation of resources, and simplified system operation and maintenance. The concept of RoF is shown in Figure 1 in an in-building network context.

RoF systems are (ideally) transparent to all signals transmitted in the optical fiber. It has been experimentally shown that RoF networks are well suited to simultaneously transport several wireless standards like wideband code division multiple access (WCDMA), IEEE 802.11 wireless local area network (WLAN) [2], global system for mobile communications

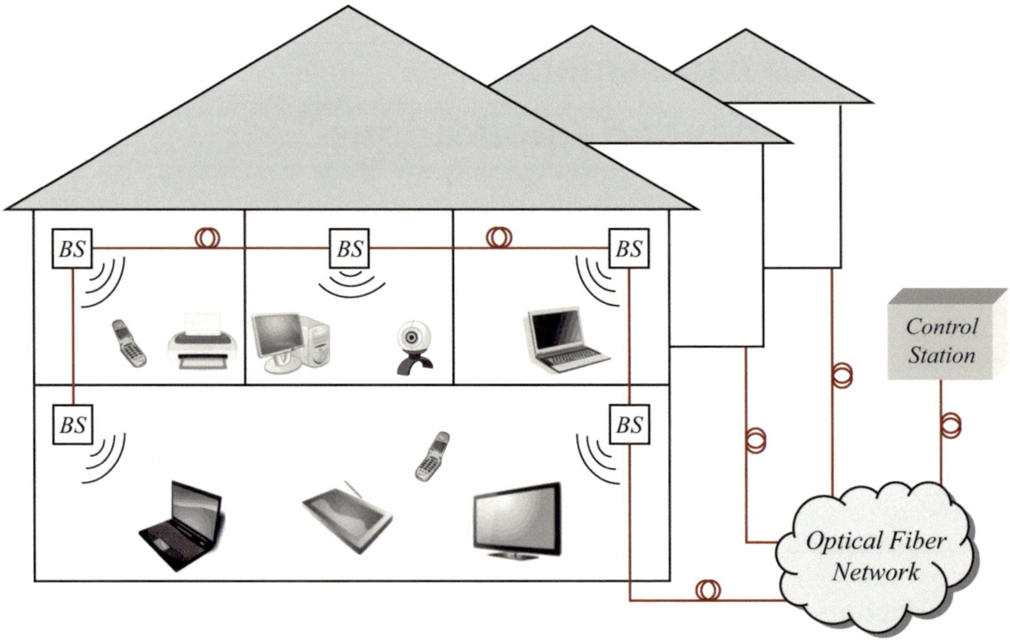

Figure 1. In-building radio-over-fiber concept.

(GSM) [3], WiMAX [4] and ultra-wide band (UWB) [5, 6]. Moreover, RoF systems also offer other attractive advantages such as low weight and immunity to electromagnetic interference. However, an optically modulated mm-wave signal can also suffer from several impairments namely nonlinear distortion, power penalty from the electric/optic/electric (E/O/E) conversion process, chromatic dispersion, attenuation from the optical fiber and phase noise from LASER sources.

Here we study the transmission performance of UWB over two distinctive optical networks. In Section 2, the packet error rate performnace in a low-cost multi-mode fiber (MMF) network composed by a 850 nm vertical-cavity surface emitting LASER (VCSEL) and a PIN photodiode connected by two different polymer optical fibers (POF) is assessed. Then, in Section 3, an analytical and experimental performance evaluation is carried out in a single-mode fiber (SMF) network composed by a reflective electro-absorption modulator (R-EAM) and a PIN photodiode.

2. UWB over perfluorinated graded index polymer optical fiber for low-cost in-building networks

The use of multimode fibers (MMF) in RoF short span networks has attracted much attention in recent years. Nowadays, the majority of building networks are based on MMF. In fact, in-building networks employing MMF topologies for high speed short-range (10 Gb/s; < 300 m) represent about 90% of all in-building networks [7]. Furthermore, it is predicted that the fastest growing part of the optical communications market will be targeting legacy MMF for installed lengths up to 300 meters.

Polymer optical fiber (POF) is an emerging medium for very short reach links. The popularity of polymer optical fiber is due to the advantages brought by its large core diameter and mechanical properties. These include connectorization simplicity due to the large numerical aperture, high tolerance to both misalignments and vibrations, low bending loss that eases installation and simple and low maintenance costs due to its robustness.

Common polymer optical fibers are based on polymethyl methacrylate (PMMA-POF). These fibers exhibit low bandwidth, multimodal dispersion and high attenuation (200 dB/km) hence are not suitable for today's high data rates or RoF systems where signals usually exhibit high bandwidths and high RF frequency carriers. Due to their relatively low bandwidth, a down-conversion of the RF signal to an intermediate frequency would be necessary, which introduces additional complexity and raises the cost of the BSs. Newer perfluorinated graded index polymer optical fiber (PF-GI-POF) from companies such as Sekisui Chemical, Chromis Fiber or Asahi Glass, solve this issue by combining a low attenuation material (about 50 dB/km @ 850 nm) with a graded index profile in their fiber construction. Bandwidth is relatively high for graded-index multimode fibers. Current PF-GI-POFs have bandwidth length products of around 1 GHz·Km, and attenuations as low as 10 dB/Km at 1310 nm [8]. In practical terms, for short links (< 100 m), it is limited by the response of directly modulated laser devices [9].

Comparing to common silica multimode fiber (SI-MMF) with respect to transmission capacity, PF-GI-POF has the potential of high bandwidth and a lower modal dispersion. Moreover, it offers lower material dispersion and higher bandwidth than standard MMF with 40 Gb/s data transmission capability for 100 m links [9].

The attenuation is not an issue for short silica-based fiber link lengths. But in the case of POF the attenuation can be as high as 20 dB for the PMMA-POF or about 5 dB for the state of the art PF-GI-POF for a 100 m length link. Large-core glass fiber shows lower attenuation than POF, however their core size is restricted to 200 μm due to the inherent inflexibility of glass. In this situation, POF again has advantages concerning easy handling and termination, tolerance to misalignments and high mechanical strength [10]. Furthermore, the typical large core of polymer fiber allows for large tolerance on misalignments that results in the possibility of using cheaper connectors. For comparison, consider the case of the power loss due to lateral (axial) misalignment of connecting two graded index (parabolic case) MMF with different core diameters. Comparing the power loss, assuming uniformly modal power distributions, for a misalignment of 25 μm, yields a loss of 1.76 dB for a 62.5 μm core diameter MMF whereas for the case of POF with a core diameter of 200 μm, the same 25 μm displacement results only in 0.48 dB loss [11]. New PF-GI-POF fibers being developed are able to withstand large temperature variation (-65 C to 125 C) and so may be suitable for applications in harsh critical environments. Their ease of installation, and tolerance to misalignment, vibration and large temperature variation operation makes these fibers suitable for short-range applications in the home environment or in critical applications such as the car and the avionics industry.

Here, we experimentally demonstrate the uplink of a MB-OFDM UWB signal (ECMA-368 standard [12]) over two different PF-GI-POFs from Chromis Fiberoptics using commercial UWB transceivers and cheap commercial off-the-shelf (COTS) components, namely an optical transceiver composed by a low-cost 850 nm VCSEL and a PIN photodiode.

2.1. A low-cost directly modulated RoF system

VCSELs are characterized by a vertical low divergence, circular beam patterns, low threshold currents (a few mA) and high bandwidths (several GHz). Their vertical wafer growth process enables in-wafer testing, and is well suited for large scale production. Their output light beam pattern enables efficient coupling to large diameter polymer optical fibers. Hence, simple plastic injection molded packages are sufficient as fiber coupling devices. These are reasons that make VCSELs desirable for low cost directly modulated systems in these types of widespread commercial applications.

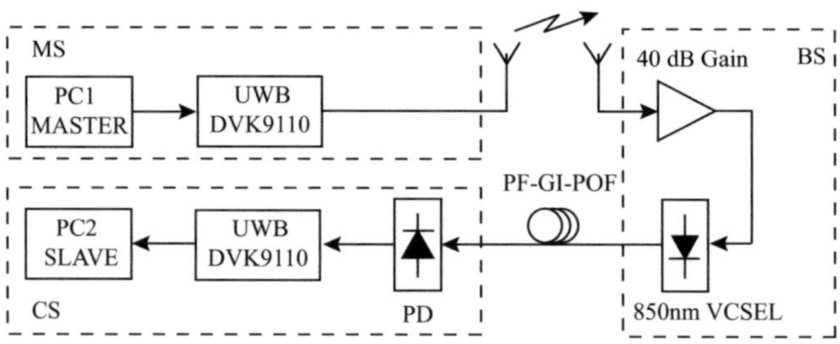

Figure 2. Schematic illustrating the RoF setup used.

A schematic of the MB-OFDM UWB over GI-POF system used is shown in Figure 2, representing a RoF communication uplink between a Mobile Station (MS) and a Control Station (CS), via a Base Station (BS). In order to generate MB-OFDM UWB signals compliant with the ECMA-368 standard, a commercially available UWB transceiver module from WisAir (DVK9110M) was used. The proposed system is based on today's commercially available low cost VCSELs and photodiodes that are not optimized for radio-over-fiber applications.

At the BS, a power amplifier (PA) amplifies the driver signal. The optical signal power (P_{OPT} in dBm) at the laser output is given by [13, 14]

$$P_{OPT} = \frac{G_{TX} + P_{RF,BS}}{2} + 10\log_{10}\left(G_M\sqrt{\frac{1000}{Z_{in}}}\right) \quad (1)$$

where G_{TX} is the PA gain in dB, $P_{RF,BS} = P_{RF,MS} - L$ is the received electrical power (in dBm) at the BS, $P_{RF,MS}$ is the transmitter electrical power at the MS, L is the wireless link loss, G_M is the VCSEL modulation gain (or slope sfficiency) in mW/mA and Z_{in} ($\sim$50 Ω) is the laser input impedance assumed constant within the band of interest.

The received electrical power at the CS, $P_{RF,CS}$, is given by

$$P_{RF,CS} = 20\log_{10}(RG_M) + 10\log_{10}\left(\frac{Z_{out}}{Z_{in}}\right) - 2OL + G_{TX} + P_{RF,MS} - L \quad (2)$$

where R is the photodiode responsivity in mA/mW, Z_{out} ($\sim$50 Ω) is the photodiode output impedance (also assumed constant within the band of interest) and OL is the optical power loss due to both the fiber attenuation and connector loss. Note the factor of 2 multiplying the optical power term which results from the quadratic optical power to electrical power conversion in the photodiode meaning that reduction on the transmitted optical power has a significant impact on the link budget.

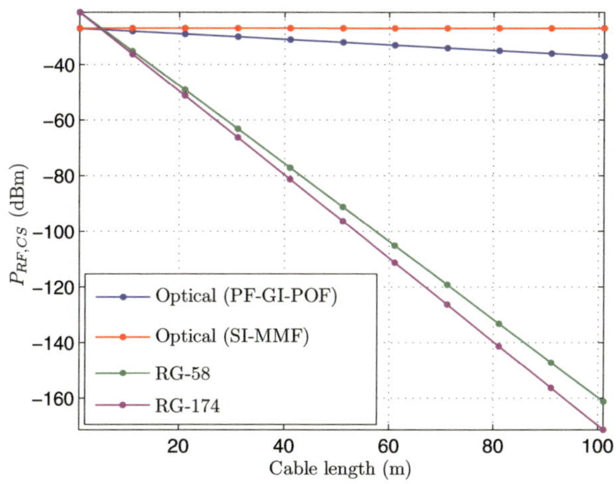

Figure 3. RF power at the CS as a function of the cable length.

Figure 3 shows the RF power at the CS as function of the cable length. For this illustrative result, we consider an UWB signal transmission over the setup given by Figure 2 with the following specifications: $P_{RF_MS} = -19.7\,\text{dBm}$, a wireless link of 1 meter ($L \approx 40\,\text{dB}$), $RG_M = 0.88$, $G_{TX} = 40\,\text{dB}$ and $Z_{out} \approx Z_{in} = 50\Omega$. We compare this result with the loss of common silica multimode fiber (SI-MMF) and RF electrical cables (RG-58 and RG-174 considering the same operating band and the same RF gain as applied to the E/O/E converter). Although there is a penalty by using the E/O/E conversion and optical transport, this is nonetheless very small when compared to the loss suffered by the signal when transported with electrical cable at distances of several tens of meters. It can be seen that the electrical to optical and optical to electrical conversion (whose efficiency is given by RG_M) jointly with the attenuation of POFs is the dominant factor reducing the link power budget of these systems. The relatively high POF attenuation can be partially overcome by post detection amplification, at the expense of some SNR degradation due to amplifier noise. In our experiments, an extra LNA was not included because the UWB DVK provides enough sensitivity.

MB-OFDM UWB radio applications make extensive use of multiple subcarriers and, hence require large dynamic range and highly linear devices. The signal transmission is mainly impaired by the laser nonlinearity, the optical loss due to the fiber, the free space loss and noise added by the system.

2.2. Experimental demonstration of concept

Figure 4 depicts the RoF conversion board based on low-cost electrical and optical components. An amplifier cascade and a polarizing circuit makes up the laser driver for the E/O conversion circuit and a photodiode with an integrated transimpedance amplifier was used for the O/E conversion circuit.

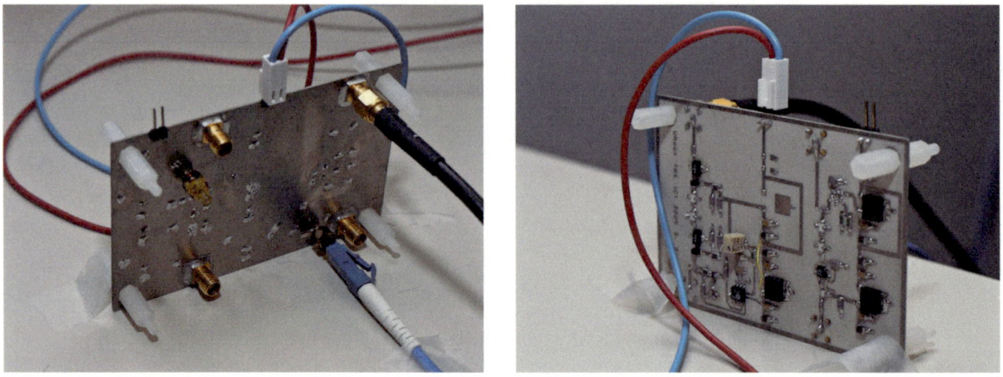

(a) Top view. (b) Bottom view.

Figure 4. RoF conversion board used as E/O/E transceiver.

The amplifiers were chosen among devices with both high IP3 (3rd Order Intercept) and low noise figure. High IP3 is essential for guaranteeing the integrity of multiple carrier ultra wideband signals. The VCSEL (HFD3180-203) and photodiode (HFE4192-581) operate at 850 nm, and have a combined 3 dB bandwidth of about 5 GHz. The laser modulation efficiency, and photodiode (PD) responsivity are 0.07 mW/mA and 12.5 mA/mW, including transimpedance amplifier gain, respectively. It was also experimentally verified that the receiver noise is much larger than the laser RIN, even with short POF lengths.

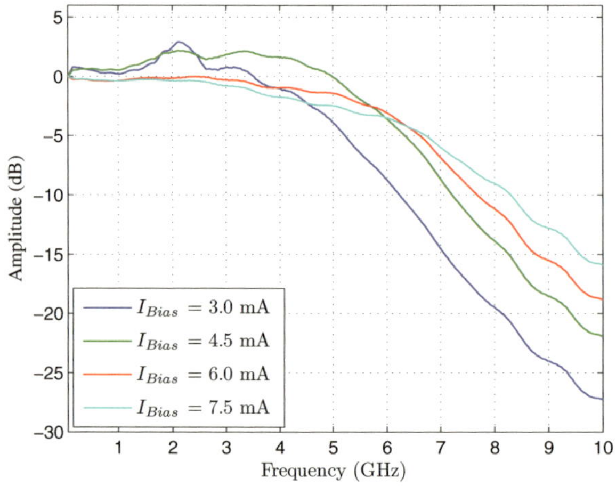

Figure 5. VCSEL transfer function vs. DC bias current.

The VCSEL bias current was also judiciously adjusted. It was found that the current that provides the maximum bandwidth (6 − 7.5 mA, see Figure 5) is not the one corresponding to the optimum operation point. Instead, the bias current was set to the lower level of 4.5 mA, which increases the output optical modulation index (the input RF intensity is fixed), without significantly compromising the bandwidth, and without significantly increasing the laser nonlinear dynamic distortion.

The total budget of this system prototype does not exceed €50. Large scale production of such a system would undoubtedly have an even lower price, meeting well in the requirements of a widespread commercial application for home or office use.

The UWB kit operates is the band group 1 (from 3.168 GHz to 4.752 GHz) and the maximum equivalent isotropic radiated power (EIRP) is −41.3 dBm/MHz using antennas with approximately 4 dBi of gain. This band group consists of three sub-bands, each occupying a bandwidth of 528 MHz and containing 128 subcarriers. Consequently, the power transmitted by the UWB DVK is approximately −19 dBm, disregarding the antenna gain. Although three subbands are available for transmission, the optical transceiver design limitations (commercial VCSEL, photodiode and amplifiers available at the moment) prevented using the entire available bandwidth. Therefore, the time-frequency code was set to TFC 5 (3.168 GHz to 3.696 GHz).

Previous experimental demonstrations have shown that a 1 meter wireless link produces similar results to the ones obtained using a 40 dB attenuator, which corresponds to the free-space air attenuation over 1 meter distance, approximately. Therefore, and for simplicity sake, the effect of the wireless link was simulated by the attenuator. In this experimental demonstration we have used two different PF-GI-POFs from Chromis Fiberoptics, namely, the GigaPOF-62LD and the GigaPOF-120LD, with core diameters of 62.5 μm, and 120 μm, respectively.

2.3. Experimental results and discussion

The experimental validation was carried out by transmitting data at bit rates of 53.3 Mbps, 200 Mbps and 480 Mbps.

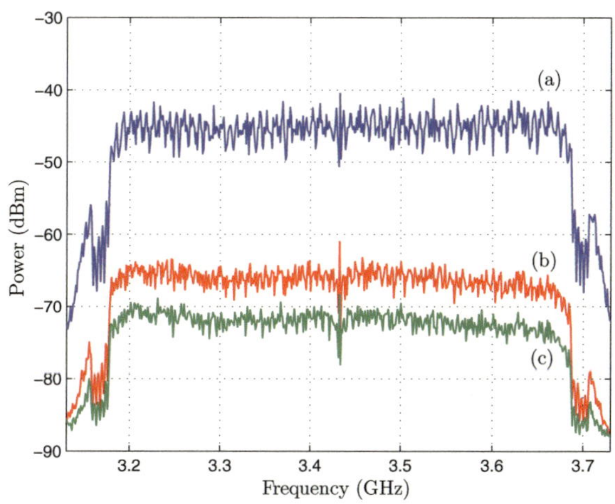

Figure 6. Spectrum of the MB-OFDM UWB signal (a) transmitted by the UWB kit and received after a 1 meter wireless link and (b) 50 meters and (c) 100 meters of GigaPOF-120LD (RBW = 1 MHz).

We also used a 64 octet packet length for PER measurements. Figure 6 shows the MB-OFDM UWB signal spectra obtained before the wireless link and after the GigaPOF-120LD. The attenuation of these links obtained from equation (2) are 20.6 dB and 25.6 dB, which agrees with the measured values of 21 dB and 26 dB from Figure 6. In addition to the wireless/optical link attenuation, it can be seen that subcarriers suffer slightly different attenuations mainly due to the photodiode and amplifier frequency response. The UWB signal spectrum shows no distortion for the tested fiber lengths, which indicates that the bandwidth-distance product is not the factor limiting the transmission on the fiber.

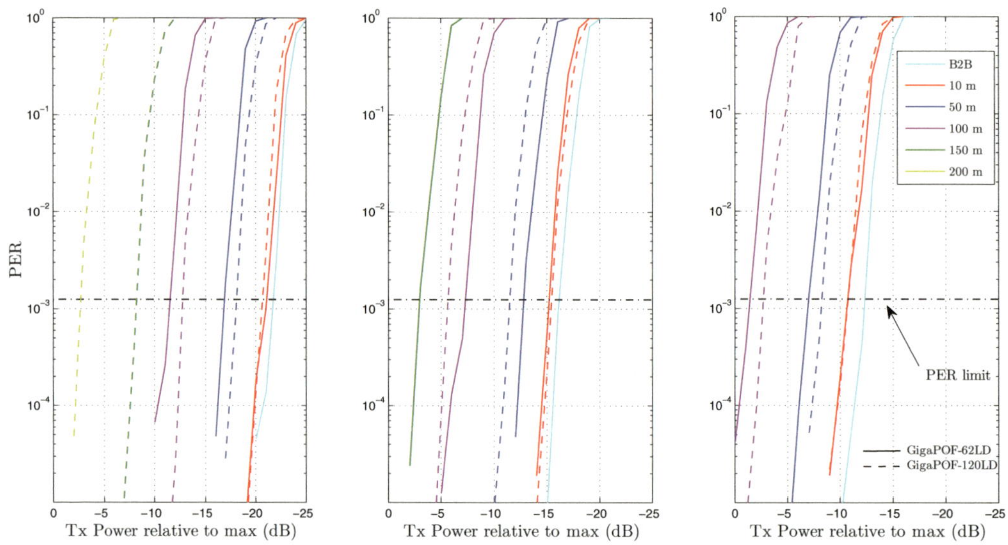

(a) 53.3 Mbps. (b) 200 Mbps. (c) 480 Mbps.

Figure 7. PER vs. transmitted power after a 1 meter wireless link for transmission rates of 53.3, 200 and 480 Mbps for both GigaPOF-62LD and GigaPOF-120LD.

Figure 7 depicts the experimental results of PER as a function of transmitted power (relative to the maximum allowed value as defined by the ECMA-368 standard [12]) for different POF lengths preceded by a 1 meter wireless link and considering transmission rates of 53.3 Mbps, 200 Mbps and 480 Mbps. As expected, results show that the PER increases when the POF length is increased. The horizontal dashed line corresponds to a PER of 0.125 % which is the maximum PER allowed for a 64 octet frame body (ECMA-368 [12]).

Figure 8 shows the minimum required signal power (relative to the maximum) corresponding to the maximum allowed PER, as a function of POF length, where a 1 meter wireless channel is included. As expected, the PER increases for larger data rates, and the required power for achieving a valid transmission also increases.

A linearly increasing transmitted signal power (in dB) is necessary to compensate for the linearly increasing POF loss with distance, keeping both the receiver SNR and the PER constant, as indicated in Figure 8. This shows that the overall noise level is constant at the output of the receiver board, and that intermodulation products are sufficiently below the receiver noise level (for the chosen link parameters). A further interesting result is

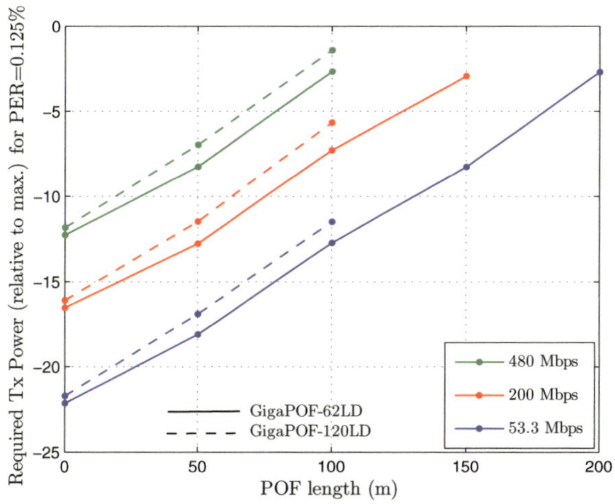

Figure 8. Required signal transmit power after 1 meter wireless link to achieve a PER of 0.125 % as a function of POF length for both GigaPOF-62LD and GigaPOF-120LD.

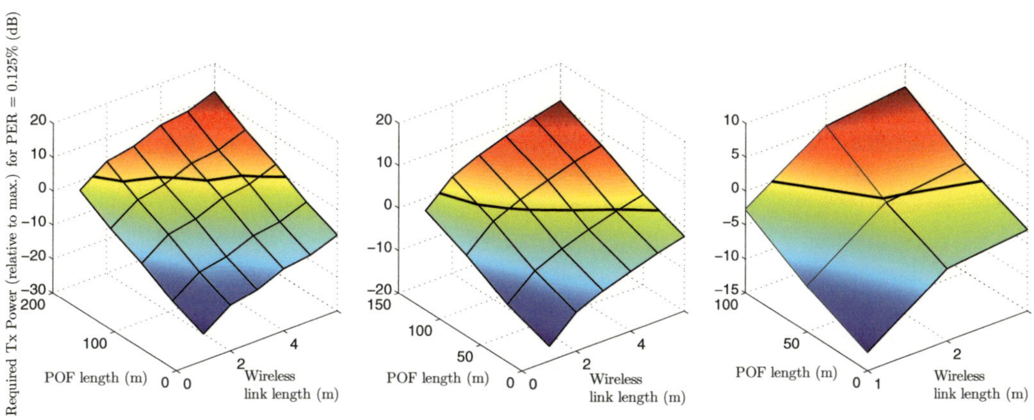

(a) 53.3 Mbps. (b) 200 Mbps. (c) 480 Mbps.

Figure 9. Required signal transmit power (relative to max.) to achieve a PER of 0.125% as a function of GigaPOF-120LD length and wireless link length for three different transmission rates.

that the slope of the plots depicted in Figure 8 correspond to the fiber attenuation, which is approximately 50 dB/Km in both cases, indicating that fiber attenuation (and not fiber bandwidth) is the dominant transmission penalty (see Figure 6).

Results also show that it is possible to transmit 480 Mbps up to 100 meters of POF when preceded by a 1 meter wireless link as well as 200 Mbps and 53.3 Mbps over 150 meters and 200 meters, respectively. The slight difference found in the back-to-back configuration shows that the large core diameter fiber has a better light coupling efficiency. This difference persists in all the POF lengths.

It was not possible to obtain results for fiber lengths longer than 100 meters, due to the unavailability of suitable GigaPOF-62LD cables at our lab. Nonetheless, by looking at the

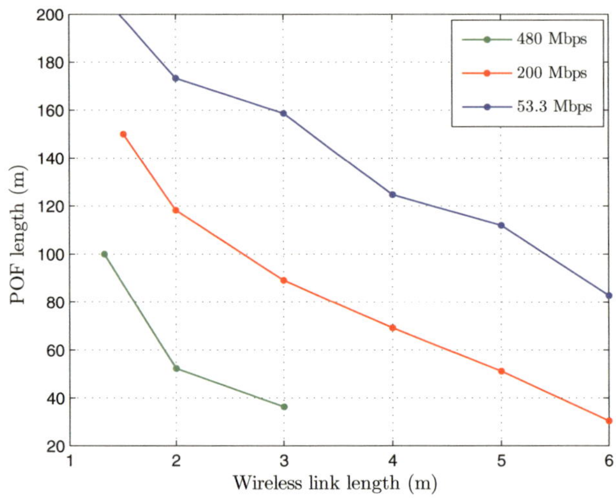

Figure 10. GigaPOF-120LD length vs. wireless link length when transmitting at the maximum allowed power for a 0.125 % PER.

plots trend, we can also infer that valid transmissions of 200 Mbps over 150 meters of GigaPOF-62LD and 53.3 Mbps over 200 meters of the same fiber, are likely achievable.

A generalization of the results plotted in Figure 8 is depicted in Figure 9 for the GigaPOF-120LD. In order to obtain these results, the required signal power for the PER of 0.125% was obtained for wireless links up to 6 meters (in one meter spans) in a back-to-back optical configuration. For the 10, 50 and 100 meters of GI-POF cases, results were extrapolated using the start point given by the back-to-back configuration and using the previous derived result that the transmission power increases linearly with the GI-POF length (see Figure 8). Note that the 1 meter wireless link results are the ones plotted in Figure 8. The solid black lines in Figure 9 indicate the intersection of the surface with the 0 dB plane, which represents the maximum transmitted power. Thus, combinations of POF and wireless links with required transmit powers above 0 dB are not allowed by the standard and their representation in the Figure result from the extrapolation used.

With this result it is possible to see that a 1 meter wireless link followed by a range extension of 150 meter of GI-POF requires the same amount of transmitted power as the 6 meter wireless link with a 30 meter GigaPOF-120LD link for a bitrate of 200 Mbps. Similarly, at 480 Mbps, we can see that 100 meters of GigaPOF-120LD preceded by a 1.3 meters wireless link gives the same performance as 3 meter wireless link followed by approximately 35 meters of fiber.

The specific cases given by the black lines in Figure 9, are represented in Figure 10. These results represent the maximum POF length as a function of the wireless link length when transmitting at the maximum allowed power for a 0.125 % PER.

In conclusion, results show that it is possible to transmit 200 Mbps over a 6 meter wireless link followed by 30 meters of GigaPOF-120LD. We have experimentally demonstrated maximum transmission distances of 150 meters and 200 meters, respectively, using GigaPOF-120LD, with data rates of 200Mbps and 53.3 Mbps. It is also demonstrated that the PG-GI-POF attenuation, and not its bandwidth, is the dominant factor limiting the fiber link length.

3. UWB transmission in a radio-over-fiber system based on reflective electro-absorption modulators

As stated earlier, phase noise from optical sources is one of the factors impairing RoF system performance. Thus, very stable narrow linewidth optical sources are mandatory but also very expensive [15]. For the downlink signal transmission, an ultra-stable and common optical source can be used since it is located at the CS. However, in the uplink, it is not attractive in terms of complexity, size, power consumption and cost to have an optical source for each BSs. Furthermore, by eliminating the need of an optical source, BSs can be colorless and the wavelength assignment can be done at the CS.

Two main BS schemes are usually used in colorless RoF systems. The first scheme is based on an external modulator, photoreceiver and utilizes optical filtering techniques and wavelength reuse, or a more convenient method in which the optical carrier is remotely provided from the CS [16–18]. Another scheme of source free BSs is based on a single electro-absorption waveguide device in which a single component acts both as a modulator for the uplink and as photoreceiver for the downlink [19, 20]. Therefore, this transceiver device is a very attractive solution for a full-duplex RoF transmission.

Although an electro-absorption transceiver (EAT) based RoF system is simple and potentially has low power consumption, alternative solutions based on a dual lightwave approach and passive EATs have also been reported [21]. Using different wavelengths for segregating the uplink from the downlink optical paths makes it possible to minimize the transceiver insertion loss for both uplink and downlink signal transmission. Therefore, optical transceivers are increasingly considered as key components for the implementation of low-cost BSs [21].

The reflective EAM where the rear facet is coated with high reflection layer is an interesting device for operating simultaneously as a modulator and photoreceiver. In the following, the R-EAM performance as a transceiver in RoF systems is discussed. Additionally, an experimental evaluation of the R-EAM in terms of its slope efficiency (SE) and responsivity (R_e) at different wavelengths, optical powers and bias points is reported. Finally, a case study of UWB signal transmission is reported, where the optimum operation points are discussed for different scenarios: the bias for maximum SE and for maximum R_e are compared to zero bias.

3.1. R-EAM performance analysis

The experimental setup for the characterization of the 60 GHz R-EAM transceiver (CIP EAM-R-60-C-V-FCA) is shown in Figure 11.

Both electrical and optical test signals are used in order to obtain the electro-optical (E/O) and optical to electrical (O/E) response, which correspond to the slope efficiency and responsivity, respectively. We assume the decibel units for these variables to be obtained by $20\log_{10}(\cdot)$, as defined in the measurements of the laboratory equipment. Figure 12 shows the EO response

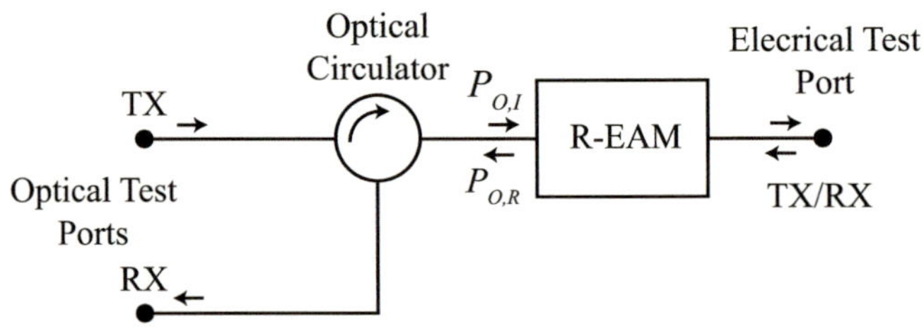

Figure 11. R-EAM characterization setup.

as a function of reverse bias voltage, for different frequencies and wavelengths. The R-EAM average optical input power, $P_{O,I}$, is varied by controlling the CW laser output power. It is apparent that the optimum bias voltage increases with both wavelength and the optical input power. It is also apparent that SE degrades slightly with frequency especially for high $P_{O,I}$, and lower wavelengths. Furthermore, it is easily seen that SE at the optimum bias voltage increases with $P_{O,I}$. Figure 13 shows the O/E response as a function of reverse bias voltage, for different frequencies and wavelengths. Due to the similarity of results, only two wavelengths are plotted, in order to allow a clear visualization. Similarly to what was observed for the E/O case, the responsivity degrades with frequency, particularly for high values of $P_{O,I}$, shorter wavelengths and decreasing reverse bias. Nevertheless, responsivity is shown to be more affected than SE. Furthermore, results also show that responsivity increases monotonically with reverse bias voltage.

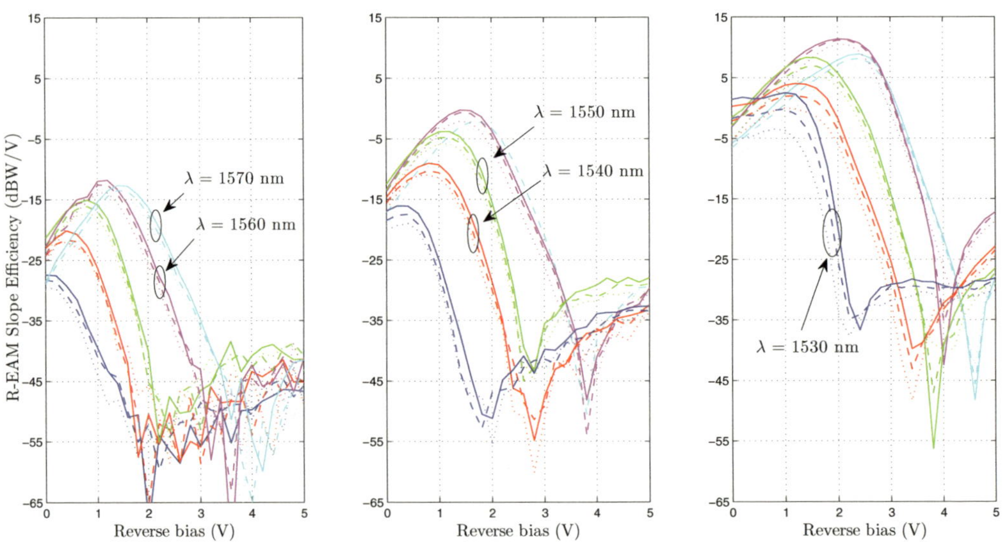

(a) $P_{O,I} = -6\,\text{dBm}$ (b) $P_{O,I} = 0\,\text{dBm}$ (c) $P_{O,I} = 7\,\text{dBm}$

Figure 12. E/O response vs. reverse bias voltage for different wavelengths and frequencies and optical input powers. Solid, dashed and dotted lines correspond to frequencies of 2.4, 5, 15 GHz, respectively.

The optimum bias points for maximum slope efficiency and responsivity have been extracted from the results of Figure 12 and 13. Considering these values, the R-EAM performance

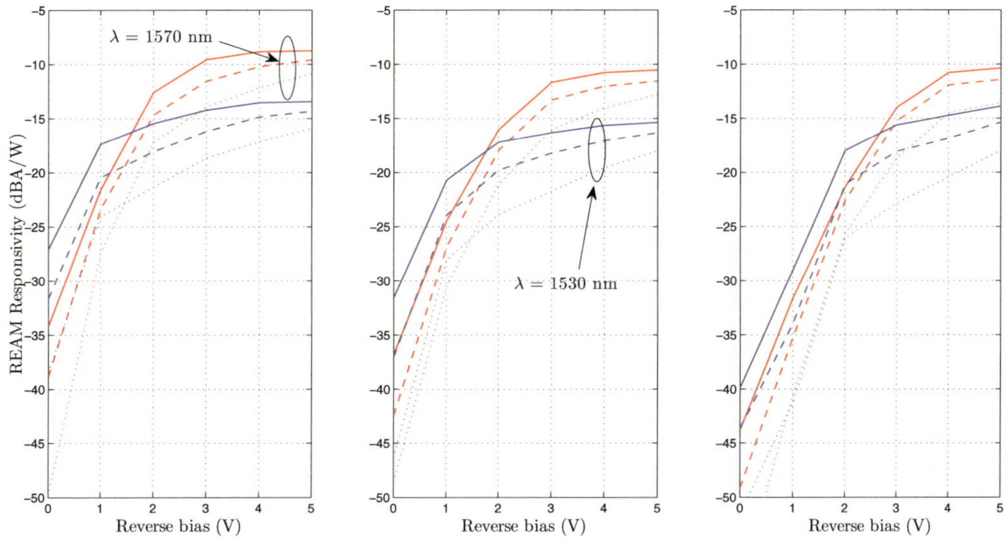

(a) $P_{O,I} = -6\,\mathrm{dBm}$ (b) $P_{O,I} = 0\,\mathrm{dBm}$ (c) $P_{O,I} = 7\,\mathrm{dBm}$

Figure 13. O/E response vs. reverse bias voltage for different wavelengths and frequencies and optical input powers. Solid, dashed and dotted lines correspond to frequencies of 2.4, 5, 15 GHz, respectively

is assessed in terms of slope efficiency and responsivity for the following cases: bias for maximum SE, bias for maximum R_e and zero bias. The results of this analysis are plotted in Figure 14 and 15.

The results in Figure 14 show that the best performance is obtained for a wavelength of 1560 nm, when the EAM is biased for maximum SE. However, when zero biased, the optimum wavelength is reduced to 1530 nm for $P_{0,I} = 7\,\mathrm{dBm}$, where a penalty of 13 dB is incurred, compared to the case of maximum SE. Finally, when the EAM is biased for maximum responsivity, the optimum wavelength is 1560 nm and the SE decreases by 15 dB, compared to the zero bias case. It has also been verified experimentally that, as expected [18], the slope efficiency is proportional to the input optical power, as seen in Figure 14 (a) and (b), until it saturates at high optical powers, as shown by Figure 14 (c). Nevertheless, high optical input powers should be used in order to maximize the SE.

Concerning the EAM responsivity, its value is optimum for $P_{O,I} = -6\,\mathrm{dBm}$, while a noticeable reduction is observed with increasing $P_{O,I}$, especially for zero bias. In both cases of biasing for maximum R_e and SE, the EAM responsivity improves with increasing wavelength, except for high $P_{O,I} = +7\,\mathrm{dBm}$, where its maximum is achieved at $\lambda = 1560\,\mathrm{nm}$.

The best responsivity values for both cases of biasing for maximum R_e and SE are $-9.6\,\mathrm{dBA/W}$ and $-18.6\,\mathrm{dBA/W}$, respectively. For the zero bias case, the responsivity decreases with wavelength for both input optical powers $-6\,\mathrm{dBm}$ and $0\,\mathrm{dBm}$, reaching a maximum of $-31\,\mathrm{dBA/W}$ for $P_{O,I} = -6\,\mathrm{dBm}$ and 1530 nm. However, for $P_{O,I} = +7\,\mathrm{dBm}$ the responsivity is relatively constant with the wavelength, except for 1570 nm where it degrades by 6 dB. The

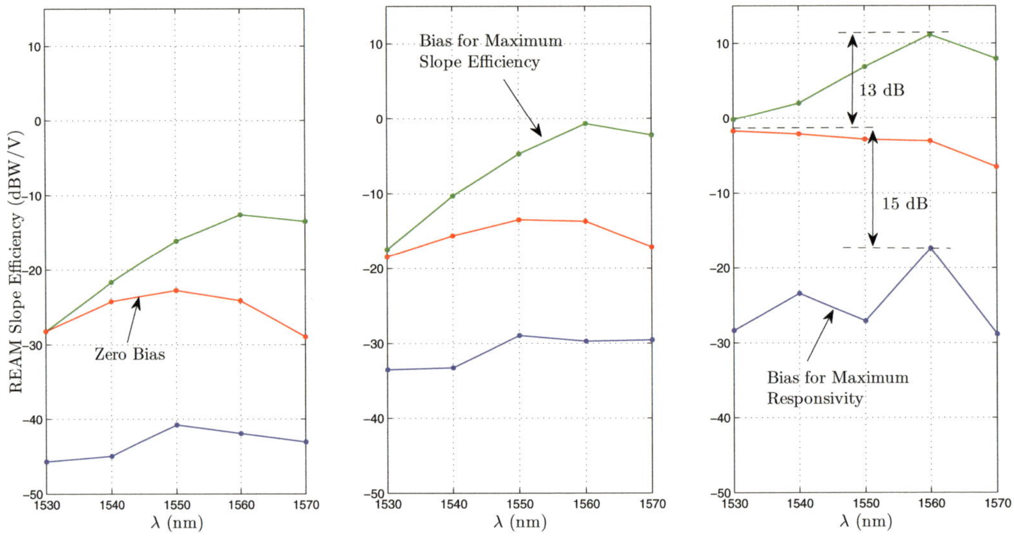

(a) $P_{O,I} = -6\,\text{dBm}$ (b) $P_{O,I} = 0\,\text{dBm}$ (c) $P_{O,I} = 7\,\text{dBm}$

Figure 14. E/O response vs. wavelength for different bias configurations and optical input powers.

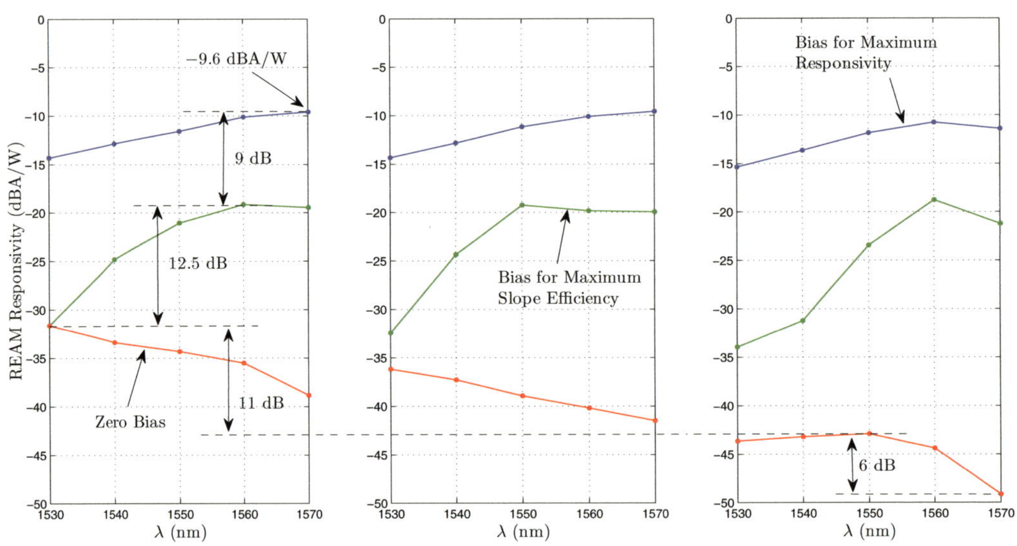

(a) $P_{O,I} = -6\,\text{dBm}$ (b) $P_{O,I} = 0\,\text{dBm}$ (c) $P_{O,I} = 7\,\text{dBm}$

Figure 15. O/E response vs. wavelength for different bias configurations and optical input powers.

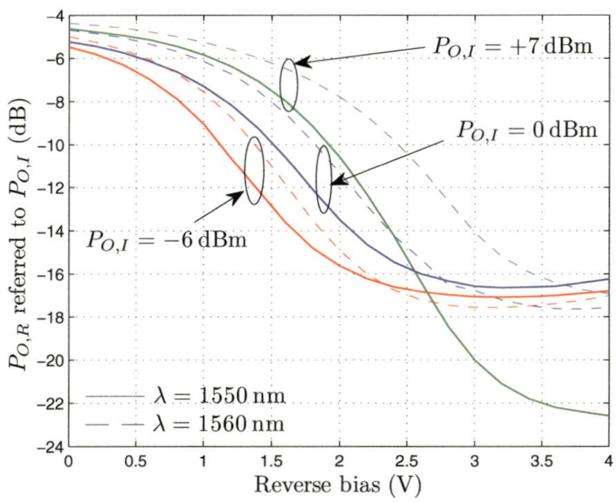

Figure 16. R-EAM reflected optical power, $P_{O,R}$ referred to the input optical power $P_{O,I}$.

responsivity degradation with increasing $P_{O,I}$, translates into a penalty of 11 dB when $P_{O,I}$ is increased from -6 dBm to $+7$ dBm.

The R-EAM performance was also assessed in terms of reflected optical power, $P_{O,R}$, for different values of input optical power $P_{O,I}$, the results of this analysis being shown in Figure 16. The reflected optical power is a relevant parameter with impact in the optical to electrical conversion. Therefore, this analysis considered the wavelengths of 1550 nm and 1560 nm, because these provide the best performance according to the measurements of slope efficiency given in Figure 15. The results given by the current analysis, indicate that the reflected optical power tends to decrease with increasing reverse bias voltage. Furthermore, the optical power reflected at the wavelength of 1550 nm is generally lower than that reflected at 1560 nm, except for higher reverse bias at low optical input powers.

3.2. System performance assessment

In this section, we consider the application of the R-EAM as a base station in a bidirectional transmission system, in a typical RoF network. A diagram of the application scenario is shown in Figure 17.

The RF downlink (DL) signal, generated by the CS transceiver (TRX) with a power of $P_{tx,DL}$, passes through an electrical circulator and amplifier with gain $G_{E,DL}$, and drives an E/O modulator, considered ideal in the present analysis. The optical downlink signal then passes through an optical circulator, and reaches the R-EAM through an optical fiber, with an incident optical power of $P_{O,I}$. The RF modulated optical signal is converted to the electrical domain by the R-EAM with a responsivity (R_e), and then reaches the mobile station through the wireless channel, which induces a signal loss of L. Note that the attenuation parameter L already includes both BS and MS antenna gains. Conversely, the RF uplink (UL) signal is generated by the mobile station transceiver with a power of $P_{tx,UL}$, and reaches the base station after being attenuated by the wireless link. The weak RF uplink signal can be amplified electrically ($G_{E,UL}$) before being converted from the electrical to the optical domain with a conversion

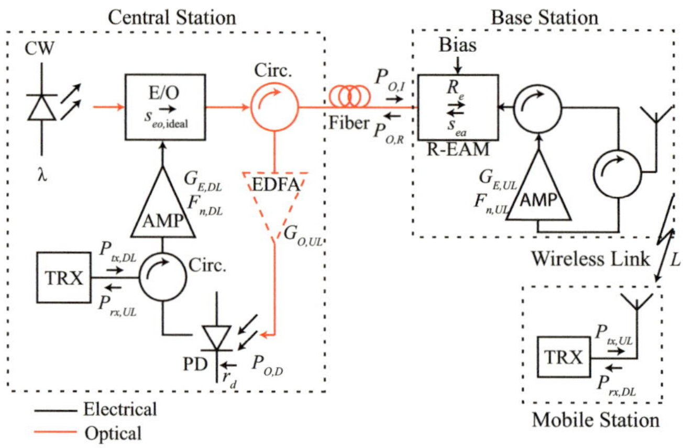

Figure 17. Considered setup for performance assessment. CW represents a continuous wave light source. Both TRX shown, represent the transceivers at the Control Station and Mobile Station. PD represents the Control Station photodiode. The EDFA depicted, represents an optional Erbium Doped Fiber Amplifier.

efficiency given by the EAM slope efficiency, s_{ea}, by modulating the reflected optical carrier with power $P_{O,R}$. The uplink signal might be further optically ($G_{O,UL}$) amplified before reaching the CS transceiver. This optical amplification is only adequate if the optical power is low (typically less than $-3\,\text{dBm}$), so that the noise added by the EDFA is still below the noise level at the receiver [22]. In the following analysis, we assume $P_{O,R} \cong P_{O,D}$, since the optical circulator loss and the use of an EDFA are not considered. The metric used to evaluate performance of the system is the signal to noise ratio (SNR).

The SNR of the RF signal arriving at the mobile station can be written as:

$$\text{SNR}_{rx,MS} = \frac{\langle I^2_{rx,DL}\rangle}{\sigma^2_{n,MS}} = \frac{\langle I^2_{tx,DL}\rangle G_{DL}}{\langle I^2_{t,MS}\rangle} \tag{3}$$

where,

$$G_{DL} = G_{E,DL} s^2_{eo,\text{ideal}} R^2_e L \tag{4}$$

$$\langle I^2_{t,MS}\rangle = (4T + T_a)k\Delta_f / R_{MS} + (F_{n,DL} - 1)kTG_{DL}\Delta_f / R_{EO} \tag{5}$$

$$P_{tx,DL} = R_{CS}\langle I^2_{tx,DL}\rangle \tag{6}$$

Therefore, the SNR at the receiver can be written as the following ratio:

$$\text{SNR}_{rx,MS} = \frac{(P_{tx,DL}/\Delta_f)G_{DL}/R_{EO}}{(4T + T_a)k/R_{MS} + (F_{n,DL} - 1)kTG_{DL}/R_{EO}} \tag{7}$$

where R_{MS} represents the impedance of the mobile station circuitry, R_{EO} the impedance of the E/O converter circuitry, Δ_f the transmission bandwidth, $s_{eo,\text{ideal}}$ the slope efficiency of the ideal E/O modulator, $F_{n,DL}$ the noise factor of the electrical amplifier that precedes the modulator, k is Boltzmann's constant, $T = 290\,\text{K}$ and $T_a = 120\,\text{K}$ is the antenna temperature. Since the link gain (G_{DL}) is low, the noise at the receiver is dominated by the thermal noise of the receiver circuitry itself. Furthermore, since the R-EAM efficiency is higher in the downlink, regardless of the bias conditions, the SNR is considerably higher than in the uplink.

The SNR of the RF signal arriving at the CS can be written as:

$$\text{SNR}_{rx,CS} = \frac{\langle I^2_{rx,UL}\rangle}{\sigma^2_{n,CS}} = \frac{\langle I^2_{tx,UL}\rangle G_{UL}}{\langle I^2_{rin}\rangle + \langle I^2_{sn}\rangle + \langle I^2_{t,CS}\rangle} \tag{8}$$

where,

$$G_{UL} = G_{E,UL} s^2_{ea} R^2_e L \tag{9}$$

$$\langle I^2_{rin}\rangle = r^2_d \langle P_{O,D}\rangle^2 10^{\text{RIN}/10} \Delta_f \tag{10}$$

$$\langle I^2_{sn}\rangle = 2q r_d \langle P_{O,D}\rangle \Delta_f \tag{11}$$

$$\langle I^2_{t,CS}\rangle = 4kT\Delta_f / R_\text{L} + (T_a + (F_{n,UL} - 1)T) k G_{E,UL} s^2_{ea} r^2_d \Delta_f / R_{EAM} \tag{12}$$

$$P_{tx,UL} = R_{MS} \langle I^2_{tx,UL}\rangle \tag{13}$$

where R_{EAM} represents the impedance of the R-EAM circuitry.

The noise terms are referred to the photodiode output, and stem from three main components: the relative intensity noise (RIN) from the laser source (proportional to the square of $\langle P_{O,D}\rangle$), the shot noise (proportional to $\langle P_{O,D}\rangle$), and the last term is the thermal noise from both the photodiode load (R_L), the antenna and the uplink electrical amplifier, where $F_{n,UL}$ is the noise factor of the amplifier at the base-station. The thermal noise from the electrical transmitter at the mobile station can be neglected, due to the low total gain of the link (G_{UL}). The optical power detected by the photodiode, $P_{O,D}$, is expected to have a significant impact on the noise contribution at the receiver. In the presented results, the wireless channel attenuation (L) was not considered, the fiber is considered ideal, $G_{E,UL/DL} = 1$, $F_{n,UL/DL} = 1$, $R_\text{L} = 1000\,\Omega$ and $R_{MS/EO/EAM} = 50\,\Omega$.

3.3. UWB-over-fiber results and discussion

The present analysis considers again the usage of typical UWB transceivers (*Wisair DVK9110*), which operate in band group 1 (from 3.168 GHz to 4.752 GHz) and have a maximum transmission power of approximately $P_{tx,UL}/\Delta_f = -45.3\,\text{dBm}$, when the MS antenna gain (considered to be 4 dB) is lumped into the wireless channel attenuation.

The UWB receiver sensitivity of −70.4 dBm at 480 Mbit/s specified in the standard [12], is not valid for an optical front-end. This value assumes a typical value of noise level in a wireless receiver of −80.5 dBm, which indicates that a receiver should be able to meet the target performance specified in the standard for a SNR of approximately 10 dB. In the present

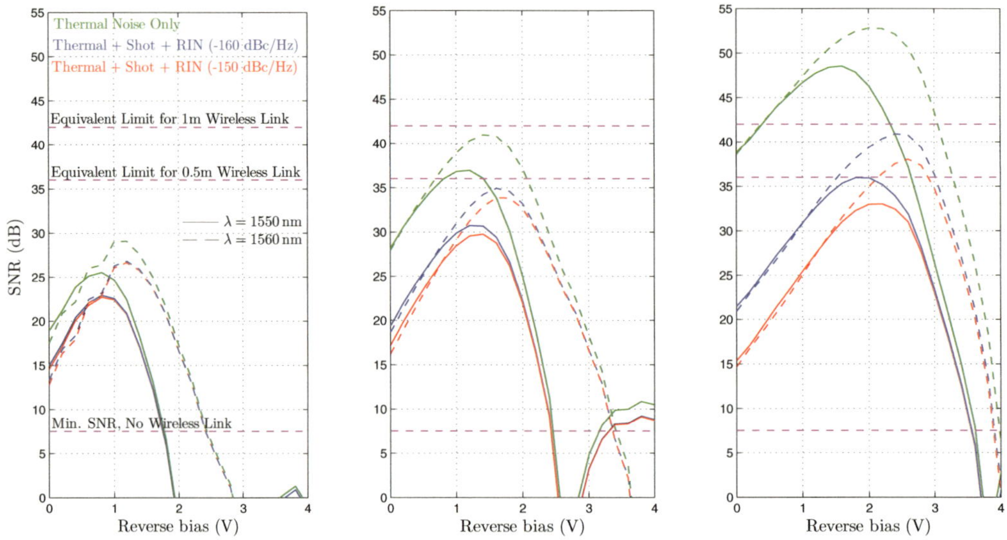

(a) $P_{O,I} = -6\,\text{dBm}$ (b) $P_{O,I} = 0\,\text{dBm}$ (c) $P_{O,I} = 7\,\text{dBm}$

Figure 18. SNR as a function of reverse bias for three different values of optical input power.

analysis, a minimum SNR of 7.5 dB is considered, which has been measured experimentally for the mentioned commercial UWB transceiver.

The results in Figure 18 show the SNR as a function of the reverse bias, for three different values of $P_{O,I}$, (a) -6 dBm, (b) 0 dBm and (c) $+7$ dBm, considering a responsivity of 1 A/W and two different wavelengths, 1550 nm and 1560 nm and RIN values, -150 dB/Hz and -160 dB/Hz. Moreover, SNR results considering only the thermal noise term are also depicted for comparisson purposes. Two additional equivalent SNR limits that consider a wireless link length of 0.5 m and also 1 m are also shown in the results, which account for a signal loss of 28.5 and 34 dB at 4 GHz, respectively, when considering a total gain of the antennas of 10 dBi, among MS and BS. Results show that for low reverse bias the impact of both shot noise and RIN increase with the incident optical power. This result comes in line with the ones of the reflected optical power plotted in Figure 16. Moreover, it is clearly seen that the SNR is optimum for a wavelength of 1560 nm with a maximum SNR 5 dB better than that at 1550 nm, although slightly worse for zero bias.

In Figure 19 the SNR is obtained as a function of $P_{O,I}$, for the following cases: (a) zero bias, (b) optimum bias for maximum SE and (c) optimum bias for maximum SNR, considering only the wavelength of 1560 nm. Again, two additional equivalent SNR limits that consider a wireless link length of 0.5 m and also 1 m are also shown in the results. The results considering optimum bias for maximum Slope Efficiency are, at the maximum points, $5 - 6$ dB worse than those for maximum SNR, essentially because there is a compromise between signal power, which is affected by the $s_{e,a}$ as a function of bias voltage, and noise power, affected by the $P_{O,D}$ which also depends on the bias voltage.

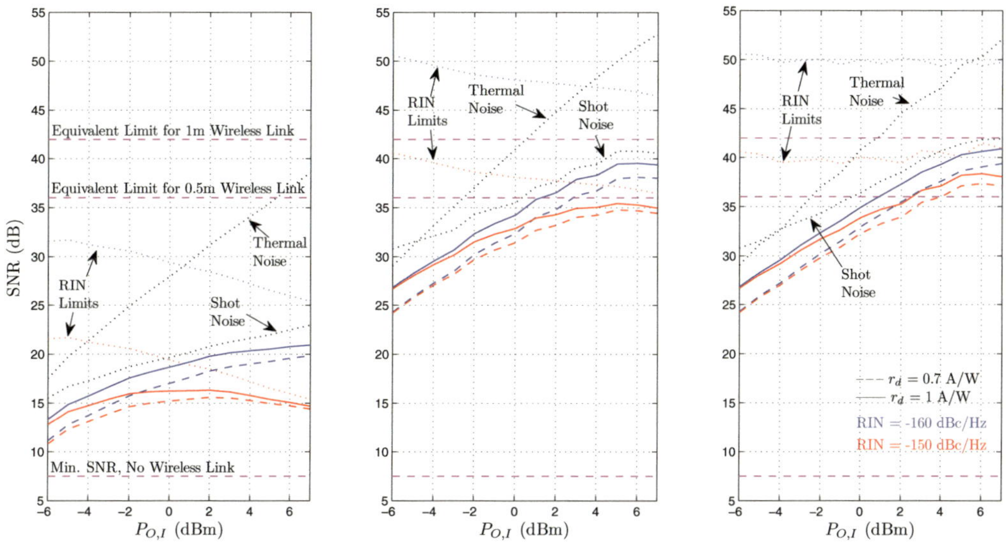

(a) Zero bias (b) Optimum bias for maximum SE (c) Optimum bias for maximum SNR

Figure 19. SNR as a function of optical input power for thre different bias cases.

From an overview perspective, it can be noted that the SNR performance increases with $P_{O,I}$ for a RIN of $-160\,\text{dB/Hz}$, whereas for $-150\,\text{dB/Hz}$ there is an optimum value of $P_{O,I}$ for maximum SNR. At zero bias, a maximum margin of 9 dB compared to the UWB SNR limit is obtained with a laser RIN of $-150\,\text{dB/Hz}$ for a $P_{O,I}$ of $\sim$0 dBm, whereas for a RIN of $-160\,\text{dB/Hz}$ the performance becomes limited by shot noise, allowing for a SNR margin of $\sim$ 13.5 dB. However, the results indicate that none of such limits are achieved at zero bias, because of both RIN and shot noise limitations, which means that a totally passive base-station is not practicable, for reasonable wireless link distances. Results for optimum bias for maximum SE indicate that the SNR for a RIN of $-150\,\text{dB/Hz}$ practically achieves the UWB SNR limit for a wireless link distance of 0.5 meters with a $P_{O,I}$ of $\sim$6dBm, whereas for a RIN of $-160\,\text{dB/Hz}$ a 1 meter link is almost achieved. Note that while the former SNR is limited by the RIN term, the latter is limited by the shot noise. At optimum bias for maximum SNR, a margin of 33.5 dB is obtained at the maximum $P_{O,I}$, for a RIN of $-160\,\text{dB/Hz}$ and $R_e = 1\,\text{A/W}$, allowing for an acceptable wireless link distance of approximately 1 meter. For a RIN of $-150\,\text{dB/Hz}$, there is enough SNR margin to allow for a wireless link distance between 0.5 and 1 meter. Furthermore, by reducing the UWB throughput, a maximum distance of 2.8 meters would be achievable at 53.3 Mbit/s, at a minimum SNR of 0 dB. Since the R-EAM slope efficiency at optimum bias for maximum responsivity is worse than that at zero bias, there is no advantage in analyzing that scenario.

In conclusion, the optimum operation point was found to be the biasing for maximum SNR using an high optical input power and a wavelength of 1560 nm. Although a zero bias configuration is an attractive technique, it is not suitable to provide a reasonable wireless link distance for UWB. Additionally, we also conclude that one of the performance-limiting factors comes from the laser RIN that imposes a limit on the achievable SNR, especially for a zero biased modulator. A laser RIN of −160 dB/Hz would be required in order to avoid the RIN limitation. In this case the performance becomes limited by shot noise. Although a zero bias configuration is an attractive technique, it is not suitable to provide a reasonable wireless link distance for UWB signals.

4. Conclusions

This chapter presented a study of the transmission performance of UWB signals over two different optical based networks. Firstly, a low-cost RoF solution based on a VCSEL and two high performance POFs for assessing the UWB signal transmission performance was presented. Experimental measurements of packet error rate (PER) and minimum transmitted powers to achieve the maximum allowed PER showed that it is possible to have a viable transmission at data rates of 480 Mbps, 200 Mbps and 53.3 Mbps over 100, 150 and 200 meters of PF-GI-POF, respectively, preceded by a 1 meter wireless link. The two PF-GI POFs have diameters of 62.5 μm and 120 μm with bandwidth distance products higher than 1 GHz·Km.

It was shown that the PF-GI-POF attenuation, and not its bandwidth, is the dominant factor limiting the fiber link length. Therefore, the PF-GI-POFs are a viable solution for transporting UWB signals. In particular, the larger core GI-POF with 120 μm diameter may find applications in both the home and office environments where easiness of installation and handling are vital.

Finally, a performance analysis of both the slope efficiency and responsivity of a R-EAM transceiver for different wavelengths, optical powers and bias points was presented. In particular, the performance of RoF transmisssion with zero bias transceivers was evaluated for the UWB ECMA-368 standard. The optimum operation point was found to be the biasing the transceivers for maximum link SNR, using a high optical input power at 1560 nm wavelength. Although a zero bias may seem limitative, the R-EAM can be optimally biased using a small battery, which can last for several months, allowing the BS to operate almost as a passive device.

Acknowledgments

This work was supported in part by EC Framework 7 (FP7) project DAPHNE (www.fp7daphne.eu) – Developing aircraft photonic networks (grant ACP8-GA-2009-233709). We acknowledge funding from FCT and program POCTI/FEDER under the National Plan for Scientific Hardware Renewal with grant REEQ/1272/EEI/2005. D. Coelho also acknowledges support from FCT through a PhD grant.

Author details

João M.B. Oliveira, Diogo Coelho and Henrique M. Salgado
Faculdade de Engenharia da Universidade do Porto
Instituto de Engenharia de Sistemas e Computadores do Porto - INESC TEC, Portugal

Luís M. Pessoa and Jorge C.S. Castro
Instituto de Engenharia de Sistemas e Computadores do Porto - INESC TEC, Portugal

5. References

[1] Federal Communications Commission. First report and order, revision of part 15 of commission's rule regarding ultra-wideband transmission system. FCC 02-48, 22:2, 2002.

[2] JE Mitchell. Performance of OFDM at 5.8 GHz using radio over fibre link. *IEEE Electronics Letters*, 40(21):1353–1354, 2004.

[3] Pak Kay Tang, Ling Chuen Ong, A. Alphones, B. Luo, and M. Fujise. PER and EVM measurements of a radio-over-fiber network for cellular and WLAN system applications. *IEEE Journal of Lightwave Technology*, 22(11):2370 – 2376, 2004.

[4] N.J. Gomes, M. Morant, A. Alphones, B. Cabon, J.E. Mitchell, C. Lethien, M. Csörnyei, A. Stöhr, and S. Iezekiel. Radio-over-fiber transport for the support of wireless broadband services [invited]. *Journal of Optical Networking*, 8(2):156–178, 2009.

[5] M. Jazayerifar, B. Cabon, and J.A. Salehi. Transmission of multi-band OFDM and impulse radio ultra-wideband signals over single mode fiber. *IEEE Journal of Lightwave Technology*, 26(15):2594–2603, 2008.

[6] M. Sauer, A. Kobyakov, and J. George. Radio over fiber for picocellular network architectures. *IEEE Journal of Lightwave Technology*, 25(11):3301–3320, 2007.

[7] C. Lethien, C. Loyez, J.P. Vilcot, L. Clavier, M. Bocquet, and P.A. Rolland. Indoor coverage improvement of mb-ofdm uwb signals with radio over pof system. *Optics Communications*, 282(24):4706–4715, 2009.

[8] O. Ziemann, H. Poisel, and J. Vinogradov. Potential of high speed, short distance optical data communication on large diameter optical fibers. In *Proc. 1st Electronics Systemintegration Technology Conf*, volume 1, pages 409–414, 2006.

[9] A. Polley, R. J. Gandhi, and S. E. Ralph. 40gbps links using plastic optical fiber. In *Proc. Conf. Optical Fiber Communication and the National Fiber Optic Engineers Conf. OFC/NFOEC 2007*, pages 1–3, 2007.

[10] T. Ishigure, Y. Aruga, and Y. Koike. High-bandwidth pvdf-clad gi pof with ultra-low bending loss. *Journal of Lightwave Technology*, 25(1):335–345, 2007.

[11] T. Kibler and E. Zeeb. Optical data links for automotive applications. In *Electronic Components and Technology Conference, 2004. Proceedings. 54th*, volume 2, pages 1360–1370. IEEE, 2004.

[12] S. ECMA. Ecma-368: High rate ultra wideband phy and mac standard, 2005.

[13] X. N. Fernando and A. Anpalagan. On the design of optical fiber based wireless access systems. In *Proc. IEEE Int Communications Conf*, volume 6, pages 3550–3555, 2004.

[14] T. Marozsak and E. Udvary. Vertical cavity surface emitting lasers in radio over fiber applications. In *Proc. MIKON-2002 Microwaves, Radar and Wireless Communications 14th Int. Conf*, volume 1, pages 41–44, 2002.

[15] C. Lim, A. Nirmalathas, M. Bakaul, P. Gamage, K.L. Lee, Y. Yang, D. Novak, and R. Waterhouse. Fiber-wireless networks and subsystem technologies. *IEEE Journal of Lightwave Technology*, 28(4):390–405, 2010.

[16] Z. Jia, J. Yu, and G.K. Chang. A full-duplex radio-over-fiber system based on optical carrier suppression and reuse. *IEEE Photonics Technology Letters*, 18(16):1726–1728, 2006.

[17] L. Chen, H. Wen, and S. Wen. A radio-over-fiber system with a novel scheme for millimeter-wave generation and wavelength reuse for up-link connection. *Photonics Technology Letters, IEEE*, 18(19):2056–2058, 2006.

[18] C.H. Cox. *Analog optical links: theory and practice.* Cambridge Univ Pr, 2004.
[19] D. Wake, D. Johansson, and DG Moodie. Passive picocell: a new concept in wireless network infrastructure. *IEEE Electronics Letters*, 33(5):404–406, 1997.
[20] A. Stohr, K. Kitayama, and D. Jager. Full-duplex fiber-optic RF subcarrier transmission using a dual-function modulator/photodetector. *IEEE Transactions on Microwave Theory and Techniques*, 47(7):1338–1341, 1999.
[21] K.I. Kitayama. Architectural considerations of fiber-radio millimeter-wave wireless access systems. *Fiber and Integrated Optics*, 19(2):167–186, 2000.
[22] G. Keiser. *Optical fiber communications.* Wiley Online Library, 2000.